APPLICATION OF VEGETATION SCIENCE TO GRASSLAND HUSBANDRY

W0225645

HANDBOOK

OF

VEGETATION SCIENCE

Editor in Chief

Reinhold Tüxen

1977

Dr. W. JUNK b.v. - PUBLISHERS - THE HAGUE

PART XIII

APPLICATION OF VEGETATION SCIENCE TO GRASSLAND HUSBANDRY

edited by

W. Krause

1977

Dr. W. JUNK b.v. - PUBLISHERS - THE HAGUE

ISBN 978-94-010-1317-8   ISBN 978-94-010-1315-4 (eBook)
DOI 10.1007/978-94-010-1315-4
Cover design: Charlotte van Zadelhoff

CONTENTS

INTRODUCTION

Werner Krause

INTRODUCTION

GENERAL COMMENTS ON THIS VOLUME

The subject "Application of vegetation science to grassland husbandry" is not limited by geographical boundaries. It is proposed to cover this subject in an internationally applicable work of reference. However, to avoid omissions it would have to embrace all the diversity of grassland from the Arctic to the Equator.

The term "application" requires that the papers contained herein take into account both generally relevant scientific knowledge and the special needs of practical farming. The latter are not satisfied solely by the laws of vegetation science; the changing effects of the market, politics, the distribution of wealth and poverty, highly developed technology and the continued use of archaic farming methods must all be considered.

To try to cover all this adequately in one volume of approximately 550 pages is a hopeless task. The editor and the authors had to limit themselves to presenting a selection of problems and attempts to solve them, with the aim that these examples should indicate how much vegetation science can contribute to grassland husbandry, and the scope of this contribution is broadened by the references quoted.

The arrangement of the sections within this volume does not lend itself to specific suggestions for the management of individual systems. However, grassland husbandry can expect to obtain from this work an insight into some overall general relationships, which could assist the adviser in the formulation of more specific improvement plans.

The geographical boundaries of the contributions

The contents of these articles offer a fascinating panoramic view to the geo-botanist resident within the narrow confines of Central Europe. The grassland vegetation of Central Europe has reached a high level of stability and uniformity due to an equable climate and management practices which are based on close observation of natural patterns of production and well established farming tradition. The plants present belong mainly to the indigenous flora, and many of those found in grassland are natives which were found in the original wood-lands or their natural openings, e.g. on river banks and rocks.

Although the current type of management applied to permanent pasture or mowing was only recently established, the A r r h e n a t h e r i o n and C a l t h i o n plant communities have spread over wide regions, with the result that widely separated communities nevertheless are of comparable composition. Their wealth of plant species permits botanico-sociological subdivisions to take account even of small local species variations. This allows consideration of special agricultural problems from the point of view of vegetation science.

These firmly established communities form a basis for suggesting improvement of grassland by management. This improvement in established swards can be brought about in many cases by management practices which favour the increase of desired minority species at the expense of those species which are less desirable.

The system of classifying permanent grassland by plant communities also provides a basis for a highly developed mapping system. The knowledge gained of the geographical distribution of the communities shows many relationships between effects of location and type of plant community.

In this context the relevant treatises for Central Europe are: Balátová-Krause: "Der Beitrag der Vegetationskunde zur Regulierung des Wasserhaushaltes im Grünland" ("The contribution of vegetation science to water economy in grassland") and Dietl: "Vegetationskundliche Grundlagen der Verbesserung des Graslandes in den Alpen" ("Improvement of grassland on the basis of vegetation science in the Alps").

Compared with this near-idyllic picture, the paper by Whyte: "Analysis and Ecological management of Tropical

X

Grazing Lands" presents a sharp contrast. Tropical grassland is mainly derived from areas that were formerly wooded. This woodland did not contain grasses in its undergrowth, thus no indigenous grasses could serve as forage plants in the newly formed permanent pasture, whose swards are mainly composed of new species, including a great variety of grasses. The mastery of the taxonomy of these new species presents difficulties, which will be further increased when their demands on location and their value as feeds are to be evaluated. It is also to be expected that some of these newly arrived species will change genetically due to the stress of their new environment. Legumes, which are a highly valued component of swards in temperate climates, are rarely present in the tropics. Shrubs and trees, on the other hand, some of which are browsed, occupy a high percentage of the area.

In the majority of cases no firmly established associations have yet been formed, and violent successions, in which plant communities are destroyed and new ones originate, are the norm.

The improvement of these pastures through management is hindered by alternating periods of drought and heavy rains, which, in the areas suitable for pastoral use, are not conducive either to the retention of swards of high ground-cover or to the development of soil. In addition, fires and irregular grazing are detrimental to the pasture, and orderly management is made more difficult by the size of the areas involved and deficiencies in infrastructure, e.g. traffic arrangements and facilities for manufacturing agricultural products. The provision of adequate drinking facilities for grazing animals sets another major problem.

With tropical grassland it does not suffice to know "what is present at this moment", but it is essential to be able to predict what will be present in the near future. Whyte therefore recommends a classification of the swards according to their place in the sequence of succession. Indications for this are given by plant species, which according to experience, are associated with certain stages of the sequence, e.g. therophytes in swards which are open to destruction. Judgments of this kind are frequently associated with practical experience gained *in situ* by specialists, and it is difficult to pass this experience on to others. These are examples of typical attempts at solving the specific problems of the tropics.

Werger's paper "Application of Zurich-Montpellier Method in African Tropical and Subtropical Range Lands" is an example of advanced vegetation science. The plant communities discussed there have all the characteristics considered desirable in the Central European method. However, technical difficulties will still occur with young unstabilised communities, which are subject to quick succession. To shed light on the latter, repeated botanical surveys are needed on permanent sites. However, these surveys are difficult to undertake over a period of years even within the narrow confines of Central Europe, all the more in the tropics.

After the growing conditions, the developmental patterns of grassland in Australia, monsoon Asia, Africa south of the Sahara and South America are described by White. Dominant species, limits of environment and trends in succession are mentioned. Extensive references to the literature refer the reader to the original papers.

Werger makes a small but significant contribution to the understanding of the tropical grassland ecosystem with his paper "Effects of Game and Domestic Livestock on Vegetation in East and Southern Africa". He covers the differential burden put on the pasture by large herds of wild animals on one hand, and by the settlers' sheep on the other hand. The wild herbivores, irrespective of their numbers, do not damage the pasture, as these herds are composed of many animal species, each of which has different grazing habits. The sheep are more stationary and put a one-sided burden on the sward, while the wild herbivores browse to a large extent on the leaves of shrubs and trees.

Following Whyte's discussion of the problems of tropical grassland, as seen from the point of view of natural science, Andreae provides an extensive general view of tropical regions according to their economic and agricultural structure. His inclusion of areas in which arable farming or bush and tree culture are predominant does not preclude the inclusion of his paper "Farming Regions in the Tropics" in the handbook on "Grassland Husbandry". This extensive survey, which will find, within the Handbook, its main distribution in temperate climates, should establish, for readers outside the tropics, first understanding for any endeavour towards agricultural progress in these alien areas.

All the stages from nomadic herding, utilizing sparse natural vegetation, to the completely different conditions created by paddock grazing systems are described. References are made to climate, suitable grazing animals and the output of saleable produce; in addition to this main subject, the aspects of the humid savannah and the rain forest and their hostility to grazing offer an insight into the environmental conditions for grassland in the main areas of tropical pasture.

The Mediterranean basin is also a world of its own, equally as unique as the tropics. This area is discussed in Le Houérou's paper "Plant Sociology and Ecology applied to Grazing Lands Research, Survey and Management in the Mediterranean Basin". Here, too, thousands of years of destructive farming have ruined the original woodlands and created secondary types of vegetation which consist of sclerophyllous shrubland and herbaceous plants. The author describes the general characteristics of the Mediterranean area and at the same time the great differences which are created by climatic conditions between Portugal and Israel, and the Riviera and the northern Sahara. The plant communities range from the M o l i n i o - A r r h e n a t h e r e t e a in the west to the *Artemisia* steppes in the east, and from the mesophile B r a c h y p o d i e t u m   p h o e n i c o i d e s in the north to the *Salsolaceae* desert communities in the south. Their stable botanical composition permits detailed studies of their sociology to be made. The differences between geographical locations are also shown by the occurrence of numerous individual species, e.g. *Quercus ilex, Pinus halepensis, Poterium spinosum, Argania spinosa*. Special consideration is given by Le Houérou to precipitation, which is the main factor in determining the vegetation types in the Mediterranean area.

Looman's paper on "Applied Phytosociology in the Canadian Prairies and Parklands" gives an introduction to temperate climates. There, a continental climate with an annual rainfall of between approximately 350 and 470 mm and temperature extremes from +35° to −40° C prevails, and this has led to the formation of chernozemic soil. The southern part of this extensive area is mainly covered by grassland, whereas the north is covered by a mosaic of grassland, shrubland and woodland. In the prevailing conditions forests cannot regenerate, and this differs from Central European conditions, where, with a more maritime climate, brown earth is formed and new forests can be created without difficulty.

This grassland contains few species and varies little in composition. Within the scope of this handbook, it seems important to state that no species of indigenous grass or legume is capable of greatly improved yield, and moreover none are suitable for cultivation as forage crops in arable leys. The necessary increase in productivity can only be obtained by sowing grass-legume mixtures of foreign origin.

The tundra in the Soviet Union is the subject of Andreyev's paper. Extreme climatic stress has created a uniformly poor environment, which is locally highly differentiated and in which the reindeer is the only large herbivore that can hold its own. The indispensible basis of successful reindeer husbandry is the knowledge of the plant communities, of which *Cladonia*, *Cyperaceae-Gramineae* or low-growing *Salix* species are the main sward components. These occur in small mosaics, which result from the diversely acting soil-forming powers of the sub-arctic climate (cryoturbation, permafrost, duration of snow cover, natural drainage). Each plant community has its own specific value as a forage source, and this value is determined not only by its yield but also by its accessibility. For example, new loose snow up to a depth of 1.5 m can be scraped away by reindeer, whereas a much thinner snow cover capped with ice can prevent access to the forage.

The author gives a general botanico-geographical impression of the tundra between the Kola peninsula and the extreme northeast of the Soviet Union.

VEGETATION HISTORY

The development of rapidly changing tropical grassland will be more predictable, the greater the knowledge of its history. Whyte contributes an extensive chapter on "Antiquity and Evolution", in which the subject is followed back to the Tertiary period. This paper is based on the assumption that the palaeoclimatic-ecological interpretation of geological history permits an estimate of the types of grassland communities and species that might or might not have been present during specific periods of history.

For example, the tolerance of the surviving grass species of shading by the forest canopy should be tested, as unbroken woodlands formerly covered large areas. Also, the pathways taken by the newly arrived plants should be examined, as relatively few species of grass changed their habitat from woodland to grassland. These pathways will often coincide with man's migration and trading routes. Another interesting topic is the man-made change in climate.

Further, the remains of old, multi-species plant communities are valuable as gene pools and as contrasting examples of genetic change to which newly imported plants may be subjected.

With these points in view, Whyte's paper endeavours to analyse the history of the vegetation in monsoon Asia, including the continental drift of the Gondwana complex and the formation of

XIV

the Himalayas. Furthermore, Australia is briefly discussed, and Africa south of the Sahara and South America are considered more fully.

According to Le Houérou, man is the most important formative force in the recent vegetation history of the Mediterranean basin. This history presents a uniform picture of over-use, of which the main feature is the disappearance of woodlands. These woodlands suffered destruction in particular during wars and by fire. Another important point is the destruction of grassland areas by year-round grazing without allowing periods for recovery. For this area, the future looks bleak due to the population explosion coupled with the continued use of archaic farming methods.

The vegetation history of the Great Plains of North America as depicted by Looman presents a surprisingly multi-coloured picture. These monotonous grassland areas were, until recently, thought to be climatic climax formation, millions of years old. These grassland areas will now have to be considered as being relatively young and to have resulted from non-climatic effects. Glacial and post-glacial changes in climate, the composition of the wild animal population and the Red Indians' settlement history are used as evidence for this hypothesis. The information available gives rise to the idea that wood and grassland found better growing conditions alternatively at short intervals. Only relatively recently did grassland become finally dominant, assisted to a great extent by the huge herds of bison, which are estimated to have numbered many thousands. An example of rapid change is that these herds were able to maintain their maximum numbers only for a short period, possibly 50–100 years, but that this was sufficient time to displace the woodlands.

The dearth of species also agrees with the assumption that the grassland of the Great Plains has only recently attained its present extent, as Whyte classified swards in the tropics with a high number of species as "old" and with a low number as "young".

APPLICATION OF VEGETATION SCIENCE

In the tropics, improvement of grassland depends to a greater extent on socio-economic factors than on natural factors. Apart from Australia, which is a relatively wealthy country, the use of fertilizers and associated modern systems of land use are generally difficult to establish in the tropics. So far, only cheap and "practicable" methods are feasible. Some of these are: regulation of herd size, control of grazing periods and nomadic grazing cycles associated with the location and temporary closure of watering places.

Recently, a polarisation of domestic animals has taken place. On one hand, animal husbandry has retreated to land of the lowest agricultural value which is not suitable for cultivation, and on the other hand production centres are formed near towns, where they are dependent on sown swards and the facilities made possible by a fixed location, e.g. sewage irrigation. Whyte's contribution is mainly concerned with extensive production methods. The utilization of shrubs and trees plays an important role in grazing in this system. They often have a growth rhythm different from that of the grasses and still offer food when the grasses have died off. Improvement of swards through preservation, as is the rule in Europe, is not practised in the tropics.

Andreae's contribution to "application" is mainly concerned with management of resources. The main aspect here is to overcome periods of drought without large losses of animals. This can be achieved by adaptation to the conditions or by overcoming periods of drought by cultivating forage crops. Means of adaptation quoted are: choice of animal species and breeds, choice of the right enterprise (milk, meat, wool), proper planning for time of year, e.g. concentration of breeding and fattening periods to coincide with peak herbage production and lastly, an increase or decrease in stock numbers to match feed supply.

In order actively to overcome periods of drought, financial investment must be made. This, in turn, is dependent on high prices for produce, i.e. on a general increase in the purchasing power of the population. All the intermediate steps from nomadic grazing to paddock systems which utilise sown fodder crops, are discussed. Without the knowledge of these basic managerial and economic requirements, vegetation science would not be in a position to propose management systems which are valid in every respect.

Le Houérou, in his contribution, maintains a close connection between vegetation science and its application. He mentions growing conditions for and the potential of forage species and their geographical distribution and usefulness under different conditions. He particularly emphasizes the relationships between vegetation science and economic factors. As soon as a plant community is identified, its potential for economic development has

XVI

to be tested in detail. This would entail research into multiple relationships of cause and effect, which are pre-determined by far-ranging and obscure causal connections. Preliminary examination of these relationships reveals only the coincidences and not their causes. In endeavouring to follow the intricate interrelationships of whole ecosystems by studying their vegetation, detailed studies of their causal chains are necessary.

The uniformity of the locations and the monotony of the vegetation in the Canadian Prairies enables Looman to cover the subject of "application" briefly and comprehensively. Due to the low yield of native forage plants when compared with imported ones, grasslegume mixtures are sown in large areas as forage crops, as production can be increased only by these means and not by the improvement of indigenous communities. Of special interest to the vegetation scientist is the fact that none of the locations of the indigenous communities differ in the growth conditions they offer for these forage crops. Another important project is the resowing of the fallow arable areas of the Prairies with pasture species. Other points considered are the correct regulation of grazing, maintaining reserve areas of pasture in addition to the calculated minimum and the grazing of open woodland and parkland.

All "application" automatically raises the question of concomitant effects of increased productivity by grassland. Each additional large animal that can be supported requires 50 l water per day, and in the Prairies this is often difficult to provide. Also, any anticipated increase in financial margins which may be attributable to fertilizer use can be thwarted by variations in meat prices.

Dietl's contribution covers the improvement of Alpine grassland, where yields were very low due to traditional methods of utilization. However, yields can be increased by moderate expenditure on fertilizers and by improved utilization. To these ends, vegetation science supplies the knowledge of the yield potential of the different communities hitherto low in yield. The potential of an area of low yield can be judged by its plant composition and an assessment made of its suitability for improvement. This is possible because the sociological classification of the Alpine plant communities is complete, and the knowledge of the natural and anthropogenous growing conditions is far advanced. Two examples of this are given to illustrate "application".

In the tundra, which is generally avoided by man, grassland

management is controlled to an even greater degree by natural forces and conditions than in the relatively highly populated Alpine regions. Andreyev's paper therefore deals with the requirements for the reindeer breeder to familiarize himself with the precise plant composition of the pastures and their micro-climatic peculiarities. In this, he can receive substantial support by extensive aerial surveys of the vegetation. This enables him to control the direction and time of migration of his herds in such a way that the best forage available is always used and that areas are reserved at times of surplus for grazing during the severe conditions of late winter. In discussing this, Andreyev's contribution reaches the essence of "Applied Vegetation Science".

The problems of regrowth of the *Cladonia* swards, vital for reindeer, are discussed extensively. Highly intensive grazing will lead to the destruction of the lichens, which grow extremely slowly. A balanced relationship between the severity of grazing and regrowth potential is therefore essential for continued productivity by the pasture.

Apart from this natural method of husbanding vegetation by management practices, experiments are being carried out which attempt to solve the problems by biochemical-technological means. For example, areas of *Cladonia* have been sprayed from the air with a growth promoter, and the results so far indicate that the rate of regeneration of the pastures can be increased. Finally, Andreyev discusses the pastures composed of phanerogamous plants, the optimal percentages of these and lichen in animal rations, breeds of reindeer and experiments carried out to break hard-frozen snow with spiked rollers in order to enable animals to reach the buried vegetation.

While the task set in the Alps and the tundra is to maintain the yield potential of low-yielding pastures or to take the first steps towards improving their yields, the main aim for the intensively managed lowland regions of Central Europe is to achieve a further increase in the already high level of production. However, this can be performed only in small, carefully considered steps.

To substantiate this, Balátová-Krause's contribution gives examples of highly differentiated ways of applying vegetation science. The basis of application is the "indicator value" of the plant cover. This value allows conclusions to be drawn from the vegetation about environmental factors. Methods of defining the indicator value are discussed, with special reference to the difference between factors which operate directly or indirectly. Also considered is the necessity for cooperation with other areas of

XVIII

science, e.g. soil science and geomorphology. Examples quoted and partly substantiated with vegetation maps are three agricultural projects, one each in Central and in East European inundated meadows and one in a mountain valley. Finally, other possibilities of application of vegetation science are listed.

THE STRUCTURE AND INNER DYNAMICS OF PLANT COMMUNITIES

When agriculture asks the advice of vegetation science, the first question asked is for increased and stabilized yields. This requires information on long, involved causal chains. One example of such a chain is the application of fertilizer and all the stages leading to the utilization of the increased production due to the fertilizer. Frequently, it is not easily recognizable at which point in the chain and by which mechanism the intervention takes effect. To avoid false conclusions, the internal structure of the vegetation has to be considered as well as structural processes. This is the main subject of Rabotnov's paper "The Influence of Fertilizers on the Plant Communities of Mesophytic Grasslands".

The author sketches an impressive picture showing the multitude of interactions which are hidden by the simple expression "fertilizer application to a sward". Two examples may be given here:

Seed production of permanent pasture plants is increased by fertilizer application, and this increase, in turn, may improve the rate of spontaneous pasture rejuvenation. However, due to the more rapid growth of the fertilized sward, which necessitates earlier cutting, the plants will not reach the seed producing stage. Thus, one biological process of increased quantitative production prevents, via management changes, i.e. earlier harvesting, the other biological process of rejuvenation of the sward by seedlings.

The different species of grass react to increased fertilizer application less in a way specific to a single species than in a way more related to their competitiveness with other species. Thus, plants which have a high growth response to fertilizer will suppress adjacent grasses of lower growth potential. If, however, the superior competitors are absent, the inferior ones become dominant.

The relationships in these systems were found to be extremely intricate. The adviser has the difficult task of selecting the appropriate solution for a specific question from a multitude of possibilities.

Geyger considers the measurement of the assimilating surface and its ratio to the covered soil area (leaf area index, LAI) to be a method by which knowledge can be gained of the pathways of organic production. Firstly, the author discusses the morphological structures, which determine the distribution of light within the plant canopy. Secondly, methods of measuring LAI are considered.

In a special part given over to the LAI of some grassland plants, their assimilative productivity and the connection between both these parameters and leaf arrangement are discussed. The results give some idea on the optimal angle of leaves and even lead to the suggestion of breeding cultivars with a particular leaf angle. However, there are still some contradictions in these observations.

The comparison of the LAI-values of five grassland communities with locations between 5 and 1500 m above sea level forms the main part of the paper, and a positive correlation between altitude, productivity and the rate of increase of LAI was found. At low altitudes, growth starts early but remains relatively slow, also it takes longer for LAI to reach its maximum value, whilst with increased altitude, growth rate increases but growth itself commences later. LAI too increases faster than at low altitudes. Only when these relationships are extrapolated to very high altitudes, which are subject to adverse micro-climates, do the predicted values greatly overestimate sward productivity and to a lesser extent LAI. Also, the production of organic matter per unit leaf area decreases with increasing altitude.

LAI is increased by fertilizer application, and this effect is greater when the original conditions were poor. In the examples quoted, the increased production is mainly due to an increase in LAI.

The last chapter of this paper deals with the comparison of the LAI values for all the forms of vegetation so far researched, from a field of maize to tropical rain forest. The results show that all these types of vegetation, widely different in structure and location, have LAI values with similar limits, the mean minimum and the mean maximum values being 1—3 and 12—15. The reasons for these variations are shown to be due to different geographical locations. Therefore, LAI is a measure which does not have a large immediate effect on production.

XX

1 ANALYSIS AND ECOLOGICAL MANAGEMENT OF TROPICAL GRAZING LANDS

ROBERT ORR WHYTE

Contents

n.b. The Latin names of plants in this chapter are those used by the respective authors. Although some may no longer conform to modern usage, it has not been possible to check them without reference to specialists in the taxonomy of the many plant families involved.

2

1   ANALYSIS AND ECOLOGICAL MANAGEMENT OF TROPICAL GRAZING LANDS

1.1   Introduction

1.1.1   OBJECTIVE AND SCOPE

The vegetation sciences, or an appreciation of botany and ecology in their widest aspects, are considered as the essential basis for economic land management and plant and animal husbandry in the tropical grazing lands.

The term, grazing land, is to be preferred to the more general and physiognomic term, grassland, which is perhaps applicable only to certain high altitude types of vegetation in the tropics. The tropical grazing lands, the present botanical composition and improvement of which are to be discussed, are comprised of a mixture, in widely varying proportions, of grasses and other herbs, shrubs and trees. For the grass assemblages within these communities, it is appropriate to use the noncommittal term, grass cover, as has already been done for Africa (Rattray, 1960) and India (Dabadghao & Shankarnarayan, 1973), to avoid involvement in the differences of opinion on terminology discussed below.

That diverse geographical entity, the biological intertropical belt, comprises the equatorial latitudes and the tropics, and to some extent also the subtropics, which are here subdivided into their relatively distinct parts:

Australia and the Pacific Islands
Asia, monsoonal and equatorial
Africa south of the Sahara, and
America: South, Central and the Caribbean.

Within each of these subregions there are, of course, extensive areas of high mountain country with temperate or alpine ecoclimates to which the criteria for tropical land environments do not apply. Their study represents a separate branch of vegetation ecology (Whyte, 1974, pp. 130–131).

The intertropical zone is not only that part of the globe lying between latitudes 30° north and south of the Equator, as proposed by some writers (Davies, W., 1960; repeated by Davies & Skidmore, 1966). Rather is it those regions of the world in which, at the lower elevations, certain families and members of the Gramineae and Leguminosae grow wild or can be cultivated. It may be a matter for discussion whether these plants in their wild communities or sown or planted crop mixtures are better indicators of a biological environment than the instruments and criteria of the meteorologists.

Subtropical and even tropical species extend down to South Africa, for possible palaeobotanical reasons discussed below (1.3.4.2). Corrientes in Argentina is included, but Rio Grande do Sul in Brazil is not (see 1.3.5.3). The brigalow lands in Australia (Coaldrake, 1970) are regarded there as subtropical, and may thus be included; it is however, found that in north-eastern Australia, the differences between the tropical and subtropical zones necessitate quite different approaches to land management. Subtropical areas with a cold season and frosts, the annual alternation on the same site of tropical and temperate conditions, are among the most difficult environments in which to develop sown pastures (Taiwan, Korea, Japan). Results of works in these areas cannot be extrapolated to fully subtropical conditions, and vice versa.

In making a resource assessment of tropical grazing lands, it is essential to accept the fact that, for technical and economic reasons, only a very small percentage of the vast area of land involved can be converted to highly productive pastures. A further percentage, still not by any means large, might be improved at lower cost under the appropriate socio-economic conditions, by reseeding following ploughing, or by the surface introduction of grasses and especially legumes into the existing sward. Fencing would then appear to be essential to provide the control of the harvesting mechanism in the ecosystem — the grazing animal or the cutting knife — and thereby to ensure the persistence of the synthetic plant communities so created.

Over the greater proportion of the tropical grazing lands, however, only those ecological and related types of management discussed in 1.7 to 1.7.6 are applicable. Those who would use these methods should therefore be fully informed of the principles and practice of succession and synecology in the widest sense, in so far as this branch of vegetation science relates to those communities which are required to provide sustenance for the wild and domestic livestock in diverse biological and economic ecosystems. It is realized that it is ecologically heterodox to speak of succession in one component only of a complex biological ecosystem of tree/shrub/gramineous ground cover, namely the grass

4

communities, but it has its practical advantages.

Most of the current research and published literature on the tropical grazing lands, natural and artificial, is concerned with what may be described as agronomic and managerial questions, with advice to the grazier or the farmer as the immediate objective. Thus it deals with management of grazing areas (frequently called range lands) by fencing, control of brush, stimulation or introduction of superior species with perhaps the application of fertilizers. The agronomists are also concerned with the establishment in appropriate environments of sown or planted grass/ legume crops and their subsequent management by optimal control and management of the grazing animal, the so-called animal agronomy, for the maximum benefit of both sward and animal.

In order to discuss this "applied" work alongside data of a more fundamental nature for the benefit of vegetation scientists in general, we consider the place, the reasons for the occurrence of grass covers in natural vegetation, beginning with the ecological history and evolution of these grass covers, with their associated tree, shrub and herbaceous species, in response to natural and anthropogenic factors; the analysis of these covers according to botanical criteria, and the assessment of their value as an actual or potential economic resource; the nature of ecological regression and progression within the grass covers themselves, and in the vegetation of which they form a part.

What is frequently called range management is the correct integration of wild or domestic livestock with the plant species available for grazing or browsing in the natural vegetation. It is the response to ecological management of the plant species which is considered in this more botanical approach. The role of fire as an ecological/historical factor and as a management tool is of major importance throughout the zone of tropical grazing lands.

1.1.2   SOCIO-ECONOMIC FACTORS

Although the ecoclimatic and biological environments may be sufficiently similar in different parts of the intertropical zone to make a particular technique such as the establishment of sown pastures a feasible proposition, there are dominant socio-economic factors which limit or actually inhibit such developments. What is economically acceptable in the high-capital land economies of Australia is not applicable to the indigenous, low-capital communities of Asia and Africa. It then becomes necessary to propose realistic but cheap methods of improvement of grazing resources, e.g. by better management of free ranging livestock and a limited amount of surface seeding of legumes.

The management of tropical grazing lands calls for the adjustment of livestock numbers (domesticated and wild), their seasons of grazing and their movements (nomadic or migratory) in such a way that the grazing lands are maintained at the most desirable botanical composition and stage of succession (Whyte, 1975a). In the drier areas, this may be achieved by appropriate location of stock watering facilities and by their periodical closure when the area they command begins to suffer through excessive pressure of livestock.

Due to the increasing awareness of certain economic factors, which differ in relation to the specific biological environment and the socio-economic structure, there is a marked tendency towards some degree of polarization in policy, research and practical advice on grassland and pasture matters. In most of the tropical zone outside Australia, it would appear that the provision of feed for livestock will be concentrated on the uncultivated and/or uncultivable land at one extreme, and on certain highly efficient areas of fodder cultivation at the other (Whyte, 1971).

Three actual or possible situations may be mentioned, as variants of this polarization in tropical and borderline conditions. These, and many others which could be quoted, indicate the great importance of correct management of the tropical grazing lands, as one component of these economic systems.

India:
a. extensive free-range grazing for sheep, goats and camels, and for the rearing of draft cattle, breeds adapted to the seasonal availability of feed on the semi-arid grass/shrub communities on the uncultivable lands;
b. no sown pastures or leys; fodder crop acreage on cultivated land insignificant and likely to decrease due to demand for food for direct human consumption;
c. village cattle fed on crop residues and concentrates;
d. cultivation of gross-feeding African grasses on sewage farms, cattle colonies and other areas where a very high level of plant nutrients and water can be maintained; green feed (supplemented with concentrates) for high-producing milch cows and buffaloes; economic return per unit area in terms of an ultimate livestock product such as milk higher than can be obtained from any other crop.

Japan:
a. sown pastures on mountains and sloping land too steep for crops, for grazing in the growing season by dairy heifers and young beef animals;
b. no sown pastures or leys and insignificant area of fodder crops in the cultivated areas (summer monsoon conditions);
c. stall or yard-feeding of productive milch cattle in zero grazing systems providing green or conserved feed, plus concentrates;
d. feeding of cattle for beef production and of growing stock from hills (in winter) in feed lots with locally produced feeds and imported grain.

Kenya:
a. free-range grazing of unimproved Boran cattle on the semi-arid lands;
b. fattening in special feed lots, using maize grain (1 per cent of total national production) and/or silage, and molasses as a by-product of the sugar-growing industry around Lake Victoria (UNDP/FAO Project).

See bibliography no. 1248 of the Commonwealth Bureau of Pastures and Field Crops for 149 references to literature on grazing systems in the tropics and subtropics.

1.1.3   CONTRASTS BETWEEN TROPICAL AND TEMPERATE LATITUDES

The scientist who has received his training and early experience on temperate grazing lands and pastures, and who comes to the intertropical zone in programmes of technical assistance, is faced with a set of entirely different conditions.

Tropical grazing lands in their natural, unimproved condition are composed of a great number of species of Gramineae, which present difficulties in recognition, taxonomy, ecology and genetic composition and behaviour. Some grasses have achieved pantropical distribution without necessarily being good pasture plants.

Specialists in the evolution of the grasses in temperate latitudes will be aware of the proposal that all angiosperms, including the Gramineae, arose in the equatorial belt. They will therefore be particularly interested in what is stated below regarding the possible antiquity, evolution and migration of grasses within the tropics. This may give a guide to the possible manner, place and time of the evolution of temperate from mesophytic tropical species, either during an earlier period of tropical climate in higher latitudes, or by long-distance migration from the tropics as delimited today. For example, the origin and evolution of the Laurasian grasses which meet the Gondwanian or their derivatives along specific contours in the Himalaya are an intriguing problem (1.3.2).

Species of the Leguminosae, so important in temperate pastures, do not acquire great significance in the tropics until the higher levels of intensification are reached. Shrubs and trees are essential components of tropical grazing lands, especially of the more arid zones.

The contrasts between wet and dry seasons are much greater over vast areas than in the temperate regions. After the original vegetation has been destroyed, the communities in the secondary stages of succession are exposed to increasing or recurrent changes in the habitat, towards greater aridity, induced mainly by man and his domestic animals on the free range. Fire has played, and continues to play, a significant role in the origin, botanical composition and management of tropical grazing lands.

As the grass covers over wide regions of the tropical grazing lands are still relatively young and not yet stabilized, as in Europe, existing as it were in a constant state of flux, the capacity to interpret climax and plant succession becomes much more important. Provided seed-producing plants of species higher in succession still exist, promotion of succession upwards may be a rapid and efficacious method for initial management, following reduction of the biotic factors associated with excessive or badly-timed grazing pressure.

Perhaps above all, the newcomer comes to accept that enough is known about most types of tropical grazing lands for correct advice to be given on measures for management and improvement, but that their application in the field is limited, if not actually prevented, by overriding social and economic factors. At the sociological level, one has first to consider the distinction between the cultivator and the pastoralist, and the low technological level and poverty of the latter. Many traditional customs and prejudices are associated with the different types of livestock and their husbandry (capital on the hoof, bride price, sanctity of bovines, etc.).

The major economic problems facing the would-be improver of the resource and of the types of animal husbandry practised thereon are the vast extent and low carrying capacity (and there-

8

Table 1.1. Gradients of increasing availability of soil moisture, soil fertility and intensification of feed and livestock production

Resource	System of use	Type of livestock	End product
1. Unimproved range, mostly arid and semi-arid	Free range grazing, mostly nomadic and/or migratory	Cattle, sheep, goats, camels, wild animals, adapted to wide seasonal fluctuations in supply of feed and water	Beef, meat, wool, hair, hides, skins, wildlife products, tourism
2. Improved range, semi-arid or savannas in humid tropics	Numbers and movement of stock controlled by herding, or by closure of watering points	Similar	Similar
3. Dryland fodder production, crop residues and stubble grazing	Stubble grazing in dry season in a monsoonal environment, with yard feeding or free-range grazing in the rainy (cropping) season	Slight improvement in quality of livestock, in line with slight improvement in total and seasonal availability of feed	Specially relevant to cattle populations in villages of India (25% of world's total), giving minute amounts of milk in a four-month lactation
4. Sown or planted synthetic stands for grazing	Rotational or fixed grazing during growing season in fenced pastures, other arrangements for dry season	Change of type of livestock and introduction of superior breeds justified, especially milch animals or cattle for fattening.	Milk, meat or beef according to social acceptability
5. Semi-intensive fodder production on cultivated land	Cut and carry, zero-grazing system for crops such as lucerne, berseem or grasses growing alone or with legumes	Production from superior animals ensured by availability of high-protein cut green fodder, or hay or silage	Similar
6. Intensive fodder production	Cross-feeding African grasses grown with irrigation water plus heavy dressings of nitrogen or with treated city sewage	Ample green feed, requiring supplementation by concentrates, for productive milch animals in specialist dairy units or milk colonies	Milk
7. Import of feeds, fodders, concentrates or coarse grains from elsewhere in same country or from abroad	Feeding as supplements to roughages and cut green fodder, in stalls, yards, or feedlots	(a) Productivity of high quality milch cows and buffaloes assured (b) Cattle brought in from range or mountain pastures for fattening	(a) Milk (b) Beef

9

fore low potential return on investment) of the grazing lands, lack of funds sufficient to cover such an area, and the problems of extension among such scattered, sparse and generally moving populations. Accessibility of the grazing lands to markets and to centres of health and education is also significant. Four levels of investment and production can be recognized, providing for a progressively more reliable seasonal availability of better quality plant material (1.7.1).

The relative composition of economic ecosystems and of the different levels or plateaux of production on a scale of increasing intensity of utilization are set out in Table 1.1. Decisions regarding the technical and economic practicability of raising an economic ecosystem from one level to the next higher, or of combining two or more ecosystems in an integrated productive system, may be made on the basis of the special economic criteria that apply to the tropical situation. The higher the level of production, the more costly is the feed. The areas involved are so vast, the climatic situations so severe and widespread, and the great livestock populations of such low productive potential, that concentration of effort on relatively small and favourable areas and enterprises must surely be economically correct.

1.1.4 TERMINOLOGY

There is still considerable difficulty in arriving at an internationally acceptable terminology for plant formations in general, and for the tropical grazing lands in particular.

Recommendation no. 5 of the UNESCO Abidjan Symposium on tropical soils and vegetation may be quoted: "Terminology for the classification of plant formation".

"In view of the serious difficulties, for all tropical vegetation students, resulting from the present confusion in vegetation terminology for the classification of plant formations, and with appreciation for the results of the meeting organized by CCTA/CSA at Yangambi in 1955, the Symposium invites the Director-General of UNESCO to convene a similar meeting of specialists on the vegetation of various tropical regions of the world in order to establish a project for a unified nomenclature of tropical plant formations on a world-wide scale, classified on the basis of the characters of the vegetation itself rather than on those of its environment."

There is considerable disagreement on the correct nomenclature to apply to the different types of tropical grazing lands. The term "savanna" is used throughout Latin America, by French, German and some British writers in Africa, particularly West Africa, and by some writers referring to certain types of forest/grass associations in Asia and Papua New Guinea. British and American workers in East Africa, however, have in general rejected the use of the term "as being a South American one seldom correctly applied outside of its continent of origin" (Pratt, Greenway & Gwynne, 1966). Some object to the use of a Spanish word of Caraib Indian origin for different physiognomic types of vegetation.

Bourlière & Hadley (1970) state: "A major factor in classifying savannas is that the physiognomy of anthropogenic savannas is very often quite similar to that of natural ones. Human influence has undoubtedly extended the range of natural savannas so that savanna communities occur under many different climatic and edaphic conditions".

Keay (1959) has discussed the origins of derived savannas. The UNESCO meeting on forest borders (Hills & Randall, 1968) concluded that savannas are 1. not a climatic climax, 2. the majority of units of this vegetation can be considered as anthropic, and 3. there are also natural savannas which can be considered as an edaphic climax.

The criteria applied in the definition of savanna in tropical South and Central America are partly geographic and partly derived from the physiognomy and structure of the vegetation (van Donselaar, 1965); the only point of agreement is that there is a ground layer, and that this ground layer is always ecologically dominant. These definitions have been made by botanists and phytogeographers; geographers such as Jaeger (1945), Troll (1952) and Lauer (1952) have adapted the nomenclature to their specialized disciplines, in particular regarding the relation between vegetation and climate.

(i) Lanjouw (1936) "savannahs are plains in the West Indian Islands and northern South America covered with more or less xeromorph herbs and small shrubs and with few trees or larger shrubs."

(ii) Dansereau (1951) "savanna: more or less closed ground layer, very discontinuous upper (woody) layer."

(iii) Beard (1953) "savannas are communities in tropical America comprising a virtually continuous, ecologically dominant stratum of more or less xeromorphic plants, of which grasses and sedges are the principal components and with scat-

11

tered shrubs, trees or palms sometimes present".

(iv) Eeckhout (1954) "savanna: open xerophytic herb formation in the tropics with scattered xero- or tropophytic trees or shrubs, isolated or in clumps."

(v) Dyksterhuis (1957) "xeromorphic grassland containing isolated trees."

(vi) Sillans (1958) "formation characterized either by a continuous herb layer, mainly consisting of more or less high and more or less dense grasses, or in addition by a layer of shrubs or by a layer of trees of very variable density."

(vii) Fosberg (1958) "closed grass or other herbaceous vegetation with scattered trees."

(viii) Heyligers (1963) "tropical vegetation that is neither swamp nor primary nor secondary forest."

(ix) Van Donselaar (1965) "a savanna or a campo is an area with a xeromorphic vegetation comprising an ecologically dominant ground layer consisting mainly of grasses, sometimes together with sedges, and with or without trees and/or shrubs either forming a more or less continuous layer, or occurring in groups, or isolated."

(x) Walter (1969) "tropical vegetation community, ecologically homogeneous, grass component dominant and woody plants dispersed among Gramineae, all growing in identical environmental conditions; savannas contain two types of plants in labile balance competing for water; to understand their interrelations, the ecophysiology of the Gramineae and the typical woody plants must be known."

The 1955 Yangambi Conference (CSA, 1956) established definitions for "formations herbeuses, savane, steppe". Discussing their applicability, Descoings (1973a) proposes a term for all the African tropical or subtropical herbaceous plant formations in which are dominant, on a physiognomic or structural basis, populations composed of gramineous species (Gramineae and Cyperaceae): formations herbeuses = grassland plant formations subjected or not to annual fires and characterized by

a. the obligatory presence of a regular herbaceous layer, discontinuous at the soil surface, composed essentially of annual or perennial species of the Gramineae and Cyperaceae, of varying height and density, with which other herbaceous plants may be mixed in relatively small and varying proportions, and

b. the non-obligatory presence of a regular population of woody or partly woody plants (shrubs, trees, palms) of

12

varying height and density, influencing the gramineous components to a varying degree but never eliminating them.

This type of vegetation cover is considered to rank with the other major subdivisions, e.g. forêts claires, forêts denses, and to embrace savanna, steppe and pseudosteppe.

1.2 Grazing Lands in Tropical Ecosystems*

1.2.1 CONCEPTS AND DEFINITIONS

A. G. Tansley (1935) first introduced the term ecosystem: "The more fundamental conception is ... the whole system (in the sense of physics) including not only the organism complex, but also the whole complex of physical factors forming what we call the environment ... We cannot separate them (the organisms) from their special environment with which they form one physical system ... It is the systems so formed which ... (are) the basic units of nature on the face of the earth ... These ecosystems, as we may call them, are of the most various kinds and sizes."

The patterns of the earth are expressed as expanses of forests, grasslands and croplands, as rivers and lakes, estuaries and oceans. "Each is physically and biologically different. Each is occupied by different organisms well adapted to the environment in which they are found. Yet, in spite of the differences, oceans and lakes, forests, grasslands and deserts all function the same. Energy fixed by plants flows through them. Nutrients are deposited in the tissues of plants and animals, cycled from one feeding group to another, released by decomposition to soil and water and recycled again. Rarely are the desert or the forest, the stream, the lake or the sea independent of one another." (Smith, 1972).

Definitions are given by McMillan (1960, 1971): a. the terms ecotype, ecotypic differentiation, ecogenic gradient, refer to genetically based variation that is correlated with habitat, b. community is applied to the sum total of organisms in a given area, c. population is applied to the one or more individuals of a close genetic lineage in a given area, d. ecosystem is applied to the sum total of organisms (the community and its included

---

* For a more detailed discussion of the ecosystem as a biological entity and an economic resource, see Land and Land Appraisal by R. O. Whyte, Junk, The Hague 1976.

populations) and to their relations with their environmental surroundings in a given area, and e. ecosystem-type results from the lumping or sorting of certain ecosystems into a particular kind.

Expanding this further, with reference to ecotypes and the functioning of ecosystems composed of grasses and other plant species, McMillan (1971) states (see also Wiens, 1972; Ellenberg, 1971); a. the role of the ecotype in ecosystem function is primarily one of allowing a community of organisms to adjust to its habitat, b. the simultaneous selection of ecotypic variants within different kinds of organisms occupying a given area results in harmonious functions of a particular ecosystem, and c. the selection of ecogenetic gradients results in the continuity of an ecosystem-type over geographic diversity.

P. Duvigneaud & J. D. Lockie (in UNESCO, 1966) gave the concepts and definitions of the ecosystem terminology at a Regional Seminar on the primary and secondary productivity of tropical savannas, convened by UNESCO in Nairobi, Kenya, in October, 1966. The following sections A—F are quoted from the preliminary report:

A *Levels of integration of biological material* (populations, communities, biomes).

1. Biological materials (principally holo- and heteroproteins and lipoids) are integrated in nature in a number of levels of organization of increasing complexity: cells, individuals, populations, communities. Ecology is particularly interested in populations and communities.

2. A population is a collection of individuals of the same species in a given area at a given moment. Examples: populations of *Pennisetum benthamii*: population of *Equus zebra*.

3. A community is an assembly of different species populations (thus of individuals belonging to different species) in a given area at a given moment.

One can distinguish (and study separately if one wishes): plant communities, phytocoenoses: animal communities, zoocoenoses: communities of microbes, microbiocoenoses: communities of fungi, mycocoenoses.

4. The collection of phyto-, zoo-, microbio- and mycocoenoses in a given place forms a biocoenose or better a biome. The innumerable living things which make up a biome are linked to one another in various ways of which the principal are food and distribution.

B *Food chains at the heart of the biome. Trophic levels and chains.*

1. *Trophic levels*

a. The producers in the biome are the organisms which, by photosynthesis (rarely chemosynthesis) accumulate potential energy in the form of organic material fashioned from minerals derived from the abiotic environment. These are for the most part macro- or micro-green plants.

b. The consumers are organisms which nourish themselves directly or indirectly from the organic materials elaborated by the producers.

We can distinguish: primary consumers, obtain nourishment directly from the producers, phytophagous organisms, herbivores, plant parasites; secondary consumers,

14

feed upon primary consumers, for example, carnivores which eat herbivores; tertiary consumers, feed upon secondary consumers: carnivores which eat carnivores, carrion eaters or scavengers, feed upon the corpses of the preceding consumers or the prey abandoned by carnivores.

c. The decomposers (or bioreducers) assure the progressive mineralization of organic material and its return to the inorganic world. It is a complex group of organisms which should be further sub-divided. They include among others, the insects, fungi, coprophagous bacteria, which feed on corpses not eaten by scavengers. In addition there is the immense population of fungi and saprophytic bacteria which change organic material in the soil to CO_2 and H_2O (respiration of the soil) assuring the continuity of the cycle of nitrogen in mineralization and by the fixing of atmospheric nitrogen; finally restoring to the soil the cations and anions necessary for living things.

2. *Trophic chains*

Food material thus passes from one level to another, the producers, consumers, and decomposers forming a trophic chain.

3. *Trophic web*

The same producer can serve as food for different kinds of herbivore or the same herbivore can feed on many producers. These herbivores can in that turn be eaten by various carnivores. This results, in the biome, in a multiplicity of trophic chains which anastomose in a trophic web.

C *Distributional relationships at the heart of the biome*

These are set up between organisms depending, among other things, on the place occupied in the biome (struggle for food, light, water; search for protection against an unfavourable environment or against an enemy). These border on the determined structure of the biome which in the terrestrial environment more or less coincides with the structure of the phytocoenoses.

D *The Ecosystem*

Each population forming the communities of the biome depends for its existence on factors of the non-living (abiotic) environment. These can be grouped in two large categories: those which depend on climate (climatope) and those which depend on the soil (edaphotope). If one integrates in the same system the collection of populations forming the biome and the various factors of the environment, one obtains the ecosystem; the ecosystem is a functional system which includes any community of living things and their environment. It consists of (Sukachev): biome: phytocoenoses, zoocoenoses, microbiocoenoses, mycocoenoses and all the food chains which unite them; ecotope: (factors of the environment) climatope and edaphotope.

It functions really like a machine with complicated wheel-cog system, utilizing the energy of the sun (light and heat), feeds on water and mineral substances in the soil, elaborates and transforms diverse materials by the action of living beings and their metabolism, making the materials circulate, or sometimes accumulates them all along the trophic chains, and breaks them down by the action of decomposers in an open or closed cycle.

E *Some important definitions*: Station: geographical site, marked on a map where a population, or a community, or a biome or an ecosystem exists. Habitat: all the conditions of the environment which affect a population, community, biome or ecosystem. Biotope: geographical site, marked on a map where a particular habitat exists. Ecological Niche: the total of functional and spatial aspects in a biotope that apply to one population; in an ecosystem in equilibrium, there is only one theoretical ecological niche for each species population as a consequence of adaptations in the struggle for existence. Example: in the *Acacia* savannas of East Africa, the possibility of feeding in a very dry environment on leaves mixed with spines several metres above the ground in trees of a parasol shape constitutes an ecological niche occupied by the giraffe.

15

1. Biomass: the weight of living organisms or of individuals forming a population or of the population forming a community. Biomass can be expressed in grams, kg. or tons of material dried at 70° centigrade (in order to avoid losses of nitrogen). In terrestrial ecology the biomass is usually expressed as so much per unit area (m², ha, kM² depending on the importance of the individual). Biomass per unit of surface particularly in dry savannas or "standing crop" is more or less the same as production. 2. Productivity (primary and secondary); each trophic level or each population making part of a trophic level has a particular biomass. Productivity of this level (or of this population) is called the biomass formed during a given time usually expressed as per second, day or year per unit surface area. Productivity is thus the speed with which new material is produced. Primary productivity is the productivity of the producers, that is, the green plants. Secondary productivity is the productivity of consumers (especially animals or decomposers such as fungi or bacteria). It is also possible to study the total productivity of an ecosystem.

1.2.2   ASSESSMENT OF PRIMARY PRODUCTION

A symposium in New Delhi on tropical ecology with particular emphasis on organic production (Golley & Golley, 1972) had as its terms of reference the production of the primary producers, the only group that receives its major energy input from abiotic sources. The other groups are the consumers, with food chains based on living plants, and the decomposers, with food chains based on dead organic material.

F. B. Golley and H. Leith (in Golley & Golley, 1972) state that the definitions of primary production are confusing, and sometimes lead to the assumption that primary production equals photosynthesis. To include nutrient uptake in the equation, one would have to say that dry matter production equals photosynthesis plus mineral uptake. Since plant dry matter may consist of more than 20 per cent ash, this distinction is considered to be reasonable. Golley & Leith distinguish the following production categories: gross primary production = net primary production plus metabolic (respiration) loss; net primary production = standing crop increase (above and below ground) plus litter; standing crop increase = yield plus waste. Net primary production is equal to the organic material available for the consumers and decomposers and for export and storage.

Golley & Leith then examine methods of measuring primary production, under the heads of gross and net production, potential and realized production, and the range of production in forests, grasslands (a doubtful distinction is drawn between "savannas" in Nigeria and "grasslands" in Bihar) and annual crops.

16

Any specialist concerned with advising graziers regarding the improved management of their grazing grounds on the basis of ecological succession must feel uneasy with some of the conclusions which arise as a sequel to assessments of primary production. One example will suffice. It is reported (Golley, summary to Golley & Golley, 1972) that *Heteropogon contortus* gives the highest net production at Varanasi, higher than *Dichanthium annulatum*, and higher than "natural" relatively undisturbed vegetation in Rajasthan, *Dichanthium annulatum* may be out of its true environment at Varanasi, and net production may be expected to be much lower than may be obtained in protected plots at Poona or Palghar in Maharashtra. A direct comparison between herbage cut at Varanasi and in Rajasthan is not possible, because of the higher protein content and therefore higher nutritive value and palatability of herbage in Rajasthan, especially if *Cenchrus ciliaris* is dominant. Further, one notes that, in the contribution of the Indian Grassland and Fodder Research Institute, Jhansi, U.P. (K. A. Shankarnarayan, P. M. Dabadghao, V. S. Upadhyay & P. Rai in Misra, 1974) to the January, 1974, Symposium at Varanasi, that the levels of primary production of species, without and with added nitrogenous fertilizer, in a *Sehima/Dichanthium* cover (Whyte, 1974, Table 5.2, p. 86), are highest in *Iseilema laxum*, followed closely by *Sehima nervosum* and far behind by *Heteropogon contortus*.

Workers at the Central Arid Zone Research Institute find that it is most difficult to draw specific conclusions regarding the above-ground productivity of grasslands at Jodhpur, Rajasthan. This varies widely because of the characteristics of the climate and the livestock management of the arid biome (Gupta, Saxena & Sharma, 1972a). The above-ground productivity and phytomass of three individual grasses, *Cenchrus ciliaris, C. setigerus* and *Lasiurus sindicus* are also highly variable under different rainfall conditions (Gupta, Saxena & Sharma, 1972b).

Sehima/Heteropogon grasslands are major constituents and have high successional status in the *Sehima/Dichanthium* cover, the largest in area of the five grass covers of India, extending over large parts of southern and central India lying south of the Great Indian Plains (Dabadghao & Shankarnarayan, 1973). The *Sehima/Heteropogon* grasslands give a stable expression to the grass cover over large tracts with medium to low rainfall with undulating topography and red-gravelly (murrum) soils. Their production ecology is important (Shankar, Shankarnarayan & Rai, 1973) because of their high potential for the maintenance of free-range livestock. *Sehima nervosum* provides the major contribution towards both photosynthetic and non-photosynthetic community structure. Comparison with grasslands elsewhere in India shows that the standing crop biomass and rate of production are especially high in the *Sehima/Heteropogon* grasslands at Jhansi, U.P. It is not stated whether the experimental area used had formerly been grazed, heavily or otherwise, or protected, and if protected, for how long, before these observations were made.

The ecosystematists also state that mean annual productivities of tropical grassland may equal or exceed that of tropical forests, but that productivity from "savannas" is less than half that of "grassland" (again, when is a grass cover a grassland and when a savanna?); also that net production from temperate grassland is half that from tropical grassland; this generalization does not recognize the higher protein content of temperate grasses and the need to compare temperate and tropical on the basis of net production of protein per hectare; nor the higher fibre content for much of the year in tropical grasses in a monsoonal environment. But above all, it is not recognized that the differences in net productivity between grass cover types in tropical environments and between grass cover types in temperate environments are probably greater than those claimed to exist between tropical and temperate grasslands.

The work of the physiologists would appear to be of more direct practical application, which must be the ultimate test; for example, on the comparative light and temperature requirements for the growth of young vegetative material of temperate and tropical grasses, the efficiency of energy conversion through photosynthesis, and hence the potential dry matter production at three levels of increasing complexity (Cooper, 1970, Cooper & Tainton, 1968, in Whyte, 1974 p. 175).

1.2.3   ECOSYSTEMS AND COMMUNITIES – TRUE AND FALSE

Following the study of the antiquity and evolution of the Gramineae of the Asian region (Whyte, 1972, 1976b), it becomes apparent that grassland workers should distinguish between true, false and artificial communities, and therefore also between true, false and artificial ecosystems. A community may be said to be true if it is composed entirely of species which are indigenous to that locality or habitat. The grass covers of Rajasthan are true, except where they are contaminated by relatively recent arrivals, such as perhaps *Panicum antidotale*, or species lower in the succession which have come from elsewhere to take up their position in the environments which have become unsuitable for the true indigenous species higher in the succession.

A community may be said to be false if it is composed entirely or largely of species which have arrived from elsewhere. The three or four species in such a community may each have arrived at different times in the vegetation history of the particular locality, for different reasons (natural migration or man-assisted movement), and from different places of origin. They come together on a site adapted to their individual autecologies. They therefore have to combine and compete with species with which they may not have been associated in their place of origin. Such false communities and ecosystems are characteristic of sites where grass covers have occupied the ground storey on land which was formerly under tropical rain forest or other grassless types of forest.

Artificial communities and ecosystems are the sown pastures and hay leys, and the cultivated fodder crops sown or planted pure or in mixtures, the composition of which is based on experimental or empirical knowledge regarding compatability of species and related characteristics.

Two groups are working on grass ecosystems in Asia, one on temperate grasslands (Chairman, M. Numata, Chiba, Japan), and one on tropical grasslands (Chairman, R. Misra, Varanasi, India). The latter group held a meeting in Varanasi on 17 to 22 January, 1974, to assemble regional information which could be used as the tropical contribution (Misra, 1974) to the International Grassland Synthesis volumes to be published by the U.S./IBP Ecosystem Analysis Studies on the Grassland Biome. The agriculturist is, according to R. T. Coupland in his opening address, devoted to producing the maximum amount of food on a land surface. The biologists, on the other hand, are somewhat concerned about this because of the increasing inputs of N, P and K that are required in order to produce this food. The biologist's approach has been to find out more about the fundamentals of this production process in plants, and the use of this material in the natural and artificial association, so that one may better understand how this food production can be maintained at a sustainable level and what that sustainable level might be.

Thus, six working groups were formed at Varanasi to consider the tropical contribution to the Synthesis volumes under the heads: site description, primary producer system, consumer system, decomposer and nutrient cycling, systems analysis, and management and conservation. The few grassland practitioners that exist per hundred thousand square kilometers in Asia may be forgiven for wondering how all this applies to the advice they are to give to the animal husbandmen; and whether the data that are laboriously and meticulously collected on biomass, litter, nutrient cycling, the higher rates of annual dry matter production in tropical grasslands as compared with temperate ones, and all the other criteria of the ecosystem specialist can really be accepted as reliable and permanent, on grass-dominant communities that vary so widely month by month and year by year, in accordance with climate, the biotic pressures, and the competition from the other constituent species. And the economic value of biomass as a nutrient resource is far greater in 100 kg. of desert grass and shrub from Rajasthan than in 100 kg. of tall jungle grass from the Terai or southeast Asia.

Supplementary to the techniques given in the earlier studies by the writer, reference should be made to the contributions from the Centre d'Etudes Phytosociologiques et Ecologiques, C.N.R.S., Montpellier, France, in the *American Journal of Range Management*, on sampling methods in dense herbaceous pasture which would be applicable in much of monsoonal and equatorial Asia (Long *et al.*, 1972, Poissonet *et al.*, 1972, 1973), also to the paper by Descoings (1971) on the concept and method of use of standardized data sheets for the analysis of the structure of intertropical herbaceous formations, also to papers on the methods of study and inventorization of the herbaceous stratum in palm savannas in Côte d'Ivoire (Descoings, 1972, Poissonet & César, 1972).

Fig. 1.1 is a simulation model of the dynamics of a grassland developed as part of the systems analysis effort of the U.S./IBP Grassland Biome study at Colorado State University, Fort Collins. The model was developed to be representative of the grassland network sites and to address the effect of net or grass primary production as influenced by level and type of herbivory, moisture and temperature, and added nitrogen and phosphorus. ELM is considered a total system model as the abiotic, producer, consumer, decomposer and nutrient (phosphorus and nitrogen) components are all represented.

The abiotic model simulates the abiotic parameters via a water flow submodel and a temperature profile submodel which are stratified through the air, plant canopy and soil profile.

The producer model considers carbon and phenological

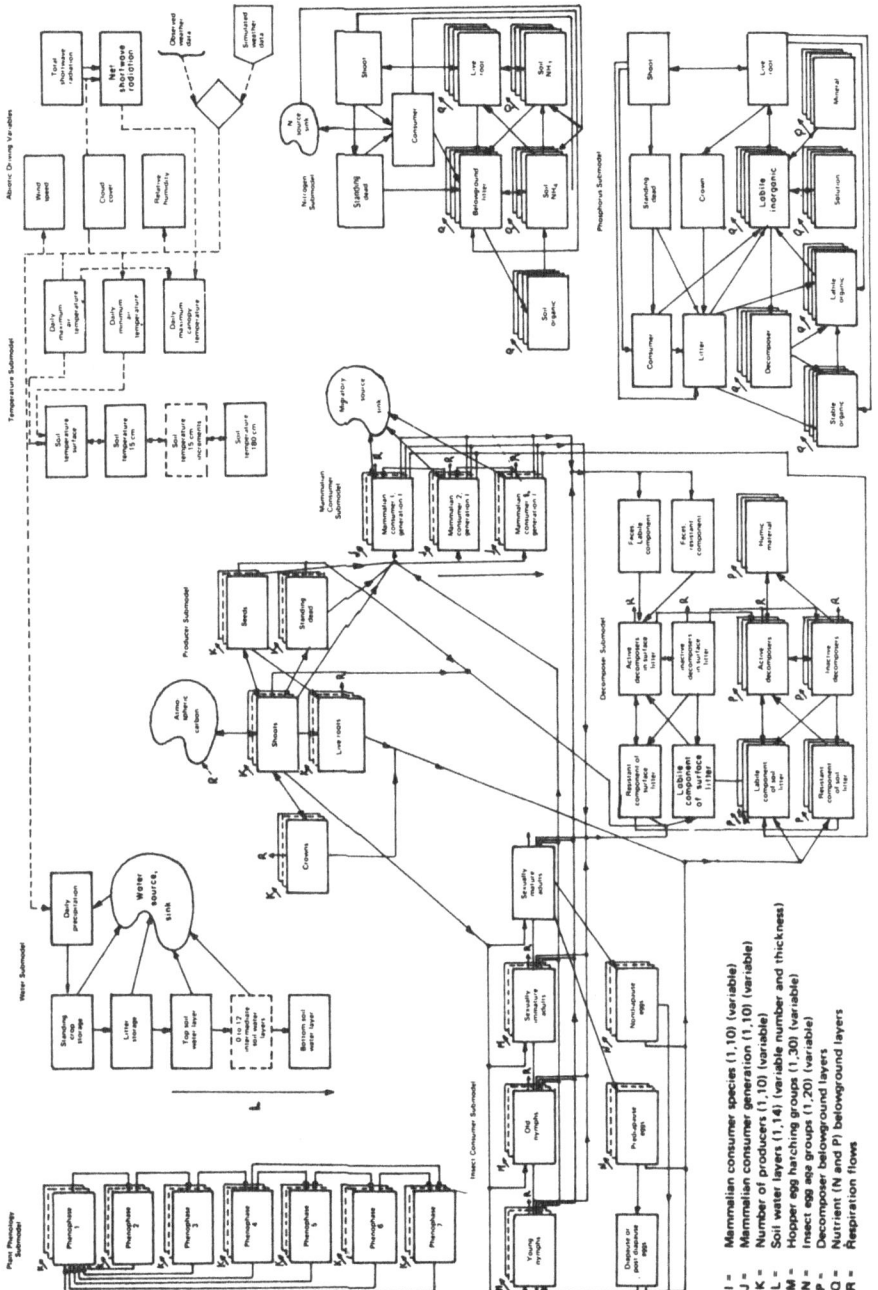

Fig. 1.1. A grassland ecosystem model constructed by US/IBP Grassland Biome Study at Colorado State University, Fort Collins.

I = Mammalian consumer species (1,10) (variable)
J = Mammalian consumer generation (1,10) (variable)
K = Number of producers (1,10) (variable)
L = Soil water layers (1,14) (variable number and thickness)
M = Hopper egg hatching groups (1,30) (variable)
N = Insect egg age groups (1,20) (variable)
P = Decomposer belowground layers
Q = Nutrient (N and P) belowground layers
R = Respiration flows

21

dynamics of both above and below ground parts of a variable number of primary producers.

The decomposer model follows decomposition rates and microbe biomass in litter and dead material both above ground and below ground.

The mammalian consumer model and the grasshopper model simulate organismal, intrapopulation and interpopulation dynamics of consumers.

The nitrogen and the phosphorus models simulate nutrient flow through the system.

Each submodel is independently operational and also interacts with the other six submodels to give the total model (from *U.S./IBP Grassland Biome Newletter* no. 14, June, 1974).

1.2.4   HYDROLOGY

The hydrological effects of methods of grassland practices (A. B. Costin & J. C. I. Dooge in FAO/UNESCO, 1973) depend on the climate and on the level of grassland management. Generally, a permanent grass cover has an effect on the elements of the hydrological cycle which is intermediate between that of a forest or dense scrub and that of bare soil or land cultivated for crops. Thus, for example the water yield and the tendency to erosion will be greater than for forest cover but less than for ordinary farm land on comparable topography. The higher the level of grassland management, the closer will the hydrological characteristics of the grazing area correspond to those of good forest cover.

"In semi-arid areas, mismanaged grassland is particularly liable to produce high flood peaks and relatively large amounts of soil erosion. In these areas the native grass cover tends to be seasonal and at times offers minimal protection to the whole of the soil surface. Thus the amount of livestock which can be supported by such grazing land on a long-term basis may be quite limited. Over-grazing of the area will intensify the problems of surface run-off and erosion. The over-grazing of grassland can also result in a hydrological change due to the compaction of a surface with inadequate cover due to the trampling of the grazing animals. This phenomenon is just as likely to occur in humid as in arid areas and it further decreases the amount of infiltration and increases the amount of surface run-off and the liability to surface erosion. Good management of the pasture is not only of

22

benefit for hydrological control but is also in most cases capable of giving better economic returns ...

"It is possible to increase stock-carrying capacity several times, and yet also reduce or eliminate flood run-off and surface erosion. The improvement of grassland or the replanting of land to forest may, by reducing the run-off, affect the water supplies downstream. The planting of land to trees or grass to prevent siltation in a reservoir may accomplish that purpose, but may at the same time reduce the chances of filling the reservoir with water. With more integrated economic development in a grass-land area, the concentration of animals in confined spaces for winter feeding or before marketing may lead to quite serious problems of waste disposal", unless the cowshed wash is distributed into nearby crops of the gross-feeding African grasses, Para, elephant and Guinea, as is done in the Milk Colony at Aarey, Bombay and other centres.

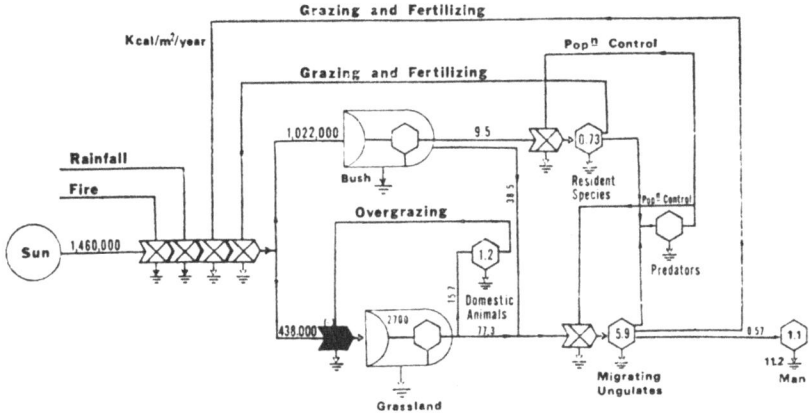

Fig. 1.2. Energy diagram for the Serengeti Mara Plains Region, Tanzania, based on data of Talbot & Talbot. The caloric value of animal biomass was assumed to be 2,000 Kcal/gm, wet weight. The caloric requirement of man was assumed to be 2,000 Kcal/day. Grazing acts as a positive feedback by maintaining the vegetation in a young productive state. Overgrazing acts as a negative feedback since it represents an exploitation of the vegetation leaving a low amount of photosynthetic tissue for recovery to occur. If only the feeding area is considered, the standing crop of ungulates may be 22 Kcal/m^2. However, this value neglects the total area of the home range which becomes very important during periods of moisture stress (Lugo, 1969).

23

It is in East Africa in particular that the end product of management systems on tropical grazing lands is wildlife, whether produced as a source of protein in the human diet, as a tourist attraction bringing in foreign exchange, or for other purposes.

Meat of all species of wild ungulates is eaten readily by various tribes in East Africa (Talbot, Ledger & Payne, 1961); many species such as eland and Thomson's gazelle are also prized by Europeans. Since most meat except that of zebra contains little fat, it can easily be converted to biltong or other dried meat products where ready outlets for fresh meat are not available. There are good markets for hides, either locally or overseas, and for other animal products such as elephant ivory, rhino horn, hippopotamus teeth, giraffe and wildebeest tails (for fly whisks).

The standing crop of wild animals on *Acacia* savanna land may be two to eight times that of domestic livestock, and the crop of wildlife on bushland may be from four to fifteen times that of goats and sheep (Talbot *et al.*, *op. cit.*).

"Since the wild animals breed earlier, reach maximum growth quicker, and achieve higher killing-out percentages than the domestic live-stock on comparable land, it can be assumed that, other things being equal, the potential harvest of protein and other animal products from wild animals should be several times that possible from domestic livestock alone on much East African rangeland."

Wise manipulation of control and use of wild animals requires a better understanding of their role in the ecosystem and

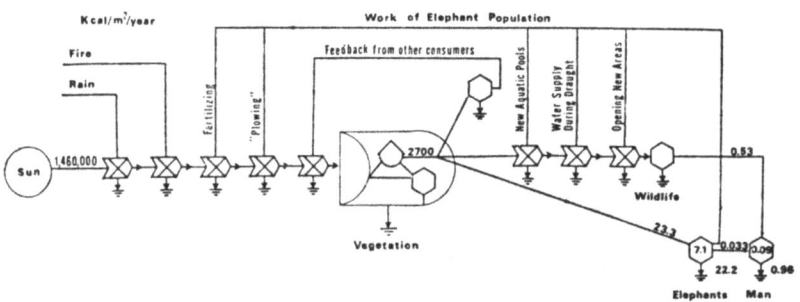

Fig. 1.3. Suggested work flows of elephant populations in the Luangwa Valley, Zambia. Wildlife and vegetation production were assumed to be equal to those reported in Fig. 1.2. (Lugo, 1969).

the energy flow stream (FAO/UNDP, 1971). An energy diagram and suggested work flows proposed by Lugo (1969) are reproduced as Figs. 1.2 and 1.3.

The most important phase of evaluation of ecosystems is an understanding of the food and feeding habits and preferences of the occupant fauna (Casebeer & Koss, 1970). Gaps in the fields of widlife biology and ecology on rangeland that also maintains domestic animals include the role of specific widlife species in disease and parasite relations, the degree of competition or compatability between domestic livestock and wildlife, the habitat requirements of designated wildlife species, and the potential economic value of wildlife in situations of multiple use.

The Fourth Session of the ad-hoc Working Party on Wildlife Management of the African Forestry Commission of FAO was held in Nairobi in February, 1972. K. Curry-Lindahl discussed ecosystems, biomes and national parks, referring to the African Convention for Conservation of Nature and Natural Resources, to the fact that most national parks are located in areas unsuitable for development, to the threats to existing national parks, and to the urgent need to conserve the components of the biomes, including the ecosystems comprising the grazing and browsing resources. For these purposes it is necessary to consider the place of wildlife in land-use planning (W. G. Swank & R. L. Casebeer, UNDP/FAO, Kenya), and the problems associated with harvesting and marketing the wildlife "crop" (P. Thresher). Any such developments will in due time increase the economic resources available for improving the initial stages in these ecosystems, that is, the condition of the rangeland itself.

Conclusive evidence is required regarding the role of wild animals (which may migrate for hundreds of miles twice each year) in the origin and present ecological status of grass covers in the tropical and equatorial zones. It is generally understood that elephants have been a primary factor in the origin of African grasslands, because of the rapid destruction of trees and shrubs now witnessed in some national parks. However, in South Asia, Thailand and Malaysia, it would appear that the damage to climax vegetation by elephants leads to the appearance of a secondary dicotyledonous community on the forest floor, with the initial gramineous flora being represented then or later by bamboos rather than grasses. It would appear that the diffusion of the grasses may have had to wait for the adventious introduction of seeds by man or other medium, and for the spread of initial grass clusters or communities by the grazing animal.

Antiquity and Evolution

1.3.1   TECHNIQUE OF ECOLOGICAL/HISTORICAL DEDUCTION

If the results of ecological analysis are considered in relation to the evidence becoming increasingly available on geological history and palaeoclimatology, it is possible to deduce which type of plant covers, communities and species could or could not have been found in the natural vegetation at any particular period of geobotanical history. Those concerned with biological species, communities and ecosystems, in relation to their past and present environments, may propose three rather distinct but interrelated categories: 1. the total aerial and global environments or macroclimates, recognized by meteorologists, 2. the land-based environments which have become superimposed locally upon these macroclimates as a result of the modifying actions of major events in geological history, such as tectonic movements and/or orogeny, and 3. the microclimates at ground level, frequently so extensive in area as to qualify for the category of induced macroclimates, in which plants and animals exist and reproduce and in which their response to physiological stress is influenced and generally greatly intensified by the use and misuse of land. Environmental changes in category (2) lead to major changes in zones of vegetation and the composition of plant communities (Whyte, 1976b). Those in category (3) may be regarded as the major cause of diverse irregularities in reproductive behaviour in the component species of tropical grazing lands (Whyte, 1975b), which lead either to extinction, to near or full speciation of new perennials, or to their replacement by annuals.

Specialists in the taxonomic geography of the Gramineae, ancient and modern, have, in association with forest ecologists, to learn to categorize forest covers in respect of their capacity to support growth and reproduction of particular members of the Gramineae. Much of the intertropical zone was originally, but most regrettably is no longer, covered with tropical rain forest (tropical wet evergreen in the terminology of Champion & Seth, 1968) and associated types of "forêts denses". The primitive grass species of the forest shade are neither adapted nor sufficiently aggressive to spread from their restricted ecological niches with the clearance of forest and the increase in light intensity and reduction in soil moisture. These new exposed areas have been colonized by light-loving, more xerophytic grasses from elsewhere. It is necessary to turn to the history of human migrations and trade, or to evidence of significant climatic change, natural or man-induced, in order to find an acceptable explanation for the present composition of many if not most of the grass covers of the intertropical zone.

Thus the results of ecological and historical analyses are essential not only for a better appreciation of the present composition of these communities, but also in the study of their generic and specific diversity (rich in tree savannas or comparable types of ancient grass covers, poor in relatively young derived savannas), and in the choice of areas for collection of grass genotypes, the desired variability having some relation to distance in time and space from centre of origin, and to the possibility of renewed speciation in an environment which exposes immigrants to physiological stress of genetical and taxonomic significance (Whyte, 1975b).

The field surveyors would find that the grass cover communities which they recognize at the present day are quite different in botanical composition and percentage frequency of species from those that existed perhaps fifty, probably one hundred, certainly five hundred years ago. The systematists must surely realize that they also are reporting on the taxonomic characters of the species of a genus as they exist now, and that the specific diversity would have been quite different in the former days of less exposure to the factors of stress. The analyses and descriptions given by taxonomists for the genera and species of the arid and semi-arid tropics apply to the plant at the date of collection, ± 25 years.

The field material now before us for recognition and classification may represent only residual populations from assemblages which have been depleted over the millenia of geobotanical history and palaeoclimatic change, and particularly since man entered the fragile ecosystems. It will be desirable to locate the habitat(s) in which those members which are accepted as more primitive (according to taxonomic and cytological criteria) are or were to be found. From this starting point, one may follow the history of their further dispersal and migration, with or without associated speciation, generally to a higher taxonomic level.

In such studies, it is essential to recognize that it is the history and evolution and the latter-day degradation of the total vegetation that are most significant. It cannot be claimed that a species was truly ancient and indigenous in a vegetation that could not have provided it with a habitat, or even an ecological niche suited to its autecology.

The history of the Gramineae and of other herbaceous families in this part of the intertropical belt necessarily begins with the study of the palaeoclimatic and geobotanical consequences of the history of Gondwana, its progressive disintegration and the movement of the Plate so created, out of the southern hemisphere, across the Equator (Whyte, 1976b). It has been postulated that the Indian or Indo-Pakistan Plate (here called the South Asian Plate to give geographical rather than political emphasis) was originally located along the eastern edge of Madagascar which, in turn, lay "against the coast of East Africa rather than South Africa and Mozambique, where it has sometimes been placed. There appears to be independent geological support for a more northerly location ... For instance, the strata of the late Palaeozoic Karroo System of Madagascar and Tanzania show close similarities and both countries lack the early Mesozoic volcanic rocks that are so widespread in southern Africa. There is also some oceanographic evidence to support the idea of southward drift of Madagascar relative to Africa ... The position of Madagascar is clearly the most enigmatic piece in the whole puzzle. Until more compelling evidence of a more southerly position comes forward than has hitherto been adduced, the Smith-Hallam reconstruction can be accepted provisionally as a reasonable approximation to the truth" (Hallam, 1973). When the South Asian Plate was in position adjacent to Madagascar, its main axis was in an east-north-east/west-south-west direction, rather than its present north/south alignment.

Mainland and insular southeast Asia has hitherto been regarded as part of Laurasia (Figs. 34 and 35 in Hallam, 1973). However, certain theories and/or substantive facts have been produced for and against the existence of a Southeast Asian Plate of Gondwanian origin, supposedly originally located between the land masses of the southern hemisphere which were later to become the South Asian Plate to the west, and the Australasian Plate to the east. This evidence had different bases, including stratigraphy, palaeobotony and palaeontology. However, Raven & Axelrod (1974) and McElhinny, Haile & Crawford, (1974) consider that it is unlikely that southeast Asia, and the Malay peninsula in particular can, on the basis of palaeontological and geological evidence, have been part of Gondwana.

The drift of the South Asian Plate was relatively rapid. Palaeomagnetic evidence shows that the central part of India moved from latitude 30°S. to 20°N. between the time of formation of the Deccan Trap and the present. 'If India had reached its present position by the time of formation of the Himalayas in, say, the Miocene, then this 50° change of latitude took place over about 50×10^6 yr. This corresponds to an average drift of India of 10 cm/yr during the Tertiary" (McElhinny, 1968). Recent palaeomagnetic data indicate that India was close to its present position during the Oligocene, having moved at $26 \, \text{cm/yr}^{-1}$ from the Cretaceous to the middle Eocene, and resumed at $16 \, \text{cm/yr}^{-1}$ from the Miocene (Blow & Hamilton, 1975). Other workers place junction during the Eocene, and calculate movement at from 10–18 cm/yr until junction, and 5 cm/yr subsequently (Molnar & Tapponnier, 1975).

During India's passage through the various climatic zones south of the Equator there occurred what Florin (1963) has called the complete annihilation of its rich *gymnosperm* flora. But Axelrod (1970) states that "viewed from the standpoint of ocean floor spreading and plate tectonics, the Indo-Australasian region, rich in ancient *angiosperms*, was composed of the eastern fragments of Gondwanaland which were rafted apart during the early Cretaceous. India was separated from Madagascar in the Cretaceous." The nature and stage of evolution reached by this flora are important in relation to its possible gramineous components before, during and after the movement of the Plate. Grass species associated with a gymnosperm flora would be quite different from those in a (tropical) angiosperm flora.

In the early Tertiary, warm wet climates extended into latitudes beyond 45° in both hemispheres in the Eocene, with a change to slightly cooler and drier conditions in the late Eocene and Oligocene. Whatever primitive species of humid tropical adaptation existed in Madagascar after its separation from the African continent in the early Tertiary may have crossed on to specialized niches on the proto-South Asian Plate before its own separation. It is possible that the South Asian Plate and Madagascar were joined longer than Madagascar and Africa — the last more or less direct migration having been possible 100 million years ago (Raven & Axelrod, 1974). The Mozambique Channel between East Africa and Madagascar later became a general but not absolute barrier to the spread of dicotyledons, but less so for the Gramineae (Exell & Wild, 1973). The indigenous gramineous flora of Madagascar may thus be relevant to the evolution of the Gramineae of western Monsoon Asia (see Africa, 1.3.4.3).

If any species of temperate adaptation did originally exist

on the Plate, they could not have persisted in that form during the northward drift, in the absence of mountains in the main segment of the Plate (Pakistan, India, Bangladesh, Sri Lanka) sufficiently high to provide a relict non-tropical habitat. Genera of temperate adaptation are, however, found above threshold elevations in west Indonesia. Furthermore, all xerophytic grasses (including the Eragrostideae and Aristideae of Brown & Smith, 1972) must have been eliminated when the South Asian Plate passed from the southern tropics with its high seasonal rainfall, through the equatorial zone with high rainfall at all seasons, and into the northern tropical zone, again with seasonal rainfall. During this period of movement across the equator, palynological evidence shows that the Plate was covered with tropical wet ever-green forests (apart from the active area of the Deccan Trap) with some slight admixture of deciduous but still grassless tree communities.

What is now peninsular India came into position astride the southwest winds which had hitherto brought the rains to the southern coast of the Asian mainland. From this stage onwards, the Western Ghats intercepted the rain-laden southwest winds. From the crest of the Ghats down to the western seaboard, rainfall became extremely heavy. Within relatively short distances from the crests to the east, however, rainfall dropped abruptly, creating a new and vast rainshadow of significance to the further evolution of the natural vegetation and its constituent herbaceous families. This event has to be considered also in relation to the post-volcanic plant cover of the Deccan.

The third geological event of significance in the botanical evolution of western Monsoon Asia was the periodic and progressive uplift of the Himalaya and their associated ranges, and the high plateaux of Tibet to 4,000 to 5,000 m. elevation. The first rise took place during the Cretaceous, but was not initially of sufficient magnitude greatly to affect the physiography, climate and flora (Vishnu-Mittre, 1969). The South Asian Plate came into position and joined the Asian mainland, along a line probably located slightly to the north of the present Gangetic Plain. Do the stony soils of the bhabar zone lying between the northern edge of the Terai and the first rise of the Himalayan foothills represent the old sea coast? The Plate then passed under the former mainland as far as the Indus suture and its eastern continuation in a belt near and roughly parallel to the Tsangpo (Brahmaputra) River (Chang Cheng-fa & Chong Hsi-lan, 1973).

The Himalaya had then reached an elevation sufficient to

cause a marked shift in climate in the north of the subcontinent but still south of the range, and resulting in an increase in the number of dry months per year. The "equable maritime climate" of the Plate which had already become influenced in the south by the rainshadow became further affected by the new monsoon pattern which evolved as a consequence of Himalayan uplift. When the rain-laden monsoons passed up the Bay of Bengal, they were deflected to the east (see maps of Wu & Chen and Dao & Chen in Chang Jen-Hu, 1971), and west by the eastern Himalaya, as soon as the range had reached an effective height. This northern monsoon pattern introduced in what was formerly a humid tropical habitat and vegetative cover the new ecological factor of aridity, of even greater intensity and variability than that produced in peninsular India by the rain shadow to the east of the Ghats.

Superimposed upon this aridity came the Neothermal following the end of the Pleistocene, the change from a cold to a warm climate which took place in the Northern Hemisphere 11,000 years ago and, to judge from evidence from foraminiferal faunas of deep-sea cores obtained off the south coast of Western Australia (Conolly, 1967), at about the same time also in the Southern Hemisphere. This new combination of aridity and rising temperatures is considered to be the main cause of the physiological stress so important in plant evolution. Where aridity (combined with variable, progressively rising temperatures) is the dominant factor of physiological stress, it is not so much aridity per se which is operative, as the highly variable conditions which obtain during the periods of abrupt change from less arid to more arid ecoclimates, and which occur particularly along fluctuating boundaries between contrasting ecoclimates and vegetation zones (Valentine, 1970).

Some nine major periods of significance in plant migration and speciation may be recognized: a. the long periods of slow geological and palaeoclimatic evolution through which the vegetation on the South Asian Plate and along the northern shores of the Tethys Sea passed before rafting began; b. during the movement of the Plate across the equator into the northern hemisphere, and the cutting off of the southwest monsoon from the northern shores of the Tethys Sea; c. the creation of the rainshadow and introduction of aridity into peninsular India, when the Plate took up its permanent position astride the southwest monsoon; d. the uplift of the plateaux of Tibet and of the Himalaya and its associated ranges; e. the increase of aridity in the

31

northwest and north of the subcontinent, due to the new pattern of monsoonal climate which followed the uplift of the mountain barrier, now to deflect the northward rain-bearing winds of the Bay of Bengal to the east into China, and to the west, as far as southern Iran, Pakistan and eastern Afghanistan; f. the geographically limited fluctuations in climate at lower altitudes in north India and Pakistan related to the four glaciations; g. the beginning and extension of the Neothermal, introducing high temperature to act in combination with the already widespread aridity as major operative factors in plant speciation; h. the period beginning some 5,000 years ago (S.K. Seth, personal communication, 1976), and particularly following the introduction of cattle and fire (and much later of iron) by the Aryans, when the actions of man first became widespread in the northwest of the subcontinent in destroying vegetative cover or reducing it in successional status, with marked effects at least on the microclimates at ground level in which the grasses live and progress through the highly vulnerable stages of meiosis (Whyte, 1975b); and i. the recent period with short-term fluctuations in climate which continue to expose the grass species at irregular intervals and in varying intensities to the action of factors operative in specific variability and change. Man has spread over the land in ever-increasing numbers, cutting fuel, burning and clearing for cultivation and grazing and browsing his domestic livestock. The intensity of these activities has increased catastrophically over the last hundred, even fifty years, thus exacerbating in the ground-level microclimates the aridity and higher temperatures and their effect on the physiology, genetics and taxonomy of the herbaceous plants.

It is in this final and most recent stage that the grasses have been given an opportunity to spread in all manner of now degraded forest types to which they do not truly belong in an ecological/historical sense, and to evolve to their present state of cytogenetical and taxonomic complexity (Whyte, 1975b). It is these types of grass covers and successions that are currently described for Pakistan, Nepal, Burma and Sri Lanka (Whyte, 1968a and 1973), and for India by the Government of India Grassland Survey (Dabadghao & Shankarnarayan, 1973). This combination of the continued operation of the factors of aridity and temperature with a great increase in the pressure of cultivation and the grazing of domestic livestock has led to the elimination over much of the region of the high caste grasses, and to their replacement by inferior perennials and by annuals.

Current attempts to map the chorology of Indian grass species on the basis of a computer-aided method of analysis of their present distribution (Clayton & Panigrahi, 1974) are unrealistic. Their maps of kingdoms, regions and endemic centres are maps of the degree of intensity of human and animal pressure on the natural vegetation which has created conditions for the (very recent) spread of species from their initial core areas or ecological niches.

The major part of the grass flora of western Monsoon Asia, comprising some 250 genera and 1,250 species (Bor, 1960), is thus of very recent origin in geobotanical history. Within the region itself, one may recognize the more ancient species to be found in specific niches in the relict areas of Gondwanian vegetation along the Western Ghats, in Sri Lanka, and in the north-east of South Asia. These shade-adapted species and the few types of Gramineae found in pollen cores of Tertiary marshy habitats represent the primitive types. At a much later date, in fact during the past 5,000 years, a great new assemblage of species adapted to xerophytic conditions and intense light advanced from the north-west of the subcontinent with the progressive opening of the forest canopy due to increasing aridity and the effects of man's actions. Much of the grass flora of this region arrived from elsewhere, became adapted to the new conditions, and proceeded to speciate, after a. immigration on the South Asian Plate, from close proximity to Madagascar in its more equatorial position on the east of the African continent, of a few types which could persist in appropriate niches in a tropical wet evergreen forest during the passage of the Plate across the equator; b. evolution of high-caste species adapted to semi-arid ecoclimates, from a small and perhaps ancient residual enclave of ancestral types in Pakistan along the foothills of the mountains of Baluchistan and in the adjacent plains, outside the influence of the southwest monsoon even before the arrival of the South Asian Plate; c. immigration of African types of semi-arid adaptation, by natural means or more probably through the actions of man, across the Red Sea and through Arabia, within the limits of the so-called Saharo-Sindian phytogeographical zone; are the monsoonal type species described by Bor (1968) for Iraq on their way to the South Asian subcontinent from Africa, or did they enter Iraq from the east during the trade between Sumer and the Indus civilizations, or at other times; d. transfer from north-east and east Africa by the maritime trade that has been conducted since time immemorial, through the stepping stones (watering points) of the islands of the Indian Ocean (Whyte 1974, p. 24; Renvoize, 1971); e. possible, but less likely, immigration into the region from further east and southeast (Whyte, 1972); f. migration into the western Himalaya of temperate grasses of Irano-Turanian and Mediterranean adaptation to the middle-elevation habitats of similar ecoclimate; and g. migrations of temperate species from continental Asia to the north and northeast, southwards through what had been the warm moist zone which extended up to 45°N. latitude before the uplift of the Himalaya and its associated ranges (Frakes & Kemp, 1972), over the plateaux of Tibet, through the rising Himalaya and down their southern slopes to the limits of their tolerance of a subtropical environment; and then to be cut off from their northern origins by the creation of the arid core of Asia.

Before man's so recent introduction of plantation agriculture in southeast Asia, its natural vegetation was composed of climax formations of forest covers, part of the continental eastern Asian forest communities which once flourished in an almost unbroken expanse over 30 degrees of latitude from the tropics to eastern Siberia (Wang Chi-Wu, 1962). They were subject only locally and spasmodically to the temporary regressive influence of natural disturbances. These formations were part of a great assemblage of plants now called the Indo-Malayan flora (Good, 1964; van Steenis, 1948), because it is said to cover the South Asian subcontinent, Burma, Thailand, Indo-China, southern China, the Malay Peninsula and Malay Archipelago, or Malesia in the terminology of the Rijksherbarium, and to extend far into the Pacific to the east of the Hawaiian Islands (van Balgooy, 1960).

In the undisturbed rain forest and its associated types of forest along the ecoclimatic rainfall gradient, all classifiable as "forêts denses", there were only a few ecological niches in which certain specialized species of the Gramineae (hygrophytic, shade-loving or littoral) could have grown. Holttum (1954) recognizes only one grass, *Leptaspis*, as occurring in undisturbed rain forest. Gilliland and others (1971), as analyzed by Whyte (1972) mention a few more adapted to true forest conditions or to one of the niches on the forest fringes. These include the small but widely distributed tropical genus, *Centosteca* (until recently incorrectly named *Centotheca* or *Ramosia*) and the small genus of tropical Asian forests, *Lophatherum*, which Soderstrom & Decker (1973) include with their Centostecoid and Bambusoid grasses, most of which are related to bamboos and are found in the shaded habitat of tropical American forests. The area of distribution of these forest grasses in southeast Asia has now become greatly reduced through the felling of their associated tree species in the original tropical forest ecosystem.

It is not possible to refer, as one apparently may with forest tree species, to the existence of an Indo-Malayan (or Indo-Malesian) gramineous flora on the basis of the presence now in the central part of the region of some 200 or more species of the Gramineae. The species included in modern floras were (and are) well-adapted to spread rapidly in the Indo-Malesian environment, but only when the forest canopy had become sufficiently open or absent altogether, and the previous close correlation between regional macroclimate and local ground microclimate had been destroyed, introducing bioclimatic conditions at ground level characterized by greater aridity and increased light intensity.

Raven & Axelrod (1974) consider that several drought periods in the Quaternary not only impoverished the rain forests of South America and Africa, but also those of southeast Asia and Australasia. They raise the question of whether tropical grasslands are man-made, and conclude that evidence of this ice-age aridity implies that at least some are native. The actual distribution of grasslands in New Guinea, West Asia and the Philippines, as judged from high-altitude aerial photos, is believed to be governed by terrain and climate. This conclusion may be accepted for grass communities composed of temperate species at elevations above the level for tropical rain forest and associated types of forest in the mountain chains from Taiwan through the Philippines and Indonesia to New Guinea, but less readily for lower altitudes, despite the requirement of a dry savanna landscape on the Sahul shelf to explain the distribution of grassland birds.

The technique of ecological/historical analysis is being extended to apply to eastern Monsoon Asia where, according to present indications, the final story will differ considerably from that given for the other subregions of monsoonal and equatorial Asia. The grass species found in the deforested monsoonal lands of China and in the transition zones between these and the Asian continental grasslands to the north and west have been listed by Whyte (1968a, p. 215). An analysis will be made of the statements of distribution and habitats of tropical and temperate species (at the southern limits of their distribution in China) given in the flora of Keng Yi-Li (1965). Many of the Gramineae in Noda's flora of Manchuria (1972) have subtropical and monsoonal affinities, and are doubtfully indigenous at these latitudes.

As in southeast Asia, the grass species now found in eastern Monsoon Asia occur largely on land from which the earlier tropical rain forests or associated types of relatively grassless tree communities have been removed. The few species of grasses and herbs originally adapted to the ecological niches in these types of forests were not able to persist in the new aerial and edaphic environments created by clearance. They had to retreat with their tree associates to relict areas, to be replaced by species adapted to full light and more arid soils, brought there primarily through the migration, colonization and trade of man. Down through this zone of false grass ecosystems (1.2.3) run the high-altitude tongues of temperate species, cold-tolerant but adapted to progressively shorter days as they approach the Equator.

1.3.3 AUSTRALIA AND THE PACIFIC ISLANDS

Raven & Axelrod (1972) write as follows: As the Australian plate was rafted north, it met the west-moving Pacific plate and underwent fragmentation ... Many new taxa evolved over the lowlands as these lands were rafted north to warmer climates. In Australia, the mixed gymnosperm-evergreen forest of Cretaceous to Eocene time was replaced progressively by taxa adapted to mediterranean and desert climates, xeric shrublands and low woodlands, savanna and tropical forest as the plate moved into lower latitudes. The collision of the Australian with the Asian plate in the Miocene established Wallace's line, and the mixing of Oriental and Australasian taxa commenced. In the late Cenozoic, the elevation of high mountains from Malaysia to New Guinea and Australia/New Guinea and Australia/New Zealand provided dispersal routes for numerous herbs from cool-temperate parts of the Holarctic and new sites for their rapid evolution.

1.3.4 AFRICA SOUTH OF SAHARA

1.3.4.1 *Africa in Gondwana times*

The monocotyledons, and perhaps the angiosperms themselves, originated in West Gondwana (Africa with South America before their rupture) — in vast arid to semi-arid tracts in tropical latitudes (Raven & Axelrod, 1974). It appears that the angiosperm flora of Australasia, including the Gramineae, could have come for a time either through links with West Gondwana —

warm temperate and subtropical (proto-South Asian Plate, Madagascar and Africa), or for much longer from South America via Antarctica. The Poaceae and Cyperaceae are stated to have passed between Madagascar and Australasia via islands of the Indian Ocean (see refs. in Whyte, 1972; also Zimmerman, 1963 and Carlquist, 1974). They have the characteristics of groups dispersed over long distances — unbalanced representation in the two floras, close relations, and occurrence in open habitats. But none of the major gramineous species of Australian tropical grasslands appear in Morat's list (1973) of autochthonous species in southwest Madagascar (1.3.4). Many of the grasses of tropical Australia are of south Asian origin; likewise they do not occur among the ancient species of Madagascar; if they had, they could not have crossed the Equator during the rafting of the Plate. Reference may also be made to Table 28 in Whyte (1968a) from Specht (1958), showing distribution outside Australia of grass species recorded for Arnhem Land aboriginal reserve, a classification which includes continental southeast Asia, peninsular India, the east African islands and tropical Africa. Bird migration may apparently not be proposed for plant dispersal from Eurasia into Australia across Wallace's Line (McClure, 1971), possibly because of the relatively recent juxtaposition of Asia and Australia — 15 m.y. B.P. (Raven & Axelrod, 1974). Many endemic families and taxa may have survived in or later evolved in such edaphic deserts, and long-distance dispersal would be favoured. As Africa and South America separated, climates became moister and more equable. The dry climate of Africa probably began near the close of the Oligocene, at which time upwarp commenced and has continued to the present day as rift valleys grow. The altitude of East Africa has increased some 2,400 m. above Miocene levels, with consequent cooler and drier climates. Other factors were the development of the Benguela Current and the fluctuations of Quaternary climate.

1.3.4.2 *Palaeoclimatic hypothesis*

Aubreville (1962) has proposed a palaeoclimatic origin for certain savannas which occur today in Africa. Where soil is not a factor and the influence of man did not early acquire a significant intensity, it is necessary to suggest other explanations for the vast areas of savanna now found in climates suitable for dense humid or dry forest. Some believe that the Quaternary glaciations would have brought humid conditions and only a slight fall in temperature in Africa and Latin America. Aubreville, on the other hand, considers that a polar shift of about 12.5° latitude south would have brought dry and very cold conditions to west and central Africa. This change would have been sudden, causing a retreat southwards of the forest, while its place was taken by grassland. The forest moved to the southern tip of the continent. The Angola coast remained covered in savanna and steppe in spite of being in the Equatorial zone. This was due to the desertifying influence of the Benguela current. The east African climate became favourable for the expansion of tropical forests on mountains and the high plateau, and their expansion towards the south where they approached the region of the Cape. (A possible ecological origin of the grass components of these derived African savannas is given in 1.5.2).

With the return of the Pole northwards and the retreat of the glaciers of the last Ice Age, Africa south of the Sahara again acquired a forest climate (Aubreville, *op. cit.*). This created conditions for the return of the many forest types of the region from the favourable mountain and coastal bastions where they had persisted, just as today there remain in favoured localities in southern Africa certain vestiges of the Guineo-Congolese forest. Forest regeneration proceeds along banks of watercourses, extending in corridors, taking over hills, coastal plateaux on young soils, more humid soils, and eroded areas protected from fire. Tentacles join isolating savannas, which are then colonized from the forest and taken over. The forest has not yet reconquered all its old domain (Aubreville, 1968). Progression in succession is always slower than regression; therefore this change back to forest was probably slower than the original change from forest to savanna; it has, of course, also been impeded or even prevented by the actions of man in his attempts to retain the economically valuable savanna cover.

If there had been an ancient indigenous grass flora in South Africa, it would probably have been eliminated at this time, its place to be taken later in the post-glaciation phase by grasses of east African origin.

When the forests returned to roughly their present position between Gabon and Uganda, between its southern borders and the limits of the position occupied during the Quaternary glaciation, a vast region was invaded by grasses. Then the semi-xerophytic flora of the present dry forests, open forests and Sudano-Zambesian wood savannas typified by *Brachystegia* also pushed north into the newly "savannized" region. This progress is, however, less rapid than the withdrawal of the humid forest. That is why there persists between the southern borders of the humid forest and the forêts claires of *Brachystegia* characteristic of Angola, Katanga and Zambia, a great expanse of grassy or poorly wooded savannas – a biological hiatus. Duvigneaud (1951) has noted that the Angolo-Rhodesian open forest is moving north across the great savannas, and has not yet established contact with the equatorial forest.

1.3.4.3   *Significance of Madagascar*

Because Madagascar was a Tertiary Gondwanian stepping-stone between mainland Africa and the South Asian and Australian Plates to the east, its flora is of great interest. In his study of the Gramineae of the pastures and crop lands, Bosser (1969) deals with about two-thirds of the 450 species of the Gramineae in the flora of Madagascar. However, he has omitted those species that would be of greatest historical relevance – those which are rare, forestal, or which grow in special habitats in a former tropical rain forest environment now largely destroyed. Bosser lists twenty species of *Panicum*, including *P. maximum*, some of which might be old enough to have provided the ancestors of some of the more mesophytic species now found in the flora of western Monsoon Asia.

Morat (1973) has described the origin, antiquity and evolution of the savannas of the southwest of the island, which became separated from the African mainland since the beginning of the Tertiary. The fragile equilibrium which had been built up before the very recent arrival of man was particularly sensitive to his very presence, and was destroyed suddenly and brutally.

The 83 typical savanna species of the western region may be classified in six groups according to their distribution (Morat, *op. cit.*); only the Gramineae are listed here:

Pantropical: *Heteropogon contortus, Hyparrhenia rufa, Imperata cylindrica, Cynodon dactylon, Chloris barbata, Schizachyrium sanguineum.*

Palaeotropical: (Madagascar, Asia and Africa): *Schizachyrium brevifolium, Eragrostis cilianensis, Tricholaena monachne, Themeda quadrivalvis.*

Afro-Malagasy American: *Hyperthelia dissoluta, Trachypogon spicatus, Schizachyrium domingense.*

Asiatic-Malagasy: *Chrysopogon serrulatus*
African-Malagasy: *Hyparrhenia cymbaria, Craspedorachis africana,*
Eragrostis chapelieri, Aristida congesta ssp. *congesta* and ssp. *barbicollis, Enneapogon cenchroides, Pogonarthria squarrosa, Perotis* aff. *patens, Eragrostis cylindriflora, Sporobolus festivus,* Endemics (limited to Madagascar and in some cases including Mascarenas):
Aristida rufescens, Loudetia simplex ssp. *stipoides, L. filifolia* ssp. *humbertiana, Panicum voeltzkowii, P. pseudovoeltzkowii, P. luridum, Isalus isalensis, Vignierella madagascariensis, Neostapfiella perrieri.*

In addition to some 20 allochthonous species of Gramineae, Morat recognizes autochthonous species, those which were originally (before the arrival of man) part of the local flora, of diverse origin: species distributed by sea or by wind currents or by birds, species arriving by slow migration along the continental connections, or descendants of floristic elements already installed in the Gondwanian age. Some of these are heliopilous, not of forest origin in the primitive vegetation, but occurring on slopes and screes: *Enneapogon cenchroides, Pogonarthria squarrosa, Perotis* aff. *patens, Sporobolus festivus,* the subspecies of *Aristida congesta, Eragrostis cylindriflora.*

1.3.4.4   *Ecological Status of Grass Covers*

"Dans les régions intertropicales le climax est de façon générale constitué par des forêts. Le fait est unanimement admis par les botanistes et les forestiers" (Schnell, 1970/1971).
Aubreville (1962) enumerates the climax tropical vegetation type as follows:
"Les climax forestiers dans la zone intertropicale.
Nous arrivons alors à ces conclusions générales très importantes d'écologie tropicale que dans les pays tropicaux la végétation primaire est sous les climats humides une végétation de forêt dense, ou plus généralement de végétation ligneuse dense et, sous les climats semi-arides, encore une végétation ligneuse, biologiquement différente des précédentes, fermée ou ouverte; lorsqu'elle est ouverte il y a transition vers des formes mixtes graminéennes et forestières. Dans les régions semi-désertiques, les formations climax sont encore des steppes à végétation ligneuse, des

sous-arbrisseaux nains et des succulents, très ouvertes, où les herbacées pérennes (graminées) n'ont plus une place préponderante, alors qu'au contraire les herbacées annuelles abondent en saison estivale pluvieuse. Tels sont à notre avis les grands types climaciques de végétation tropicale."

Wooded savanna and open forests are climax only in very arid country, where due to poor soil moisture, the roots of shrubs and trees spread, leaving gaps between them which are occupied by grass. Elsewhere a mixture of shrub/grass is an anomaly. The forest types will eliminate the grass, in the absence of repeated bush fires and clearing. In actual fact, however, the forest rarely takes over savanna, and thus a fire subclimax acquires an apparent stability in the form of wooded savanna. It has been shown that protection from fire in Ivory Coast, Katanga, Zambia, Zaire and Nigeria leads to the regeneration of tree cover (Aubreville, 1966a). Adejuwon (1971) has suggested that the climax vegetation in the savanna zones of Nigeria (Sudan, Sahel, north and south Guinea) was tropical, xerophytic woodland, tropical deciduous woodland and tropical rain forest respectively.

In their survey of the land resources of Lesotho, Bawden & Carroll (1968) state that this is a grassland country and there is an almost complete absence of natural tree growth. Acocks (1953) has suggested that conditions are too dry and frosty for tree growth and that most of the grasslands of Lesotho are therefore "climax" types. Other southern African botanists have preferred to regard them as a fire sub-climax grassland (Rattray, 1960). A sub-climax succession to shrubs may occur, in sufficiently sheltered spots at any altitude if fire is excluded. Many years of overpopulation, over-grazing and veld fires have greatly influenced the development of the present vegetative cover and most of the shrubs that now occur are unpalatable to stock. Aubreville (1962) notes that in the east, rainfall 730 to 1,500 mm. and dry season of four to five months, the climax should be dry forest or at least open forest. More to the west, with rainfalls down to 450 mm., with fairly high saturation deficit and coldest average month 4 to 7° C, there are a number of mountain tropical woody formations that are adapted.

Walter (1969) claims that in South-West Africa climate is the dominant factor governing the distribution of grass and shrub communities. At 100 mm. annual rainfall, the grasses take up all the available soil moisture. At 200 mm. the same situation occurs, but with taller grasses. Only at 300 mm. is sufficient water

Table 1.2. Woody types of vegetation in difficult environments in Africa (Aubreville, 1962).

	Climate	Rainfall mm.	Duration of ecologically dry season months	Annual saturation deficit and regime of monthly deficit
Wooded savannas and Sudanian open forests	Sudano-Guinean	medium or high, 950–1750	4–5	high: low or medium rainy season, very high in dry
Open forest and wooded savanna of Upper Katanga and Zambia	Upper Katangan	medium 1000–1400	5–6	medium: high at end of dry
– id – Zambia and Rhodesia	Rhodesian	low or medium 600–1200	6–7	medium: high at end of dry
Steppe. Acacia and Sahélo-Sudanian wooded savanna	Sahélo-Sudanian	400–1000	6–7	very high, sometimes excessive
Thickets. South Madagascar, Didieraceae and Euphorbeae	South Madagascar	500–550	6–7	medium

available to permit small woody species to enter the communities. Shrubs become taller with increase in rainfall, until tree savanna is reached. In all cases, grasses are dominant, with non-woody plants using the residual moisture (Walter, *op. cit.*). Between 500 and 600 mm. rainfall, trees develop until their canopies touch, in a type of deciduous tropical forest from which grasses are eliminated by shading. Aubreville (1962) considers that the south-west African coast and the Karroo remained desertic during the palaeoclimatic shift of the Quaternary, under the influence of the cold Benguela current. This explains why today the biological spectrum of the vegetation is so different from the Sahara, which is a young desert having undergone frequent changes since the Tertiary.

Workers familiar with conditions in Africa would expect to find more emphasis on the soil poverty which is said to inhibit tree growth and which they may regard as a primary cause for much of the open grassland near the equator. There are marked localized differences in soils which may explain variants in composition and succession within grass covers. As the supply of nitrogen is a main limiting factor, the free-living organisms which fix nitrogen in tropical soils assume importance.

1.3.4.5   *The Forest/Savanna border*

Throughout the intertropical region, one is impressed by the sharp break which occurs between tropical forest and savanna; the absence of intermediate forms of forest or shrubby communities appears as an ecological anomaly (Hills & Randall, 1968). Aubreville (1966a) has considered this problem with special reference to West Central Africa, the Guineo-Congolese vegetation zone. Here the climatic gradient is gradual without sharp changes in ecoclimate which could account for this phenomenon. Quite distinct soil types occur over small areas, but these are found in both forest and savanna. A phytogeographical hiatus is found between the dense humid tropical forest on the Congo-Guinea zone and the open forest and woody savanna of the Sudano-Zambezian (see Letouzey, 1968, for a phytogeographical study of these types in Cameroon). Even in its arid interior, the Sudano-Zambezian zone has a relatively rich flora, with genera and species which may be endemic although related to the Guineo-Congolese flora. Between these two chorological regions there exists a gap, which is occupied by sparsely wooded savanna with a limited number of species.

Fire has in most areas eliminated the transition from dense humid to dense dry forest, and all stages in the succession to the semi-arid, sub-desertic tree savanna. Annual fires generally prevent colonization from fire-sensitive forest pioneers, while the dense humid forest forms an impassable barrier to savanna vegetation. Hence the comparative floristic poverty of the savanna border (Aubreville, 1966a).

1.3.4.6   *Anthropogenic Factors*

The outstanding factor in the evolution and maintenance of the grass cover of tropical Africa is, of course, fire. Anthropologists report the practice of burning vegetation among the customs of primitive communities from the earliest times. Clark (1970) states that man has used fire for well over 50,000 years, while during Neolithic times it was a common tool in land management in north and sub-Saharan Africa. However, dense humid tropical forest could not have been destroyed by fire until primitive peoples possessed iron tools for felling, and these must have come from arid regions (Whyte, 1972).

The plant geographers recognize that fire has played a predominant role in the creation of the vegetation of Africa as it occurs today:

"Il paraît aujourd'hui indiscutable que la végétation climacique, c'est-à-dire le manteau végétal naturel, est en général forestier. Les savanes, les prairies dans l'ensemble, paraissent bien n'être que des paysages secondaires, issus d'une déforestation et entretenus dans leur état actuel par l'action directe ou indirecte de l'homme" (Schnell, 1970/1971).

The managers of grazing lands have adopted fire for the maintenance of a grass cover in a vegetation which is always struggling to revert upwards in succession to a forest climax, through scrub, shrub, thicket, thorn forest, etc. Fire has been used for several purposes:

a. to concentrate wild animals for easier slaughter by hunters;
b. to control ligneous growth and so reduce competition;
c. to promote regeneration of grass growth in tall grass savanna, and
d. to clear and fertilize land for shifting cultivation. (see also 1.7.4.).

Scott (1970) quotes earlier writers as saying that veld fires have been a feature of the African landscape since time immemorial. Early Portuguese explorers saw that the interior of South Africa was covered by a pall of smoke due to veld burning, and called the country Terra dos Fumos. Ramsay & Rose-Innes (1963) have reviewed the literature on grass and bush burning in West Africa. They describe the principal types of subclimax and climatic climax vegetation in the northern Guinea savanna zone of northern Ghana, and discuss the physical and biotic factors which have influenced this. Quantitative analyses have been made of the effect on degraded Guinea savanna vegetation of annual early dry-season fires, annual late dry-season fires, and complete protection. Adejuwon (1970) states that the coastal savannas of Nigeria are not the result of human activities, but then admits that they are maintained as a subclimax by annual fires.

Lemon (1968a, b) has considered fire on the Nyika plateau in Malawi in relation to the pattern of wildlife grazing. Fire is an important agency of control, if the objective is to prevent animals from wandering off the central grasslands during the dry season into areas where they will certainly be subject to poaching.

Lamotte (1967) states that fires are prevented to promote comparative studies of unburnt and burnt grasslands in an area around the Lamto Ecological Field Station in Ivory Coast. Annual rainfall varies from 1078 to 1657 mm. with two dry seasons, a long one from November to March and a short one in August. Gallery forests are found along the Bandama River and its tributaries; elsewhere the open savanna is interspersed with numerous *Borassus* palms.

There are two possible types of savanna:
1. those forming in old, dry forests, including species partially resistant to fire; more densely wooded and floristically rich; and
2. those appearing after cultivation on the borders of the humid forest; derived savannas later colonized by ubiquitous species from drier regions; floristically poor.

With time the differences may become less marked, since a dry forest flora has a tendency to spread into all savannas, in spite of fire (Aubreville, 1966a).

Where human activities (cultivation and burning) are absent, forest regeneration is rapid, e.g. in protected enclosures in the Congo, Gabon, the High Sangha, North Cameroon and Uganda. In the process of regeneration, the trees of the savanna (e.g. *Terminalia* in Uganda) are completely eliminated as the true forest vegetation advances.

Aubreville (1966a) has summarized his conclusions on the savannas of Africa: "Sauf le cas de savanes herbeuses édaphiques, les savanes herbeuses ou peu boisées, ayant une flore forestière pauvre ne sont pas des formations climaciques. Elles ont une origine anthropique contemporaine ou peut-être une origine paléoclimatique récente, disons du Quaternaire le plus proche".

1.3.5 AMERICA: SOUTH, CENTRAL AND CARIBBEAN

1.3.5.1 *Origins of Savannas*

It is appropriate to give the impressions of a specialist on African savannas after studying the savannas of tropical America, and to compare these with a presentation of a specialist in South American savannas (Blydenstein, 1968), made on the basis of local experience and a study of African literature.

Aubreville (1962, 1965) believes that, where savannas are found in areas of very differing soil and relief, with a forest climax climate, where the forest has disappeared and where population is low, the same palaeoclimatic origins as those noted in Africa apply. With the Quaternary polar shift and the movement south of the Equator some 8° in Latin America, the pine forests of south Brazil were eliminated and their place taken by savannas. In Central America (Costa Rica, Nicaragua and Honduras), pine savannas occur in an area with a wet tropical climate which should maintain tropical rain forest. Aubreville (1966b) states that the indigenous peoples of these areas could not have burnt the tropical rain forest. Therefore the reason for the occurrence of a pine forest in a tropical rain forest climate is palaeoclimatic in the first instance, perpetuated by fires to produce pine savannas over the subsequent millenia (Taylor, 1962).

The Sabana Grande took the place of humid tropical forest in the Guyanas. The northern coastal areas within the intertropical zone still received abundant rainfall, and here the dense tropical forest remained. Amazonian Hylaea moved southwards, joining the Atlantic forest of the east and south, which persisted under maritime influence. The zone between the Guyana forest and Hylaea, marked today by predominantly xerophytic vegetation (Llanos, the savannas of Ampapa, caatingas of Piauhy and Ceara, the pseudo-caatingas of Upper Rio Negro, the campos of upper Rio Branco, the campos and dry forests of the middle Amazon) was probably drier at that time, while the basin of the middle Amazon was more or less deforested.

In a study of avian speciation in tropical South America (Haffer, 1974), it is stated that during Pleistocene dry phases, the rain forest survived in high-rainfall areas, but was replaced by savanna and other open habitats in low-rainfall areas. As climate fluctuated (possibly correlated with glacial phases in the temperate zones), these forest "islands" alternately expanded and rejoined and contracted and became separate again, while the changes in the savannas took the opposite course.

Faced today with clear evidence of the progressive destruction by fire of the Venezuelan forests and the extreme poverty of the ligneous flora of the Llanos, Aubreville (1965) considers it inconceivable that the savannas are climax, as many phytogeographers still believe.

"The tropical wet and dry climate, with its alternating rainy and dry seasons, is a prerequisite for good burning conditions ... The typical wet and dry climate has been called a "savanna climate" because the long dry season following a continuously rainy wet season was considered "hostile to woodland" (Schimper, 1903–1904). In actual fact, the major part of the region with a tropical wet and dry climate is covered by deciduous forest, or at least its visible remains ... Although there is little argument about the dominant role of fire in the environment of savannas, once these are formed, the role of fire in the destruction of forests must be studied in the forests which are supposedly destroyed ... trained observers have been able to recognize the signs of forest relicts within these grasslands and the general tendency to vegetation succession in the direction of forest ... such degraded forest is common, and is usually associated with cultivation and forest clearing practices, among which fire plays an important, but not exclusive role" (Blydenstein, 1968).

The floods of the western Llanos discourage forest; higher areas support forests, however, and dense semi-deciduous forests exist in the western plains, even in more or less flooded areas.

The origin of the Llanos is frequently attributed to edaphic causes, due to the great extent of the impermeable layer or hardpan, and a succession of alternating drought and flood which are inimical to tree growth. In Surinam, Teunissen & Wildsehut (n.d.) state that the hydrology of the soil, as governed by the presence or absence of a more or less impermeable soil layer at shallow depth, is a principal habitat factor influencing the location of xerophytic forest, savanna scrub, savanna bushes and open vegetation, and exclusively open vegetation.

Van Donselaar (1965, 1968, 1969) states that the climate of northern Surinam is a "savanna climate" characterized by a certain difference between the precipitation in dry and wet seasons, independent of absolute values. A savanna vegetation is natural, i.e. determined edaphically, if the upper layer of the soil is alternately desiccated and saturated, in wet and very wet localities. A savanna vegetation occurs in dry localities only if fires prevent the formation of a closed layer of shrubs or trees. The origin of savanna formations on white sands in Surinam is discussed by Heyligers (1963) (see 1.6.5 for grass communities).

The presence of trees has been attributed to their possible ability to penetrate the hardpan (Walter, 1969). Blydenstein (1962) has noted that the watertable at the beginning of the rains is—575 cm. and at the end—385 cm., and that the hardpan cannot therefore be entirely impermeable. However, during trials at the Estación Biológica de Los Llanos near Calabozo, Venezuela, it was discovered that tree and isolated bush roots do not penetrate the layer, while they may travel laterally as far as 15 m. from a bush only 6 m. high. Moreover Aubreville (1965) has observed dense dry forest or at least thickly wooded savannas and open forests in Africa, India and Cambodia on laterite shields with very thin soils, in climates drier than that of the Llanos. Copses may be seen in the Llanos during the dry season which are still green, showing that adequate soil humidity is available. Here again, fire appears to be the dominant ecological factor determining the savannas of the Llanos.

As in Africa, the poverty of the floristics of the Llanos presents an anomaly, contrasting sharply with the rich, ancient, autochthonous flora of arid and semi-arid South America — for example, the caatingas and the campos cerrados of Brazil (600 species belonging to 336 genera and 83 families — Goodland, 1970). The shrubs, *Curatella americana, Bowdichia virgilioides, Byrsonima* and *Roupala* occur in all poorly shrubbed savannas of French Guyana, Ampapa, Upper Rio Branco and in the Llanos; they are recent colonizers, adapted to the now mediocre soil conditions of the deforested lands. There are indications that the germination of *Curatella* seeds is improved after passing them through a flame (Blydenstein, 1968).

The presence of dry deciduous forest in Guarico and Anzoategui is taken as confirmation of the fact that, several thousand or tens of thousands of years ago, the Llanos was occupied by a dry forest formation of the type which remains today in thickets, grading towards the east into more humid forest at the foot of the Andean Cordillera, in the Orinoco delta and in Guyana (Aubreville, 1965).

The origin and present status of the Rupununi savannas in former British Guyana (now Guyana) have been studied by Fanshawe (1952), and later by the Regional Research Centre of the British Carribbean at the Imperial College of Tropical Agriculture, Trinidad (Loxton, Rutherford & Spector, 1958; Stark, Rutherford, Spector & Jones, 1959). Particulars are given of climate, geology, geomorphology and soils. The vegetation of the northern, southern and central savannas is similar; treeless stretches of grassland are common: *Trachypogon plumosus* dominant, with *Aristida setifolia, Axonopus aureus, A. compressus, Andropogon angustatus, Mesosetum loliiforme*; on higher, better-drained soils the fire-resistant tree, *Curatella americana*, in open formation, with occasional *Byrsonima crassifolia; Anacardium occidentale* common around derelict house sites, ground cover of *Axonopus compressus*. The Good Hope-Annai Mountains carry what is considered by Myers (1936) to be a deciduous monsoonal type of forest, similar in composition to fringing forests along rivers — great mixture of shrubs, small trees and lianas, with many Myrtaceae and Leguminosae.

Eden (1970) states that climatic fluctuations during the Quaternary and resultant changes of soil water conditions must be considered as possible causes of vegetational change in the Rupununi. Pollen data suggest that savanna or dry *Byrsonima* forest has existed in the northern Rupununi from late glacial times to the present, possibly pre-dated by a more extensive, semi-deciduous forest now present as relic "bush islands". One has the impression that Eden does not give enough discredit to adverse soils and the much more recent burning which have contributed to the establishment and maintenance of the present savanna formation.

The savannas of Venezuela have been described by Ramia (1961, 1964, 1968), Blydenstein (1961, 1962, 1963; see 1.6.5 for subtypes of *Trachypogon* savanna) and Goodland (1966). The Llanos Orientales of Colombia (FAO/UNSF 1966; Blydenstein, 1967) contain, in addition to forests, savannas in which annual grasses are dominant: savannas with residue of forests (i) *Melinis minutiflora*; (ii) *Paspalum carinatum*; (iii) *Trachypogon ligularis/ Paspalum carinatum*; savannas subject to flooding (iv) *Andropogon*; (v) *Mesosetum*; humid savannas (vi) *Leptocoryphium lanatum*; (vii) *Trachypogon ligularis*; (viii) *Trachypogon vestitus/Axonopus purpusii*; (ix) *Paspalum pectinatum*; (x) *Trachypogon vestitus*. A great number of species are shared by the last seven types.

Blydenstein (1967) discusses theories on the origin of tropical savannas in relation to the environment (climate, landscape, including geological history and geomorphology, soil, fire and man). Pollen analyses have shown a dominance of Cyperaceae and Gramineae in an herbaceous vegetation to the northern edge of the Llanos in Pleistocene times. Cole (1960) postulates that grass savannas in Brazil represent the older vegetation formations, which are now gradually being replaced by forest on dissected terrain as the level peneplain becomes eroded. It is now considered that man arrived on the American continent between 40,000 and 15,000 years ago, and that the palaeo-Indian epoch in the Llanos was from 15,000 to 5,000 B.C. (Blydenstein, 1967).

A modern comparison has been made in Costa Rica between environments of a tropical semi-deciduous forest with a contiguous area of derived *Hyparrhenia rufa* savanna which has been burned annually (Daubenmire, 1972b). During the five-month dry season the savanna soil becomes hard and cracks. Compaction by zebu cattle and horses, coupled with a very sparse, invertebrate soil fauna, reduces porosity. During the season of maximum leaflessness, the forest still intercepts about 33 per cent of solar radiation; the savanna soil is rendered bare and black by fire during this season. Higher soil temperatures and the annual burning probably account for much of the striking reduction in burns in the savanna soil profile. Soils become dried to wilting point to a depth of over 90 cm. in the savanna, but remain moist below 30 cm. in the forest. No downward movement of clay or other irreversible changes in the profile have been noted. There is, however, good evidence of a loss of 11 cm. of soil during the 22 years since deforestation; this is regarded as the most significant aspect of environmental deterioration.

The most characteristic herbaceous plants of the savannas in Cuba (Voronov, 1970) are grasses of the genera *Panicum, Paspalum, Arundinella* and *Arthrostylidium*. On the poorer soils, palm savannas (genus *Sabal*) occur, possibly originating from palm groves due to anthropogenic influences. Relations are not quite clear between the savannas and the thorny shrub communities, which occur mainly in tropical conditions and generally indicate increasing aridity. Savannas with low or stemless palms (*Copernicia* spp. with a herb/low shrub layer) are formed on serpentines, a markedly deviant, edaphically conditioned group of formations rich in endemic species. Gradual salinization of the soil in savanna swamps, usually stagnant, leads to the formation of solonchak meadows; grasses with rigid stalks and short leaves − *Distichlis spicata, Sporobolus domingensis, Cynodon dactylon* and *Paspalum vaginatum*.

1.3.5.2   *Cerrados of Brazil*

The vegetation of almost all Central Brazil is cerrado (meaning in Portuguese half-closed, dense or ajar), an appropriate name for this vegetation which is neither open nor closed, that contains every physiognomic type between these two extremes, herbaceous, grassy, shrubby, orchard, woodland, almost to forest (Grossman, Aronovich & Campello, 1966; Lima, 1966; Santiago, 1970; Goodland, 1970, 1971a and 1971b).

Different authors have different opinions regarding the

origin of the cerrados, for example: a. all represent climaxes; b. all are anthropogenic; c. some are climaxes, but man, by felling and burning, has created conditions for the migration of cerrados from their original habitats to other places formerly covered by forests. If a cerrado is protected for about 30 years, there will be significant modifications of the vegetation, both physiognomic and floristic. If protection continues for long enough, the land will revert to forest (Ferri, 1973).

Eiten (1972) argues that the cerrado is a climax vegetation, not a stage or a disclimax in succession to forest or any other vegetation type. Both in its central position in the country and on its margins when it is in contact with other vegetation provinces, it is an original, primitive vegetation — a whole floristic province of semi-deciduous, xeromorphic vegetation. It has not arisen following the destruction of mesophytic forests by burning.

Goodland has studied in particular the oxisoils and the plant species and vegetation types of the Triângulo Mineiro. The physiognomic gradient parallels a soil fertility gradient; basal area per hectare is correlated with P, N and K (Goodland & Pollard, 1973). Soil fertility is not necessarily the cause of the physiognomic gradient; differences in physiognomy and fertility may have arisen because of different histories of burning and cutting. The cerrado situation seems to contradict the statement of Beard (1953): "Savannas occur in ill-drained country ... the chemical status (of the soils) is of little account."

In the cerrados, several species of grass may reach 3 m. in height: *Tristachya leiostachya, Axonopus compressus, Tristachya chrysothrix, Hyparrhenia rufa, Andropogon lateralis, A. condensatus* and *Melinis minutiflora,* in order of decreasing height. Grasses are reduced in number and height in other types, probably due to shading. The fact that woody vegetation is dominant in most Brazilian cerrados may distinguish them from the floristically related vegetation of the Llanos of Venezuela and Colombia discussed above. The rich flora of this area in Brazil would suggest great venerability, following Aubreville. See bibliography no. 1268 of the Commonwealth Bureau of Pastures and Field Crops for 116 references on the llanos of Colombia and Venezuela and the cerrados of Brazil (1939–70).

The natural region called "Gran Chaco" occupies large areas in Bolivia, Paraguay and Argentina. Morello (1968) has divided the Argentine Chaco into "great units of vegetation and environment," considering vegetation "as the most sensitive and simplest indicator for registering and integrating environmental variations." The model concept (roughly a group of recurrent patterns of vegetation) and "three levels of perception" have been developed. The names of the grass genera of the Argentine Chaco would suggest at least a borderline subtropical environment: *Aristida adscensionis, Elyonurus adustus, E. muticus, Panicum prionitis, Paspalum intermedium, Setaria* and *Sorghastrum* (Marlange, 1971). The general state of specific degradation and low nutritive value of pastures in Argentina has been confirmed by enclosures (Bragadín, 1959) for the region of Los Llanos in the province of La Rioja. The species of Gramineae noted in the various closures were: *Aristida adscensionis, A. medocina, Bouteloua aristoides, Cottea pappophoroides, Chloris polydactyla, Digitaria californica, Diplachne dubia, Eragrostis cilianensis, E. longipila, E. virescens, Gouinia paraguayensis, Pappophorum* (= *Enneapogon*) spp., *Paspalum unispicatum, Setaria argentina, S. geniculata, Sporobolus pyramidalis, Tragus racemosus, Trichloris crinita, T. pluriflora* and *Tripogon spicatus*.

It is in the Chaco that human and geomorphological disturbances have led to creation of very unstable and non-adjusted marginal ecosystems with regard to structure and niches. Morello & Saravia Toledo (1959) have examined the changes caused by extensive livestock husbandry in the woods and grasslands of the Chaco. The primitive woods were of two types: a. Quebracho stands of *Schinopsis quebracho-colorado* and *Aspidosperma quebracho-blanco*, and b. Palosanto stands of *Bulnesia sarmienti* with either *Aspidosperma quebracho-blanco* or *A. triternatum*.

The primitive grasslands were composed of *Pennisetum frutescens, Elyonurus tripsacoides, Trichloris crinita, T. pluriflora, Setaria argentina, Gouinia paraguayensis* and *G. latifolia*. In the upland savannas there is always evidence of fire; they are therefore considered to be pyrogenic in origin, and caused by lightning rather than man.

In the first quarter of this century, animal husbandry was limited to the grasslands, which were transformed into shrub communities of *Acacia, Mimosa, Mimozyganthus* and *Celtis*. Where these pastures were exhausted, the cattle began to feed in the interior of the wooded areas, inducing many floristic and plant sociological changes, e.g. propagation of species hitherto unknown in virgin forest, invasion by species largely restricted to the major drainage bottoms, invasion of arborescent Cactaceae, and the formation of a number of anthropogenic communities. Morello & Saravia Toledo (1959) continued their study by re-

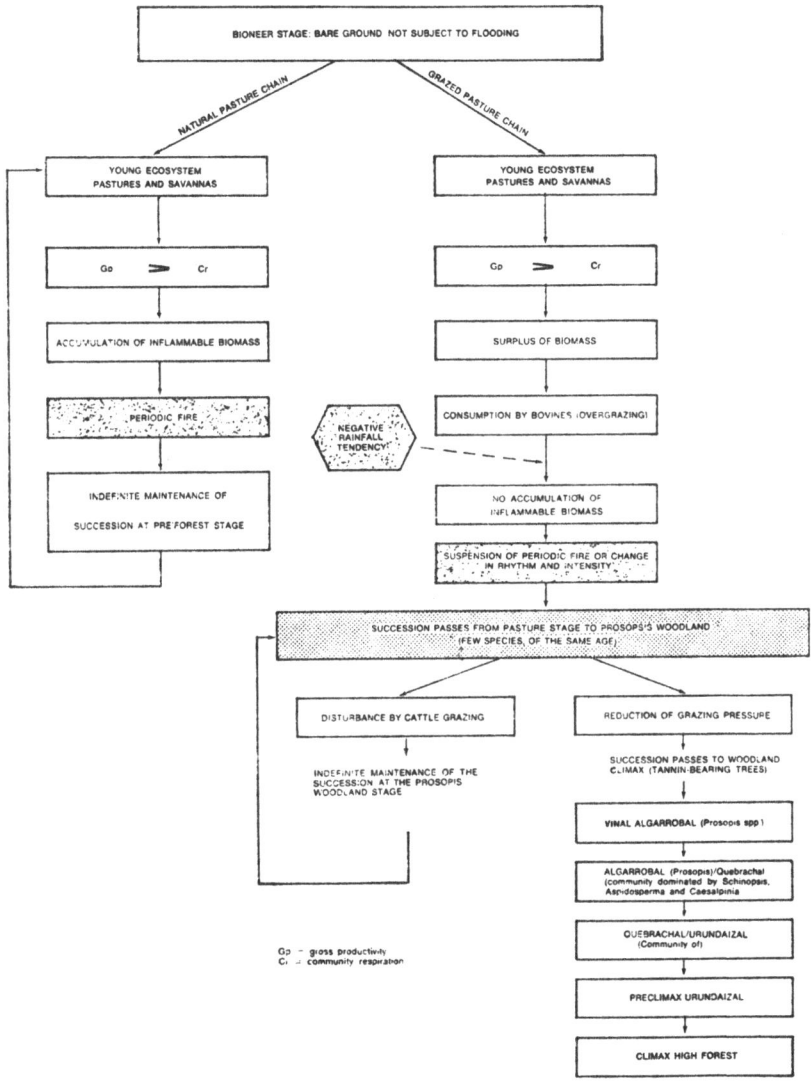

Fig. 1.4. Model for the evolution of an ecosystem on land not subject to flooding, with and without cattle (Morello, 1970).

cording the present seasonal diets and grazing patterns of domestic animals in the forest communities of the Chaco, and the role of rodents as a destructive element in the peri-domestic areas, where one of the ten plant communities recognized is characterized by grasses of ephemeral type.

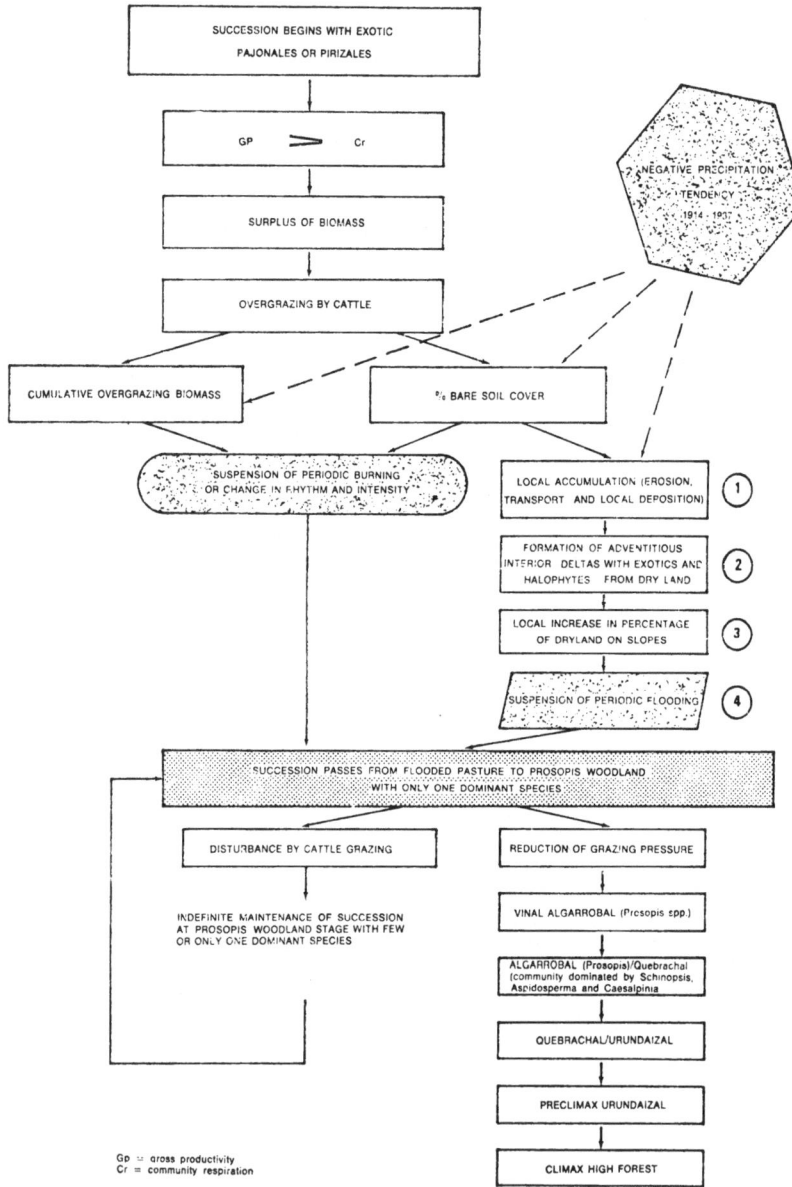

Fig. 1.5. Model for the evolution of an ecosystem on land subject to flooding and with overgrazing (Morello, 1970).

It is into some of the primitive systems that the woody colonizing species, *Prosopis ruscifolia* or vinal, has entered (Morello, Crudeli & Saraceno, 1971). Morello (1970) has evolved a conceptual model of the relations between grassland and vinal (Fig. 1.4-1.7). This model considers that, before the pressure of cattle on the Chaco, the grasslands were regularly submitted to two types of periodic pulsations — fires and floods. Their disappearance would have started "the explosive evolution of the woody plants in environments of grassland and tall grassland."

Plant formations of potential value in Paraguay include the typical grasslands, the marsh vegetation and the palm savannas of *Copernicia australis, Acrocoma totai* and *Butea yatay*, the thorn savannas with various species of *Prosopis*, the tree savannas of *Tecoma argentea*, and the mosaic savanna of the "montes", tropical, semi-deciduous, semiarid steppes (Fretes, Samudio & Gay, 1970). The Gramineae are present in greater number of genera, species (128) and individuals and occupy a much greater area than herbage legumes (33 species); the latter are present solely as "interstitial" components and have little effect on the quantity and quality of the forage (limited by acid soils, deficiency of phosphorus, calcium and possibly some essential minor elements). Woody species of *Prosopis, Mimosa* and *Acacia* are invaders in some alluvial grasslands. The genera of the Gramineae that are predominant in respect of area are *Paspalum, Andropogon, Sorghastrum, Elyonurus* and *Axonopus*; in respect of number of species, *Andropogon, Eragrostis, Panicum, Paspalum, Aristida*.

Walter (1967a and 1967b) considers that: "the pampa is a transitional zone between the temperate grasslands caused by aridity and the tropical grasslands caused by topographical factors. The grasses can withstand the change from humidity to aridity during a year, but not the trees".

But, according to Troll (1968) all months of the year have a precipitation far in excess of potential evaporation. Ellenberg (1962) concludes that many native and exotic tree species would spread spontaneously "if the omnipresent grazing cattle did not destroy the tree seedlings." Hueck (1953) reaches similar conclusions regarding the Chaco and parklands of Tucumán Province; only human influence could have created the park landscape from former dense forest.

In her review of literature on the grasslands of Latin America, Roseveare (1948) discussed the work of Rosengurtt and his colleagues (1939; 1944) on the grass covers of Uruguay; this country is also borderline in the present context, since the grass covers are admixtures of tropical and subtropical (summer-grazing) and temperate (winter) species. The Mesopotamia region of Argentina has a transitional climate (Van der Sluijs, 1971) towards a cool, steppe climate in the south and a cool, tropical

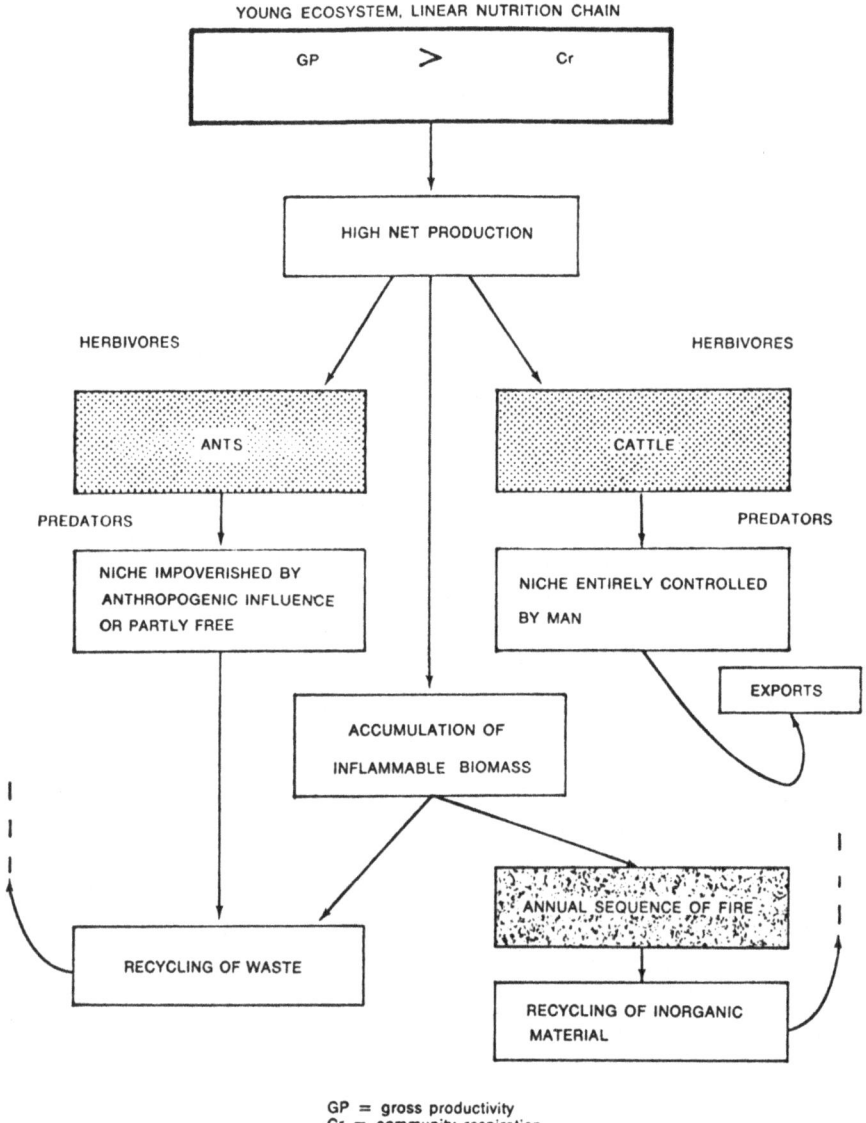

GP = gross productivity
Cr = community respiration

Fig. 1.6. Simplified model of present grazing systems in the Chaco pastures not subject to flooding (Morello, 1970).

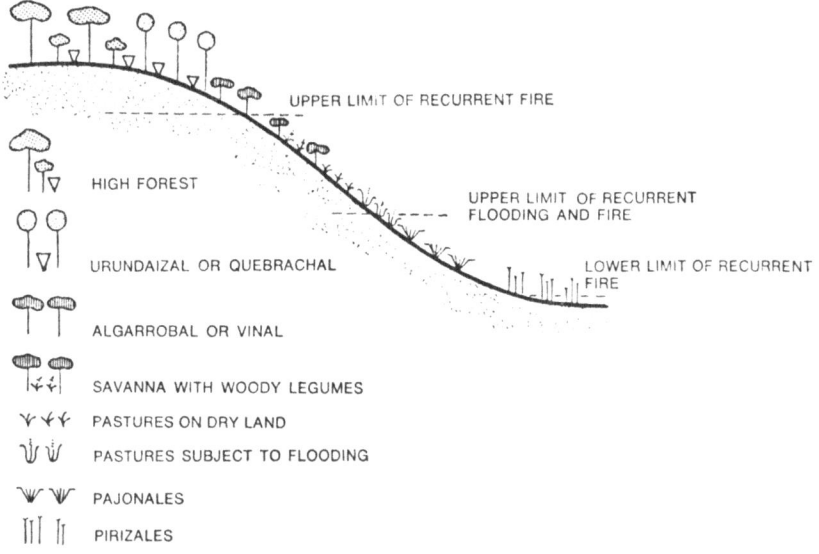

HIGH FOREST

URUNDAIZAL OR QUEBRACHAL

ALGARROBAL OR VINAL

SAVANNA WITH WOODY LEGUMES

PASTURES ON DRY LAND

PASTURES SUBJECT TO FLOODING

PAJONALES

PIRIZALES

UPPER LIMIT OF RECURRENT FIRE

UPPER LIMIT OF RECURRENT
FLOODING AND FIRE

LOWER LIMIT OF RECURRENT
FIRE

Fig. 1.7. Generalized cross-section of an interfluvial slope in tropophilous woodland in Formosa, Argentina (Morello, 1970).

climate in the north.

Transitional types and zones of vegetation in Mexico are described by Leopold (1950), and Hernandez Efrain (1963). Rzedowski (1975) has made an ecological and phytogeographical analysis of the grasslands of Mexico, tropical and transitional. Beard (1944) considers that the study of communities in tropical America may offer the key to the understanding of African types of vegetation, particularly where fire and grazing are operative. Pennington & Sarukhan (1968) describe the various biomes in the Mexican tropics.

1.4 Survey and Assessment of Resources

1.4.1 THE INTEGRATED OR HOLISTIC APPROACH

Specialists concerned with the evaluation of the total environment of tropical grazing lands accept the fact that their own detailed studies should be a part of or a sequel to an integrated survey, as defined by Christian & Stewart (1968) — a survey of the physical and biological features, climate, water, soil, vegetation, minerals, terrain and natural environment. (See also annual reports of C.S.I.R.O. Division of Land Research.) Only indirectly are the most important natural resource factors considered, the human (rural) occupants of a country who are expected to follow up the recommendations arising from the survey, for the ultimate benefit of themselves and the urban populations who are dependent upon them.

"Of the physical resource factors, climate has the most direct significance for its general characteristics set the broad pattern of biological activity. Although land use may modify the microclimate, the broader characteristics of climate are not disturbed by man and the climatic pattern may be regarded as a permanent feature. The broad features of climate often change zonally in an orderly manner according to major global influences but this zonation is interrupted by topographic effects and the interrelations of land and water masses ..."

"Each resource factor has its own spatial distribution and some have a temporal distribution as well. Thus, although the broader aspects of climate may vary geographically within a country, they will also differ from season to season ... stream flows will vary seasonally according to rainfall, run-off and seepage. The grazing value of native vegetation such as natural grasslands will also be subject to season to season variations ... Survey and assessment methods and techniques must be adapted to the nature and form of each resource factor".

The Australian workers have adopted the term "land" to denote the complex of climate, topography, soil and vegetation, since it is the combined effect of all and the interactions between them which are important. Christian (1963) has stated: "Land must be considered as the whole vertical profile at a site on the land surface from the aerial environment down to the underlying geological horizons, and including the plant and animal populations, and past and present activity associated with it. There are many features in this total profile, some easily observable like the soil and vegetation, some measurable such as the rainfall and surface slope, but many not so readily observed or measured ... The many features of this total profile vary from site to site and their many combinations and interactions result in a vast array of land types, each with its own potential and limitations for agriculture and forestry, each presenting its own specific barriers to the achievement of maximum plant or animal production. Rarely does one feature alone determine productivity. It is the combination of all that is important, and if we are to understand land we must think of it in terms of this complex rather than only of the individual components of it".

The holistic and separationist approaches are discussed by Zonneveld (1972) in Volume VII, Chapter VII; 4 Lectures on land science, land survey and land evaluation. For an international symposium, Zonneveld's discussion on the use and comparability of terms in different languages, and of the careless use of the word: "ecosystem", is of particular value. Reference is made to the assertion that vegetation is a holistic expression of all environmental factors, that a vegetation classification unit is really a characterisation of the land as a whole, and can thus be used as the sole basis for land classification.

While agreeing with Zonneveld that such "attempts are often too sweeping", one may make a case for specialized studies of vegetation in general, and of grazing in particular, to follow an integrated or holistic survey. After the more general survey has defined land within a given land system or unit as being actual or potential grazing land or sown pasture land, the subsequent detailed analysis and assessment of that resource are the job of specialists in botany, ecology and vegetation history and evolution.

The concept of integration as seen by the Montpellier plant sociology school is stated by Long (1972). One may approach from two directions:

from elementary disciplines or from the nature of variables, or

the ecosystematic, or global multidisciplinary approach. In the first, integration is made *a posteriori* by trial and error. The second or holistic integration proceeds from *a priori* hypotheses (CSIRO and Montpellier) and *a posteriori* interpretations. The phytoecological approach of Montpellier is especially well developed (vertical versus horizontal integration). Verified integration is that which proceeds from mathematical models, from historical data, or from experimentation. Total integration takes into account contributions from "naturalists" as well as "humanists".

Over vast areas of the tropics and subtropics, the vegetation as defined by the land scientists is rarely climax. The grass, shrub and tree cover on tropical grazing lands is already at a low stage in ecological regression, so much so that a semi-arid vegetation may be found growing in a sub-humid climate, and grass savanna in a forest climate. Thus it is necessary to query the Australian statement, that land use may modify only the microclimate, in view of the presence of widespread desiccation of the total environment following haphazard or even planned devegetation for land "development".

The place of vegetation, grassland and land use in the overall definition of land systems is demonstrated in the survey made by the British Land Resources Division in North-East Nigeria (Vol. 3) (Bawden, Carroll & Tuley, 1972), and in the accompanying statement (Vol. 4) on present and potential land use. A *land system* is here defined as an area or group of areas throughout which can be recognized a recurring pattern of topography, soils and vegetation. The components of this pattern are *land facets*. The land systems (123 in all) are grouped into *land regions*, and these again into *land provinces* (five defined). In spite of the failings of the land-system concept, it is regarded as admirably suitable for rapid reconnaissance surveys based on aerial photography.

In recognizing twenty grass cover communities, both physiognomy and species composition are closely related to the main agro-ecological zones. In the northern sector with a mean annual rainfall of 500 mm. or less, the short grass cover is typically annual (*Aristida, Chloris, Digitaria*). The few perennials are highly xerophytic in adaptation. In the central Sudan Zone, *Loudetia* spp. are conspicuous with shorter members of the Andropogoneae. When the rainfall exceeds 750 mm., the tall grass covers are dominated by annual and perennial members of the Andropogoneae.

1.4.2   THE APPLICATION OF AERIAL PHOTOGRAPHY AND SPACE TECHNOLOGY

The use of aerial photographs, ground checks and subsequent interpretation is a fundamental technique in integrated surveys of land resources (Christian & Stewart, 1968; Bawden, 1965; Whyte, 1976a). The cost of survey can be reduced by interpretation from aerial photographs, by the establishment of a correlation between patterns on aerial photographs and land features (Perry, 1964). As the patterns on aerial photographs are dependent on the integration of all land features, they can be correctly understood only in terms of all these factors. Conversely, the most efficient use of aerial photographs can be made only through the integrated approach, e.g. for surveys of pastoral areas in northern Australia (Perry, 1967). Much information can be extracted from aerial photographs, making it possible to cover

60

large areas with a minimum of ground control.

Similar methods have been adopted for surveys in African and other countries by the Land Resources Division of the British Directorate of Overseas Surveys (Rains, 1970). Stereoscopic examination of pairs of photographs facilitates the recognition of differences in land form, soil and vegetation, and boundaries may be marked on photographs. For example, in the Land Use Survey of Malawi, it has been possible to apply statistical sampling to air photo coverage, to provide adequate information economically, about the distribution of types of cultivated, uncultivated and uncultivable land.

New techniques involving colour and infra-red colour make it possible for a person with normal vision to distinguish a large number of colours compared with a few grey tones (more grey tones can be distinguished with transmitted than with reflected light). The Land Resources Division finds that results of analysing vegetation, soil and landuse patterns on true and infra-red colour air photographs are only marginally better than results from black and white prints. The use of infra-red colour facilitates remote sensing in the determination of the grass cover and/ or the quantity of herbage present; it is also of value in range inventory studies. The recent development of high-altitude photography and the use of multispectral sensors to identify natural formations is unlikely to supersede conventional photography for some years (Rains, 1970). Most workers will continue to use films recording the visible and near infra-red bands of the spectrum exposed at medium altitudes in fixed-wing aircraft.

Grassland, as defined by Pratt, Greenway & Gwynne (1966) can be recognized on aerial photographs by its smooth grey textureless tone and its location (seasonally flooded plains and sites with impeded drainage, apart from high altitudes). Although grassland can be recognized and the density of tall shrubs and trees determined on 1 : 60,000 scale photography (Rains, 1970), it is most difficult to assess the density of woody plants in low shrub savanna at this scale. The problem was resolved for resource surveys of Botswana (Rains & Yalala, 1972) by analysing the distribution and type of the burning patterns. The characteristic burning pattern may appear as black or shades of grey. A pattern from fires over a period of seven years is an indication of good forage production. Patterns from deliberate burning suggest that deterioration is occurring. One must, however, be careful in interpreting fire patterns until the nature of the forage and density of the shrub can be checked.

The International Institute for Aerial Survey and Earth Sciences, Enschede, has reported on vegetation patterns in a savanna region of Northern Nigeria, during the UNDP/FAO survey of the soil and water resources of the Sokoto Valley. The interpretation of the relation between these and other patterns, clearly visible on aerial photographs, to the edaphic conditions played an important part in speeding up the reconnaissance survey (Zonneveld, de Leeuw & Sombroek, 1971). Both the main patterns, the "pseudo-dune and gully-pattern" and the "black and white dot pattern" could be subdivided into sub-patterns in which various pattern "elements" were distinguished.

Zonneveld (personal communication, 1972) would agree with British workers that the ordinary, old-fashioned black and white panchromatic film will remain the cheapest and still the most valuable tool far into the future. Thermal sensing (far infrared) will never be useful for general grassland surveys. Radar will certainly be interesting for small-scale survey, but so far there is little experience with it. The more conventional remote sensing like near infrared (preferably combined with visible radiation in false colour) is of particular value on vegetation types that have much open space, as in arid or semi-arid zones. The contrast between bare soil and patches covered with vegetation is sharp, and enables a good study of patterns to be made.

No study has been seen dealing specifically with the application of space technology to analysis of tropical grazing lands, apart from their place in the general vegetation (Whyte, 1976a). However, an interagency meeting on the establishment of an African regional remote sensing centre of satellite ground receiving and the processing of data held in Addis Ababa in October 1975 recommended that a mission of experts should make a thorough study of the plans for training programmes in the technique in a number of African countries. This should lead to the assembly of data relating to the grazing lands of tropical Africa.

The NASA satellite photo of the Sahel reproduced by Wade (1974) shows dramatically the effect of overgrazing, the striking difference between the badly managed open range and the curiously shaped green pentagon of the Ekrafane ranch, 100,000 hectares, inside a barbed-wire fence established five years previously when the drought began, divided into five sectors, with the cattle allowed to graze one sector each year.

UNITED NATIONS PROGRAMMES

International Programme for Grazing Lands

Following discussions between the International Biological Programme and FAO, a project has been drawn up for a programme of training, analysis and synthesis of data and distribution of information on the world's grazing lands (UNESCO, 1971). It is considered to be unrealistic that countries should individually attempt to start separate, full-scale research programmes. U.N. Agencies are conducting studies and supporting training. The I.B.P. terrestrial productivity group has, through its grassland, tundra and arid land groups, started research on grazing lands, and established a network of communication. However, there has not yet been any integration, nor any detailed analysis and synthesis of results.

A systems analysis approach will be taken in the organization of data, information, experience, and ideas to develop models and analyses concerning important practical and theoretical problems. The programme will include development of a data bank of information on grazing lands and a programme of both informal and formal training for participants from developing nations.

FAO Resource Survey in South America

A regional project for the evaluation of grassland resources has been in operation in the Regional Office of FAO in Santiago, Chile since 1970. The objectives for the project, which covers the tropical and temperate regions of Latin America from Mexico to Patagonia, are (Blydenstein, 1972): description of correct systems of management, collection of basic information on the distribution and characteristics of grasslands, identification of problems in optimal use of the resources, establishment of priorities for research, and rapid surveys of grazing resources in selected countries.

The methodology of the rapid surveys consists of the collection and manipulation of existing statistics on livestock production in relation to available information on grazing resources. Studies are completed or are in progress for Venezuela (Blydenstein, 1971), Peru, Chile, Argentina, Colombia and the Caribbean Islands. These can be no better that the statistics upon which they are based. Estimates are also obtained on grazing resources, extent of grasslands used, types of natural grassland or improved pasture, availability of additional forage and feed resources, carrying capacity with its seasonal fluctuations.

Numerous grassland surveys are concerned with the floristic composition and ecological status of the grass cover, whereas an evaluation of resources calls, according to Blydenstein, for a different set of data of which the botanical information is only a part. Yet it is admitted that it is botanical and ecological factors which impose restrictions on the degree of livestock development. This is seen particularly in the seasonality of growth in the dry tropics. Seasonal variations in forage availability in the humid tropics are also important; growth of *Pennisetum purpureum* is depressed during the very wet season due to insufficient sunlight; *Digitaria decumbens* has two seasons of lower forage availability, one when it is in flower, another in the drier part of the year, even if rainfall then exceeds 50 mm. per month. Mineral imbalances are significant in livestock production throughout Latin America.

Survey in Kenya

A UNDP/FAO Range Management Project became operational in Kenya in 1966, at the request of the Government. A credit of U.S. $ 7.2 million was also provided for rangeland development in specified areas, negotiated with the International Development Association and the Kingdom of Sweden. The three main objectives of the UNDP/FAO project are: land-use survey and socio-ecological planning, investigations on problems of forage and animal production, and education of specialist staffs.

Casebeer (1969) has described the plans for that part of the project dealing with integrated survey, through the six phases: review of available data and literature; reconnaissance; field studies; assessment and coordination; projections and recommendations; and reporting.

1.4.4   THE APPROACH OF THE PLANT SOCIOLOGIST

The Centre d'Etudes Phytosociologiques et Ecologiques at Montpellier has developed a method of identification inventory and mapping of vegetation and environment (Long, 1966). Plant ecology is distinct from phytogeography, which is the study of natural vegetation or modified plant communities, when the units studied are large areas of physiognomic homogeneity, the "plant formations" and "vegetative types" of classical plant geography. Plant ecology has as its aim the simultaneous study of vegetation and environment, the main object being an understanding of the biological processes of ecosystems.

The first stage leading to the preparation of the large-scale maps required by range specialists is a sampling of plant communities in their natural habitat, using special forms on which are assembled the basic data, coded for transference to punched cards. Descoings (1971) has discussed the use of standardized data sheets devised for analysis of the structure of intertropical herbaceous formations, including the formations classified in the Montpellier "Code écologique" (Godron *et al.*, 1968) and especially those currently called savannas, steppes and pseudo-steppes. Correlations between species and environmental factors are attempted and statistical tests of significance are made (Long, 1966). The result is the establishment of "ecological groups" which are indicators for the respective environments. These groups are used in the compilation of vegetation/environment maps which are accompanied by maps of the factors of the environment and of land occupation.

In a separate paper Long (1969) discusses these techniques and their cartographical expression in greater depth. A general model of the perception of biogeographical facts and their reciprocal relation is termed the pyramid of perception; this comprises five levels, from the small to the largest scales. The application of this concept to plant ecology and plant geography is important.

Efficient and objective sampling methods in areas used for grazing and browsing of wild and domestic livestock will, according to Long, help developing countries to make up for lost time in the survey and analysis of their natural resources. The cost of phyto-ecological studies (1965 figures) including mapping but not printing, can vary from 4 cents U.S. per hectare for small-scale mapping of arid areas with open vegetation to U.S. $2.00 per hectare for large-scale mapping in temperate regions with a closed vegetation.

The concepts and methods of plant sociology have been applied to the tropical conditions of Brazil (Cain & Castro, 1959), in studies of the structure and distribution of vegetation as they relate to medical ecology (malaria, yellow fever, schistosomiasis), all limiting factors in the evolution of ecosystems based upon animal husbandry in forest/savanna environments.

1.4.5   RANGE CONDITION ANALYSIS

This is considered as a tool for the study of unclassified grazing lands in general, and in Brazil in particular (Humphrey, 1966). The application of this North American technique requires a knowledge of the nature of ecological succession of vegetation and the potential of specific sites for forage production. Ranges are classified for their potential for maximum production into four classes: excellent, good, fair, poor, on the basis of five criteria: forage composition (botanical), ground cover, plant vigour, litter, erosion. Condition trend is also determined on the basis of criteria that indicate an upward or a downward trend, these being the kind of plants that are reproducing, their vigour and seed-setting characteristics, and degree of erosion.

In Rhodesia, it has been agreed (Southern Rhodesia, 1962) that the most important criteria for the classification of condition in African veld are the relative abundance of trees and shrubs, of perennial grasses, of annual grasses and herbs, and of bare ground. Because, to the grazier, the best veld is that which provides the most forage for domestic animals, especially cattle, year in and year out, the highest condition class is veld which consists of perennial grasses without trees or shrubs, with few annual grasses, and a minimum of bare ground.

African workers have to distinguish between a classification of condition class that could be used in a succession advancing towards bush, and condition as understood in the American range in a succession advancing towards grass. Condition in American range is usually related to the climax species present (76 per cent means condition excellent, for example). In Africa, the situation is reversed. In vegetation with 76 percent of climax species, condition for grazing is very low for anything other than the browsing species of game. Condition classification here, and probably in the Indian subcontinent as well, should be based on a "theoretical condition of perfection, that is, on grass from which the trees have been excluded by management" (Southern Rhodesia, *op. cit.*; Vincent, 1962; Rattray, 1962).

Little is known of the potential for forage production of many million hectares of brush- or tree-covered lands of South America (Humphrey, 1966); in Brazil, for example, is it possible to transform them to grass range, as has been done on the chaparral and pine/juniper woodlands of Arizona? The three types of brush land in the arid north-east of Brazil, for example, serido, sertào and caatinga, are very extensive and yet produce relatively little forage.

1.4.6   GRASSLAND SURVEY OF INDIA

In 1953, the Indian Council of Agricultural Research decided to undertake a rapid reconnaissance survey of the grass covers of India (Dabadghao & Shankarnarayan, 1973; Whyte, 1968a). Two members of the survey team (P.M. Dabadghao & B.D. Patil) received training under FAO fellowships in the United States of America and Great Britain, and others were trained at the Forest Research Institute, Dehra Dun. Up to that time, Indian and British botanists had done invaluable work on the recognition, naming and classification of the great number of grass species in the flora of the subcontinent, and N.L. Bor (1960) has produced a major work on the subject. Comparatively little attention had, however, been given to the grass communities, to their botanical composition, and to the nature of the succession that occurred under the influence of biotic factors.

The objectives of the survey and the techniques adopted on selected one-acre plots on over 500 sites throughout India, were:
a.   Floristic composition, (i) listing and classification of species; (ii) estimates of percentage composition by the pace-transect method
b.   Density of plant cover, judged by the square foot density method from a circular sample enclosing an area of 100 sq. feet (Stewart & Hutchings, 1936)
c.   Forage production, estimated by cutting and weighing the individual species from the 100 sq. ft. density circles
d.   Plant vigour assessed by the study of plant height, leaf length, length of seed-stalk, number of tillers, basal area and reproduction of species
e.   Plant succession, data obtained by inference and by a comparison of floristic composition on undisturbed and adjacent overgrazed areas
f.   Associated forest types, listing of tree species and classification of forest types
g.   Soil types, data collected by methods adopted in soil surveys,

for colour, texture, water-holding capacity, pH, total soluble salts, and soluble phosphate and potash.

1.4.7 INSTITUT D'ELEVAGE ET DE MÉDECINE VÉTÉRINAIRE DES PAYS TROPICAUX, MAISONS ALFORT, FRANCE

The Institut d'Elevage et de Médecine Vétérinaire des Pays Tropicaux has evolved its own sequence of techniques for the study and mapping of tropical grazing lands (Boudet & Baeyens, 1963; Boudet, 1966; Bille, Lebrun & Rivière, 1968/9; Pagot, 1971):

a. detailed classification is made of phytosociological surveys of minimal areas located in different vegetation groups;

b. the floral components of each type of pasture so identified are obtained from a grid survey of a few examples of each characteristic group; a statistical survey of these results facilitates a more exact assessment of species frequency and basic cover if perennial species are abundant;

c. the forage value of each pasture type is determined by random sampling; aliquot samples are made to determine nutritive value and trace-element content; and

d. the survey is completed by the construction of a map based on aerial photographs.

In the Sahelian zone, this mapping permits the assessment of optimal head of cattle per region, the adoption of a hydrological programme in harmony with pasture potential, the establishment of a rotational system of grazing by regular closure of wells in relation to optimal season for grazing the dominant vegetation, and the probable development of pastures under intensive grazing, controlled again by closing wells at critical times.

In the Sudanian and sub-Guinean savanna, the maps show the stock-breeding potentialities of specific areas. The problems associated with the development of temporary pastures may be solved by assessment of productivity and control of pasture development by periodical grazing of 2,500 m². areas, and by making annual grid surveys.

Boudet & Ellenberger (1971) describe the techniques used in an agrostological study in the Republic of Mali:

a. definition of natural pastures following inventory of abundance/dominance of species;

b. establishment of synoptic tables of the principal vegetation communities;

c. biological or growth forms of species with a significant distri-

68

bution (mesophanerophytes, nanophanerophytes, geophytes, etc.);

d. relation between types of pastures, and the major physiognomic groupings recognized by Trochain (1957): forêt dense sèche, forêt claire, savane, savane boisée, prairie (aquatic or marshy)

e. subdivision of types of vegetation other than forest according to the characteristics of the woody cover (arbustif, arboré, boisé), and

f. primary productivity of the herbaceous cover.

1.5   Synecology

1.5.1   GRASS COVERS IN FOREST CLIMAXES

There are no climax grass covers in the intertropical zone, except perhaps in small and localized areas of climatic or edaphic climax. Thus, the grass cover ecologist is concerned with what may be called ground cover subclimaxes in a wide range of shrub and tree formations; most of these are themselves subclimaxes in regressive forest succession, due to the operation of anthropogenic factors. This was recognized in Rattray's map of the Grass Cover of Africa (1960), in which the existing tree cover was introduced as a backdrop to the grass map; also by the team conducting the Grassland Survey of India, who always noted the tree species and forest types when recording the total site cover (1.4.6).

Some writers have suggested that there is a degree of parallelism between forest type and the associated ground grass cover (Fig. 1.8., Shankarnarayan, 1963; Whyte, 1968a). Grassland workers are fully aware of the type of tree and shrub species that they find regularly associated with their grass covers, without necessarily referring to this as parallelism. It may be true to say that, in sites undisturbed by anthropogenic influences, tree canopies and grass ground covers are parallel, in that one would expect to find certain grass species and, in the more open forests, grass communities associated with certain tree or forest types. When either the forest covers or the grass covers or both come to be used and misused, that degree of parallelism may be lost due to differential rates of regression.

Bhatnagar (1960) shows that clear felling of moist *Shorea robusta* forests in north India tends to raise the watertable, and grass communities include *Narenga porphyrocoma, Arundo donax* and *Phragmites maxima*; in dry forests, pioneer grasses are *Saccharum spontaneum, Erianthus munja* and *Neyraudia arundinacea* (Fig. 1.9).

Some examples of tree or shrub/grass cover associations for two of the vegetation zones south of the Sahara are given by Boudet & Baeyens (1963). It is to those types of vegetation that the sylvo-pastoral methods of husbandry are applied (1.1.6).

Steppe zone — Mauritania:

a. *Ziziphus mauritiana* and tufts of *Cymbopogon schoenanthus*.

b. *Acacia seyal* dominant, *A. raddiana, Balanites aegyptiaca* and ground layer dominated by *Andropogon gayanus* var. *genuinus*

c. *Acacia seyal* and *Cordia gharaf*, tufts of *Cymbopogon schoenanthus*, some *Schoenefeldia gracilis*

d. trees of *Acacia nilotica* var. *tomentosa*, shrubs of *Mitragyna inermis* and *Feretia canthioides*, with *Oryza breviligulata* and *Echinochloa colonum*.

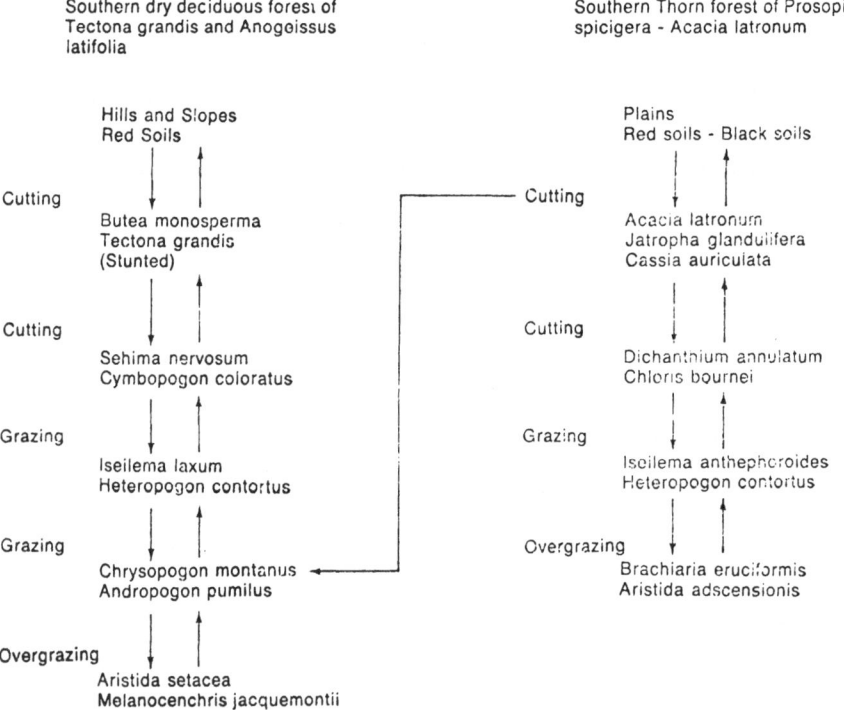

Fig. 1.8. Succession in two types of vegetation at Bellary, Mysore, India (Shankarnarayan, 1963).

70

Sudanian zone — Mali:

a. shrubs of *Parinari curatellaefolium* and *Pterocarpus lucens* with *Ctenium newtonii*

b. forêt claire with *Pachystela pobeguiniana, Bombax costatum*, with dominant shrubs *Acacia macrostachya, Boscia angustifolia* and *Combretum molle*

c. very open tree layer with *Parkia biglobosa* and *Butyrospermum parkii* (species protected for their fruits), a shrub layer of *Terminalia avicennioides* and *Guiera senegalensis*, wooded savanna cultivated with sorghum and groundnut, with *Andropogon gayanus* and *A. pseudapricus* on old fallows

d. rare *Terminalia macroptera* with *Hyparrhenia rufa* and *Brachiaria fulva*.

1.5.2   AUTHENTICITY OF INDIGENOUS SPECIES

It has been proposed in Chapter 1.3 that many of the grass communities of the intertropical and especially the equatorial

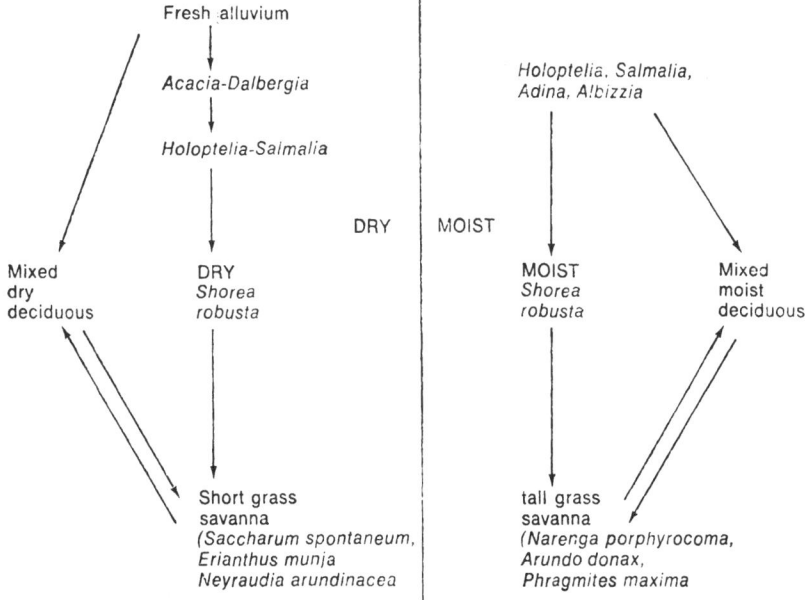

Fig. 1.9. Succession in forests of *Shorea robusta* in Uttar Pradesh, India (Bhatnagar, 1960). The border between dry and moist forest is taken as 1,143 mm rainfall. The xerophytic stage of dry *Shorea robusta* forest is mainly dominated by *Anogeissus latifolia, Acacia catechu* and *Aegle marmelos*, with *Eulaliopsis binata* most common grass.

zone are composed entirely or partly by species which are not indigenous to that site. From the academic, if not necessarily from the practical point of view, it is desirable to classify grass covers into those that are truly indigenous and those which may be called derivative or evolved. In those parts of the intertropical zone formerly and until quite recently, in terms of geobotanical time, covered with tropical rain forests and other types of forêts denses, the few truly indigenous grasses to be found in their ecological niches could not have come out into the light and formed grass communities. Those heliophilous types that now exist in the cleared or burnt-over land must have come from elsewhere. A number of species may have arrived from different habitats and are meeting for the first time in new conditions of environment and interspecific competition.

Thus in south-east Asia and southern China species from Africa and the Indian subcontinent are found taking over after forest clearance where the few local indigenous and endemic species were unable to do so. In Latin America we have the phenomenon which Parsons (1970) has called the "Africanization" of the tropical grasslands of the New World, either by intentional introductions, or by accidents associated with two major events in American history, the slave trade (Holdridge, 1947), or the introduction of the Zebu cattle and the Asian buffalo. In Central America, over 75 per cent of the total pasture area is planted to, or has been colonized by introduced African species (Horrell, 1972; Bazan, 1972). *Hyparrhenia rufa* accounts for 70 to 80 per cent, or 3.5 to 4 million hectares of this total, on infertile land and under the local type of ranching; *H. rufa* is the only herbaceous plant in the derived savanna of Costa Rica that prospers under annual burning (Daubenmire, 1972a); *Panicum maximum*, 5 to 15 per cent on more fertile soils; *Brachiaria mutica* 5 to 15 per cent on waterlogged sites; also *Pennisetum purpureum, Melinis minutiflora, Eriochloa polystachya* and, latterly, *Digitaria decumbens*.

Within Africa itself (chapter 1.3.4), one can trace a trend of species into the areas of climax forest as recognized by Aubreville (1962), from two directions in respect to their mesophytic and xerophytic origins. The mesophytic perennial species would progress from their ecological niches in the tropical forest zone (coastal and river flats, etc.) or from the Central African lakesides and marshes (Thomas, 1966) into cleared forest land, to the limits of their genetical tolerance of aridity. Conversely, from the zones of the dry forêt claires came the xerophytic perennial

72

grasses, advancing into the cleared forest zone to the limits of their genetical tolerance of humidity. Subsequent speciation and intercrossing in those new habitats has provided the gramineous flora of tropical Africa which is found today.

1.5.3   RELATIVE AGGRESSION

Grass species of various origins within the intertropical belt exhibit differential aggressiveness. When African grasses were introduced intentionally or adventitiously into Central and South America, and where they found the environment to their liking, they entirely eliminated whatever indigenous grasses may have existed on the cleared forest land. The Indian and African grasses had virtually no competition from other grasses when they invaded south-east Asia, and until they began to compete with each other. When mesophytic African grasses were introduced into the Indian subcontinent, they were generally and still are "cultivated" under rather special conditions, protected from competition with the Indian grass flora. But when dryland African grasses such as *Chloris gayana* and *Cynodon plectostachyus* are introduced into the Indian subcontinent, they are unable to compete with and are duly eliminated by the Indian species. Thus degree of competitive ability is governed at least partly by environment of place of origin; a grass that has evolved and can persist in its own rugged monsoonal, semi-arid environment is superior in that respect and in that environment to grasses from easier, summer-rainfall habitats.

1.5.4   PLACE IN SUCCESSION

If certain grass species are found to be low in ecological succession in one part of the tropics, does this mean that they will be found in a similar position in successions elsewhere? Species of the genera *Aristida* and *Eragrostis* are low in succession in the grass covers of the Indian subcontinent; it is known that, following protection from regressive ecological influences, these covers will proceed upwards through several stages to one of the few recognized sub-climax grass cover types with little or no *Aristida* or *Eragrostis*. Reports of vegetation surveys from Africa contain references to or show illustrations of grass stands dominated by different species of these and other genera. Can one extrapolate by suggesting that these African stands have been long exposed to similar regressive influences, and that they also would progress upwards towards a higher grass cover type, provided that these influences were removed, and also that seed sources of the ecologically superior species were still available? Critics of the method of ecological manipulation refer to the slowness with which this change takes place, but this is not by any means always so.

1.5.5   THE PANTROPICAL UBIQUITARIES

Imperata cylindrica, Cynodon dactylon, Heteropogon contortus – these are the star performers for migration and aggression throughout all or much of the tropics. They call for a special study, involving geobotanical and ecological history, the latter in relation to anthropological factors, trade contacts and the zoological promoters of plant migration.

Harlan, De Wet & Rawal (1970a) have stated, with reference to the eight species of *Cynodon* L.C. Rich that they: "fall clearly into four groups according to geographical distribution: South Asia and Indian Ocean – South Pacific Islands (*C. arcuatus, C. barberi*); East Africa (*C. plectostachyus, C. aethiopicus, C. nlemfuensis*); South Africa (*C. incompletus, C. transvaalensis*); Cosmopolitan with endemic varieties (*C. dactylon*). The fragmented geographic patterns, with limited distribution and complete genetic isolation of some species, and the narrow endemism of some varieties of *C. dactylon*, all imply a considerable antiquity and long evolutionary history" (see also Table 1 in Harlan 1970).

The same authors (1970b) describe plants of a seleucidus race of *C. dactylon* var. *dactylon* as being strikingly coarse, vigorous, bluish in colour, often hairy, strongly winter-hardy, spreading aggressively by coarse stolons and rhizomes that are interconvertible.

"The center of distribution corresponds closely to the original Seleucid Empire (Pakistan to Turkey), and germ plasm from the race has infiltrated into eastern and southern Europe. It is probably an introgression product involving *C. dactylon* var. *afghanicus* and a temperate race of var. *dactylon*".

It is not known whether there has been a consensus of opinion on the taxonomic age, possible centre or centres of origin and history of distribution of *Imperata cylindrica* (and the

74

other named species of this genus of doubtful taxonomic status). Santiago (1966) refers to complex interactions between aerial and subterranean parts and between genotypic and environmental variations, promoting the evolution of existing ecotypes and of new ecotypes. Swards dominated by *Imperata* frequently cover vast areas of land, for example, in the green cogon deserts of insular south-east Asia. They may become ecologically permanent at this low level of succession, when the controlling factors remain in operation, or in the absence of seed sources of species higher in the succession. With considerable labour and expense, agronomists plough and sow or plant superior species, which, however, have to be very vigorous and well-managed to prevent regrowth of the *Imperata* — for example at Keningau, Sabah, and at several centres in the Philippines. Kellman (1969) has compared *Imperata* grass covers with other types of vegetation and crop covers in respect of run-off, sediment loss and mineral loss from soils under shifting cultivation.

Grasslands are an integral part of the economy and social structure of all villages of shifting cultivators visited by Seavoy (1975) in Kalimantan, whether Dayak or Malay. A small field of *Imperata cylindrica* is created intentionally by repeated burning, because wild animals, mostly deer and wild pigs, graze on the young shoots and so are readily available for hunting. The burnt areas become enlarged from year to year, damaging the secondary forest in the shifting cultivation cycle, but because of the high social value of hunting the burning goes on. Fire hunting then becomes the general practice over the now extensive areas of *Imperata*. Residual populations engaged in grazing after the agricultural villages have moved because periodic hunger occurs too frequently have further enlarged the grasslands. It would be possible to replace the *Imperata* on the village grasslands by controlled and continued grazing favouring the growth of *Axonopus compressus* and *Chrysopogon aciculatus*, but this is not practised because of the constant supervision that would be required.

There are several reports of succession away from the *Imperata* phase. In the Sudano-Zambezian and Guinean areas of South Kasai and Katanga, Risopoulos (1966) reports that *Imperata cylindrica* rapidly becomes dominant after cropping; the species colonizes adjacent areas by its particularly dense rhizomatous root system, and is well adapted to withstand bush fires. In some sandy areas, one finds the tall caespitose, less aggressive species, *Urelytrum thyrsioides*. If a bush fallow is not brought back to cultivation, species of *Hyparrhenia* appear: *H. rufa,*

H. filipendula, H. dissoluta, followed soon afterwards by *H. familiaris, H. diplandra, Panicum phragmitoides* and *Elymandra androphila*. In the least humiferous soils, *Loudetia arundinacea* and *Andropogon shirensis* predominate.

It has been said that, due to shifting cultivation, 40 per cent of the Philippine archipelago is covered by grass (predominantly *Imperata cylindrica* and *Saccharum spontaneum*, but never together), 16 per cent secondary growth forest, and only 10 per cent cultivated (Whitford, 1906). There is little doubt that most of the grassland areas in the Philippines, ranging from treeless through various stages of regeneration, are predominantly *Imperata* grasslands (J.M. Rattray, personal communication 1969) which have developed following shifting cultivation, assisted by fire. In the early stages of succession, common grasses of undisturbed areas such as *Paspalum conjugatum, Digitaria sanguinalis, Eleusine indica, Cynodon dactylon* and *Chrysopogon aciculatus* are found, but it is not long before *Imperata cylindrica, I. exaltata* or *Saccharum spontaneum* appear and eventually take over completely. Small areas of *Themeda triandra* have been noted, of doubtful ecological status and place in succession. *Imperata* seems to favour moister conditions and is less widespread in the drier parts of Luzon. Extensive *Imperata* grasslands are found in the islands of Masbate and Burias, in the west of Mindoro, and the central part of Bohol; in Panay; much of Negros Oriental; and most extensive of all in Mindanao, provinces of Bukidnon and Cotabato.

Fire-resistant plants with large underground structures (the orchid *Eulophia exaltata* and the shrub *Blumea balsamifera*) and the fire-resistant trees (*Antidesma ghesaembilla, Bauhinia malabarica* and *Acacia farnesiana*) can gradually eliminate the grass in their immediate neighbourhood (Brown, 1919). When fire is excluded from a grass area, shrubs, vines and small trees can invade, leading in a few years to second-growth forest of small, rapidly growing trees. But the shade-intolerant *Imperata* and *Saccharum* are first replaced by other grass species forming taller, less dense stands.

1.5.6   INDICATORS OF METAL-BEARING SOILS

The gramineous species of tropical grazing lands have considerable value as indicator plants. An experienced observer can find evidence in the occurrence of a certain genus or a certain type of grass cover for differing combinations and intensities of management practices (grazing, cutting, burning) and of burning with and without shifting cultivation; also for different soil types, soil moisture status, microclimate, presence of sub-surface hardpan, and so on.

Gramineae also play an important part as members of the distinctive floras of heavy-metal-bearing soils. Duvigneaud (1958) and Duvigneaud & Denaeyer De Smet (1963) recognized types within the Katangan copper flora on an ecological basis – cuprophytes and plants growing on soils with the highest copper values, and which may sometimes be restricted to copper soils. Gramineae and Cyperaceae predominate in various life forms on soils with the highest values (5,000 to 10,000 ppm.) – *Eragrostis boehmii, Monocymbium ceresiiforme* and *Andropogon dummeri.* Cuprifuge species, those which never occur on copper and by their absence indicate its presence, include species of *Hyparrhenia.* Wild (1969a) has compared the copper flora of Rhodesia with that of Katanga; the ecological pattern rather than the individual species may be used for the identification of copper soils, and include the short graminoid zone sedges and the grasses *Loudetia simplex*, species of *Andropogon, Danthoniopsis, Microchloa, Schizachyrium* and *Schmidtia pappophoroides.* Wild has also studied graphitic soils (1969b) and nickel-bearing soils (1970). The nickel flora is similar to the serpentine flora (Wild, 1965); the genera *Acacia, Aristida, Combretum* are common on both; the familiar Leguminosae, Gramineae and Compositae are equally important on both. The most frequent grasses are *Loudetia simplex, Andropogon gayanus, Aristida leucophaea* and *Danthonia intermedia.*

In the cerrados of Brazil (1.3.5.2), there is a significant negative correlation between aluminium and the critical nutrients Ca + Mg and K. This suggests that aluminium toxicity may occur in these senile latosols which average 75 ppm. Al, and that cerrado plants must be extraordinarily Al-tolerant (Goodland & Pollard, 1973). The world flora contains a small, taxonomically unrelated group of plants that actually accumulate aluminium to several thousand ppm. in their leaves, dry weight. Aluminium accumulators contribute greatly to cerrado biomass, and are actually dominant in over 30 per cent of the stands.

Cole (1973) has assessed the role of geobotany and biochemistry in mineral exploration in the sclerophyllous woodland and shrub associations of Western Australia. An understanding of the complex relations between plant distributions and environmental parameters is essential in the use of these techniques in exploration for nickel and other metals. Anomalies due to nickel concentrations have been distinguished from those due to chromium concentrations and have established the species associated with each (no Gramineae).

1.6   Climax and Succession in Grass Covers

Types of grass covers are generally classified in relation to ecoclimatic zones, edaphic conditions or altitude. Only rarely is any indication given of the nature and trend of succession within an individual type. It is, therefore, not possible to say whether the grass cover type as recognized and described is the true grass subclimax for that particular site in that particular type of present or former forest climax.

When one has begun to look at grass cover types in the field and attempted to diagnose their place in succession, it becomes obvious that many if not most of the types now in existence are seral stages in a succession, or rather regression, caused by the operation of one or more of the major anthropogenic factors: fire, grazing, clearing for cultivation. Thus one comes to look with ecological suspicion at some of the grass cover types recognized by those who map merely what is there now. As already stated, the dominance of certain genera suggests a low place in the succession, as does a high percentage of annuals, because there are no undisturbed subclimax associations in which annuals provide a significant part of the cover: other species are indicators of excessive or badly-timed overgrazing, or of excessive or badly-timed burning, or of a combination of extreme pressures of shifting cultivation, burning and concomitant loss of soil fertility over a long period.

It is currently fashionable in certain quarters to denigrate the significance of the concepts of succession (regression and progression) and of climax or subclimax in studies on grazing lands. Nevertheless, these terms appear frequently in the literature, and they belong to the vocabulary of field workers on tropical grazing lands. A University Department of Pasture Science in South Africa states "The teaching and application of veld management leans heavily on the principles of plant succession, indicator plants and ecosystem dynamics ... Knowledge of potential climax condition is presupposed in the use of 'present condition' score sheets ... for the assessment of condition based on five criteria, namely, cover density, botanical composition, vigour, soil surface condition and insect damage".

As a basis for further discussion, one might carry the proposition further, and state that the concept of succession, or something approaching it, applies at all levels of study: in types of vegetation, along climatic gradients, or by regression on the same site (Morello, 1968); in types of grass covers, again along climatic gradients, or on the same site under the influence of anthropogenic and biotic factors; within genera of grasses, as in *Hyparrhenia*, for example, from those members of the genus high in succession to those at the lower levels, and within the species, from those ecotypes/phenotypes/genotypes adapted to one end of the humidity scale to those adapted to the opposite.

With movement in continual progress at these four levels and in opposite directions, is it a wonder that the taxonomists say that the members of the Gramineae are in a state of flux and active evolution? Unfortunately, following the excessive devegetation of recent and modern times, leading to progressive desiccation of habitats, the trend is all in one direction. Many of the superior genera, species and character groups within species will be lost.

Although it may be the ideal of the vegetation ecologist to maintain a climax flora of vegetation and grass cover, the top level in grass covers is not always the best objective of the practitioner. It was stated in the First Edition of Whyte (1964) that, although the top level of grass succession is the best for arid and semi-arid Rajasthan, the top level in the high-humidity Terai (rainfall plus runoff from the Himalayan foothills) is of no economic value to livestock other than wildlife. The tall *Sorghum/ Themeda/Narenga* grass stands of the Terai would have to be brought very low in the scale of succession before they would be of much value to domestic animals.

After allowing for a continuing state of imbalance and heterogeneity in grass covers, it is nevertheless possible to recognize stages in regression or progression on any given site (Pandeya, 1961). The basic environmental factors that govern the composition of a grass community are soil, climate, exposure, altitude and slope. In addition to these relatively permanent factors, there are others which are in a constant state of change in nature and intensity. Under their influence, some species become rare or absent, and new ones more adapted to the new conditions take their place. It is the object of the practising ecologist to be able to recognize these seral stages, to diagnose their origins, and to indicate what seral stages might be expected to appear if the governing factors were reduced or increased in intensity.

Although the overall picture is still patchy due to lack of sufficient analyses over the whole intertropical zone, some preliminary indicators are already appearing. There are certain species which are usually found high in the succession, in the absence of intensive operation of use factors. As grazing intensity increases, the more palatable species will be reduced in percentage cover; the bunch grasses of the climax will be damaged because of their slow growth habit, which becomes exposed to defoliation at an earlier, more critical stage than the other, more rapidly growing perennials in the same community. Then there is a group of species that always appears to belong to lower stages in succession, tough and less palatable and adapted to greater soil aridity. These are the pyrophytes that are resistant to fire, or may even be stimulated by it. Finally, there are the annuals representing the last-ditch attempt of a genus to perpetuate itself in conditions in which its perennial ancestors can no longer persist. A categorization of the major species of intertropical grass covers in this way would be of great value, in both taxonomic geography and taxonomy.

1.6.1 AUSTRALIA

The tropical and subtropical grasslands of Australia are described in the special volume, *Australian Grasslands* (Moore, 1970), prepared for the XIth International Grassland Congress. For the grazing lands of this region, which comprise the grassy understorey of the woodlands which have been modified to a greater or lesser extent by the influence of man and his grazing animals, Shaw & Norman (1970) quote Perry (1960) and other publications of the C.S.I.R.O. Division of Land Research with reference to the Northern Territory and Western Australia. Perry (1960) records under each type of pasture the actual or possible reaction to grazing, leading to the increase or decrease of desirable or undesirable species.

In the Tropical Tallgrass of Queensland (see Chapter 6 of Moore, 1970), *Heteropogon contortus* is now dominant, with *Themeda australis*, *Bothriochloa bladhii* syn. *B. intermedia*, *B. ewartiana*, *B. decipiens* and *Aristida* spp. *Themeda triandra* may have been the original dominant. Botanical composition varies considerably, even over short distances; other genera include *Chrysopogon, Cymbopogon, Chloris, Dichanthium, Eria-chne, Schizachyrium, Sorghum* and *Eragrostis*. The species composition of this grass cover on any specific site is due to edaphic and use factors. *Themeda australis* and *Heteropogon contortus* are also found in the *Melaleuca* woodlands to the south and east of the Gulf of Carpentaria and in Cape York peninsula.

Cull & Ebersohn (1969) have studied the dynamics of semi-arid plant communities in western Queensland, more especially in relation to population shifts of two invaders, *Cenchrus ciliaris* and *Heteropogon contortus*.

The location of the main Laboratories and field stations of the CSIRO Division of Tropical Crops and Pastures in Queensland is shown in Figs. 1.10 and 1.11.

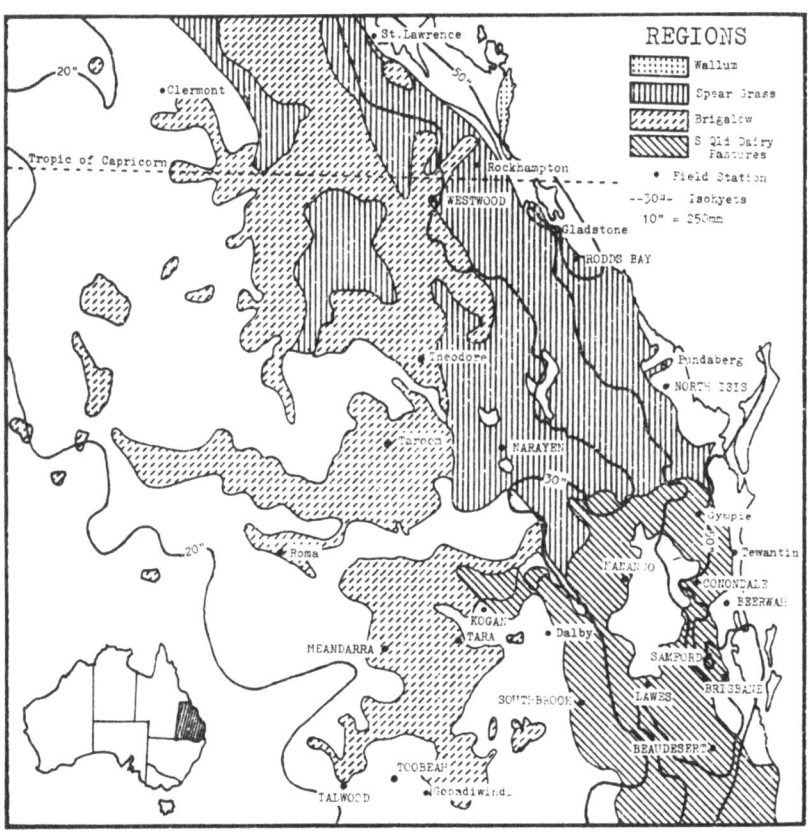

Fig. 1.10. Pasture environments and field stations where CSIRO Division of Tropical Crops and Pastures is conducting research based on the Cunningham Laboratory, St. Lucia.

81

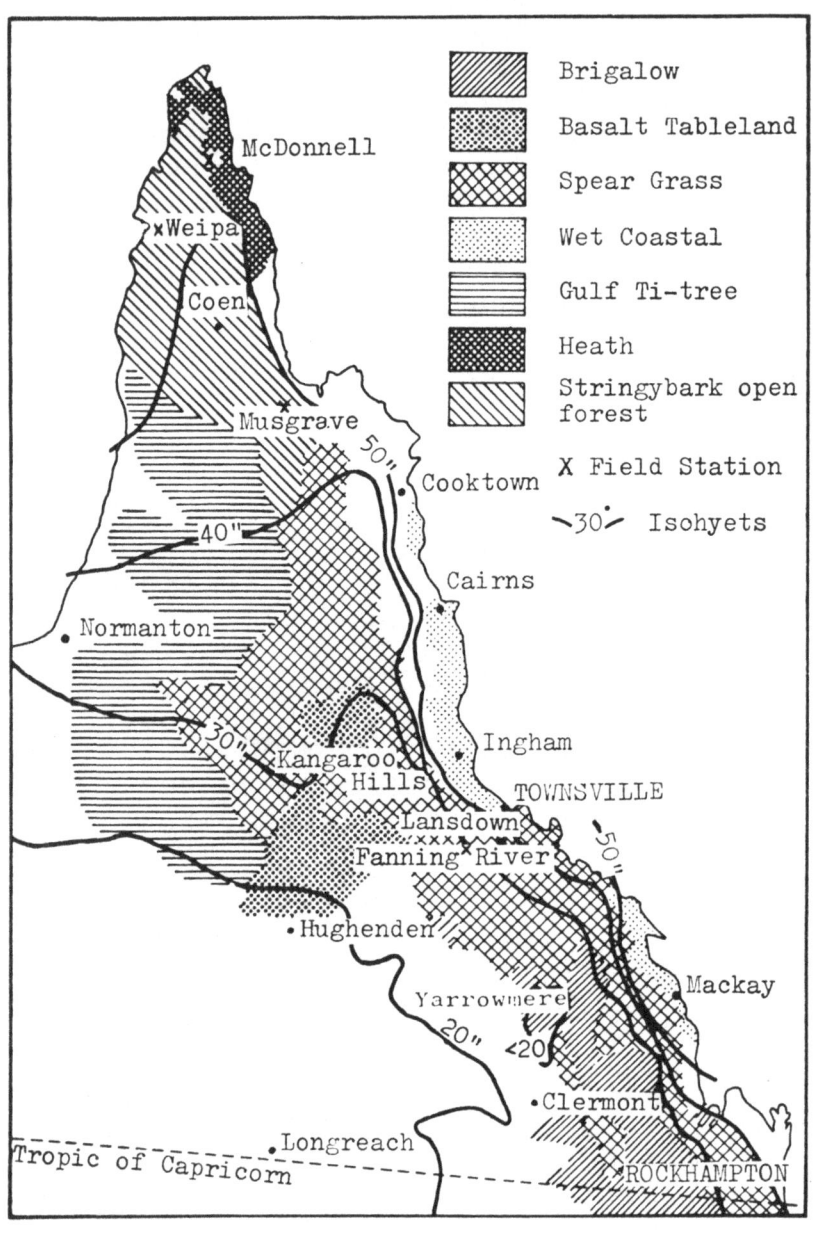

Fig. 1.11. Pasture environments and field stations where CSIRO Division of Tropical Crops and Pastures is conducting research based on the Laboratory at Townsville.

Studies on grass cover ecology relate to the monsoonal savannas at lower elevations and to the significance of the high-altitude grasslands. Earlier work by CSIRO Division of Land Research is reviewed by Whyte (1968a).

Of the 255 genera in the flora of semi-deciduous forest and scrub and of eucalypt savanna in the Port Moresby area, 8 per cent are centred in Australia and/or the Pacific, 12 per cent in Asia and/or Malesia, and 80 per cent occur in and often beyond Australia and Malesia (Heyligers, 1972). Monsoonal conditions have not led to a stronger representation of Australian elements. The large percentage of widespread genera will decrease if one considers the whole of monsoonal south-eastern Papua; a comparison with Pacific and Malesian islands shows that the relation between the numbers of widespread genera and the total number of genera can be expressed by a logarithmic series. It is surmised that most of the widespread species reached south-

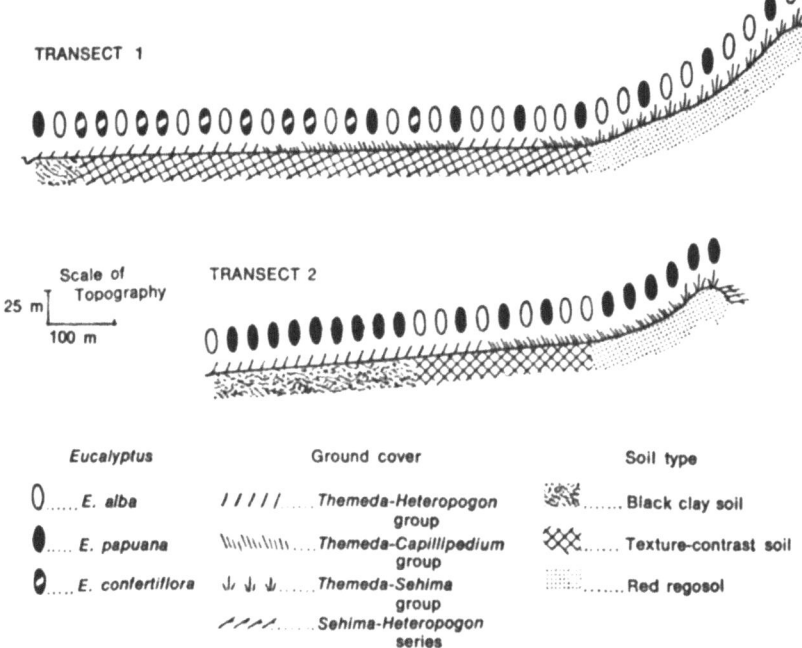

Fig. 1.12. Transect in a *Themeda australis/Eucalyptus* savanna, in Papua, showing the distribution and mutual proportions of the *Eucalyptus* species, and the distribution of the ground cover groups and of the soil types (Heyligers, 1966).

eastern Papua before the Pliocene via drier areas on the Asian and Australian continents.

In Papua, a *Themeda australis/Eucalyptus* savanna has a tall stratum of one or more species of *Eucalyptus*, some scattered *Albizzia procera*, and a grass cover dominated by *Themeda australis*, with *Heteropogon contortus* commonly associated, especially on stonier soils (see Fig. 1.12). Fires are common throughout the dry season, at the end of which most of the ground cover has been burnt.

Johns & Stevens (1971) have listed the (mostly temperate) grasses of the national park on Mt. Wilhelm (4,510 m.). Paijmans & Löffler (1972) have studied the high-altitude forests and grasslands of Mt. Albert Edward, in the Owen Stanley Range, the mountain backbone of East Papua. The major dominant on the most common soils, relatively deep and well-drained, is *Danthonia archboldii*, with local co-dominants *Danthonia vestita* and *Deyeuxia* spp. Some 70 per cent of the summit plateau would still be under forest, but for destructive fires associated with hunting.

The gramineous flora of these mountains should be considered alongside those of the arc of high mountain areas lying along the western Pacific (Whyte, 1972, 1973; Liu, 1971).

1.6.3   MONSOONAL AND EQUATORIAL ASIA

The types of grass covers and the succession to be found within them in south-east Asia and in western monsoon Asia (especially India, Pakistan and Sri Lanka) are described by Whyte (1968a, 1972 and 1973). The outstanding example is the recognition of the sub-climax and seral types of grass cover within India, as a result of the eight-year Grassland Survey (Dabadghao & Shankarnarayan, 1973). Sequels to this have been the studies of *Iseilema, Sehima* and *Heteropogon* communities within the *Sehima/Dichanthium* zone (Dabadghao & Shankarnarayan, 1970); also of the grass cover types in the western Himalaya (Gupta & Nanda, 1970). The grass covers of Maharashtra State have been classified into seven major habitat patterns and 24 minor sub-patterns, on the basis of soil, topography, hydrography and other phenological considerations (Oke, 1972).

The grass constituents in the last column headed "vegetation" in the tabulated analyses of twelve land systems in the Nagarparkar peninsula in Pakistan (between the Tharparkar dune

Table 1.3.   India: Succession in types of grass covers.

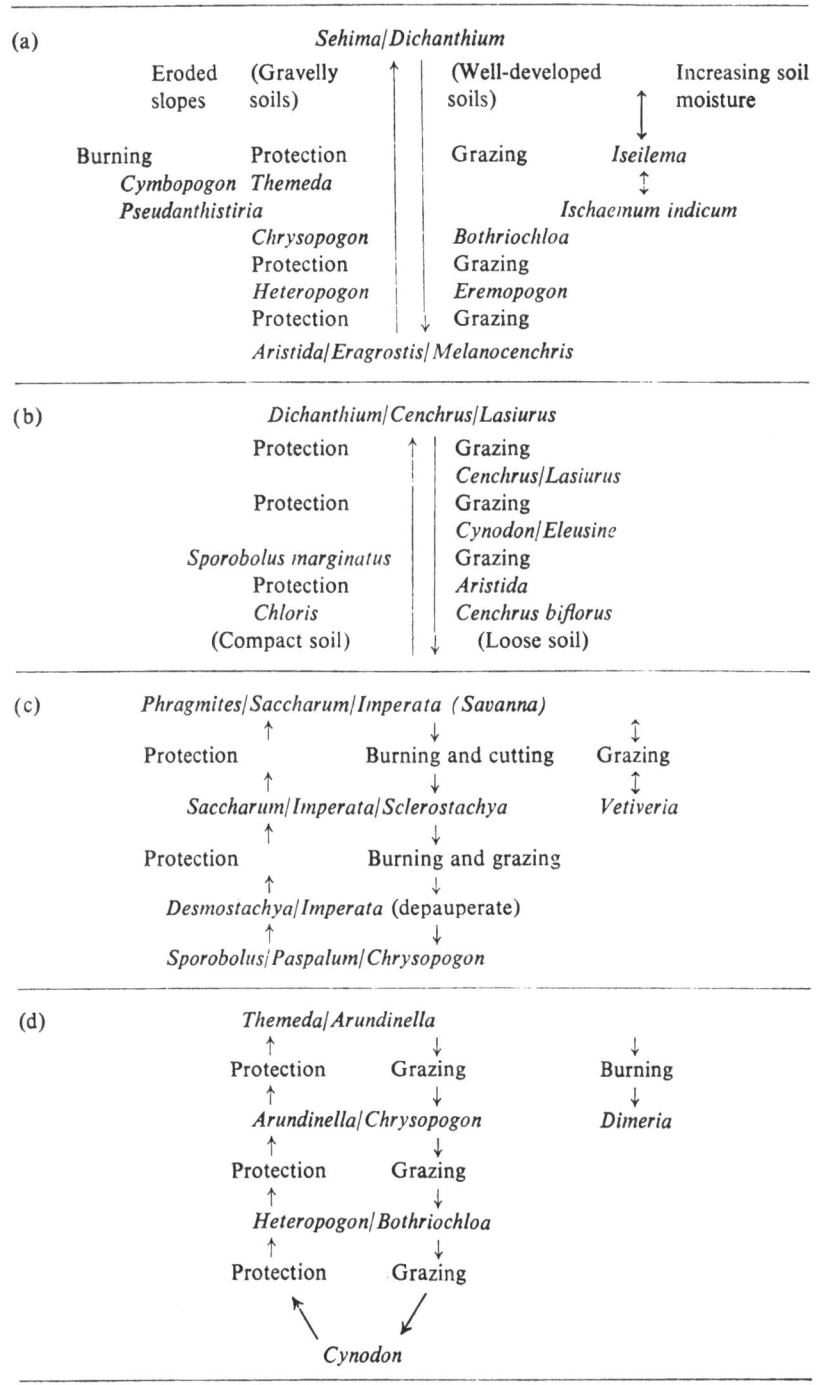

(a)   *Sehima/Dichanthium*

Eroded   (Gravelly   ↑   (Well-developed   Increasing soil
slopes   soils)   soils)   ↕   moisture

Burning   Protection   Grazing   *Iseilema*
Cymbopogon Themeda   ↕
Pseudanthistiria   *Ischaemum indicum*
           Chrysopogon   *Bothriochloa*
           Protection   Grazing
           Heteropogon   *Eremopogon*
           Protection   ↓   Grazing
           Aristida/Eragrostis/Melanocenchris

(b)   *Dichanthium/Cenchrus/Lasiurus*

           Protection   ↑   Grazing
                                  Cenchrus/Lasiurus
           Protection   Grazing
                                  Cynodon/Eleusine
    Sporobolus marginatus   Grazing
           Protection   *Aristida*
           Chloris   *Cenchrus biflorus*
           (Compact soil)   ↓   (Loose soil)

(c)   *Phragmites/Saccharum/Imperata (Savanna)*
                ↑   ↓   ↕
      Protection   Burning and cutting   Grazing
                ↑   ↓   ↕
        Saccharum/Imperata/Sclerostachya   *Vetiveria*
                ↑   ↓
      Protection   Burning and grazing
                ↑   ↓
        Desmostachya/Imperata (depauperate)
                ↑   ↓
        Sporobolus/Paspalum/Chrysopogon

(d)   *Themeda/Arundinella*
                ↑   ↓   ↓
        Protection   Grazing   Burning
                ↑   ↓   ↓
          Arundinella/Chrysopogon   *Dimeria*
                ↑   ↓
        Protection   Grazing
                ↑   ↓
          Heteropogon/Bothriochloa
                ↑   ↓
        Protection   Grazing
                  ↖   ↙
                Cynodon

Table 1.3. (continued).

(e) Temperate Alpine Type (Himalaya)

The grass vegetation of this primarily non-monsoonal cover is composed of the following species:

Agropyron canaliculatum	Chrysopogon gryllus
Agrostis canina	Dactylis glomerata
A. filipes	Danthonia jacquemontii
A. munroana	Koeleria cristata
A. myriantha	Phleum alpinum
Andropogon tristis	Poa pratensis
Calamagrostis epigejos	Stipa concinna

Associated perennial species with a contribution of more than 10 per cent in the composition are:

Agrostis pilosula	Festuca valesiaca
Brachypodium sylvaticum	Helictotrichon asperum
Bromus ramosus	Muhlenbergia sp.
Calamagrostis emodensis	Poa alpina
Eragrostis nigra	Trisetum sp.
Festuca lucida	

(f) Nilgiri grasslands

Andropogon polyptychus/Eulalia
↕
Eulalia/Themeda/Ischaemum
↕
Chrysopogon/Ischaemum/Themeda
↕
Ischaemum/Chrysopogon
↕
Ischaemum/Tripogon
↕
Tripogon/Eragrostis/Arundinella

Source: (a) to (e) – P. M. DABADGHAO, Grassland Survey of India.
(f) – S. C. GUPTA, S. CHINNAMANI & N. D. REGE, Personal communication

---

fields and the seasonally flooded, saline littoral plains of the Rann of Kutch) could be expressed quite differently by the grass cover ecologist (Wright, 1964). It is first necessary to know the history of the present shrub/grass associations, to attempt to define the original climax covers and to analyze the subsequent effects of use and misuse over millenia of time. On that basis, it would then be possible to relate the vegetation (past and present, climax and disclimax) to geomorphology, slope and the texture and moisture-holding capacity of the soil.

Pandeya (1969) has concluded his series on the ecology of the grasslands of Sagar, Madhya Pradesh, by considering edaphic factors in the distribution of associations. Singh & Misra (1969) show that species diversity increases productive efficiency of the grassland ecosystem, while dominance makes the system stable, though less efficient for production, contrary to the conclusion of McNaughton (1967), that diversity on Californian grassland reduces efficiency and generates stability in the community.

The study of stages in regression within the major grass covers on different soils and under different management influences was an essential part of the Grassland Survey of India (Table 1.3). It is to be hoped that some of the surveyors may be able to return to selected sites from time to time to record any trend that may become apparent. It is becoming increasingly obvious that botanists and ecologists everywhere have little idea what their vegetations and grass covers were like a hundred, fifty or even twenty-five years ago.

Verboom (1968) has studied grass associations and succession in Pahang, Malaysia, in a tropical rain forest area of the type discussed in Chapter 1.3.2. Seven distinct grass covers are recognized on the basis of place of occurrence. These are among the secondary successions which would ultimately revert to forest. The shade-tolerant species noted in 1.3.2 are actually forest fringe grasses. The forest clearings have the usual pioneer associations — *Paspalum conjugatum, Cynodon dactylon, Digitaria* spp.,

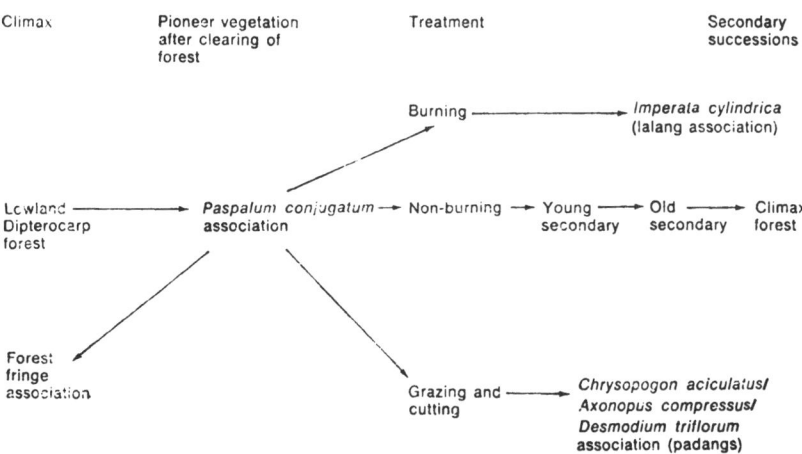

Fig. 1.13. Relation between land management and plant succession in Malaysia (Verboom, 1968).

Echinochloa colonum, Eleusine indica, etc. Species of wet padang and roadsides are listed. See Fig. 1.13.

1.6.4   AFRICA

Rattray (1960) mapped the grass covers of Africa on the basis of information available at that time. Since that date, the Land Resources Division of the British Directorate of Overseas Surveys has included the grassland resource in their overall analyses. For example, an ecological survey of Western Province, Zambia, related particularly to fodder resources (Verboom & Brunt, 1970); appendix 2 to volume 2 of the report gives the main vegetation communities (based on Trapnell, Martin & Allan, 1950, and Trapnell & Clothier, 1957), the grass cover types and their species composition. Astle (1968–9) has described the four types of grassland (virtually treeless areas) found on a ranch in Luapula Province, Zambia; because of a distinct succession of flowering in any one area, species composition appears to vary according to season of survey. Hopkins (1968, 1970a and 1970b) also refers to seasonal changes in Nigeria. Vesey-Fitzgerald (1970) describes the origin and distribution in relation to edaphic factors of valley grasslands in East Africa. Van Rensburg (1969) reviews the management and utilization of pastures in Kenya, Tanzania and Uganda.

The programme "Afrika-Kartenwerk" of the Deutsches Forschungsgemeinschaft, initiated in 1963 following a recommendation of C. Troll, was planned to cover the following regions (Kayser, Manshard, Mensching & Schulze, 1966);

North Africa (Tunis-Sfax)
Tropical West Africa (Southern Nigeria)
Tropical East Africa (Lake Victoria)
South Africa (Lourenço-Marques – north-east Transvaal).

Bader (1967a and 1967b) discusses the problems which have arisen in connection with parts of the Lake Victoria sheet, especially the *Cymbopogon afronardus* grassland and the *Combretum/Cymbopogon* savanna.

For sub-Saharan West Africa, Boudet & Rivière (1968) relate the tree/grass associations to the zones between specific isohyets: 150–200; 200–300; 300–400 and 400–550 mm. respectively; this begins with a ground storey of tufts of *Aristida plumosa, A. pungens* and *Panicum turgidum*, and reaches a steppe with *Aristida mutabilis* and *Eragrostis tremula*, giving place to a cover of tufts of *Hyparrhenia dissoluta, Cymbopogon giganteus* and *Andropogon gayanus*, with the annuals *Aristida mutabilis* and *Schizachyrium exile*. Species distribution within the recognized distributional zones of West Africa south of the Sahara is given as follows (Bille, Lebrun & Rivière, 1968/9; see also Rattray, 1960).

Sahelian

> *Andropogon gayanus, A. pseudapricus, A. schirensis, Diheteropogon hagerupii, Loudetia* spp.
> *Hyparrhenia rufa, Hyparrhenia* spp.
> *Paspalum orbicularis, Leersia hexandra*
> *Echinochloa stagnina, E. crus-pavonis*

Guinean

> *Hyparrhenia diplandra, H. chrysargyrea, H. confinis, Andropogon tectorum, A. macrophyllus, Loudetia* spp. *Panicum phragmitoides, Brachiaria brizantha, Pennisetum purpureum, Imperata cylindrica.*

A detailed study of an area in Mali (Boudet & Ellenberger, 1971) recognizes grass cover types of *Andropogon pseudapricus, Schizachyrium sanguineum* and *Andropogon gayanus*; four types of pasture flooded in the wet season contain respectively *Hyparrhenia rufa/Schizachyrium platyphyllum, Andropogon canaliculatus/Anadelphia afzeliana, Andropogon africanus/Sorghastrum trichopus* and *Paspalum polystachyum* or *Echinochloa stagnina.*

The climate of western Madagascar is monsoonal, with a long dry season; over much of this region there existed a dry semi-deciduous forest (Bosser, 1969), which with its various subtypes has been replaced by:

a. "savanes herbeuses" on deep and rich alluvial soils: *Hyparrhenia rufa, Sorghum brevicarinatum, Rottboellia exaltata, Panicum maximum.*

b. "savanes arborées" on drier soils, often eroded and poor. Better soils well supplied with water: *Hyparrhenia rufa, Bothriochloa glabra* frequent, *Imperata cylindrica* locally dominant. Poorer and drier soils: *Heteropogon contortus* the main constituent. Very eroded and very poor soils: *Aristida rufescens.*

In places, *Chrysopogon serrulatus* and, in the north, *Themeda quadrivalvis* assume some importance.

In the eastern zone,

c. "savanes herbacées": *Hyparrhenia rufa, H. cymbaria, H. variabilis, Imperata cylindrica*; poor eroded soils: *Aristida similis*; unconsolidated coastal dunes: *Panicum umbellatum, Digitaria didactyla*; plantations: *Stenotaphrum dimidiatum.*

regression stages from forests on plateaux:

soils not too degraded: *Hyparrhenia rufa, Heteropogon contortus, Imperata cylindrica*

degraded soils: *Aristida rufescens*.

along forest edges in eastern zone:

Loudetia simplex ssp. *stipoides, Sporobolus centrifugus*.

high plateaux with poor drainage: *Loudetia simplex* ssp. *stipoides, Trachypogon spicatus*.

high altitude ericoid vegetation replaced by grassland:

Pentaschistis perrieri, Loudetia madagascariensis, Andropogon trichozygus, with species of the temperate genera *Anthoxanthum* and *Poa*.

French workers in Madagascar and West Africa have adopted a dynamic approach to the survey and planned use of grazing lands (Piot, 1966; Valenza, 1970; Granier, 1965; Garnier, Lahore & Dubois, 1968). Descoings (1973b, 1974, 1976) has reported on floristic and structural analyses and estimation of pastoral value of the savannas in different regions of Gabon. In delimiting the main natural regions of the Congo Republic, the same worker (Descoings, 1975) recognizes seven out of a total eleven regions as having dominant herbaceous formations, three as heterogeneous and four as forests. The main criteria that govern the distribution of plant formations are related to the geological nature of the soil.

Lansbury, Rose-Innes & Mabey (1965) note that the "grasslands" of Ghana are mostly tree or shrub savannas or savanna woodlands, as defined at Yangambi (C.S.A., 1956). They consist of a continuous grassy ground cover, interspersed with trees and shrubs of variable height and density; the grass covers are seral in character and proclimax in status, progression to a predominantly woody climax being prevented by frequent fires and other disturbances.

Dominant Species

Sandy soils	Clay soils
Schizachyrium schweinfurthii	*Vetiveria fulvibarbis*
Andropogon canaliculatus	*Andropogon canaliculatus*
Vetiveria fulvibarbis	

Jacques-Félix (1968) returned to an area in the Cameroons that he had visited in 1939. At that time he had noted vestiges of a forest of *Khaya grandifolia*; now the region has "une allure très naturelle" and one would never suspect the earlier forest climax.

On certain plateaux, *Urelytrum fasciculatum* was the dominant grass; now the development of animal husbandry has led to its replacement by *Sporobolus pyramidalis*, with rare stands of *Brachiaria brizantha*.

In Somalia, Glover (1947) and other British officers reported the presence of considerable areas of perennial grass; now they are absent (Box, 1968) or present only in relict areas, to indicate the potential (*Themeda triandra, Chrysopogon aucheri, Bothriochloa insculpta, Hyparrhenia hirta* and other tall grasses). In the high plateau country (the Haud), grazing has been less destructive, and *Cenchrus ciliaris, Sporobolus* sp. and, near water, dense stands of *Aristida kelleri* are found.

Owen & Brzostowski (1966) have discussed the possibilities of ranching development on the Kongwa Plain, Tanzania, on the basis of the proportions of annual and perennial species in the grass cover, and the tendency for progression from a cover of annual grasses and herbs to perennial shrubs and to taller bush species and the common components of bush thickets. In those parts of Uganda (Buganda and Busoga bordering Lake Victoria), there are a number of ecological vegetation zones in an area occupied by twenty-two dairy farms, but pressure of cultivation has left little more than relict areas of the climax. Reference is made to the possible climax communities (Langdale-Brown, Osmaston & Wilson, 1964), in order to assess the true dairy potential (Thornton, Long & Marshall, 1969; Long, Thornton, Ndyanabo, Marshall & Ssekab, 1970; see also Thornton & Marshall, 1971). Nutritive value in relation to specific composition and succession in grass covers has been studied in Ankole and the Queen Elizabeth National Park, Uganda (Thornton, Long & Marshall, 1968; Long, Thornton & Marshall, 1969; Long, Ndyanabo, Marshall & Thornton, 1969). A similar study of nutritive value of seral stages and a good perennial grass cover (*Cynodon dactylon, Eragrostis superba, Digitaria scalarum* and *Cymbopogon* sp.) has been made in the browse and grass range pastures of the semi-arid upland area of Kenya — 1,575 m. elevation in Rift Valley (McKay & Frandsen, 1969).

The British Land Resources Division (Verboom & Brunt, 1970) notes in Western Zambia the change in the composition of the open grassland, compared with the grass cover found in woodland under a fire regime. The tendency for many pastures to revert to a *Sporobolus*-dominated cover when overgrazed has been noted. On Lake-dune Barotse sands and Lake Basin soils, where grassland often has a high proportion of *Brachiaria dura*, overgrazing may lead to dominance of relatively unpalatable *Aristida* species; if the shrub growth of *Bauhinia mucronata* and *Baphia obovata* is not slashed, it will slowly increase and shade out much of the grass cover. In low-lying areas, succession and hence composition are functions of depth of flooding; the most nutritious sward may be maintained by controlling depth of flooding and level of watertable. Verboom (1966) finds that the grass covers of Barotseland differ so much from the remainder of Zambia that one may regard it as a special environment with affinities with south-east Angola, the Caprivi Strip and north Botswana.

Succession in a high-altitude tropical "grassland" has been studied by Compère (1968) in Rwanda and Burundi (Fig. 1.14). Bille (1965) has studied succession in natural pastures of the high plateaux of the Central African Republic (Tables 1.4 and 1.5).

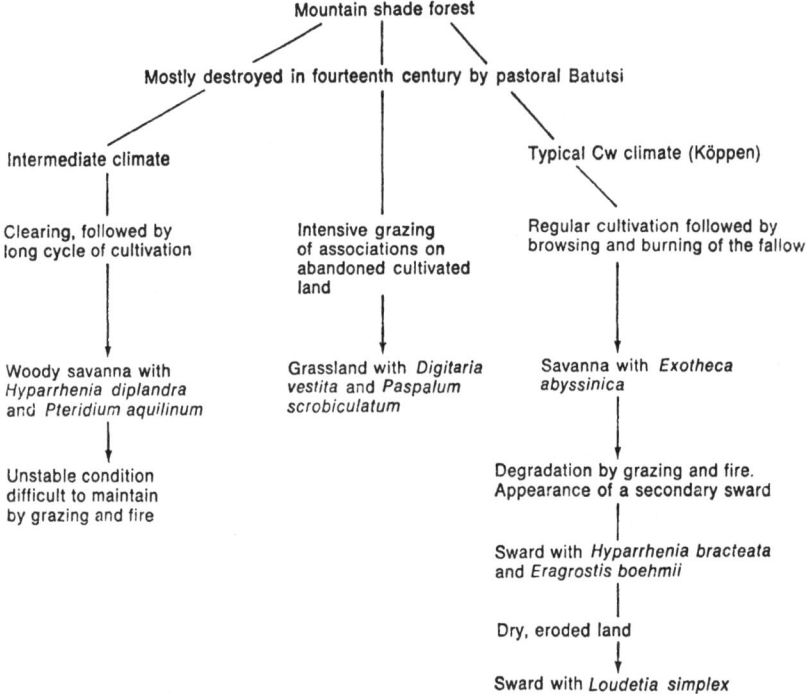

Fig. 1.14. The floristic evolution of a permanent grass cover of *Exotheca abyssinica* and *Eragrostis boehmii* in Rwanda and Burundi at altitudes of 1,800 to 2,500 m. (Compère, 1968).

Table 1.4. Central African Republic: Grass succession in regions of semi-permanent occupation, classing the undergrazed and overgrazed pastures on a scale 1–11 (Bille, 1965).

	1	2	3	4	5	6	7	8	9	10	11
Andropogon tectorum	+										
Hyparrhenia diplandra	+	+									
Panicum phragmitoides	+	+	+	+	+						
Hyparrhenia chrysargyrea		+	+	+							
Hyparrhenia rufa		+	+								
Urelytrum thyrsioides			+	+	+						
Brachiaria brizantha				+	+	+	+				
Paspalum commersonii											
Setaria sphacelata					+	+	+	+			
Eleusine indica											
Dactyloctenium aegyptiacum							+	+			
Eragrostis spp.											
Sporobolus spp.						+	+	+			
Adventices+ sol nu								+	+	+	+
Solanum spp.											+

Table 1.5. Central African Republic: Succession on pastures on laterite soils grazed by migratory stock, classing undergrazed and overgrazed pastures on a scale 1–8 (Bille, 1965).

	1	2	3	4	5	6	7	8
Hyparrhenia dissoluta	+							
Hyparrhenia filipendula	+							
Andropogon gabonensis	+	+						
Hyparrhenia gracilescens	+	+						
Hyparrhenia soluta	+	+	+	+				
Andropogon schirensis		+						
Hyparrhenia notolasia		+	+	+				
Loudetia arundinacea		+	+	+	+	+		
Panicum phragmitoides			+	+	+	+		
Loudetia kagerensis				+	+			
Hyparrhenia rufa				+	+			
Hyparrhenia chrysargyrea				+	+	+		
Eragrostis – Sporobolus						+	+	
Adventices+Sol nu							+	+
Solanum distichum								+

Three main grassland dominions of South America are recognized by Burkart (1975), based on the climatic groupings of South American grass tribes and genera as mega-, meso- and microthermic grass groups, and somewhat independent of current geobotanical subdivisions (see map no. 1: Main vegetational divisions and grassland regions in Burkart, *op. cit.*).

A. Tropical and subtropical grasslands — savanna-type, dominated by megathermic grasses with a variable admixture of woody plants, shrubs, trees or palms, which never form dense groves:

1. the Llanos of the Orinoco River system in Colombia and Venezuela

2. the Campos Cerrados of Central Brazil, when of the more open type

3. the swamp grasslands with hydrophilous or floating grasses — Marajo Island at mouth of Amazon, Gran Pantanal in Matto Grosso where the Paraguay River has its source, and the Delta of the Paraná.

4. the South Brazilian Campos, northern parts, including the humid lower Campinhas or small grassland areas surrounded by forest, mainly with acid soils

5. the tropical and subtropical east Andean grasslands, which extend on a narrow tract at median altitudes on the humid eastern slopes of the Andes, above timberline, from Venezuela and Colombia to Bolivia and northwestern Argentina.

"Everywhere in this domain there is strong competition between woody plants, tending to the natural forest climax, and grasses and herbs, which build up the savanna. Climatic or edaphic conditions have determined the existence of savannas. Grasses are better adapted to a long dry period than are trees, and the latter require better and deeper soils. Through human activities (grazing, burning, etc.) the natural savannas have been extended, favoring the grasses, and forests are becoming more and more restricted."

Reference has been made to the transitional vegetation of the Gran Chaco (1.3.5.3). Morello & Adamoli (1973) have made a macro-ecological study of the Province, stressing the relation between the type of basic rock, the relief, the types of vegetation, the specific floristic composition and the modifying actions of man.

Spatial succession of grass communities in the Alto Llano of

Venezuela is given by Blydenstein (1962). The main physiognomic types are:

a. wet grassland in gallery forest — *Paspalum pulchellum, P. chaffanjonii, Eragrostis acutiflora* and other humid grasses.

b. low Llano — *Mesosetum rottboellioides, Paspalum* spp., *Sorghastrum parviflorum, Andropogon selloanus, Axonopus purpusii*

c. palmares: *Copernicia tectorum, Leersia hexandra, Panicum laxum*

d. morichales: *Mauritia minor, Jussiaea lithospermifolia, Panicum stenodes, Andropogon virgatus, Rynchospora velutina.*

e. high Llano — *Trachypogon* savanna — *T. vestitus, T. montufari*, with trees, *Curatella americana, Byrsonima crassifolia, Bowdichia virgilioides.*

Five variations of *Trachypogon* savanna (e) are recognized, mainly on the basis of soil characters.

The vegetation of the white-sand savanna in Surinam comprises (a) forests and woods, (b) scrub and bushes, and (c) ground covers consisting of a single stratum (Heyligers, 1963). Of the last, there are five main types of vegetation, characterized respectively by:

(i) *Bulbostylis conifera*
(ii) *Axonopus attenuatus/Lagenocarpus weigelti*
(iii) *Trachypogon plumosus*
(iv) *Rynchospora tenuis*
(v) *Lagenocarpus tremulus*

The last type has four variants, with algae or *Sphagnum antillarum* covering from 10 to 90 per cent of the surface, grasses the remainder (*Paspalum pulchellum, Panicum micranthum, P. polycomum, P. nervosum*).

The cleared pasture lands of the Caribbean region may be grouped into ten main classes on the basis of soil moisture and topography. Twenty six grass communities (eleven major and fifteen minor) have been recognized in Puerto Rico by Molinari (1952). By studies of dominant, principal and secondary species, the types have been correlated with soil groups and types of rainfall. Most are also found in Santo Domingo, Cuba, Haiti and Jamaica; some occur in the Virgin Islands and Lesser Antilles (Molinari, *op. cit.*). *Panicum purpurascens* occurs in moist lowlands throughout, *Stenotaphrum secundatum* along moist limestone coastal plains on the windward sides; *Cynodon dactylon, Sporobolus virginicus* and *Chloris inflata* are the dominant pastures of the dune formations and pure sandy soils fringing the ocean; *Sporobolus indicus, Axonopus compressus* and *Andropogon* spp. occur on the moist mountainous regions.

In a coarse type of grassland among characteristic trees on true laterite soils of serpentine origin in Puerto Rico and Cuba, the dominant grasses are *Leptocoryphium lanatum* and *Andropogon semiberbis. Bothriochloa pertusa* (Oakes, 1968) and *Andropogon intermedium* var. *acidula* are important in the Virgin Islands but not in Puerto Rico or other Greater Antilles. *Andropogon* (= *Dichanthium?*) *caricosus* forms extensive pastures in Cuba, but is sporadic in Puerto Rico.

Most of the present native pasture lands in Puerto Rico has been under crops at one time or another; some areas have been undisturbed for 40 years, others for a few months to several years (Molinari, 1949). In most upland areas of the north interior mountain region, trees start reversion to the forest stage; the grasses of the open parkland disappear and are replaced under the forest canopy by shade-tolerant grasses, *Lasiacis divaricata, L. ruscifolia, Pharus glaber, Ichnanthus pallens, Olyra latifolia, Oplismenus setarius* and others.

Studies of succession show the role of annuals as first pioneers and as last relicts under adverse conditions, and the relative place in succession of perennial species.

Northern Coastal Plain Grassland

1st Stage: Annuals occupying land for only a short period — *Echinochloa colonum, Eriochloa punctata, Digitaria horizontalis, Eleusine indica, Cenchrus echinatus, Paspalum fimbriatum.*

2nd Stage: Long-lived annuals forming dense sod — *Digitaria horizontalis, D. sanguinalis*; scattered seedings of perennials — *Paspalum conjugatum, P. notatum, Axonopus compressus, Sporobolus indicus, Stenotaphrum secundatum*; first perennials following annuals in heavily grazed areas — *Paspalum conjugatum, Sporobolus indicus.*

3rd Stage: These last species are dominant at this stage; being less palatable than the annuals, they grow undisturbed for one to three years after abandonment. Ultimately, depending on soil and climatic factors and intensity of grazing, the grasses are: *Axonopus compressus, Paspalum notatum, Stenotaphrum secundatum* and *Sporobolus indicus.*

Other successions are recorded for:

a. the abandoned cotton lands of the Playa Plains, sandy lands near the seashore:

1st Stage: *Panicum adspersum, Eragrostis ciliaris*

2nd Stage: *Dactyloctenium aegyptiacum, Digitaria horizontalis, D. sanguinalis*

and on sand dunes: dominant *Sporobolus virginicus* with *Philoxerus vermicularis* and some *Cenchrus pauciflorus, C. myosuroides, Chloris inflata* and *Tricholaena repens*;

b. the fresh water lagoons and swamps, in the submerged, floating, reed-swamp and sedge-meadow stage, the last composed of many species of sedges with *Paspalum distichum, Sporobolus virginicus, Sacciolepis striata, Leersia hexandra* and *Paspalum millegranum*; and

c. the dry grasslands of the south and south-west:

1st Stage: *Panicum reptans, P. adspersum, Dactyloctenium aegyptiacum, Digitaria horizontalis, Eleusine indica, Paspalum fimbriatum* and *Cenchrus echinatus*

2nd Stage: *Chloris inflata, Cynodon dactylon* leading to a climax of *Bouteloua heterostega* and *Sporobolus indicus*.

1.7   Ecological Management of Natural Resources

1.7.1   LEVELS OF PRODUCTION

In operations directed towards improving the output from native grasslands and their supplementation through fodder production on arable lands, one may consider four production levels. These involve progressive increases in capital expenditure to provide a more reliable seasonal availability of better quality plant material, leading through the fodder production and utilization chain to the end objective, production of specific requirements from an efficient livestock industry:

Production levels:

primary: using the resource as it stands, by adjusting stock pressures and movements, adoption of sylvo-pastoral technique, selective opening and closing of watering places;

secondary: improving the ground cover resource by removing competitive trees and shrubs;

tertiary: introducing foreign species of grasses, and especially
               legumes, into the existing resource; and
quaternary: removing the natural resource and replacing it with
               artificial or synthetic plant communities for use as
               pasturage or for fodder production.

1.7.2   MANAGEMENT ON BASIS OF CONDITION ANALYSIS

Humphrey (1966) introduced the botanical and ecological
components of the technique of range condition analysis adopted
in the United States of America to an audience of tropical specialists at the Ninth International Grassland Congress in Brazil.
The following paraphrased summary may suffice in the present
context (see 1.4.5).

The ecology of a region must be thoroughly known before
making range condition analyses on which management practices
may be founded. However, a range condition analysis is not a
purely ecological classification. It is rather a practical means of
rating to determine whether, under the current type of management, they are producing the crop of forage that they should.
Five condition classes may be used: excellent, good, fair, poor,
and very poor or depleted. To recognize an excellent condition,
one must know the ecology of the area, its vegetation and the
indicator value and forage value of its component species; also
the climate, the production potential of the soils, and their genesis. One must assess the history of the vegetation in relation to
past climatic fluctuations and crises, to use and misuse.

The five criteria of condition to be adopted by the range
research worker and adviser have been noted in 1.4.5: a. forage
composition: b. ground cover; c. plant vigour; d. litter; e. erosion,
with a possible sixth, reproduction.

The vegetation or forage types are usually separated on a
life-form basis, e.g. forest, woodland, savanna, chaparral, desert
scrub, grassland. Each type may, if necessary, be further subdivided on a floristic basis. This initial mapping into vegetation
types is necessary because each occupies its own ecological niche.
Condition classes are delimited within each vegetation type.

Trend in range conditions is also important in assessing
whether present management should be changed, whether the
range cover is improving, deteriorating or remaining static. Trend
may be determined by two methods. The ideal would be a combination of the two.

98

The interim method records the vegetation growing on an area at a given time, records it again in a later year, and analyzes the changes. This gives a permanent record but does involve at least two surveys. It also assumes that the analyst knows the successional ecology of his vegetation types. 'There is a general human tendency when working with noncultivated species to assume a high degree of permanence to the plants one encounters under natural conditions' (Humphrey, *op. cit.*).

The current appraisal method observes and evaluates certain characteristics of the soil and vegetation, whether erosion is increasing, static or decreasing, and what plants are maintaining themselves and reproducing. The general objective of management on range lands should be maximum forage production for a particular kind of animal, domesticated or wild, at a particular time of the year or for the full year. A range is analyzed as to present condition and direction of trend as a kind of diagnosis to assist in developing effective plans for management, leading to maximum production of livestock products in the legion of local ecosystems that exist, for example, throughout the intertropical zone.

Swartzman & Singh (1974) propose five criteria which relate to range utilization and range condition on *Dichanthium annulatum/Desmodium triflorum* grassland near Varanasi, Uttar Pradesh: a. diversity of major perennial species; b. utilization of herbage; c. percentage of bare space; d. percentage of area occupied by legumes (*Desmodium triflorum, Alysicarpus monilifer, Indigofera linifolia*; e. percentage cover of *Dichanthium annulatum*. A combination of moderate grazing (herbage harvested during October or November — grazing during February at 15 cows per ha. — total herbage removal about 55 per cent of total biomass) with an occasional light grazing year is the best strategy over a 12-year period.

1.7.3   THE ECOLOGICAL COMPONENT IN RANGE MANAGEMENT

A review of the ecological bases for the management of intertropical grazing lands would call for a detailed study of an extensive and scattered literature of a scientific and practical nature. It would be necessary to compare the indigenous systems which have evolved over the years in many different ecoclimates and in many different seral stages in vegetation succession.

The East African Range Classification Committee (Pratt,

Greenway & Gwynne, 1966) considers that the arid zone of northern Kenya comprises the semi-desert together with contiguous lands of higher rainfall, nearly one half of the total area of Kenya (East African Royal Commission, 1955). Mt. Kulal (2290 m.) and Mt. Marsabit (1430 m.) both carry forest at their upper elevations and are flanked by a ring of semi-arid country, characterized by woodland species of *Acacia* and perennial grasses such as *Themeda triandra* (Pratt, 1968, 1969). The vegetation of the basement plains is dwarf-shrub grassland; an open scatter of larger shrubs including *Acacia reficiens* ssp. *misera*, with dwarf shrubs, species of *Indigofera* (especially *I. spinosa*), *Sericocomopsis, Barleria* and *Disperma*. The commonest grasses are annual species of *Aristida* and *A. papposa*, but there is much variation; in some areas *Cenchrus pennisetiformis* may be dominant, or even the perennial species of *Dactyloctenium, Latipes senegalensis*, and species of *Enneapogon* normally classed as perennials.

Pratt (*op. cit.*) finds that the traditional systems of management of the Hamitic or Nilo-Hamitic pastoralists are sound and no radical change is proposed. The present grazing units are the tribal areas, and these are often too large for development purposes. Research on range management at the East African Agricultural and Forestry Research Organization has been described by McKay (1970; by Adegbola, Onayinka & Eweje (1968) for Nigeria; FAO (1970) for Malawi; Garnier, Lahore & Dubois (1968) for Madagascar.

In Rwanda and Burundi, within the pattern of succession at altitudes of 1,800 to 2,500 m. shown in Fig. 1.14 the maintenance of a favourable floristic equilibrium in which *Exotheca abyssinica* and *Hyparrhenia bracteata* are dominant is favoured by the subdivision of the pastures into eight equal plots, and by the adoption of a grazing plan ensuring a rest period of 30 to 40 days (Compère, 1968).

The botanical and ecological bases of the management of grazing lands in India have been fully described (Whyte, 1968a, 1968b). Policy in India is governed by socio-economic/religious factors that prevent the evolution of a scientific system of management based upon the full knowledge of grass succession which is now available. Khan (1970, 1971a, 1971b) has described the current situation in Pakistan, for which Johnston & Hussain (1963) have already proposed grass covers comparable to those in western India. In both countries will be seen an increasing tendency to polarization in grassland and fodder development (Whyte, 1971).

Perhaps it is particularly in Africa that those responsible for ecological management of natural grazing lands are faced with the problem of bush encroachment, which is an expression of a natural ecological tendency of the vegetation to progress upwards through successive stages of shrubs, bush and small trees to the local woodland or forest climax. Burning is the main tool adopted to control such growth, and also to remove old, dead unpalatable herbage that smothers new growth of the grasses.

A discussion whether bush is to be regarded as an infestation or an asset depends on the type of livestock being produced – cattle, goats or wildlife (Ivens, 1972). Some argue that cattle obtain nourishment by browsing, especially during dry seasons, others that carrying capacity of grass from which bush has been cleared is doubled or trebled. A certain proportion of bushy growth may be retained for shade.

Two reviews discuss the problem of "to burn or not to burn" in general terms, in relation to West Africa and South Africa respectively (1.3.4.6). Ramsay & Rose-Innes (1963) report on some quantitative observations on the effects of fire on the Guinea savanna vegetation of northern Ghana over a period of eleven years. They suggest widespread planned experimentation on similar lines, an international effort on a continental scale, at intervals of about 300 to 400 miles, along one or more transects stretching from the southern limits of the savanna country northwards to the Sahara.

"Whatever the merits or demerits of grass burning may be, the fact remains that, at the present time and in present socioeconomic conditions, fire is the only effective tool available to the peasant farmer for clearing and keeping land free of encroaching woody growth; for producing a uniform grass cover free of harsh, unpalatable stubble at each new grazing season, and for promoting a quick flush of off-season green growth for his animals. The practice will undoubtedly continue for a long time to come and the immediate problem is how to use it to the best advantage ... interests which have to be catered for are chiefly those of forestry, agriculture and soil conservation."

To the north of the proposed transect, French workers make frequent reference to the use of fire, for example, Peyre de Fabrègues (1965) refers to running fires in the traditional management of pastures in the Republic of Niger; such a fire at the beginning of the dry season on an annual-dominant cover destroys the reserve of feed for the year.

The second review is that by Scott (1970) who records that veld fires have been a feature of the African landscape since time immemorial (1.3.4.6). In terms of ecological management, when fire has been prevented, plant succession has resulted in bush encroachment in the savanna or in fynbos encroachment in others (fynbos next stage in succession above grassland in potential forest areas, and quite useless for grazing). Fire results in much old grass being wasted, but if this is not removed, grazing deteriorates. Old grass may be removed by fire, or mowing, or by the grazing of animals fed high protein supplements. All methods other than mowing may be equally deleterious.

Other papers from Africa include Thomas & Pratt (1967) on effects of controlled burning of secondary thicket in upland *Acacia* woodland in the drier areas of Kenya; Pratt & Knight (1971) on effects of controlled burning and grazing management on *Tarchonanthus/Acacia* thicket; Pratt (1971) on the effects of fenuron pellets in which a correlation is reported between quantity of combustible grass, fire temperature and percentage kill. Subsequent burning may have general application in the use of arboricides; Ross & Harrington (1969) on the practical aspects of implementing a controlled burning scheme in a national game park in Uganda. Harrington (1974) has studied the fire ecology of the grasslands of Ankole, Uganda, where the vegetation (Langdale-Brown, Osmaston & Wilson, 1964) is dry *Acacia* savanna with the herb layer dominated by *Themeda triandra* in the valleys and by *Cymbopogon afronardus* on the hills. The implications for range management are that pasture on which *C. afronardus* is dominant needs to be burnt when it becomes choked with litter, but the burn should not take place until the rains have started. Pasture free of *C. afronardus* should be controlled by the grazing animal; burning encourages flowering and reduces the quantity of herbage.

In Australia, the views of Tothill (1971a, 1971b) on the burning of vegetation in Queensland, with reference to *Heteropogon contortus*, have been summarized in Rural Research in C.S.I.R.O., No. 74, 1971, and appropriate advice is given to farmers. Five factors favour dominance of this species in annually burnt pastures in Queensland.

— of primary importance, the seed becomes buried in the top half inch of the soil, gaining protection from fire and the elements;

— fire removes ground cover — exposing the soil to increased

102

solar radiation which raises its daily maximum temperature;
– given adequate soil moisture, this can bring about germination when temperatures are otherwise too low;
– fire destroys some of the weaker perennial plants – reducing the level of potential competition from well-established plants;
– fire destroys much of the litter and seed at the soil surface – minimizing the potential competition from seeds that may otherwise germinate.

Norman (1969) has reported on the effect of burning and seasonal rainfall on native pasture at Katherine, Northern Territory, and Perry (1970) on the effects on grass and browse production of the felling by various methods of *Acacia aneura* with a ground cover of *Aristida contorta* (more than half the field), *Digitaria brownei, Sida cardiophylla, Enneapogon polyphyllus, Bassia costata* and *Psilotus helipteroides.*

While recognizing the dominant role of fire in the evolution of tropical American grasslands and in their maintenance in their present form, Blydenstein (1968) states that:

"Precious little information exists on the effect of fire on the tropical vegetation of Latin America ... With the importance of fire in the ecological balance of these grasslands, the control of fire, as opposed to its elimination, should be a basic subject for investigation. We know too little about the effect of season or interval of burning on forage production, brush control, soil conditions and all the other factors which contribute to the ecological balance in producing grasslands. The studies done in other environments, with different species, provide an excellent guide as to the methods and designs which could be adopted to carry out the studies, but the applicable knowledge must come from local studies."

The Tropical Ecology Group of the British Ecological Society discussed savanna burning in Africa at its Spring Meeting in London on 9th April, 1973, under the chairman, B. Hopkins. Contributors were R. Rose-Innes (definition of communities most at risk; deliberate use of fire as management tool), D. H. N. Spence (attempts to restore a diverse habitat in Uganda), M. J. Delany & B. R. Neal (effect of burning on pattern of regrowth of grasses and recolonization by rodents), R. P. Moss (distinctive patterns of savanna burning revealed by aerial photographs), R. M. Lawton (woodland is maintained if burnt early, but destroyed if burnt late, after the trees have flushed; woodland grasses are suppressed by a succession of tree canopies, starting with the most fire-hardy species and finally creating conditions suitable for the regeneration of fire-sensitive canopy species), R. H. Kemp (research to increase forestry production in savanna lands could reduce demands on complicated tropical high forest to meeting increasing world demand for forest products), and A. Blair Rains (in semi-arid regions burning patterns persist for several years; on small-scale aerial photographs these patterns can be used to estimate grazing resources; satellite imagery could be of value in savanna management).

1.7.5   SHRUBS AND TREES IN TROPICAL GRAZING LANDS

Although the main emphasis in this book is on the herbaceous members of the Gramineae and Leguminosae, the woody plants that are browsed by wild and domestic livestock must not be overlooked. The whole subject of the biology and utilization of wildland shrubs was considered at an international symposium at Utah State University, July 1971 (McKell, Blaisdell & Goodin, 1972).

"Shrubs range from some of the highest mountain elevations to the lowest. They extend from the foothills out into the drier desert areas where most grasses fail to accompany them. Only the ephemeral opportunistic annual grasses and forbs are the associated species for many shrub communities in desert and saline areas. Even under such conditions, shrubs offer certain advantages because of their productivity, palatability, nutritional qualities, value as wildlife habitat, cover for the soil, and general role in ecosystem functioning."

Geographical reviews relate to the Indian subcontinent (K. M. M. Dakshini), Australia (S. L. Everist), Asia (M. P. Petrov), South America (A. Soriano), and North America, including Mexican tropics (W. G. McGinnies). Other papers relevant to the present context are: browse and cover for wildlife (W. L. Robinette); evolution and diversity of arid-land shrubs (G. L. Stebbins); genetic improvement in crop species as contrasted with possibilities in shrubs (H. C. Stutz); salt desert shrub response to grazing use (R. C. Holmgren & S. S. Hutchings); the role of shrubs in nutrient cycling (J. L. Charley); papers on physiological and salinity stress, gas exchange, carbohydrates; nutritive value and animal use; regeneration; future of shrubs in arid lands.

104

"Browsing preceded grazing by a good many millions of years" (Gray, 1970). While in some parts of the intertropical belt, the primary objective is to provide a pure grass sward, in others a balance between tree and shrub cover and grass ground cover is recommended. The latter arises more particularly in the former French territories in sub-Saharan West Africa, where the tree/shrub/grass communities are found (1.5.1). Under these conditions it becomes debatable whether the use of the land is the responsibility of the forester and soil conservationist or the specialist in grazing resources. In Cameroon, Piot (1969, 1970) has studied communities under the widely different conditions obtaining on granitic and basaltic soils. Practical considerations relate to the possibilities of complementary feeding, the correct treatment of the woody species and the increasing necessity to keep animals permanently on the grazing lands. Barttha (1970) discusses the specific ecological characteristics of the Sahelian zone, and the nutrition of the nomadic herds based on bush, shrub and herbaceous vegetation.

In regions of West Africa where nomadic grazing ensures maximum utilization of all types of forage available, seasonal use of the woody plants on the cultivated land is adopted since they stay green longer. But with a doubling of the human population every 25 years, and the development of waste land for arable cropping and the swamps for rice, even this resource will become exhausted. It may be desirable to plant edible shrubs such as *Acacia albida* and *Pterocarpus erinaceus* as windbreaks, for lopping during the dry season (Boudet, 1970).

In the Northern Province of Zambia, the Research Branch of the Ministry of Rural Development (Zambia, 1970) has studied the management of Miombo woodland. The system adopted utilizes resources to the maximum, providing for increased production of grass and of protein from browse. It appears to be relatively cheap and well suited to the local peoples, who follow similar clearing methods in the Chitemene system. Further work is necessary to achieve a qualitative balance between ground herbage and browse on the trees, and to improve the sward further by introducing perennial legumes.

Of the small tree and shrub vegetation of western Rajasthan, *Ziziphus nummularia, Prosopis spicigera, Calligonum polygonoides, Capparis aphylla, C. decidua* and *Balanites roxburghii* represent valuable sources of fodder, the first-named being by far the most important (Kaul & Ganguli, 1963). The leaves of *Ziziphus* are rich in digestible crude protein and minerals; leaves are still abundant when the associated grasses on these depleted ranges have dried. The shrubs provide a microhabitat in which better perennial grasses can grow, and from which they can spread under favourable management. By ocular estimate, it is found that fully, medium and poorly stocked stands of this shrub correspond to 18, 14 and 11 per cent density respectively.

SUMMARY

It is generally accepted that the angiosperms in general, and the Gramineae, Leguminosae and other plant families now characteristic of the grazing lands of the world, had their origin in specific ecoclimates and types of vegetation in the equatorial and tropical zones, which were formerly much more extensive than now. It is therefore essential for the taxonomic historian to examine the evidence for the antiquity and evolution of the species which are components of the plant cover of tropical grazing lands at the present time, particularly in relation to the ecological history of those types of vegetation in which these species could have existed at earlier stages of geobotanical evolution.

In order to evolve appropriate systems of management of tropical grazing lands for economic production, the vegetation scientist should be able to define the present ecological status in the overall vegetation of these plant communities within access of grazing and browsing wild and domestic livestock, and also to indicate the types of ecological succession to be expected in the grass and shrub associations themselves.

Bibliography

Acocks, J. P. H., – 1953 – Veld types of South Africa. Botanical Survey of South Africa Memoir no. 28. 191 pp.
Adegbola, A. A., B. O. Onayinka & J. K. Eweje, – 1968 – The management and improvement of natural grassland in Nigeria. *Nigerian Agric. J.* 5: 5–6.
Adejuwon, J. O., – 1970 – The ecological status of coastal savannahs in Nigeria. *J. Trop. Geogr.* 30: 1–10.
Adejuwon, J. O., – 1971 – A biogeographical survey of the dynamics of savanna vegetation in Nigeria. *Nigerian Geogr. J.* 14: 31–48.
Astle, W. L., – 1968–69 – The vegetation and soils of Chishinga Ranch, Luapula Province, Zambia. *Kirkia* 7: 73–102.

Aubreville, A., – 1962 – Savanisation tropicale et glaciations quaternaires. *Adansonia* 2: 16–84.

Aubreville, A., – 1965 – Les étranges savanes des Llanos de l'Orénoque. *Adansonia* 5: 3–13.

Aubreville, A., – 1966a – Les lisières forêt-savane des régions tropicales. *Adansonia* 6: 175–187.

Aubreville, A., – 1966b – Le Costa Rica; quelques aspects du pays, de son climat, de sa végétation et de sa flore. *Adansonia* 7: 29–54.

Aubreville, A., – 1968 – Les étranges mosaïques forêt-savane du sommet de la boucle de l'Ogooué au Gabon. *Adansonia* 7: 13–22.

Axelrod, D. I., – 1970 – Mesozoic palaeogeography and early angiosperm history. *Bot. Rev.* 36: 277–319.

Bader, F., – 1967a – Probleme der Erfassung und Darstellung der Vegetation an ausgewählten Teilgebieten des Kartenblattes Lake Victoria. *Die Erde* 98: 142–149.

Bader, F., – 1967b – Die Vegetationsgeographie auf dem Blatt Lake Victoria in Afrika-Kartenwerk der Deutschen Forschungsgemeinschaft. *Berichten der Deutschen Botanischen Gesellschaft* 80: 291–297.

Balgooy, M. M. J. van, – 1960 – Preliminary plant geographical analysis of the Pacific, as based on the distribution of phanerogam genera. *Blumea* 10: 385–430.

Barttha, R., – 1970 – Plantes fourragères de la Zone sahélienne d'Afrique. Weltforum Verlag, Munich. 306 pp.

Bawden, M. G., – 1965 – A reconnaissance of the land resources of eastern Bechuanaland. *J. appl. Ecol.* 2: 357–365.

Bawden, M. G., & D. M. Carroll, – 1968 – The Land Resources of Lesotho. Land Resource Study no. 3. Land Resources Division, Directorate of Overseas Surveys, Tolworth, Surrey. 89 pp.

Bawden, M. G., D. M. Carroll & P. Tuley, – 1972 – The Land Resources of North East Nigeria. 3. The Land Systems. Land Resource Study no. 9. Overseas Development Administration, Foreign and Commonwealth Office, Surbiton, Surrey, 466 pp.

Bazan, R., – 1972 – Informe de Instituciones Internationales que participen en el desarrollo y fomento de Pastos y Forrajes tropicales: Instituto Interamericano de Ciencias Agricolas, Turrialba. FAO/ICEA-CPAT mimeo. 12 pp.

Beard, J. S., – 1944 – Climax vegetation in tropical America. *Ecology* 25: 127–158.

Beard, J. S., – 1953 – The savanna vegetation of northern tropical America. *Ecol. Monogr.* 23: 149–215.

Bhatnagar, H. P., – 1960 – Succession in some *Shorea robusta* forests of U.P. *J. Ind. Bot. Soc.* 39: 22–29.

Bille, J. C., – 1965 – Evolution des pâturages naturels des hauts plateaux de la République Centrafricaine en exploitation traditionnelle Bororo. *Rev. d'Elev. Méd. vétér. Pays tropicaux* 18: 313–316.

Bille, J. C., J. P. Lebrun & R. Rivière, – 1968/69 – Pâturages et Cultures fourragères en Afrique intertropicale. Institut d'Elevage et de Médecine Vétérinaire des Pays Tropicaux, Maisons Alfort/Secretariat d'Etat aux Affaires Etrangères chargé de la Coopération, Paris. 218 pp.

Blow, R. A. & N. Hamilton, – 1975 – Palaeomagnetic evidence from DSDP cores of northward drift of India. *Nature* 257: 570–572.

Blydenstein, J., – 1961 – La vegetación de la Estación Biologica de los Llanos. *Bol. Soc. Venez. Cienc. Natur.* 22: 208–212.

Blydenstein, J., – 1962 – La sabana de *Trachypogon* del Alto Llano. *Bol. Soc. Venez. Cienc. Natur.* 23: 139–206.

Blydenstein, J., — 1963 — Cambios en la vegetación despues de protección contra el fuego. 1. El aumento anual en material vegetal en varios sitios quemados y no quemados en la Estación Biologica. *Bol. Soc. Venez. Cienc. Natur.* 23: 233–244.

Blydenstein, J., — 1967 — Tropical savanna vegetation of the Llanos of Colombia. *Ecology* 48: 1–15.

Blydenstein, J., — 1968 — Burning and tropical American savannas. In "Proceedings Annual Tall Timbers Fire Ecology Conference no. 8," pp. 1–14.

Blydenstein, J., — 1971 — Recursos Forrajeros de Venezuela. AGP: PFC/19. FAO, Rome. 18 pp.

Blydenstein, J., — 1972 — Evaluación de recursos naturales. FAO/ICA-CPAT/6. 6 pp.

Bor, N. L., — 1960 — Grasses of Burma, Ceylon, India and Pakistan (except Bambuseae). Pergamon Press, Oxford, London, New York, Paris. 767 pp.

Bor, N. L., — 1968 — Flora of Iraq, Vol. 9. Gramineae. Ministry of Agriculture, Baghdad. 588 pp.

Bosser, J., — 1969 — Graminées des Pâturages et des Cultures à Madagascar. Mem. ORSTOM no. 35, Paris. 440 pp.

Boudet, G., — 1966 — Investigation and cartographic method for tropical pastures applied at the Breeding and Veterinary Medicine Institute of Tropical Countries. In "Proceedings of Ninth International Grassland Congress", pp. 475–477. Departamento da Produção Animal da Secretaria da Agricultura do Estado de São Paulo.

Boudet, G., — 1970 — Management of savannah woodland range in West Africa. In "Proceedings of Eleventh International Grassland Congress," pp. 1–3. University of Queensland Press, St. Lucia.

Boudet, G. & F. Baeyens, — 1963 — Une méthode d'étude et de cartographie des pâturages tropicaux. *Rev. d'Elev. Méd. vét. Pays trop.* 16: 191–219.

Boudet, G. & J. F. Ellenberger, — 1971 — Etude agrostologique du Berceau de la Race N'Dama dans le Cercle de Yanifolila (République du Mali). Etude agrostologique no. 30. I.E.M.V.T., Maisons-Alfort. 175 pp.

Boudet, G. & R. Rivière, — 1968 — Practical use of fodder analysis to assess the value of tropical pastures. *Rev. d'Elev. Méd. vét. Pays trop.* 21: 227–66.

Bourlière, F. & M. Hadley, — 1970 — The ecology of tropical savannas. *Ann. Rev. Ecol. Syst.* 1: 125–152.

Box, T. W., — 1968 — Range resources of Somalia. *J. Range Mgmt.* 21: 388–392.

Bragadín, E. A., — 1959 — Las pasturas en la Región de los Llanos. *Revista Agronomica Noroeste Argentino* 3: 289–333.

Brown, W. H., — 1919 — Vegetation of Philippine Mountains: the Relation between the Environment and Physical Types at different Altitudes. Bureau of Science, Manila. 434 pp.

Brown, W. V. & B. N. Smith, — 1972 — Grass evolution, the Kranz syndrome, $^{13}C/^{12}C$ ratios and continental drift. *Nature* (Lond.) 239: 345–346.

Burkart, A., — 1975 — Evolution of grasses and grasslands in South America. *Taxon* 24: 53–66.

Cain, S. A. & G. M. de O. Castro, — 1959 — The Manual of Vegetation Analysis. Harper and Brothers, New York. 337 pp.

Carlquist, S., — 1974 — Island Biology. Columbia University Press, New York. 660 pp.

Casebeer, R. L., — 1969 — Integrated resource surveys for rangeland development in Kenya. FAO Ad Hoc Technical Conference on Grassland Production and Fodder Management in Africa south of Sahara, Nairobi, 1969. Paper PL: GFA/69/2. FAO, Rome. 5 pp.

Casebeer, R. L. & G. G. Koss, — 1970 — Food habits of wildebeest, zebra, hartebeest and cattle in Kenya Masailand. *East Afric. Wildlife J.* 8: 25–36.

Champion, H. G. & S. K. Seth, – 1968 – A Revised Survey of the Forest Types of India. Manager of Publications, Delhi. 404 pp.

Chang, Cheng-fa & Chong Hsi-lan, – 1973 – Some tectonic features of the Mt. Jolma Lungma area, southern Tibet, China. *Scientia Sinica* 16: 257–265.

Chang Jen-hu, – 1971 – The Chinese monsoon. *Geogr. Rev.* 61: 370–395.

Christian, C. S., – 1963 – The use and abuse of land and water. In "The Population Crisis and the Use of World Resources", Vol. 2, – World Academy of Art and Science, W. Junk, The Hague 387–406.

Christian, C. S. & G. A. Stewart, – 1968 – Methodology of integrated surveys. In "Aerial Surveys and Integrated Studies: Proceedings of the UNESCO Conference, Toulouse," pp. 233–280. UNESCO, Paris.

Clark, J. D., – 1970 – The Prehistory of Africa. Thames and Hudson, London. 302 pp.

Clayton, W.D. & G. Panigrahi, – 1974 – Computer-aided chorology of Indian grasses. *Kew Bulletin* 29: 669–686.

Coaldrake, J. E., – 1970 – The Brigalow. In "Australian Grasslands," ed. R. M. Moore, pp. 123–140. Australian National University Press, Canberra.

Cole, M. M., – 1960 – Cerrado, caatinga and pantanal: the distribution and origin of the savanna vegetation of Brazil. *Geogr. J.* 126: 168–179.

Cole, M. M., – 1973 – Geobotanical and biochemical investigations in the sclerophyllous woodland and shrub associations of the eastern goldfields area of Western Australia, with particular reference to the role of *Hybanthus floribundus* (Lindl.) F. Muell. as a nickel indicator and accumulator plant. *J. appl. Ecol.* 10: 269–320.

Compère, R., – 1968 – Exploitation rationnelle de la prairie permanente d'altitude à *Exotheca abyssinica* Anders. et *Eragrostis boehmii* Hack. au Rwanda et au Burundi. *Bull. Rech. agron. Gembloux* 3: 583–604.

Conolly, J. R., – 1967 – Postglacial-glacial change in climate in the Indian Ocean. *Nature* (Lond.) 214: 873–875.

Conseil Scientifique pour l'Afrique au sud du Sahara (CSA), – 1956 – Rapport final de la Réunion de Spécialistes en Phytogéographie/Final Report of the Specialist Meeting on Phytogeography, Yangambi, Belgian Congo. Joint Secretariat, Bukavu. 379 + 32 pp.

Cooper, J. P., – 1970 – Potential production and energy conversion in temperate and tropical grasses (106 references). *Herbage Abstracts* 40: 1–15.

Cooper, J. P. & N. M. Tainton, – 1968 – Light and temperature requirements for the growth of tropical and temperate grasses. (71 references). *Herbage Abstracts* 38: 167–176.

Cull, J. K. & J. P. Ebersohn, – 1969 – Dynamics of semi-arid plant communities in Western Queensland. 1. Population shifts of two invaders: *Cenchrus ciliaris* cv Gayndah and *Heteropogon contortus*. *Queensland J. Agric. Anim. Sci.* 26: 193–198.

Dabadghao, P. M. & K. A. Shankarnarayan, – 1970 – Studies of *Iseilema*, *Sehima* and *Heteropogon* communities of the *Sehima-Dichanthium* zone. In "Proceedings of Eleventh International Grassland Congress," pp. 36–38. University of Queensland Press, St. Lucia.

Dabadghao, P. M. & K. A. Shankarnarayan, – 1973 – The Grass Cover of India. Indian Council of Agricultural Research Scientific Monograph, New Delhi. 713 pp.

Dansereau, P., – 1951 – Description and recording of vegetation upon a structural basis. *Ecol.* 32: 172–229.

Daubenmire, R., – 1972a – Ecology of *Hyparrhenia rufa* (Nees.) in derived savanna in north-western Costa Rica. *J. appl. Ecol.* 9: 11–23.

Daubenmire, R., – 1972b – Some ecologic consequences of converting forest to savanna in northwestern Costa Rica. *Tropical Ecology* 13: 31–51.

Davies, W., – 1960 – Plenary paper. In "Proceedings of Eighth International Grassland Congress," pp. 1–7.

Davies, W. & C. L. Skidmore (eds.), – 1966 – Tropical Pastures. Faber and Faber, London. 215 pp.

Descoings, B., – 1971 – Méthode de description des formations herbeuses intertropicales par la structure de la végétation. *Candollea* 26: 223–257.

Descoings, B., – 1972 – Note sur la structure de quelques formations herbeuses de Lamto (Côte d'Ivoire). *Ann. Univ. Abidjan sér. E. (Ecologie)* 5: 7–30.

Descoings, B., – 1973a – Les formations herbeuses africaines et les définitions de Yangambi, considerées sous l'angle de la structure de la végétation. *Adansonia* 13: 391–421.

Descoings, B. – 1973b – Les savanes du Moyen-Ogooué, région de Booué (Gabon). Conditions générales, analyse floristique, analyse structurale, valeur pastorale. CNRS–CEPE, Montpellier.

Descoings, B. – 1974 – Notes de phyto-écologie équatoriale. 2. Les formations herbeuses du Moyen-Ogooué (Gabon). *Candollea* 29: 13–37.

Descoings, B. – 1975 – Les grandes régions naturelles du Congo. *Candollea* 30: 91–120.

Descoings, B. – 1976 – Notes de phytoécologie équatoriale. 3. Les formations herbeuses de la vallée de la Nyanga (Gabon). *Adansonia* 15: 307–329.

Donselaar, J. van – 1965 – An ecological and phytogeographic study of northern Surinam savannas. *Wentia* 14: 1–163.

Donselaar, J. van, – 1968 – Phytogeographic notes on the savanna flora of Southern Surinam (South America). *Acta Bot. Neerl.* 17: 393–404.

Donselaar, J. van – 1969 – Observations on savanna vegetation-types in the Guianas. *Vegetatio* 17: 271–312.

Duvigneaud, P., – 1951 – Les savanes du Bas-Congo: essai de phytosociologie topographique. *Lejeunia, Rev.-Bot. Liège* 10: 1–192.

Duvigneaud, P., – 1958 – La végétation du Katanga et de ses sols métallifères. *Bull. Soc. Roy. Bot. Belg.* 90: 127–286.

Duvigneaud, P. & S. Denaeyer de Smet, – 1963 – Cuivre et végétation au Katanga. *Bull. Soc. Roy. Bot. Belg.* 96: 93–231.

Dyksterhuis, E. J., – 1957 – The savanna concept and its use. *Ecol.* 38: 435–442.

East African Royal Commission, – 1955 – 1953–1955 Report. H.M.S.O., London, 482 pp.

Eden, M. J., – 1970 – Savanna vegetation in the northern Rupununi, Guyana. *J. Trop. Geogr.* 30: 17–28.

Eeckhout, L. E., – 1954 – Contribution à l'uniformisation de la terminologie phytogéographique. In "Rapp. et Comm. 8me Congrès Int. Bot., Paris, Sect. 7 et 8," pp. 69–74.

Eiten, G., – 1972 – The cerrado vegetation of Brazil. *Bot. Rev.* 38: 201–341. (14 pp. bibliography).

Ellenberg, H., – 1962 – Wald in der Pampa Argentiniens? *Veröff. d. Geobotanischen Inst. d. E.T.H. Stiftung Rübel in Zürich* (Fest. F. Firbas) 37: 39–56.

Ellenberg, H., – 1971 – Ecological Studies. Analysis and Synthesis. Vol. 2. Integrated Experimental Ecology. Lange and Springer, Berlin, Heidelberg, New York. 214 pp.

Exell, A. W. & H. Wild, – 1973 – A statistical analysis of a sample of the Flora Zambesiaca. II. Gramineae and Pteridophyta. *Kirkia:* 9: 87–93.

Fanshawe, D. B., – 1952 – The Vegetation of British Guyana: A Preliminary Review. Institute Paper no. 29. Imperial Forestry Institute, Oxford. 96 pp.

Ferri, M. G., – 1973 – Sobre a origem, a manutenção e a transformação dos cerrados. *Ecologia* 1: 5–10.

Florin, R., – 1963 – The distribution of conifer and taxad genera in time and space. *Acta Hort. Berg.* 20: 121–312.

Food and Agriculture Organization of the United Nations, – 1970 – Improvement of Livestock and Dairy Industry, Malawi. Report for Government of Malawi, based on work of H. H. Heady. AGA:SF/MLW/5. Technical Report 3. FAO, Rome. 15 pp.

FAO/UNDP, – 1971 – Wildlife and Land Use in Kenya. Range Management Division of Ministry of Agriculture, Kenya, based on work of R. L. Casebeer. AGP: SF/ KEN/11. Technical Report no. 3. FAO, Rome. 22 pp.

FAO/UNESCO, – 1973 – Irrigation, Drainage and Salinity: An International Source Book. Hutchinson/FAO/UNESCO, London, Rome, Paris. 510 pp.

FAO/UNSF, – 1966 – Reconocimiento edafologico de los Llanos Orientales: Colombia. Tomo 6. La vegetación natural y la ganaderia en los Llanos Orientales. Seccion primera. La vegetación. FAO/SF: 11/COL. FAO, Rome. 233 pp.

Fosberg, F. R., – 1958 – On the possibility of a rational general classification of humid tropical vegetation. In "Proceedings of UNESCO Symposium on Humid Tropics Vegetation, Tjiawi, Indonesia," pp. 34–59.

Frakes, L. A. & E. M. Kemp., – 1972 – Influence of continental positions on early Tertiary climates. *Nature* (Lond.) 240: 97–100.

Fretes, R., R. Samudio & C. Gay, – 1970 – Las praderas naturales del Paraguay. 1. Classificación y descripción. Programa Nacional de Investigación y Extensión Ganadera, Asunción. 86 pp.

Garnier, P., J. Lahore & P. Dubois, – 1968 – Etude du pâturage naturel à Madagascar: productivité, conséquences pratiques. *Rev. Elev. Méd. vét. Pays trop.* 21: 203–218.

Gilliland, H. B., with contributions by R. E. Holttum, N. L. Bor and H. M. Burkill, – 1971 – A Revised Flora of Malaya. Vol. 3. Grasses of Malaya. Botanic Gardens, Singapore. 319 pp.

Glover, P. E., – 1947 – A Provisional Checklist of British and Italian Somaliland Trees, Shrubs and Herbs. Crown Agents, for the Government of Somaliland, London. 446 pp.

Godron, M., P. Daget, L. Emberger, G. Long, E. Le Floc'h, J. Poissonet, C. Sauvage & J. P. Wacquant, – 1968 – Code pour le Relevé méthodique de la Végétation et du Milieu. C.N.R.S., Paris. 292 pp.

Golley, P. M. & F. B. Golley (compilers), – 1972 – Tropical Ecology. Athens, Georgia. 413 pp.

Good, R., – 1964 – The Geography of the Flowering Plants (3rd. ed.). Longmans Green, London. 518 pp.

Goodland, R. J. A., – 1966 – On the savanna vegetation of Calabozo, Venezuela, and Rupununi, British Guiana. *Boln. Soc. Venez. Cienc. Nat.* 26: 341–359.

Goodland, R. J. A., – 1970 – Plants of the cerrado vegetation of Brazil. *Phytologia* 20: 57–78.

Goodland, R. J. A., – 1971a – A physiognomic analysis of the "cerrado" vegetation of central Brasil. *J. Ecol.* 59: 411–419.

Goodland, R. J. A., – 1971b – The cerrado oxisols of the Triângulo Mineiro, Central Brazil. *An. Acad. brasil. Cienc.* 43: 407–414.

Goodland, R. J. A. & R. Pollard, – 1973 – The Brazilian cerrado vegetation: a fertility gradient. *J. Ecol.* 61: 219–224.

Granier, P., – 1965 – Le rôle de l'élevage extensif dans la modification de la végétation à Madagascar. *Rev. Elev. Méd. vét. Pays trop.* 18: 293–305.

Gray, S. G., – 1970 – The place of trees and shrubs as sources of forage in tropical and subtropical pastures. *Tropical Grasslands* 4: 57–62.

Grossman, J., S. Aronovich & E. do C. B. Campello, – 1966 – Grasslands of Brazil. In "Proceedings of the Ninth International Grassland Congress," pp. 39–80. Departamento da Produção Animal da Secretaria da Agricultura do Estado de São Paulo.

Gupta, R. K. & P. C. Nanda, – 1970 – Grassland types and their ecological succession in the western Himalayas. In "Proceedings of Eleventh International Grassland Congress," pp. 10–13. University of Queensland Press, St. Lucia.

Gupta, R. K., S. K. Saxena & S. K. Sharma, – 1972a – Aboveground productivity of grasslands at Jodhpur, India. In "Tropical Ecology," ed. P. M. Golley & F. B. Golley, pp. 75–93. Athens, Georgia.

Gupta, R. K., S. K. Saxena & S. K. Sharma, – 1972b – Aboveground productivity of three promising desert grasses at Jodhpur under different rainfall conditions. In "Eco-physiological Foundation of Ecosystems Productivity in Arid Zone," Soviet National Committee for IBP, International Symposium, pp. 134–137. Nauka, Leningrad.

Haffer, J., – 1974 – Avian Speciation in Tropical South America. With a Systematic Survey of the Toucans (Ramphastidae) and Jacamars (Galbulidae). Publications of the Nuttall Ornithological Club no. 14. (c/o Museum of Comparative Zoology, Harvard University). Cambridge, Mass. 390 pp.

Hallam, A., – 1973 – A Revolution in the Earth Sciences: from Continental Drift to Plate Tectonics. Oxford University Press, London. 127 pp.

Harlan, J. R., – 1970 – *Cynodon* species and their value for grazing and hay. *Herbage Abstr.* 40: 233–238.

Harlan, J. R., J. M. J. de Wet & K. M. Rawal, – 1970a – Geographic distribution of the species of *Cynodon* L. C. Rich. (Gramineae). *East Afric. Agric. For. J.* 36: 220–226.

Harlan, J. R., J. M. J. de Wet & K. M. Rawal, – 1970b – Origin and distribution of the Seleucidus race of *Cynodon dactylon* (L.) Pers. var. *dactylon* (Gramineae). *Euphytica* 19: 465–469.

Harrington, G. N., – 1974 – Fire effects on a Ugandan savanna grassland. *Tropical Grasslands* 8: 87–101.

Hernandez Efrain, M. F., – 1963 – Los tipos de vegetación de México y su clasificación. Instituto de Biología de la U.N.A.M. y al Escuela Nacional de Agricultura, Chapingo. 29–195.

Heyligers, P. C., – 1963 – Vegetation and soil of a white-sand savanna in Suriname. Mededelingen van het Botanisch Museum en Herbarium van Rijksuniversiteit te Utrecht no. 191. 148 pp.

Heyligers, P. C., – 1966 – Observations on *Themeda australis-Eucalyptus* savannah in Papua. *Pacific Science* 20: 477–489.

Heyligers, P. C., – 1972 – Analysis of the plant geography of the semi-deciduous scrub and forest and the Eucalypt savannah near Port Moresby. *Pacific Science* 26: 229–241.

Hills, T. L. & R. E. Randall (eds.), – 1968 – The Ecology of the Forest/Savanna Boundary. Proceedings of I.G.U. Humid Tropics Symposium, 1964. Savanna Research Series. McGill University, Montreal. 128 pp.

Holdridge, L. R., – 1947 – Determination of world plant formations from simple climatic data. *Science* 105: 367–368.

Holttum, R. E., – 1954 – Plant Life in Malaya. Longmans, London and Kuala Lumpur. 254 pp.

Hopkins, B., – 1968 – Vegetation of the Olokemeji Forest Reserve, Nigeria. 5. The vegetation on the savanna site with special reference to its seasonal changes. *J. Ecol.* 56: 97–115.

Hopkins, B., – 1970a – Vegetation of the Olokemeji Forest Reserve, Nigeria. 6. The plants on the forest site with special reference to their seasonal growth. *J. Ecol.* 58: 765–793.

Hopkins, B., – 1970b – Vegetation of the Olokemeji Forest Reserve, Nigeria. 7. The plants on the savanna site with special reference to their seasonal growth. *J. Ecol.* 58: 795–825.

Horrell, C. R., – 1972 – Informes regionales sobre problemas, actividades y programas crecientes de desarrollo en el campo de pastos y forrajes tropicales: Región Centroamericana. FAO/CIAT-CPAT/10. 15 pp.

Hueck, K., – 1953 – Urlandschaft, Raublandschaft und Kulturlandschaft in der Provinz Tucumán in nordwestlichen Argentine. *Bonner Geogr. Abh.* 10: 102 pp.

Humphrey, R. R., – 1966 – Range condition analysis – a practical ecological approach to the classification of rangelands. In "Proceedings of Ninth International Grassland Congress," pp. 1437–1442. Departamento da Produção Animal da Secretaria da Agricultura do Estado de São Paulo.

Ivens, G. W., – 1972 – The problem of bush in rangeland. *Span* 15: 23–26.

Jacques-Félix, H., – 1962 – Les Graminées d'Afrique tropicale. I. Généralités, Classification, Description des Genres. Bull. Sci. no. 8, IRAT, Paris. 345 pp.

Jacques-Félix, H., – 1968 – Evolution de la végétation au Cameroun sous l'influence de l'homme. *J. Agric. trop. Bot. appl.* 15: 350–356.

Jaeger, F., – 1945 – Zur Gliederung und Benennung des tropischen Graslandgürtels. *Verh. Naturforsch. Ges. Basel* 56: 509–520.

Johns, R. J. & P. F. Stevens, – 1971 – Mount Wilhelm Flora: A Check List of the Species. Botany Bulletin no. 6, Department of Forests, New Guinea. 61 pp.

Johnston, A. & E. Hussain, – 1963 – Grass cover types of West Pakistan. *Pak. J. For.* 13: 239–247.

Kaul, R. N. & B. N. Ganguli, – 1963 – Fodder potential of *Zizyphus* in the scrub grazing lands of arid zone. *Indian For.* 89: 623–630.

Kayser, K., W. Manshard, H. Mensching & J. H. Schultze, – 1966 – Das Afrika-Kartenwerk. Ein Schwerpunkt-Program der Deutschen Forschungsgemeinschaft. *Die Erde* 97: 85–95.

Keay, R. W. J., – 1959 – Derived savanna: derived from what? *Bull. d'IFAN Sér. A.* 21: 427–438.

Kellman, M. C., – 1969 – Shifting cultivation in upland Mindanao. *J. Trop. Geogr.* 28: 40–56.

Keng, Yi-Li, – 1965 – Flora Illustrata Plantarum Primarum Sinicarum: Gramineae. Nanking University/Academia Sinica, Peking. 1178 pp. (in Chinese).

Khan, C. M. Anwar, – 1970 – Range management – a challenge in West Pakistan. *Pak. J. For.* 20: 329–350.

Khan, C. M. Anwar, – 1971a – Rainfall pattern and monthly forage yields in Thal ranges of Pakistan. *Pak. J. For.* 21: 93–102.

Khan, C. M. Anwar, – 1971b – Range management strategy in Pakistan. 1. *Pak. J. For.* 21: 313–324.

Lamotte, M., – 1967 – Recherches écologiques dans la savane de Lamto (Côte d'Ivoire): Présentation du milieu et du programme de travail. *La Terre et la Vie* 21: 197–215.

Langdale-Brown, I., H. A. Osmaston & J. G. Wilson, – 1964 – The Vegetation of Uganda and its Bearing on Land Use. Government Printer, Entebbe. 147 pp.

113

Lanjouw, J., – 1936 – Studies on the vegetation of the Suriname savannahs and swamps. *Ned. Kruidk. Arch.* 46: 823–851.

Lansbury, T. J., R. Rose-Innes & G. L. Mabey, – 1965 – Studies on Ghana grasslands: yield and composition on the Accra Plains. *Trop Agric. Trin.* 42: 1–18.

Lauer, W., – 1952 – Humide und aride Jahreszeiten in Afrika und Südamerika und ihre Beziehung zu den Vegetationsgürteln. *Bonner Geogr. Abhandl.* 9: 15–98.

Lemon, P. C., – 1968a – Effects of fire on an African plateau grassland. *Ecol.* 49: 316–322.

Lemon, P. C., – 1968b – Fire and wildlife grazing on an African plateau. In "Proceedings of Annual Tall Timbers Fire Ecology Conference no. 8," pp. 71–88.

Leopold, A. S., – 1950 – Vegetation zones of Mexico. *Ecology* 31: 501–518.

Letouzey, R., – 1968 – Etude phytogéographique du Cameroun. Ed. P. Lechevalier, Paris. 511 pp.

Lima, D. de A., – 1966 – Vegetation of Brazil. In "Proceedings of Ninth International Grassland Congress," pp. 29–38. Departamento da Produção Animal da Secretaria da Agricultura do Estado de São Paulo.

Liu, T., – 1971 – Studies on the classification of the climax vegetation communities of Taiwan. 3. *Proc. Nat. Sci. Council Taiwan* no. 4: 1–36.

Long, G., – 1966 – Modern phytosociological and ecological methods and their use in range research inventory. In "Proceedings of Ninth International Grassland Congress," pp. 1349–1352. Departamento da Produção Animal da Secretaria da Agricultura do Estado de São Paulo.

Long, G., – 1969 – Perspectives nouvelles de la cartographie biogéographique végétale intégrée. *Vegetatio* 18: 44–63.

Long, G., – 1972 – Le concept d'intégration en écologie appliquée. *Canadian J. Bot.* 50: 533–541.

Long, G., P. S. Poissonet, J. A. Poissonet, P. M. Daget & M. P. Godron, – 1972 – Improved needle point frames for exact line transects. *J. Range Management* 25: 228–229.

Long, M. I. E., W. K. Ndyanabo, B. Marshall & D. D. Thornton, – 1969 – Nutritive value of grasses in Ankole and the Queen Elizabeth National Park, Uganda. 4. Mineral content. *Trop. Agric. Trin.* 46: 201–209.

Long. M. I. E., D. D. Thornton & B. Marshall, – 1969 – Nutritive value of grasses in Ankole and the Queen Elizabeth National Park, Uganda. 2. Crude protein, crude fibre and soil nitrogen. *Trop. Agric. Trin.* 46: 31–42.

Long. M. I. E., D. D. Thornton, W. K. Ndyanabo, B. Marshall & H. Ssekab, – 1970 – The mineral status of dairy farms in the parts of Buganda and Busoga bordering Lake Victoria, Uganda. 2. Nitrogen and mineral content of pastures. *Trop. Agric. Trin.* 47: 37–49.

Loxton, R. F., G. K. Rutherford & J. Spector, – 1958 – Soil and Land-Use Surveys, no. 2. British Guiana. The Rupununi Savannas. Regional Research Centre of the British Caribbean at the Imperial College of Tropical Agriculture, Trinidad. 33 pp.

Lugo, A. E., – 1969 – Energy flow in some tropical ecosystems. In *Proc. Soil Crop Soc. Fla* 29: 254–64.

McClure, H. E., – 1971 – Some aspects of bird migration in Asia. Proc. XII Pac. Sci. Congr. 1: 219–220.

McElhinny, M. W., – 1968 – Northward drift of India – examination of recent palaeomagnetic results. *Nature* (Lond.) 217: 342–344.

McElhinny, M. W., N. S. Haile & A. R. Crawford, – 1974 – Palaeomagnetic evidence shows the Malay Peninsula was not a part of Gondwanaland. *Nature* 252: 641–645.

McKay, A. D., – 1970 – Range management research at EAAFRO. *E. Afr. agric. For. J.* 35: 346–349.

McKay, A. D. & P. E. Frandsen, – 1969 – Chemical and floristic components of the diet of Zebu cattle (*Bos indicus*) in browse and grass range pastures in semi-arid upland area of Kenya. 1. Crude protein. *Trop. Agric. Trin.* 46: 279–292.

McKell, C., J. P. Blaisdell & J. R. Goodin (technical eds.), – 1972 – Wildland Shrubs – Their Biology and Utilization. Intermountain Forest and Range Experiment Station, USDA Forest Service General Technical Report INT-1, Ogden, Utah. 494 pp.

McMillan, C., – 1960 – Ecotypes and community function. *Amer. Naturalist* 94: 245–255.

McMillan, C., – 1971 – Ecotypes and ecosystem function. In "Ecology, Foundations for Today," eds. R. S. Leisner & E. J. Kormondy, Vol. 3, pp. 38–41. American Institute of Biological Sciences. Brown and Co., Dubuque, Iowa.

McNaughton, S. J., – 1967 – Relationship among functional properties of Californian grassland. *Nature* (Lond.) 216: 168–169.

Marlange, M., – 1971 – Caractères écologiques généraux du Chaco argentin. C.N.R.S. doc. no. 58. Centre d'Etudes phytosociologiques et écologiques, Montpellier. 172 pp.

Misra, R. (ed.), – 1974 – Proceedings of IBP Symposium and Synthesis Meetings on Tropical Grassland Biome. Banaras Hindu University, Varanasi. 174 pp.

Molinari, O. Garcia, – 1949 – Succession of grasses in Puerto Rico. *Rev. Agric. Puerto Rico* 39: 199–217.

Molinari, O. Garcia, – 1952 – Pasture and forage grass species and their ecology in the Caribbean area. Paper presented at Sixth International Grassland Congress. 18 pp.

Molnar, P. & P. Tapponnier, – 1975 – Cenozoic tectonics of Asia: effects of a continental collision. *Science* 189: 419–426.

Moore, R. M., (ed.), – 1970 – Australian Grasslands. Australian National University Press, Canberra. 473 pp.

Morat, P., – 1973 – Les Savanes du Sud-Ouest de Madagascar. Mém. ORSTOM no. 68. ORSTOM, Paris, 235 pp.

Morello, J. H., (with J. Adamoli), – 1968 – La vegetación de la Republica Argentina: Las grandes unidades de vegetación y ambiente del Chaco Argentino. 1. Objectivos y Metodologia. Instituto de Botanica Agricola, Ser. Fitogeográfica no. 10, Buenos Aires. 125 pp.

Morello, J. H., – 1970 – Modelo de relaciones entre pastizales y leñosas colonizadoras en el Chaco Argentino: (Plan Ecología, Difusión y Control del Vinal). *Idia* no. 276: 31–52.

Morello, J. & J. Adamoli, – 1973 – Subregiones ecológicas de la Provincia del Chaco. *Ecologia* 1: 29–33.

Morello, J. H., N. E. Crudeli & M. Saraceno, – 1971 – Los Vinalares de Formosa (Republica Argentina). (La Colonizadora Leñosa *Prosopis ruscifolia* Gris). Instituto Nacional de Tecnologia Agropecuaria, Serie Fitogeográfica no. 11. Buenos Aires, 111 pp.

Morello, J. H. & C. Saravia Toledo, – 1959 – El bosque chaqueño. 1. Paisaje primitivo, paisaje natural y paisaje cultural en el oriente de Salta. 2. La ganaderia y el bosque en el oriente de Salta. *Revista Agronomica del Noroeste Argentino* 3: 5–81, and 209–258.

Myers, J. G., – 1936 – Savanna and forest vegetation of the interior Guiana plateau. *J. Ecol.* 24: 162–183.

Noda, M., – 1972 – Flora of the North-Eastern Province (Manchuria) of China (in Japanese). Kazama Bookshop Co., Tokyo. 1613 pp.

Norman, M. J. T., – 1969 – The effect of burning and seasonal rainfall on native pasture at Katherine, N.T., *Austral. J. Exper. Agric. Anim. Husb.* 9: 295–298.

Oakes, A. J., – 1968 – Replacing of hurricane grass in pastures of the dry tropics. *Trop. Agric. Trin.* 45: 235–241.

Oke, J. G., – 1972 – Studies in grassland ecology of the Maharashtra State. 1. Ecological classification of grassland patterns found in different ecological habitats and their botanical characterization. *Indian For.* 98: 86–106.

Owen, M. A. & H. W. Brzostowski, – 1966 – A grass cover for the upland soils of the Kongwa Plain, Tanganyika. *Trop. Agric. Trin.* 43: 303–314.

Pagot, J. R., – 1971 – Pâturages naturels, Cultures fourragères et Elevage dans les Régions tropicales de l'Afrique francophone. Institut d'Elevage et de Médecine Vétérinaire des Pays Tropicaux, Maisons Alfort. 11 pp.

Paijmans, K. & E. Löffler, – 1972 – High-altitude forests and grasslands of Mount Albert Edward, New Guinea. *J. Trop. Geogr.* 34: 58–64.

Pandeya, S. C., – 1961 – On some new concepts in phytosociological studies of grasslands. 1. Dominance diagrams. 2. Community coefficient (F x C)ICC. *J. Indian bot. Soc.* 40: 263–266; 267–270.

Pandeya, S. C., – 1969 – Ecology of grasslands of Sagar, Madhya Pradesh. 3. Edaphic factors in the distribution of grassland associations. *Trop. Ecol.* 10: 163–182.

Parsons, J. J., – 1970 – The "Africanization" of the New World tropical grasslands. *Tübinger Geographische Studien* 34: 141–153.

Pennington, T. D. & J. Sarukhan, – 1968 – Arboles tropicales de México. Instituto Nacional de Investigaciones Forestales, Mexico/FAO, Rome. 413 pp.

Perry, R. A., – 1960 – Pasture Lands of the Northern Territory, Australia. Land Research Series, no. 5. CSIRO, Melbourne. 55 pp.

Perry, R. A., – 1964 – Value of ecological surveys. In "Proceedings and Papers of the IUCN Ninth Technical Meeting, 1963," Part IV, "Ecological Research and Development," pp. 303–312.

Perry, R. A., – 1967 – Integrated Surveys of Pastoral Areas. S20 ITC/UNESCO, Delft. 27 pp.

Perry, R. A., – 1970 – The effects on grass and browse production of various treatments on a mulga community in central Australia. In "Proceedings of Eleventh International Grassland Congress," pp. 63–66. University of Queensland Press, St. Lucia.

Peyre de Fabrègues, B., – 1965 – Etudes et principes d'exploitation de pâturage de steppe en République du Niger. *Rev. Elev. Méd. vét. Pays trop.* 18: 329–332.

Piot, J., – 1966 – Etudes pastorales en Adamaoua Camerounais. *Rev. Elev. Méd. vét. Pays trop.* 19: 45–62.

Piot, J., – 1969 – Végétaux ligneux et pâturages de savanes de l'Adamaoua au Cameroun. *Rev. Elev. Méd. vét. Pays trop.* 22: 541–559.

Piot, J., – 1970 – Pâturage aérien au Cameroun. Utilisation des ligneux par les bovins. *Rev. Elev. Méd. vét. Pays trop.* 23: 503–517.

Poissonet, J. & J. César, – 1972 – Structure spécifique de la strate herbacée dans la savane à palmier Ronier de Lamto (Côte d'Ivoire). *Ann. Univ. Abidjan sér. E. (Ecologie)* 5: 577–601.

Poissonet, P. S., P. M. Daget, J. A. Poissonet & G. Long, – 1972 – Rapid point survey by bayonet blade. *J. Range Management* 25: 313.

Poissonet, P. S., J. A. Poissonet, M. P. Godron & G. Long, – 1973 – A comparison of sampling methods in dense herbaceous pasture. *J. Range Management* 26: 65–67.

Pratt, D. J., – 1968 – Rangeland development in Kenya. *Ann. arid Zone* 7: 177–208.

Pratt, D. J., – 1969 – Management of arid rangeland in Kenya. *J. Brit. Grassland Soc.* 24: 151–157.

Pratt, D. J., – 1971 – Bush-control studies in the drier areas of Kenya. 6. Effects of fenuron (3-phenyl-1, 1-dimethylurea). *J. appl. Ecol.* 8: 239–245.

Pratt, D. J., P. J. Greenway & M. D. Gwynne, – 1966 – A classification of East African rangeland. *J. appl. Ecol.* 3: 369–382.

Pratt, D. J. & J. Knight, – 1971 – Bush-control studies in the drier areas of Kenya. 5. Effects of controlled burning and grazing management on *Tarchonanthus/Acacia* thicket. *J. appl. Ecol.* 8: 217–237.

Rains, A. B., – 1970 – The evaluation of African rangeland using aerial photographs. In "Proceedings Eleventh International Grassland Congress," pp. 99–101. University of Queensland Press, St. Lucia.

Rains, A. B. & A. M. Yalala, – 1972 – The Central and Southern State Lands, Botswana. Land Resource Study no. 11. Land Resources Division, Tolworth Tower, Surbiton, Surrey. 118 pp.

Ramia, M., – 1961 – Sabanas llaneras. *El Farol* no. 197 (unpaginated).

Ramia, M., – 1964 – Distribución de Sabanas en Venezuela. *Rev. Venez. Geogr.* 4: 25–34.

Ramia, M., – 1968 – Nuestras sabanas. *El Farol* 24: 8–12.

Ramsay, J. M. & R. Rose-Innes, – 1963 – Some quantitative observations on the effects of fire on the Guinea savanna vegetation of Northern Ghana over a period of eleven years. *African Soils* 8: 41–85.

Rattray, J. M., – 1960 – The Grass Cover of Africa. FAO Agricultural Studies no. 49. FAO, Rome. 168 pp. with map.

Rattray, J. M., – 1962 – Criteria used for the measurement of condition and trend in grazing resources. In "Interdepartmental Conference of Pasture Workers," pp. 44–67. Government of Southern Rhodesia, Bulawayo.

Raven, P. H. & D. I. Axelrod, – 1972 – Plate tectonics and Australasian paleobiogeography. *Science* 176: 1379–1386.

Raven, P. H. & D. I. Axelrod, – 1974 – Angiosperm biogeography and past continental movements. *Ann. Missouri Bot. Gdn.* 61: 539–673.

Rensburg, H. J. van, – 1969 – Management and Utilization of Pastures: East Africa. I. Kenya. II. Tanzania. III. Uganda. Pasture and Fodder Crop Studies no. 3. FAO, Rome. 118 pp.

Renvoize, S. A., – 1971 – Miscellaneous notes on the flora of Aldabra and neighbouring islands. 1. Five new species of grasses. *Kew Bulletin* 25: 417–422.

Risopoulos, S. A., – 1966 – Management and use of grasslands. Democratic Republic of the Congo. Pasture and Fodder Crop Studies no. 1. FAO, Rome. 150 pp.

Rosengurtt, B., – 1944 – Estudios sobre praderas naturales del Uruguay. 4a. Las formaciones campestras y herbaceas del Uruguay. *Agros* no. 134: 1–45.

Rosengurtt, B., J. P. Gallinal, L. Bergalli, L. Aragone & E. F. Campal, – 1939 – Estudios sobre praderas naturales del Uruguay. La variabilidad de la composición de las praderas. *Rev. Asoc. Ing. Agron.* 11(3): 28–33.

Roseveare, G. M., – 1948 – The Grasslands of Latin America. Imperial Bureau of Pastures and Field Crops. Bulletin no. 36, 291 pp. Publisher now: Commonwealth Agricultural Bureaux, Farnham Royal, Slough.

Ross, I. C. & G. N. Harrington, – 1969 – The practical aspects of implementing a controlled burning scheme in the Kedepo Valley National Park (second year of operation). *East Afr. Wildl. J.* 7: 39–42.

Rzedowski, J., – 1975 – An ecological and phytogeographical analysis of the grasslands of Mexico. *Taxon* 24: 67–80.

Santiago, A. A., – 1966 – Studies in autecology of *Imperata cylindrica* (L.) Beauv. In "Proceedings of Ninth International Grassland Congress," pp. 499–502. Departamento da Produção Animal da Secretaria da Agricultura do Estado de São Paulo.

Santiago, A. A., – 1970 – Pecuária de Corte no Brasil Central. Instituto de Zootecnia, Agua Branca. 635 pp.

Schimper, A. F. W., – 1903–1904 – Plant Geography upon a Physiological Basis. Oxford.

Schnell, R., – 1970–1971 – Introduction à la phytogéographie des Pays tropicaux. Vol. 1. Les Flores, les Structures. 500 pp. Vol. 2. Les Milieux, les Groupements végétaux. pp. 501–952. Gauthier-Villars, Paris.

Scott, J. D., – 1970 – Pros and cons of eliminating veld burning. *Proc. Grassland Soc. South Africa* 5: 23–26.

Seavoy, R. E., – 1975 – The origin of tropical grasslands in Kalimantan, Indonesia. *J. trop. Geogr.* 40: 48–52.

Shankar, V., K. A. Shankarnarayan & P. Rai, – 1973 – Primary productivity, energetics and nutrient cycling in *Sehima-Heteropogon* grassland. 1. Seasonal variations in composition, standing crop and net production. *Tropical Ecol.* 14: 238–251.

Shankarnarayan, K. A., – 1963 – Parallelism between grassland and forest types of Bellary District. *Ann. Arid Zone* 1: 132–141.

Shaw, N. H. & M. J. T. Norman, – 1970 – Tropical and sub-tropical woodlands and grasslands. In "Australian Grasslands," ed. R. M. Moore, pp. 111–122. Australian National University Press, Canberra.

Sillans, R., – 1958 – Les savanes de l'Afrique central française. Encyclopédie Biol. 55. Paris. 423 pp.

Singh, J. S. & R. Misra, – 1969 – Diversity, dominance, stability and net production in the grasslands at Varanasi, India. *Canad. J. Bot.* 47: 425–427.

Sluijs, D. H. van der, – 1971 – Native grasslands of the Mesopotamia region of Argentina. *Neth. J. agric. Sci.* 19: 3–22.

Smith, R. L., 1972 – The Ecology of Man: An Ecosystem Approach. Harper and Row, New York, London. 436 pp.

Soderstrom, T. R. & H. F. Decker, – 1973 – *Calderonella*. A new genus of grasses, and its relationships to the Centostecoid genera. *Ann. Missouri Bot. Gdn.* 60: 427–441.

Southern Rhodesia, Government of, – 1962 – Proceedings of the First Interdepartmental Conference of Pasture Workers. Bulawayo. 118 pp.

Specht, R. L., – 1958 – The geographical relationships of the flora of Arnhem Land. In "Records of the American-Australian Scientific Expedition to Arnhem Land. Vol. 3. Botany and Plant Ecology," (eds.) Specht & Mountford, pp. 415–478. Melbourne University Press.

Stark, J., G. K. Rutherford, J. Spector & T. A. Jones, – 1959 – Soil and Land-Use Surveys no. 6. British Guiana. 1. The Rupununi Savannas (contd.). 2. The Intermediate Savannas. 3. General Remarks. Regional Research Centre of the British Caribbean at the Imperial College of Tropical Agriculture, Trinidad. 24 pp.

Steenis, C. G. G. J. van (ed.), – 1948 – Flora Malesiana (1948–onwards). Rijksherbarium, Leyden and Kebun Raya, Bogor, Indonesia.

Stewart, G. & S. S. Hutchings, – 1936 – The point-observation-plot (sq. foot density) method of vegetation survey. *J. Amer. Soc. Agron.* 28: 714–726.

Swartzman, G. L. & J. S. Singh, – 1974 – A dynamic programming approach to optimal grazing strategies using a succession model for a tropical grassland. *J. appl. Ecol.* 11: 537–548.

Talbot, Lee M., H. P. Ledger & W. J. A. Payne, – 1961 – The possibility of using wild animals for animal production in the semi-arid tropics of East Africa. In "Festschrift zum VIII Intern. Tierzuchtkongress in Hamburg, 1961," pp. 205—210. Verlag Eugen Ulmer, Stuttgart.

Tansley, A. G., — 1935 — The use and abuse of vegetational concepts and terms. *Ecology* 16: 284—307.

Taylor, B. W., — 1962 — The status and development of the Nicaraguan pine savannas. *Caribbean Forester* 23: 21—26.

Teunissen, P. A. & J. T. Wildschut — (n.d.) — Vegetation and flora of the savannas in the Brinckheuvel Nature Reserve, Northern Suriname. Verhandelingen der Koninklijke Nederlandse Akademie van Wetenschappen, afd. Natuurkunde, 59, no. 2. 60 pp.

Thomas, A. S., — 1966 — The importance of African aquatic grasses. In "Proceedings of Ninth International Grassland Congress," pp. 1397—1399. Departamento da Produção Animal da Secretaria da Agricultura do Estado de São Paulo.

Thomas, D. B. & D. J. Pratt, — 1967 — Bush-control studies in the drier areas of Kenya. 4. Effects of controlled burning on secondary thicket in upland *Acacia* woodland. *J. appl. Ecol.* 4: 325—335.

Thornton, D. D., M. I. E. Long & B. Marshall, — 1968 — Nutritive value of grasses in Ankole and the Queen Elizabeth National Park, Uganda. 1. Collection sites, soils and grasses. *Trop. Agric. Trin.* 45: 257—267.

Thornton, D. D., M. I. E. Long & B. Marshall, — 1969 — The mineral status of dairy farms in the parts of Buganda and Busoga bordering Lake Victoria, Uganda. 1. General ecology and soils. *Trop. Agric. Trin.* 46: 269—277.

Thornton, D. D. & B. Marshall, — 1971 — The mineral status of dairy farms in east Uganda. 1. Ecology and soils. *Trop. Agric. Trin.* 48: 217—224.

Tothill, J. C., — 1971a — A review of fire in the management of native pasture with particular reference to north-eastern Australia. *Tropical Grasslands* 5: 1—10.

Tothill, J. C., — 1971b — Grazing, burning and fertilizing effects on the regrowth of some woody species in cleared open forest in south-east Queensland. *Tropical Grasslands* 5: 31—34.

Trapnell, C. G. & J. N. Clothier, — 1957 — The soils, vegetation and agricultural systems of north-western Rhodesia. Government Printer, Lusaka. 87 pp.

Trapnell, C. G., J. D. Martin & W. Allan, — 1950 — Vegetation-Soil Map of Northern Rhodesia, with accompanying Memorandum. Government Printer, Lusaka.

Trochain, J. L., — 1957 — Accord interafricain sur la définition des types de végétation de l'Afrique tropicale. *Bull. Inst. Et. Centraf.* (13—14): 55—93.

Troll, C., — 1952 — Das Pflanzenkleid der Tropen in seiner Abhängigkeit von Klima, Boden und Mensch, Tagungsber. wiss. Abhandl. Deutscher Geogr. Tag. Frankfurt, 1951. pp. 35—66. Remagen.

Troll, C., — 1968 — Das Pampaproblem in landschaftsökologischer Sicht. *Erdkunde* 22: 152—155.

United Nations Educational, Scientific and Cultural Organization, — 1966 — Regional Seminar of Tropical Ecology: Primary and Secondary Productivity of Tropical Savannas. Preliminary Report. UNESCO Regional Centre for Science and Technology for Africa, Nairobi. mimeo. 39 pp. + bibliography.

UNESCO, — 1971 — International grazing lands programme. *Nature and Resources* 7: 21.

Valentine, D. H., — 1970 — Evolution at zones of vegetational transition. *Feddes Repert.* 81(1—5): 33—39.

Valenza, J., — 1970 — Survey of different types of natural pasture land in the Senegal Republic. In "Proceedings Eleventh International Grassland Congress," pp. 78—82. University of Queensland Press, St. Lucia.

Verboom, W. C., – 1966 – The grassland communities of Barotseland. *Trop. Agric. Trin.* 43: 107–115.

Verboom, W. C., – 1968 – Grassland successions and associations in Pahang, central Malaya. *Trop. Agric. Trin.* 45: 47–59.

Verboom, W. C. & M. A. Brunt, – 1970 – An Ecological Survey of Western Province, Zambia, with Special Reference to the Fodder Resources. Volume 2. The Grasslands and their Development. Land Resource Study no. 8. Land Resources Division, Directorate of Overseas Surveys, Tolworth, Surrey. 133 pp.

Vesey-Fitzgerald, D. F., – 1970 – The origin and distribution of valley grasslands in East Africa. *J. Ecol.* 58: 51–76.

Vincent, V., – 1962 – Means and criteria for establishing vegetation type, condition and trend. In "Interdepartmental Conference of Pasture Workers," pp. 23–43. Government of Southern Rhodesia, Bulawayo.

Vishnu-Mittre, – 1969 – Some evolutionary aspects of Indian flora. In "J. Sen Memorial Volume," (eds.) H. Santapau, A. K. Ghosh, S. Chanda, S. K. Roy & S. K. Chaudhuri, pp. 385–394. J. Sen Memorial Committee and Botanical Society of Bengal, Calcutta.

Voronov, A. G., – 1970 – The effect of geographical factors on the distribution of vegetation in tropical countries. *Forestry Abst.* 31: 446–452.

Wade, N., – 1974 – Sahelian drought: no victory for Western aid. *Science* 185: 234–237.

Walter, H., – 1967a – Das Pampaproblem in Vergleichend Okologischer Betrachtung und seine Lösung. *Erdkunde* 21: 181–203.

Walter, H., – 1967b – The Pampa problem and its solution. In "Proceedings of Second International Seminar on Integrated Surveys of Natural Grazing Lands," pp. 3–18. ITC-UNESCO Centre, Delft. (now Enschede).

Walter, H., – 1969 – El problema de la sabana: investigaciones ecofisiológicas en el Africa Sur-Occidental en comparación con las condiciones existentes en Venezuela. *Boln. Soc. Venez. cienc. nat.* 28 (115–116): 123–144.

Wang Chi-wu, – 1962 – The development of forest communities in eastern Asia. In "Proceedings IX Pac. Sci. Congr., Bangkok, 1957. Botany," pp. 103–113. Secretariat, Department of Sciences, Bangkok.

Whitford, H. N., – 1906 – The vegetation of the Lamao Forest Reserve. *Philippines J. Sci.* 1: 373.

Whyte, R. O., – 1964 – The Grassland and Fodder Resources of India (2nd ed.). Indian Council of Agricultural Research, Scientific Monograph no. 22. New Delhi. 553 pp.

Whyte, R. O., – 1968a – Grasslands of the Monsoon. Faber and Faber, London/ F. A. Praeger, New York. 325 pp.

Whyte, R. O., – 1968b – Land, Livestock and Human Nutrition in India. Praeger, New York, 309 pp.

Whyte, R. O., – 1971 – Grazing in the land ecosystems of India. *Ann. Arid Zone* 10: 111–119.

Whyte, R. O., – 1972 – The Gramineae, wild and cultivated, of monsoonal and equatorial Asia. 1. Southeast Asia. *Asian Perspectives* 15: 127–151.

Whyte, R. O., – 1973 – Grasses and grasslands. In "UNESCO Review of Research: The Natural Resources of Humid Tropical Asia," pp. 239–262. UNESCO, Paris.

Whyte, R. O., – 1974 – Tropical Grazing Lands: Communities and Constituent Species. Junk, The Hague. 222 pp.

Whyte, R. O., – 1975a – The nature and utilization of the grazing resources. In "Nomads and Pastoralists in South Asia," (eds.) L. S. Leshnik & G. D. Sontheimer. Südasien-Institut, University of Heidelberg.

Whyte, R. O., – 1975b – The geography of abnormal meiosis in plants. *The Nucleus* 18: 183–203.

Whyte, R. O., – 1976a – Land and Land Appraisal. Junk, The Hague. 370 pp.

Whyte, R. O., – 1976b – Bioclimatic and taxonomic consequences of tectonic movement and orogeny. *Annals of Arid Zone* 15: (in press).

Wiens, J. A., – 1972 – Ecosystem Structure and Function. Proceedings 31st Annual Biology Colloquium, Corvallis. Oregon State University Press. Corvallis. 176 pp.

Wild, H., – 1965 – The flora of the Great Dyke of Southern Rhodesia with special reference to the serpentine soils. *Kirkia* 5: 49–86.

Wild, H., – 1968–69a – Geobotanical anomalies in Rhodesia. 1. The vegetation of copper bearing soils. *Kirkia* 7: 1–71.

Wild, H., – 1968–69b – Geobotanical anomalies in Rhodesia. 2. A geobotanical anomaly occurring in graphitic soils. *Port. Acta biol.* (b) 9: 291–299.

Wild, H., – 1970 – Geobotanical anomalies in Rhodesia. 3. The vegetation of nickel-bearing soils. *Kirkia* 7: suppl. 1–62.

Wright, R. L., – 1964 – Land system survey of the Nagarparkar Peninsula, Pakistan. *Arid Zone* (UNESCO), no. 25: 5–13.

Zambia, Republic of – 1970 – Annual Report of the Research Branch. Ministry of Rural Development. 140 pp.

Zimmerman, E. C., – 1963 – Nature of land biota. In "Man's Place in the Island Ecosystem," (ed.) F. R. Fosberg, pp. 57–63. Bishop Museum Press, Honolulu.

Zonneveld, I. S., – 1972 – Land evaluation and land(scape) science. In "ITC Textbook of Photo-Interpretation. 7. Use of Aerial Photographs in Geography and Geomorphology," chapter 7:4, 106 pp. ITC, Enschede.

Zonneveld, I. S., P. N. de Leeuw & W. G. Sombroek, – 1971 – An ecological interpretation of aerial photographs in a savanna region in northern Nigeria. ITC, Enschede. Series B, no. 63. 41 pp.

2   APPLICABILITY OF ZÜRICH-MONTPELLIER METHODS IN
AFRICAN TROPICAL AND SUBTROPICAL RANGE LANDS

M. J. A. WERGER

In extensive areas of Africa with natural and semi-natural pastures, varying from open woodlands and savannas to grasslands, grass steppes and dwarf shrub steppes, the main source of subsistence is domestic livestock husbandry (Rattray, 1960). Although this type of husbandry occupies such an important place in the economy of most African countries, little progress has been made in establishing scientifically based management programmes or even in understanding the ecology of the areas. Ecological knowledge is a necessary precondition to the formulation of a management programme which should be based on the recognition of the various biogeocoenoses or ecosystems in the area as structural parts or functional entities of the landscape respectively. It is necessary to understand their general and peculiar features as complex natural entities of organisms and habitat mutually influencing one another and, although being an open system, maintaining an equilibrium (Tansley, 1935; Sukachev, 1944 as translated by Raney & Daubenmire, 1958; Major, 1969; Ellenberg, 1973; Whittaker, 1975). Recognition of these complex natural entities should provide mappable units which are ecologically meaningful and, therefore, also suitable for management purposes. As plants are excellent indicators of habitat factors and vegetation forms an integral and characteristic part of an ecosystem, the area occupied by an ecosystem can be delimited accurately by its plant community (Braun-Blanquet, 1951; Major, 1969; Edwards, 1972; Küchler, 1973; Mueller-Dombois & Ellenberg, 1974). Thus a vegetation map provides an invaluable basis for a management programme. Once the fundamental units are delineated, it is useful to study experimentally the relationships between plant species and environmental factors including animals and mutually between plant species. It is of particular importance to include this latter aspect in experimental studies since plants react very differently in pure cultures instead of in co-occurrence with other species (Ellenberg, 1952; Knapp, 1967; Holzner, 1973). Also, evaluation of the various constituent species of the plant community as fodder plants and their palatability are appropriate objects of study in a later stage of ecosystem research and so are many other functional relationships within the complexity of the ecosystem. In order to carry out these studies in a meaningful way, however, it is necessary to delineate the ecosystems or biogeocoenoses in an area first, and, therefore, to first undertake a vegetation survey that results in a meaningful classification of the plant communities (Rattray, 1960; Knapp, 1965; Bommer, 1965; Major, 1969; Greenway & Vesey-Fitzgerald, 1969; Edwards, 1972; Küchler, 1973).

A classification of plant communities can be based on three types of criteria: physiographic, physiognomic or floristic characters (cf. Gounot, 1969). Of these three types of criteria only the latter two deal directly with plants. Since it is logical to base a classification of plant communities on characters of the plants themselves, physiographic characters are not entirely suitable.

Physiognomic characters, which for the sake of this paper are meant to include structural-functional characters, are attributes of plants, but are not as precisely definable as floristic characters. Furthermore, in vegetation types where fire and man play an important role in determining its structural composition, as is the case in most range land vegetation in Africa, the physiognomic appearance at a particular site is often dependent upon the time interval since fire last passed through the vegetation or on the need of fuel by man in that area (cf. for example Leippert, 1968). Physiognomically based classifications are consequently often rather crude approximations and are inconsistent over a relatively short time span. In some instances, however, such as the structural-functionally based classification of North American forest vegetation by Knight & Loucks (1969), apparently good results were obtained although the effort expended on recording the characters can be prohibitive for primary surveys.

The use of floristic criteria as a basis for classification of plant communities thus has strong advantages. As stated above, taxa are good indicators of habitat factors although the occurrence of a plant species in a community need not necessarily indicate its optimal physiological requirements. Taxa with narrow ecological amplitudes in the area under study indicate more specific ecological factors, whereas taxa with wide amplitudes in the area indicate more general ecological factors operative in the area (Werger, 1973a; Coetzee & Werger, 1975). Thus a classification based on total floristic composition can be expected to be the most significant ecologically (Braun-Blanquet, 1951; Ellenberg, 1956; Knapp, 1965; Major, 1969). The units of such a classification system should, however, not only be floristically defined but also physiognomically and physiographically meaningful (Werger, 1973a, 1974b; cf. Von Glahn, 1968; Segal, 1969; Westhoff & Van der Maarel, 1973). In order to arrive at such a classification, various sampling and analysis or synthesis strategies may be followed (cf. Williams, 1971; Lambert, 1972; Coetzee & Werger, 1975). To allow consideration of total floristic composition, plots of sufficient size to provide a representative description of the vegetation in any one sampled stand must be laid out through the area. Sampling can be accomplished by systematic, random, stratified random or representative distribution of plots. Systematic and random sampling have the disadvantage of being inefficient in that certain vegetation types may be undersampled while others are oversampled. Stratified random sampling prevents this shortcoming but both strategies do not prevent the sampling of heterogeneous plots. In a representative sampling strategy, such as often used by the followers of the Zürich-Montpellier School, plots are selected to be more or less homogeneous in floristical, structural and environmental characters and to be representative, in those characters, of the vegetation to be sampled. As many plots are sampled as are required to cover properly all variation in the vegetation of the study area. Apart from noting all species present, an estimate of importance of each species is usually made. This procedure is often severely critized by statistically or numerically orientated ecologists (cf. for example, Lambert & Dale, 1964; Lambert, 1972), who fear that in this way *a priori* assumptions about the nature of the vegetation are made. The only prior assumption is, however, that vegetation consists of natural entities, which generally contact one another along narrow boundaries, and which can be abstracted to phytosociological units by comparing representative samples of such entities (Whittaker, 1962; Westhoff & Van der Maarel, 1973; Werger 1973a, 1974a). Besides, as has been pointed out by Ivimey-Cook & Proctor (1966), it will be difficult to collect any substantial body of phytosociological data to support a conclusion seriously at variance with the facts. Although stratified random sampling and other sampling strategies may also lead to useful results, it is accepted by many plant ecologists that a representative sampling strategy is the most effective to arrive at scientifically sound conclusions (for example Dahl, 1957; Daubenmire, 1968; Moore *et al.*, 1970; Werger, 1973a; 1974b; Westhoff & Van der Maarel, 1973; Mueller-Dombois & Ellenberg, 1974).

The various procedures to be followed for analysis or synthesis of the data have been discussed extensively in the literature, and have often proved rather controversial (cf. Whittaker, 1973). Again, statistically and numerically orientated ecologists often objected to the procedures of what they called the "traditionalists", and in particular ecologists of the Zürich-Montpellier School. However, comparative studies on the same data using both "traditional" and computer-based phytosociological approaches simultaneously often produced similar or compatible results (e.g. Van Groenewoud, 1965; Ivimey-Cook & Proctor, 1966; Kershaw, 1968 a, b; Moore *et al*, 1970; Kartawinata & Mueller-Dombois, 1972; Bouxin, 1973; Werger, 1973b; Stanek, 1973; Coetzee & Werger, 1973; Coetzee, 1974a). This also indicated the objectivity of the "traditional" phytosociological approach, and gradually made the approach more acceptable and appreciable to opposing ecologists (compare Dale & Anderson, 1972; Lambert, 1972). It was realised by the "traditional" ecologists that it was necessary to reconstruct their procedures and define them exactly before their approach would be entirely acceptable. Such efforts were started during the sixties and at present are nearing completion so that the objections against the "traditional" approach, as recently formulated by Lambert (1972), are nearly overcome. Several multivariate vegetation analyses of tropical or subtropical African vegetation, (e.g. Grunow, 1967; Grunow & Lance, 1969) have been successful in arriving at a meaningful classification for management purposes but often these successes have only been achieved by very intensive sampling. In general these multivariate approaches are less efficient than "traditional" approaches, particularly that of the Zürich-Montpellier School, both in effort expended and results obtained (Moore *et al.*, 1970; Coetzee, 1974a). This is especially true for reconnaissance and semidetailed surveys, mapping at scales between 1:500 000 and 1:100 000, and between 1:100 000 and 1:10 000 respectively. These scales are of major importance in delineating, classifying and studying community and habitat relations and in providing the main classificatory reference framework of plant communities from which predictions and extrapolations can be made as a basis for management programmes (Edwards, 1972; Young, 1973). Other drawbacks in using the multivariate approaches are that they only produce *ad hoc* classifications which hinder the comparison of geographically separated but chorologically related areas. They also require extensive computer facilities and highly-trained scientists, neither of which are always readily available. These difficulties are not encountered in using the Zürich-Montpellier approach, which takes total

128

floristical composition into account and aims at a taxonomy of ecologically meaningful vegetation units by emphasizing differentiating species at all levels of the classification. It thus achieves "a coherence and consistency denied other systems" of vegetation classification (Whittaker, 1962: 135). The differentiating species need not be dominants. They are species with narrow ecological amplitudes in the area studied and are, therefore, the best indicators of the complex of ecological factors that determine the pattern in the distribution of the species and consequently the delimination of the communities. In the badly overgrazed H e r m a n n i o   c o c c o c a r p a e — N e s t l e r e t u m   c o n - f e r t a e , a steppe community of dwarf shrubs and grasses occurring in the plains of the Karoo region in South Africa, Werger (1973a) found that the character species were all inconspicuous and never dominant and that most of them were not even constant. These character species were, nevertheless, together with three groups of differential species typifying three subassociations, completely adequate for delineation of the community and its subassociations in the field into three meaningful ecological entities, each with their specific degree of erosion as a result of prolonged overgrazing, and their other specific habitat characteristics.

It has sometimes been stated that application of Zürich-Montpellier methods would be impossible in floristically very rich vegetation, as often encountered in the tropics (e.g. Van Donselaar, 1965; Westhoff, 1967; Vareschi, 1972). Although in such floristically rich areas the extensive floristical knowledge required to carry out a phytosociological survey taking total floristic composition into account should be recognized as imposing a special problem for the investigator, the species number as such does not prevent a successful application of Zürich-Montpellier methods. This was clearly shown by Mullenders (1954) in a study of Central African vegetation and by Werger et al. (1972) in a study of Cape Fynbos vegetation, which is according to Mullenders (1954) floristically as rich as tropical forest vegetation.

Although the species richness of an area can always be coped with, the extent of the regions to be surveyed often make some practical adaptations of the sampling method desirable. Because time does not always allow a researcher to return to a sampling point in various seasons, sometimes only permanently recognizable species are taken into account, omitting most geophytes and therophytes from the data. This does not impose a serious drawback, since, firstly, the species omitted are few and the permanently recognizable species are usually so numerous that they allow a clear characterization of the communities on the basis of differentiating species; secondly, annuals which are usually not included in the group of permanently recognizable species, are often reported not to have such restricted ecological amplitudes as perennials, so that the former are usually not important as differential species (cf. Werger, 1974b). Obviously, when plots are only sampled once it is of utmost importance that sampling is carried out at an appropriate time, which is normally late in the growing season, when most species are mature. This is, however, not an issue specific to sampling tropical vegetation but one that applies to sampling vegetation everywhere in the world.

As species richness also influences the size of plots to be sampled in order to be representative of the phytocoenoses, suitable plot sizes are often larger in floristically rich areas than in most temperate areas. In the savanna and woodland vegetation of Africa a larger plot size is often required to adequately represent the vegetational structure than the plot size necessary to represent the floristic composition of the phytocoenosis, and it will sometimes be necessary to sample plots of 10 x 20 m or even larger in size in order to represent the structure of the vegetation. Plots of this size are, nevertheless, perfectly manageable for a researcher. In most tropical grasslands plots of about 5 x 5 m till 5 x 10 m are sufficiently large to sample both floristic composition and structure representatively.

In various parts of Africa, Zürich-Montpellier methods have been successfully applied to classify the vegetation of range lands, first in Central Africa (e.g. Lebrun, 1947; Duvigneaud, 1953; Mullenders, 1954; Devred, 1956; Germain, 1965; Schmitz, 1963) and West Africa (e.g. Schnell, 1952a, b, c, 1961; Adjanohoun, 1962; Kershaw, 1968a; Schmidt, 1973) but eventually also in East Africa (e.g. Leippert, 1968; Schmidt, 1974, 1975) and southern Africa (e.g. Myre, 1962, 1964, 1971, 1972; Volk & Leippert, 1971; Monteiro, 1970; Leistner & Werger, 1973; Werger, 1973a; Coetzee 1974b, 1975; cf. Werger & Edwards, 1976). Lebrun (1947) and Leistner & Werger (1973) studied the vegetation of national parks in Zaire and South Africa respectively. As the vegetation of these nature reserves was well developed without showing a high degree of degradation over extensive areas as a result of overgrazing and trampling the communities could be delimited clearly by differentiating species and each community was defined by specific habitat characteristics. Part of the results of Leistner & Werger (1973) are shown in Table 1, which repre-

TABLE 1: Part of a phytosociological table of grass communities on Kalahari sands. (Details in Leistner & Werger, 1973)

Relevé number		1	2	3	4	5	6	7	8	9	10	11	12	13	14	15	16	17	18	19	20	21	22	23	24	25	26	27	28	29	30	31	32	33	34	35	36
Total cover (%)		20	25	9	7	12	10	8	7	9	7	8	10	18	20	40	14	7	10	11	15	12	7	5	6	20	18	20	30	45	20	7	10	6	10	15	30
Total number of species		15	22	22	19	27	22	20	31	22	31	16	22	28	22	28	31	38	28	44	29	21	32	37	32	36	35	26	21	17	25	25	19	22	23	36
differential species of Stipagrostietum amabilis																																					
Stipagrostis amabilis	N	2	1	2	1	2	2	2	2	2	2	2	2	2	2																						
Eragrostis trichophora	H	+	+				+	+	+	+	+	1	+	r	2				+																		
Crotalaria spartioides	N	+	1		+		+	+	1	+	+	+	+		+																						
differential species of Hirpicio echini-Asthenatheretum																																					
Crotalaria sphaerocarpa	T					+											r		r	+	+	+	+	+	+	+	+	+		1							
Hirpicium echinus	GH						+									+	+	+	+	+	+	+	+	+	+	+	+	+						+			
Dicoma schinzii	GH															+	+	r	+	+	+	+	+			+	+							+			
Cassia italica	GH															+		r	r	+	+		+									+					
Merremia verecunda	T																+		+						+	+	+										
Portulaca kermesina	T																+		+	+																	
Neuradopsis austro-africana	GH																		+	+			+		+						+						
Dimorphotheca polyptera	T																+		+				+		+												
Cyamopsis serrata	T																					+				+	+			+							
Melhania burchellii	GH															+							+					+					+				+
species of the red sand																																					
Asthenatherum glaucum	H	1	1	+	1	1	1	+	+	+	+	+	+	+	+	1	1	2	2	2	1	2	+	+	1	2	+	+								+	+
Oxygonum delagoense	T	2	+	+	+	+	+	1	1	+	1	+	+	+	+	+	+	+	+	+	2	+	+	+	2	+	+	+	+							+	+

TABLE 1 Continued).

Relevé number		1	2	3	4	5	6	7	8	9	10	11	12	13	14	15	16	17	18	19	20	21	22	23	24	25	26	27	28	29	30	31	32	33	34	35	36
Total cover (%)		20	25	9	7	12	10	8	7	9	7	8	10	18	20	40	14	7	10	11	15	12	7	5	6	20	18	30	45	20	7	10	6	10	15	23	36
Total number of species		15	22	22	19	27	22	20	31	22	16	22	28	22	28	31	38	28	44	29	21	32	32	37	32	36	35	26	21	17	25	19	22	23	36
Limeum arenicolum	T	2	+	+	1	+	1	+	+	+	1	+	+	+	+	+	+	+	+	+	+	+	+	+			+	+	+								
Stipagrostis uniplumis	H	+	+	+	+	+	+	+	+	+	+	+	+	+	+	1	1	1	1	1	1	1	1	1		+	+	+					+				
Hermannia tomentosa	GH	+	+	+	+	+	+	+	+	+	+	+	+	+	+	+	+	+	+	+	+	+	+	+		+	+	+						+			
Plinthus sericeus	Ch	+	+	+	+			+	+	+	+	+	+	+	+		+		+	+	+	+		+													
Sericorema remotiflora	GH	+	+	+		+	+		+	+	+	+		+	+	+	+	+	+	+	+	+	+	+		+											
Aristida meridionalis	H	+	+	+						+		+	1	+	+	+	+		r	+	+	+	+	+	+												
Chascanum pumilum	GH	+	+	+	+	+	+	+	+	+	+	+	+	+	+	+	+	+	+	+	+	+	+	+	+	+	+	+			+						
Citrullus lanatus	T	+	+	+	+	+	+	+	+	+	+	+	+	+	+	+	+	+	+	+	+	+	+	+	+	+	1					+					
Indigofera flavicans	GH	+	+	+	+	+	+	+	+	+	+	+	+	+	+	+	+	+	+	+	+	+	+	+	+	+	+	+			+			+			
Acanthosicyos naudinianus	GH							+		+		+	+	+	+	2	+	+	1	+	+	1	1	1	1			+	+								
Sesamum triphyllum	T	+	+	+	+									+	+	+	+		+	+	+	+	+	+		+	+	+									
Merremia tridentata	T	+	+	+				+	+	+	+	+	+	+	+	+	+	+	+	+	+	+	+	+		+	+						+				
Indigofera daleoides	GH	+				+	+	+	+	+	+	+				+		+	+	+	+	+	+	+	+	+	+	+									
Hermannia burchellii	N				+	+	+	+	+	+		+	+				+	+	+	+	+	+	+	+	+												
Cleome kalachariensis	T	+														+	+	+	+							+	+				+						
Ipomoea hackeliana	T												+			+	+	+	+	+			+			+	+										
Limeum vicosum	T	+		+												+	+							+		+	+										
Phyllanthus omahekensis	T								+	+			+	+	+	+							+										+				
differential species of Monechma incanum- Indigofera alternans community																																					
Monechma incanum	Ch			+												2										2	+	+	+	+		1	1	1	+	+	1

Species	Life form
Rhigozum trichotomum	N
Stipagrostis obtusa	H
Indigofera alternans	GH
Dicoma capensis	GH
Aptosimum albomargi-natum	Ch
Chrysocoma polygalifolia	Ch
Triraphis fleckii	T

Species of the red and pink sand

Species	Life form
Brachiaria glomerata	T
Eragrostis lehmanniana	H
Requienia sphaerosperma	GH
Fimbristylis hispidula	T
Limeum fenestratum	T
Limeum sulcatum	T
Heliotropium ciliatum	GH
Acacia haematoxylon	MN
Jatropha erythropoda	GH
Cynanchum orangeanum	GH
etc.	

Fig. 1. Klein Skittery Pan in the southern Kalahari. Dune crests, on which the Stipa-grostietum amabilis occurs, are seen in the foreground and on the horizon. Principal species are *Acacia haematoxylon, Stipagrostis amabilis, Crotalaria spartioides, Era-grostis trichophora, Asthenatherum glaucum* and, creeping over the sand, *Limeum arenicolum.* In the centre a dune valley with the Hirpicio echini – Asthenatheretum is visible. Across the pan, below the dune, the Monechma incanum – Indigofera alternans Community occurs.

sents three communities occurring on the stabilized dune sand of the southern Kalahari, covering about 75 per cent of the area. They are open communities (Fig. 1), dominated by bunch-grasses, mixed with dwarf shrubs, shrubs, low trees, annual herbs and perennials, which are of a life-form in between geophytes and hemicryptophytes and which are typical of sandy, arid habitas. The three communities, the S t i p a g r o s t i e t u m   a m a b i l i s, the H i r p i c i o   e c h i n i — A s t h e n a t h e r e t u m and the *Monechma incanum – Indigofera alterans* C o m m u n i t y, are each characterized clearly by a number of differentiating species, which include character species. At the same time the table shows that the first two communities have a considerable number of species in common, which are lacking in the other community. This means that the S t i p a g r o s t i e t u m   a m a b i l i s and the H i r p i c i o   e c h i n i — A s t h e n a t h e r e t u m are floristically more related to one another than either is to the third community. It also indicates that the habitats of the first two communities are more similar to one another than to the habitat of the third. There are also, however, a fair number of species which occur in all three communities. The floristically delimited communities are restricted to certain habitats and therefore ecologically significant entities. The S t i p a g r o s t i e t u m   a m a b i l i s occurs on the dune crests and the slightly undulating loose sandy plains of the southern Kalahari; the H i r p i c i o   e c h i n i — A s t h e n a t h e r e t u m is found in the dune valleys and on the lower slopes of dunes, whereas the *Monechma incanum – Indigofera alternans* C o m m u n i t y is encountered on more compact calcareous sand in deep dune valleys with an internal drainage and where a thick calcrete sheet is close to the surface. The soils on which these three communities occur differ markedly in total percentage of sand, in sorting, in pH and in quantity of water-soluble Ca-ions. Soils of the dune crests are best sorted (compare Leistner & Werger, 1973). Apart from the overall composition of the communities and their floristic interrelationships, as well as the average percentage cover for each community, the phytosociological table (Table 1) shows the variation in species composition and cover within each community. The table thus shows the synthetic characters of the community without obscuring the individual characters of each stand. It shows, for example, that wide differences in cover occur more often in the *Monechma incanum – Indigofera alternans* C o m m u n i t y than in the other two communities, and that the high total cover value in relevé 15 should be regarded as more unusual than that in relevé 29. At the same time the table shows at a glance which species are the most important in each community and how variable their importance is in different stands of the community. The grass *Stipagrostis amabilis* is, for example, in all stands of the S t i p a g r o s t i e t u m a m a b i l i s the most important species, while in the H i r p i c i o e c h i n i — A s t h e - n a t h e r e t u m the most important species is often, but not always, the grass *Asthenatherum glaucum*. In the *Monechma incanum — Indigofera alternans* C o m m u n i t y the dwarf shrub *Monechma incanum*, the forb *Indigofera alternans*, the dwarf shrub *Aptosimum albomarginatum* or the low shrub *Rhigozum tri-*

Fig. 2. General view of Bankenveld showing *Acacia caffra—Combretum molle* savanna communities on dolomite in the foreground and grassland and *Protea caffra* savanna communities on shale in the centre and background. The grasslands on shale in the centre are interrupted by two diabase dykes carrying an *Acacia caffra* savanna community.

(Photo: B.J. Coetzee)

TABLE 2. Part of a summarized phytosociological table of the grassland and savanna communities of the South African Bankenveld. (for community names and details see Coetzee, 1974b)

Community	1	2	3	4	5	6a	6b	6c	7a	7b	8	9	10	11	12	13	14	15
number of relevés	6	7	7	19	16	8	12	5	12	9	4	3	10	3	12	10	10	23
no. of species entered in table (Coetzee, 1974b)	27	48	61	81	84	59	73	47	57	52	42	47	69	47	49	47	56	66
geology	dolomite					chert									shale			
physiogomy (s = savanna; g = grassland)	s	s	g	g	g	g	g	g	g	g	s	s	s	s	s	s	s	g
Chrysopogon montanus	V																	
Enneapogon scoparius	V	II																
Kalanchoë paniculata	IV	I			I													
Setaria lindenbergiana	III	I																
Loudetia fluvida	I	III			I													
Trichoneura grandiglumis	I	I	III	III	II							II						
Triraphis andropogonoides			III	III	I													
Euphorbia cf. pseudotuberosa		I	III	III	I													
Chaetacanthus costatus			V	I	I													
Scabiosa columbaria			III	II	I													
Indigofera malacostachys			III	I	II													
Hypoxis rigidula			I	II	I													
Kalanchoë thyrsiflora			I	I	I								I					
Aster muricatus			V	IV		I			I	II				I		I		I
Callilepis leptophylla			III	II														
Fingerhuthia sesleriaeformis	I		V	I														
Euphorbia rhombifolia			III															
Sutera burkeana			III	I														
Euphorbia inaequilatera			III															
Eustachys mutica	IV	V	IV	II	IV													
Acacia caffra	IV	IV																
Anthephora pubescens	V	III			I													

137

Community	1	2	3	4	5	6a	6b	6c	7a	7b	8	9	10	11	12	13	14	15
number of relevés	6	7	7	19	16	8	12	5	12	9	4	3	10	3	12	10	10	23
no. of species entered in table (Coetzee, 1974b)	27	48	61	81	84	59	73	47	57	52	42	47	69	47	49	47	56	66
geology	dolomite					chert									shale			
physiognomy (s = savanna; g = grassland)	s	s	g	g	g	g	g	g	g	g	s	s	s	s	s	s	s	g
Microchloa caffra	IV	III	I	I	III													
Phyllanthus parvulus	III	IV	II	I	II													
Kohautia lasiocarpa	I	II	III	III	III													
Ipomoea obscura	II	I	III	I														
Sporobolus stapfianus	III	I		I														
Mariscus capensis	II	I	I	I	I													
Vellozia viscosa	I	II	I															
Monocymbium ceresiiforme						V	V	II	III	II	II	IV	I		I			
Leucas neuflizeana						IV	II	III	III	IV	I	IV	IV					I
Digitaria brazzae						III	II	V	IV	IV	II	I						
Dianthus mooiensis						III	IV		III	II			II					
Pygmaeothamnus zeyheri							II	III			II	II	I	II				
Crassula transvaalensis						II	III	III	III	II	IV	I						
Hemizygia pretoriae						II	I	III			II							
Sporobolus eylesii						I	III		II				I	II				
Babiana hypogea						IV	I				I	I						
Hermannia lancifolia						III	III											
Asclepias stellifera						II	I	I	I	I			I					
Oxygonum dregeanum						II	I	I	I	III								
Helichrysum galpinii								III			I	I						
Psammotropha mucronata						I	I	I	I	I	II	I						
Helichrysum setosum						I	I	I	I	I	V	I						
Bulbostylis schoenoides										II	I	II	V		IV	III	IV	I
Indigofera pretoriana										I			II	II	II	II	III	II

138

Species									
Vernonia natalensis				III III II	I			I I I II	I
Polygala amatymbica			III I	III I II	I			I I I	I
Digitaria diagonalis				I V				I V	I I
Lasiosiphon burchellii				I		I		II I	I
Graderia subintegra				II I	I			II	I
Senecio erubescens				III	II				III
Polygala sp.				II				I	II
Phyllanthus incurvis	I			I	I I			I I	III
Cyperus obtusiflorus	II			II	II			I I	II
Cassia biensis	I	II	II II	III II II II	I II IV V	I II II IV		IV V IV V	V III II II
Bulbostylis burchellii	II	I IV	I IV	V III II II	V III V V	II V V II		II II III V	III III II II
Crabbea angustifolia	II	IV V III	IV V III	II IV II II	III I III II	II IV III IV		III II III II	II II II I
Cyanotis speciosa		I	I	III IV III II	V IV V IV	V V V V		IV IV III II	II II I
Tristachya rehmannii		II V	II V	V IV IV II	V V V V	II V IV V		V IV IV V	II II I
Sporobolus pectinatus		V II IV	V II IV	V IV III IV	V V V V	II V V V		V III III IV	V III III V
etc.									
Schizachyrium sanguineum	I	V	V	V IV IV IV	II V II V	II IV III V		IV V IV IV	V III II II
Tephrosia longipes	I	I	I	III V II II	II I I III	II I I I		II II V V	III III II I
Loudetia simplex		II	II	IV V V II	IV V V II	IV IV II V		III II II III	III II II II
Sphenostylis angustifolia	I	I III	I III	IV IV III II	II III I IV	II III I IV		IV IV I II	II I I I
Parinari capensis	I	II II	II II	IV IV III II	IV IV V V	IV IV III V		IV V V II	I I
etc.									
Rhynchelytrum setifolium	II III	II I	II I	III V V IV	IV V V V	IV V IV IV		II II IV V	V V V V
Bracharia serrata	II	V V	V V	V V V IV	V V V V	V V V V		V V V V	V V V V
Diheteropogon amplectens	III	V V	V V	V V IV V	V IV II V	V IV II V		V V V V	V V V V
Eragrostis racemosa	II II	I IV	I IV	V V V IV	V V V V	V V V V		V V V V	V V V V
Trachypogon spicatus	I	IV V	IV V	III IV V V	III V III IV	V V IV V		III IV V V	III IV III V
Senecio venosus	I	II IV	II IV	III IV III IV	III III V V	V IV IV III IV		III IV III IV	III IV IV V
Elyonurus argenteus	V	V IV	V IV	IV IV IV V	IV V V V	V V V V		IV IV IV V	IV IV V V
etc.									

chotomum can each be the most important in various stands of the community. It is important that cover and species composition are shown, since they should both be considered before arriving at yield and grazing potentials for a community (cf. Myre, 1964, 1971; Herbel *et al.*, 1972). Furthermore, exceptionally low cover values and floristic poorness of a stand can show its status and degree of degeneration through overgrazing or other forms of mismanagement. Finally, the table shows that not all stands can be classified unequivocally into a particular community, but that transitions do occur. Relevé 27 and particularly relevés 28 and 34 are examples of such transitional stands, which can, however, still be classified as being more typical for one of the communities than for the other. Such transitional stands usually occur on sites of which the habitat has a mixture of features characteristic of the habitat of each of the communities.

The findings of Coetzee (1974b), who used a stratified random sampling strategy and a constant plot size in a study of part of the floristically rich Bankenveld of South Africa, are also instructive. Bankenveld occurs on the transition from tropical and subtropical woodland and savanna to subtropical grassland (Fig. 2). Table 2 is a summarized phytosociological table of part of Coetzee's (1974b) results representing the savanna and grassland communities. The floristic differences between the communities as well as their similarities are clearly shown. Most communities are characterized by species which, within the study area, are virtually restricted to them. Others are characterized by differentiating species combinations. Certain species occur in a small number of related communities, while others occur in most or all communities. This demonstrates a hierarchy of relationships between the vegetation types and allows for various levels of generalization in mapping the results. For example, communities 3, 4 and 5 can be mapped separately on large scale maps, but if that amount of detail is unnecessary for management purposes, it is also possible to lump them together and map them as one community on the basis of the species restricted to the three communities. The table also shows the constancy of each species within each community and, if summarized cover-abundance values had been entered, would also in this respect have shown the importance of each species in each community. Similarly, when field observations established that relevés summarized in a particular community always represent overgrazed stands, whereas closely related but less overgrazed stands are summarized in another community, the table would clearly show that the first community is just a degenerated version of the second. The table would show which floristic differences correspond to a particular degree of overgrazing and, thus, to the degeneration status of the community. Several such examples may be found in Schmidt (1975) and Werger (1973a). The communities in table 2 are all distinctly characterized physiognomically as well as in habitat characters, which improves their mapping feasibility. Geological substrate largely accounts for the main divisions in the table, with soil factors, rockiness, slope angle and exposure being correlated with the finer divisions of the table. Because of these unequivocal characterizations of the communities in terms of floristics, physiognomy and habitat at various levels of generalization, Coetzee concluded that they represent ecologically homogeneous units which should serve excellently as a basis for management purposes. Similar examples could be taken from the works cited above as well as from many other studies.

It may be concluded that in the Zürich-Montpellier approach, the ecological amplitudes of the plant species, as indicators of environmental factors, are used as criteria for the delimitation of plant communities. Communities are consequently ecologically characterized and are excellent indicators of the biogeocoenoses in an area. Phytosociological studies result not only in the delimitation of the plant communities and an expression of their floristical and thus of their ecological interrelationships but also indicate which plant species are the best indicators of specific ecological factors and should therefore be selected for experimental research. They also indicate which species contribute most to the biomass of each community and should, as a result, be emphasized in further studies on yield, nutrient value and palatability. The proper recognition of plant communities should, thus, not only form the basis of a general management programme, but also the basis for the planning of experimental ecological work and yield and related studies, in range lands in tropical and subtropical Africa as much as in the temperate regions, where the Zürich-Montpellier approach has, in this respect, proved its usefulness.

References

Adjanohoun, E. J., – 1962 – Etude phytosociologique des savanes de Basse Côte d'Ivoire (savanes lagunaires). *Vegetatio* 11: 1–38.

Bommer, D., – 1965 – Möglichkeiten der Weideverbesserung. In: Knapp, R. (ed.): Weide-Wirtschaft in Trockengebieten. pp. 115–126. Stuttgart, Fischer.

Bouxin, G., – 1973 – Étude quantitative de quelques aspects de la végétation du Rwanda. Dr. thesis Univ. Liège. (unpubl.).

Braun-Blanquet, J., – 1951 – Pflanzensoziologie. Wien, Springer, 2. Aufl.

Coetzee, B. J., – 1974a – Improvement of association-analysis classification by Braun-Blanquet technique. *Bothalia* 11: 365–367.

Coetzee, B. J., – 1974b – A phytosociological classification of the vegetation of the Jack Scott Nature Reserve. *Bothalia* 11: 329–347.

Coetzee, B. J., – 1975 – A phytosociological classification of the Rustenburg Nature Reserve. *Bothalia* 11: 561–580.

Coetzee, B. J. & M. J. A. Werger, – 1973 – On hierarchical syndrome analysis and the Zürich-Montpellier table method. *Bothalia* 11: 159–164.

Coetzee, B. J. & M. J. A. Werger, – 1975 – On association-analysis and the classification of plant communities. *Vegetatio* 30: 201–206.

Dale, M. B. & D. J. Anderson, – 1972 – Qualitative and quantitative information analysis. *J. Ecol.* 60: 639–653.

Dahl, E., – 1957 – Mountain vegetation in south Norway and its relation to the environment. *Skr. Norske Vid. – Akad. i Oslo. I. Mat. – Naturv. Kl.* 1956 (3): 1–374.

Daubenmire, R., – 1968 – Plant communities. A textbook of plant synecology. New York, Harper & Row.

Devred, R., – 1956 – Les savanes herbeuses de la région de Mvuazi (Bas-Congo). *Publ. I.N.E.A.C. Sér. Scient.* 65: 1–115.

Duvigneaud, P., – 1953 – Les savanes du Bas-Congo. *Lejeunia* 10: 1–192.

Edwards, D., – 1972 – Botanical survey and agriculture. *Proc. Grassld. Soc. Sth. Afr.* 7: 15–19.

Ellenberg, H., – 1952 – Physiologisches und ökologisches Verhalten derselben Pflanzenarten. *Ber. Dtsch. Bot. Ges.* 65: 350–361.

Ellenberg, H., – 1956 – Aufgaben und Methoden der Vegetationskunde. *In*: Walter, H.: *Einführung in die Phytologie.* IV. 1. Stuttgart, Ulmer.

Ellenberg, H., – 1973 – Ziele und Stand der Ökosystemforschung. *In*: Ellenberg, H. (ed.): *Ökosystemforschung.* pp. 1–31. Berlin, Springer.

Germain, R., – 1965 – *Les biotopes alluvionnaires herbeux et les savanes intercalaires du Congo équatorial.* Bruxelles, Acad. Roy. Sc. d'Outre-mer.

Gounot, M., – 1969 – *Méthodes d'étude quantitative de la végétation.* Paris, Masson.

Greenway, P. J., & D. F. Vesey-Fitzgerald, – 1969 – The vegetation of Lake Manyara National Park. *J. Ecol.* 57: 127–149.

Grunow, J. O., – 1967 – Objective classification of plant communities: a synecological study in the sourish mixed bush veld of Transvaal. *J. Ecol.* 55: 691–710.

Grunow, J. O. & G. N. Lance, – 1969 – Classification of savanna by information analysis. *S. Afr. J. Sci.* 65: 341–348.

Herbel, C. H., F. N. Ares, & R. A. Wright, – 1972 – Drought effects on a semidesert grassland range. *Ecology* 53: 1084–1093.

Holzner, W., – 1973 – Forschungsergebnisse der modernen Ökologie in ihrer Bedeutung für Biologie und Bekämpfung der Unkräuter. *Die Bodenkultur* 24: 61–74.

142

Ivimey-Cook, R. B. & M. C. F. Proctor, – 1966 – The application of association-analysis to phytosociology. *J. Ecol.* 54: 179–192.

Kartawinata, K., & D. Mueller-Dombois, – 1972 – Phytosociology and ecology of the natural dry-grass communities on Oahu, Hawaii. *Reinwardtia* 8: 369–494.

Kershaw, K. A., – 1968a – A survey of the vegetation in Zaria province, Northern Nigeria. *Vegetatio* 15: 244–268.

Kershaw, K. A., – 1968b – Classification and ordination of Nigerian savanna vegetation. *J. Ecol.* 56: 467–482.

Knapp, R., – 1965 – Pflanzenarten-Zusammensetzung, Entwicklung und natürliche Produktivität der Weide-Vegetation in Trockengebieten in verschiedenen Klima-Bereichen der Erde. *In:* Knapp, R. (ed.): *Weide-Wirtschaft in Trockengebieten.* pp. 71–97. Stuttgart, Fischer.

Knapp, R., – 1967 – *Experimentelle Soziologie und gegenseitige Beeinflussung der Pflanzen.* Stuttgart, Ulmer. 2. Aufl.

Knight, D. H., & O. L. Loucks, – 1969 – A quantitative analysis of Wisconsin forest vegetation on the basis of plant function and gross morphology. *Ecology* 50: 219–234.

Küchler, A. W., – 1973 – Problems in classifying and mapping vegetation for ecological regionalization. *Ecology* 54: 512–523.

Lambert, J. M., – 1972 – Theoretical models for large-scale vegetation survey. *In:* Jeffers, J. N. R. (ed.): *Mathematical models in ecology.* pp. 87–109. Oxford, Blackwell.

Lambert, J. M. & M. B. Dale, – 1964 – The use of statistics in Phytosociology. *Adv. Ecol. Res.* 2: 59–99.

Lebrun, J., – 1947 – La végétation de la plaine alluviale au sud du Lac Edouard. *Expl. Parc. Nat. Albert* Fasc. 1: 1–800. Bruxelles.

Leippert, H., – 1968 – *Pflanzenökologische Untersuchungen im Masai-Land Tanzanias.* München, IFO-Inst.

Leistner, O. A. & Werger, M. J. A., – 1973 – Southern Kalahari phytosociology. *Vegetatio* 28: 353–399.

Major, J., – 1969 – The historical development of the ecosystem concept. *In:* Van Dyne, G. M. (ed.): *The ecosystem concept in natural resource management.* pp. 9–22. London, Acad. Press.

Monteiro, R. F. R., – 1970 – Estudo da Flora e da vegetação das Florestas abertas do planalto do Bié. Luanda, Inst. Inv. Ciên. Agr.

Moore, J. J., P. Fitzsimons, E. Lambe, & J. White, – 1960 – A comparison and evaluation of some phytosociological techniques. *Vegetatio* 20: 1–20.

Mueller-Dombois, D. & H. Ellenberg, – 1974 – Aims and methods of vegetation ecology. New York, Wiley.

Mullenders, W., – 1954 – La végétation de Kaniama. *Publ. I.N.E.A.C. Sér. Scient.* 61: 1–500.

Myre, M., – 1962 – A grassland type of the South of the Mozambique Province. *C.R. IV. Réun. A.E.T.F.A.T.* pp. 337–362.

Myre, M., – 1964 – A vegetação do extremo sul da Provincia de Moçambique. *Estudos, ensaios e documentos* 110: 1–145.

Myre, M., – 1971 – As pastagens da região do Maputo. *Memórias Inst. Inv. Agron. Moçambique* 3: 1–181.

Myre, M., – 1972 – Reconhecimento pascícola ao Vale do Save. *Communicacões Inst. Inv. Agron. Moçambique* 75: 1–171.

Raney, F. & R. Daubenmire, – 1958 – On the principles of genetic classification in biocenology. *Ecology* 39: 364–367.

Rattray, J. M., − 1960 − The grass cover of Africa. *FAO Agric. Studies* 49. Rome: FAO.

Schmidt, W., − 1973a − Vegetationskundliche Untersuchungen im Savannenreservat Lamto (Elfenbeinküste). *Vegetatio* 28: 145−200.

Schmidt, W., − 1974− Plant communities on permanent plots of the Serengeti Plains. *Vegetatio* 30: 133−145.

Schmidt, W., − 1976 − The vegetation of the northeastern Serengeti National Park, Tansania. *Phytocoenologia* 3: 30−82.

Schmitz, A., − 1963 − Aperçu sur les groupements végétaux du Katanga. *Bull. Soc. Roy. Bot. Belg.* 96: 233−447.

Segal, S., − 1969 − Ecological notes on wall vegetation. Junk, The Hague.

Schnell, R., − 1952a − Contribution à une étude phytosociologique et phytogéographique de l'Afrique occidentale: les groupements et les unités géobotaniques de la région guinnéenne. *Mém. IFAN* 18: 41−236.

Schnell, R., − 1952b − Végétation et flore de la région montagneuse du Nimba. *Mém. IFAN* 22: 1−604.

Schnell, R., − 1952c − Végétation et flore des Monts Nimba. *Vegetatio* 3: 350−406.

Schnell, R., − 1961 − Contribution à l'étude botanique de la Chaîne de Fon (Guinée). *Bull. Jard. Bot. Bruxelles* 31: 15−54.

Stanek, W., − 1973 − A comparison of Braun-Blanquet's method with sum-of-squares agglomeration for vegetation classification. *Vegetatio* 27: 323−338.

Tansley, A. C., − 1935 − The use and abuse of vegetational concepts and terms. *Ecology* 16: 284−307.

Van Donselaar, J., − 1965 − An ecological and phytogeographic study of northern Suriname savannas. *Wentia* 14: 1−163.

Van Groenewoud, H., − 1965 − Ordination and classification of Swiss and Canadian coniferous forests by various biometric and other methods. *Ber. Geobot. Inst. ETH. Stift. Rübel* 36: 28−102.

Vareschi, V., − 1972 − El problema de la vegetacion optima. *Mém I. Congr. Lat.-am. Bot.* 437−449, Caracas.

Volk, O. H. & H. Leippert − 1971 − Vegetationsverhältnisse im Windhoeker Bergland, Südwestafrika. *J.S.W.A. Wissensch. Ges.* 25: 5−44.

Von Glahn, H., − 1968 − Der Begriff des Vegetationstypes im Rahmen eines allgemeinen naturwissenschaftlichen Typenbegriffes. *In:* Tüxen, R. (ed.): Pflanzensoziologische Systematik. Ber. Int. Symp. Stolzenau 1964. pp. 1−14. Junk, Den Haag.

Werger, M. J. A., − 1973a − Phytosociology of the Upper Orange River valley, South Africa. Pretoria: V & R.

Werger, M. J. A., − 1973b − On the use of association-analysis and principal component analysis in interpreting a Braun-Blanquet phytosociological table of a Dutch grassland. *Vegetatio* 28: 129−144.

Werger, M. J. A., − 1974a − The place of the Zürich-Montpellier method in vegetation science. *Folia Geobot. Phytotax.* 9: 99−109.

Werger, M. J. A., − 1974b − On concepts and techniques applied in the Zürich-Montpellier method of vegetation survey. *Bothalia* 11: 309−323.

Werger, M. J. A. & D. Edwards, − 1976 − Progress with vegetation studies in South Africa. *Boissiera* 24 (in press).

Werger, M. J. A., F. J. Kruger & H. C. Taylor, − 1972 − Pflanzensoziologische Studie der Fynbos-Vegetation am Kap der Guten Hoffnung. *Vegetatio* 24: 71−89.

Westhoff, V., − 1967 − Problems and use of structure in the classification of vegetation. *Acta Bot. Neerl.* 15: 495−511.

Westhoff, V. & E. van der Maarel, − 1973 − The Braun-Blanquet approach. *In:* Whittaker (1973), pp. 617−726.

144

Williams, W. T., – 1971 – Principles of clustering. *Ann. Rev. Ecol. Syst.* 2: 303–326.

Whittaker, R. H., – 1962 – Classification of natural communities. *Bot. Rev.* 28: 1–239.

Whittaker, R. H. (ed.), – 1973 – Ordination and Classification of Communities. Handbook of Vegetation Science, part 5. Junk, The Hague.

Whittaker, R. H., – 1975 – Communities and Ecosystems. 2nd. ed. New York, MacMillan.

Young, A., – 1973 – Soil survey procedure in land development planning. *Geogr. J.* 139: 53–64.

3 EFFECTS OF GAME AND DOMESTIC LIVESTOCK ON VEGETATION IN EAST AND SOUTHERN AFRICA

M. J. A. WERGER

EFFECTS OF GAME AND DOMESTIC LIVESTOCK ON
VEGETATION IN EAST AND SOUTHERN AFRICA*

The mutual interactions between plants and animals and between these organisms and the environment form the basic relationships in an ecosystem. In the non-forested regions of east and southern Africa, which have always been famous for their richness of large mammals, animals have played a particularly important role in the functioning of ecosystems. Their impact on the vegetation and habitat has often been decisive in either maintaining the equilibrium or in inducing changes. Two other factors of importance in this respect have been fire and man, the latter directly and as the manipulator of both fire and animal populations. Man has markedly changed the ecosystems of the savanna and grassland regions of east and southern Africa, most noticably by his agricultural activities, involving the hunting and eradication of game animals and replacing them by his domestic livestock.

Whereas vegetation changes induced by wild animals have usually led to new, stable ecosystems suitable for the existence of the same or a modified game spectrum, influence of domestic livestock has most often led either to bare surfaces with a very high incidence of erosion, or to the development of impenetrable thickets, or to an abundant growth of unpalatable species both of which are useless for the subsistence of animals. This undesirable situation may also result when an area is overstocked with game, but when compared with domestic livestock a considerably larger quantity of game is necessary.

In areas with a rainfall high enough to support a dry forest or dense woodland type of vegetation, wild animals have often initiated the development and furthered the preservation of open grasslands. These changes are generally begun by elephants (*Loxodonta africana*) which, in a wooded area, particularly where water is unavailable, destroy the trees by ring-barking them or pushing them over (Fig. 1 and 2) (Vesey-Fitzgerald, 1960, 1973; Buechner & Dawkins, 1961; Glover, 1963; Lamprey, 1963; Napier Bax & Sheldrick, 1963; Wager, 1963; Lamprey *et al.*, 1967;

* Acknowledgements: I wish to thank D. Mason for drawing my attention to several relevant references and J. W. Morris for his comments on the text prior to publication.

Fig. 1. In a formerly densely wooded area in the Kruger National Park, South Africa, near a waterhole, many trees have been pushed over by elephant. Trees are mainly *Colophospermum mopane, Acacia nigrescens* and *Adansonia digitata*. The tree in the center right has been ring-barked. Note the heavily overgrazed, bare surface showing sheet erosion. (Photo N. G. Jarman).

Van Wyk & Fairall, 1969; Laws, 1970; Lawton, 1971; Austen, 1972; West, 1972; Myers, 1973; Croze, 1974; Harrington & Ross, 1974). The enhanced grass growth on clearings thus formed attracts grazing ungulates which maintain it as a grassland (Austen, 1972; Vesey-Fitzgerald, 1969, 1972). If there is not a large enough animal population to maintain the grassland, litter accumulates. Owing to such an accumulation of fuel, hot fires, frequently caused by lightning (Komarek, 1972; West, 1972), or by man, can occur, which will maintain and extend the grasslands by burning seedlings and saplings of woody plants and cutting back the margin of the surrounding woody vegetation (Buechner & Dawkins, 1961; Lamprey, 1963; Vesey-Fitzgerald, 1972; Knapp, 1973). Large areas of grassland, originating from the action of elephants and maintained and enlarged by grazing ungulates and fire, exist throughout east and southern Africa. These areas, with over 500 mm of annual precipitation, would normally support dense woodland or dry forest. If game and fire were excluded from these areas, woody elements would rapidly increase and the original woodland or forest vegetation would soon be restored (cf. Spencer & Angus, 1971). Introduced domestic livestock can replace game and maintain these grasslands, provided that the grazing pressure is not too high. If the latter condition is not met, the vegetation deteriorates rapidly, resulting in destruction typical of overgrazed areas to be discussed below.

In areas with an annual precipitation of less than 500 mm and non-rocky, somewhat sandy soils, natural vegetation consists of an equilibrium between grasses and woody plants in which grasses are the dominant growth form. Grasses have intensive,

Fig. 2. Small baobab (*Adansonia digitata*) in Tsavo National Park, Kenya, severely damaged by elephant, which sometimes eat the soft, juicy wood. (Photo: D. R. M. Stewart).

151

Fig. 3. Thicket of *Acacia karroo* which has encroached following prolonged over-grazing in Cape Province, South Africa. (Photo: Dept. Agric. Techn. Serv.).

shallow root systems, whereas woody plants possess extensive, deep root systems. The precipitation in regions where these savanna types of vegetation occur is always strongly seasonal and occurs in summer. Rains early in the season wet only the top soil and benefit the growth of the shallow rooted grasses. The growing season of the grasses is soon past and the moisture remaining in the soil together with that falling later in the season becomes available to and invigorates the growth of the woody plants with their deeper root systems (cf. Roux, 1966). Overgrazing and trampling impairs grass growth and reduces the amount of water transpired by grasses. Thus, at the end of their growing season, more moisture remains in the soil to increase the growth of woody plants. Heavy grazing and trampling over consecutive years has an accumulative effect and the original savanna vegetation changes rapidly into a dense thicket, often of thorny *Acacia* spp., which cannot be used for grazing (Fig. 3) (Walter, 1939, 1954, 1962; Walter & Volk, 1954; Scott, 1967; cf. Vesey-Fitz-gerald, 1973). Other harmful effects of overgrazing and trampling on the vegetation, particularly in the semi-arid and arid regions, can include extensive denudation of the surface, resulting in se-

152

vere erosion and the sparse growth of virtually only ephemerals. This results in a desiccation of the soil, particularly in loamy soils, since the infiltration rate of rain water is much lower in bare loamy soils than in loamy soils covered by litter and caespitose, perennial grasses. Revegetation of such bare soils consequently is hampered seriously (Walker, 1974). In other cases a strong increase in poisonous and unpalatable plants occur with overgrazing (Fig. 4) (Werger & Leistner, 1973; Werger, 1976). The vegetation seems to be particularly vulnerable when grazed and trampled shortly after fire (Walter, 1962). Herds of game attracted by regrowth of vegetation have also been recorded to cause damage to the vegetation shortly after fire (Pratt, 1967).

The effects of overgrazing are however usually associated with domestic livestock, such as cattle, and particularly sheep and goats (Shaw, 1875; Acocks, 1953; Phillips, 1956-57; Volk, 1966; Scott, 1967; Myre, 1970; Werger, 1973, 1976). There are several reasons why, in Africa, the influence of domestic livestock on vegetation is often so much more devastating than that of game. In the first place, the usual farming practice is important in that it restricts the movement of domestic livestock. Game are usually mobile and, although often concentrated in herds, wander over a large area. They stay nowhere for any length of time and only graze here and there. Economic considerations often not only tempt the farmer to over-stock his ranch but also force him to keep his animals in camps. For this reason, and because domestic animals are less mobile than game, part of the vegetation is more thoroughly grazed by them, and with poor management, easily overgrazed (Walter, 1962; Vesey-Fitzgerald, 1969; Walker, 1974). This becomes particularly noticable at water points. Game usually stay only for short periods in the immediate vicinity of a drinking place, probably owing to the increased danger of large predators. Domestic livestock, which generally need to drink more often than game (Hill, 1972; Strang, 1973), stay close to the water points. This results in a strong zonation from a totally bare surface to lesser degrees of overgrazing and trampling around a water point used by domestic livestock. In South West Africa, according to Walter & Volk (1954), overgrazing was no longer an important factor in the vegetation at an average distance of five to eight km from a water point.

In the higher rainfall areas, where grassland is promoted by grazing and fire, grazing by cattle, being largely restricted to grass, reduces the amount of fuel for a hot, late season burn. This can lead to regeneration of the woody vegetation (Vesey-Fitz-gerald, 1972).

From several studies it has become apparent that the diet of game includes far more plant species than that of domestic animals (Jewell, 1969; Hill, 1972). Liversidge (1970), comparing the number of grasses grazed by merino sheep and springbok (*Antidorcas marsupialis*) in the semi-arid parts of South Africa, found that the latter ate 33 species as against 20 species eaten by sheep. The quantity of the various food plants eaten also differed considerably. Whereas grass formed about 39 per cent of the dry mass intake of springbok, the balance being obtained from bushes, merino sheep took 75 per cent grass in their feed (Liversidge, 1972). Comparable results are reported by Taylor (1974) in a comparative study of game and cattle in semi-arid Rhodesian range land (cf. Walker, 1974). Several other studies show the wide range of species included in the diets of wild, East African animals (e.g. for black rhinoceros (*Diceros bicornis*) by Goddard, 1968; for Cape eland (*Taurotragus oryx*) by Hofmeyr, 1970, and for lesser kudu (*Tragelaphus imberbis australis*) by Leuthold, 1971). Casebeer & Koss (1970) found that in Kenya, white-bearded wildebeest (*Connochaetus taurinus albojubatus*), zebra (*Equus burchelli*) and kongoni or Coke's hartebeest (*Alcelaphus buselaphus cokii*) had a greater diet variation between seasons than zebu cattle grazing in the same area. In South Africa, Van Zyl (1965) showed a strong seasonal change in diet of the black wildebeest (*Connochaetus gnou*), blesbok (*Damaliscus dorcas phillipsi*), Cape eland, Cape oryx (*Oryx gazella*), impala (*Aepyceros melampus*) (cf. also Stewart, 1971), red hartebeest (*Alcelaphus caama*) and springbok. A similar seasonal change in diet was demonstrated for large mammals in the Zambezi Valley (Jarman, 1971, 1972) and in Uganda (Field, 1972) and for buffalo (*Syncerus caffer*) in Tanzania (Vesey-Fitzgerald, 1969). Talbot (1962) reported that in Kenya and Tanzania five of the most common ungulates, whitebeared wildebeest, Thomson's gazelle (*Gazella thomsoni*), Grant's gazelle (*Gazella granti*), topi (*Damaliscus korrigum*) and impala apparently prefer plant species which are recorded locally as being relatively unpalatable to domestic livestock, or which invade or become dominant in places over-grazed by domestic livestock. Except for the wider variety in species eaten and the strong seasonal changes in their diets, wild animals are often complementary to each other in the species they eat, rather than overlapping. When they do overlap they often eat the species common to their diets at different times of the year. They thus fill different ecological niches in the sense of Whittaker *et al.* (1973) (Vesey-Fitzgerald, 1960; Talbot, 1962; Lamprey, 1963; Bourlière & Hadley, 1970; Bothma, 1972; Hill, 1972). Many studies have confirmed the complementary feeding habits by demonstrating a grazing sequence of animals resulting in the use of different pastures in rotation during the year. This sequence is started by heavy animals, such as elephant, hippopotamus (*Hippopotamus amphibius*) or buffalo, who open up the vegetation and are then followed by the lighter species (Vesey-Fitzgerald, 1960, 1965; Lamprey, 1963; Gwynn & Bell, 1968; Bell, 1970; Stewart, 1971).

Anatomical differences in the feeding mechanisms of the various species of animals are probably contributing to the differences in diet. The ability of certain animal species to separate pieces from the whole plant of a certain species will to some degree, influence their choice of plant species (Gwynne & Bell, 1968). At the same time, the way in which pieces of the plant are removed will influence the damage caused to the plant. Although several species of game feed similarly to domestic livestock, many only nibble the plants here and there, thus only slightly damaging

Fig. 4. Effect of overgrazing on the South African Highveld: on the right of the fence an overgrazed grassland shows strong encroachment of the unpalatable dwarf shrub *Chrysocoma tenuifolia*, while left of the fence the better managed grassland is still in good condition. (Photo: Dept. Agric. Techn. Serv.).

the plant (cf. Van Zyl, 1965). Differences in diet are also correlated with differences in the digestive system of the animals (Hofmann & Stewart, 1972).

Once an area is covered by thickets through overgrazing, it is virtually useless, since, mechanical clearing or the use of arboricides often are too expensive and labour intensive (cf. Scott, 1967). Burning is usually not effective, as hot fires do not occur owing to lack of grass litter which usually provides fuel, and because most encroaching species coppice vigorously from buds just below the soil surface. Special burning techniques which may be effective as, for example, proposed by Donaldson (1967) often cannot be realized because of the costs. Denuded and severely eroded areas are useless too and require complete protection. Planting or seeding of desirable plant species, sometimes attended by ripping of the top soil, may be necessary. For selectively- and overgrazed areas in the semi-arid regions of South Africa, which have become dominated by unpalatable species, Acocks (1966) recommended a non-selective grazing system for reclamation to a more useful species composition. His system consists of a large number of small camps, which are rotationally grazed for short periods of about two weeks at a time. Intensive stocking ensures that grazing pressure is high enough to accomplish heavy grazing of all species. After a short grazing period, the camps are rested for a long period of at least six weeks. Acocks (1966) claims that palatable species will increase under such a grazing system. It may be possible to devise a grazing system for domestic livestock which prevents overgrazing but in natural or semi-natural vegetation, the most successful in preventing overgrazing and its harmful consequences is apparently a fairly large, mixed population of wild animals with browsers complementing grazers ecologically and perhaps mixed with a small quantity of domestic livestock. An optimum utilization of the vegetation, without it leading to degradation is then ensured and, according to Jewell (1969), Hill (1972) and Strang (1973), it results in a more efficient use of vegetation and habitat than achieved by domestic livestock alone.

References

Acocks, J. P. H, – 1953 – Veldtypes of South Africa. *Mem. Bot. Surv. S. Afr.* 28: 1–192.
Acocks. J. P. H., – 1966 – Non-selective grazing as a means of veld reclamation. *Proc. Grassld. Soc. Sth. Afr.* 1: 33–39.

Austen. B., – 1972 – The history of veld burning in the Wankie National Park, Rhodesia. *Proc. Ann. Tall Timbers Fire Ecol. Conf.* 11: 277–296.

Bell. R. H. V., – 1970 – The use of the herb layer by grazing ungulates in the Serengeti. *In:* Watson, A. (ed.): Animal populations in relation to their food resources. Oxford, Blackwell. pp. 111–124.

Bothma, J. du P., – 1972 – Short-term response in ungulate numbers to rainfall in the Nossob River of the Kalahari Gemsbok National Park. *Koedoe* 15: 127–133.

Bourlière, F. & Hadley, M., – 1970 – The ecology of tropical savannas. *Ann. Rev. Ecol. Syst.* 1: 125–152.

Buechner, H. K. & Dawkins, H. C., – 1961 – Vegetation change induced by elephants and fire in Murchison Falls National Park, Uganda. *Ecology* 42: 752–766.

Casebeer, R. L. & Koss, G. G., – 1970 – Food habits of wildebeest, zebra, hartebeest and cattle in Kenya Masailand. *E. Afr. Wildl. J.* 8: 25–36.

Croze, H., – 1974 – The Seronera bull problem. II. The trees. *E. Afr. Wildl. J.* 12: 29–47.

Donaldson, C. H., – 1967 – Further findings on the effects of fire on blackthorn. *Proc. Grassld. Soc. Sth. Afr.* 2: 59–61.

Field, C. R., – 1972 – The food habits of wild ungulates in Uganda by analyses of stomach contents. *E. Afr. Wildl. J.* 10: 17–42.

Glover J., – 1963 – The elephant problem at Tsavo. *E. Afr. Wildl. J.* 1: 30–39.

Goddard, J., – 1968 – Food preferences of two black rhinoceros populations. *E. Afr. Wildl. J.* 6: 1–18.

Gwynne, M. D., & Bell, R. H. V., – 1968 – Selection of vegetation components by grazing ungulates in the Serengeti National Park. *Nature* 220: 390–393.

Harrington, G. N., & Ross, I. C., – 1974 – The savanna ecology of Kipedo National Park. I. *E. Afr. Wildl. J.* 12: 93–105.

Hill, P., – 1972 – Grass foggage – food for fauna or fuel for fire, or both? *Proc. Ann. Tall Timbers Fire Ecol. Conf.* 11: 337–375.

Hofmann, R. R., & Stewart, D. R. M., – 1972 – Grazer or browser: a classification based on the stomach-structure and feeding habits of East African ruminants. *Mammalia* 36: 226–240.

Hofmeyr, J. M., – 1970 – A review of the food preferences and feeding habits of some indigenous herbivores in the Ethiopian faunal region and some studies on animal: plant relationship. *Proc. S. Afr. Soc. Anim. Prod.* 9: 89–99.

Jarman, P. J., – 1971 – Diets of large mammals in the woodlands around Lake Kariba, Rhodesia. *Oecologia* 8: 157–178.

Jarman, P. J., – 1972 – Seasonal distribution of large mammal populations in the unflooded middle Zambezi valley. *J. app. Ecol.* 9: 283–299.

Jewell, P. A., – 1969 – Wild mammals and their potential for new domestication. In P. J. Ucko & G. W. Dimbleby (ed.): The domestication and exploitation of plants and animals. London, Duckworth, pp. 101–109.

Knapp, R., – 1973 – Die Vegetation von Afrika. Stuttgart, Fischer.

Komarek, E. V., – 1972 – Lightning and fire ecology in Africa. *Proc. Ann. Tall Timbers Fire Ecol. Conf.* 11: 473–511.

Lamprey, H. F., – 1963 – Ecological separation of the large mammal species in the Tarangire Game Reserve, Tanganyika. *E. Afr. Wildl. J.* 1: 63–92.

Lamprey, H. F., Glover, P. E., Turner, M. I. M., & R. H. V. Bell, – 1967 – Invasion of the Serengeti National Park by elephants. *E. Afr. Wildl. J.* 5: 151–166.

Laws, R. M., – 1970 – Elephants as agents of habitat and landscape change in East Africa. *Oikos* 21: 1–15.

Lawton, R. M., – 1971 – Destruction or utilization of a wildlife habita? *In:* E. Duffey & A. S. Watt (eds): The scientific management of animal and plant communities for conservation. Oxford, Blackwell. pp. 333–336.

Leuthold, W., – 1971 – Studies on the food habits of lesser kudu in Tsavo National Park, Kenya. *E. Afr. Wildl. J.* 9: 35–45.

Liversidge, R., – 1970 – Identification of grazed grasses using epidermal characters. *Proc. Grassld. Soc. Sth. Afr.* 5: 153–165.

Liversidge R., – 1972 – Grasses grazed by springbok and sheep. *Proc. Grassld. Soc. Sth. Afr.* 7: 32–38.

Myers, N., – 1973 – Tsavo National Park, Kenya, and its elephants; an interim appraisal. *Biol. Cons.* 5: 123–132.

Myre, M., – 1970 – A invasão dos arbustos nas áreas de pastagens. Primeiras observações de uma experiencia em curso. *Agron. Moçamb* 4: 115–138.

Napier Bax, J., & Sheldrick, D. L. W., – 1963 – Some preliminary observations on the food of elephant in the Tsavo Royal National Park (East) of Kenya. *E. Afr. Wildl. J.* 1: 40–53.

Phillips, J., – 1956-57 – Aspects of the ecology and productivity of some of the more arid regions of southern and eastern Africa. *Vegetatio* 7: 36–68.

Pratt, D. J., – 1967 – A note on the overgrazing of burned grassland by wildlife. *E. Afr. Wildl. J.* 5: 178–179.

Roux, P. W., – 1966 – Die uitwerking van seisoenreënval en beweiding op gemengde Karooveld. *Proc. Grassld. Soc. Sth. Afr.* 1: 103–110.

Scott, J. D., – 1967 – Bush encroachment in South Africa. *S. Afr. J. Sci.* 63: 311–314.

Shaw, J., – 1875 – On the changes going on in the vegetation of South Africa through the introduction of the merino sheep. *J. Linn. Soc. London* 14: 202–208.

Spence, D. H. N., & Angus, A., – 1971 – African grassland management – burning and grazing in Murchison Falls National Park, Uganda. *In:* E. Duffey & A. S. Watt (eds): *The scientific management of animal and plant communities for conservation.* Oxford, Blackwell. pp. 319–331.

Steward, D. R. M., – 1971 – Food preferences of an impala herd. *J. Wildl. Mgmt.* 35: 86–93.

Steward, D. R. M. & J., – 1971– Comparative food preferences of five East African ungulates at different seasons. *In:* E. Duffey & A. S. Watt (eds): The scientific management of animal and plant communities for conservation. Oxford, Blackwell. pp. 351–366.

Strang, R. M., – 1973 – Bush encroachment and veld management in South-central Africa: the need for a reappraisal. *Biol. Cons.* 5: 96–104.

Talbot, L. M., – 1962 – Food preferences of some East African wild ungulates. *E. Afr. Agric. For. J.* 27: 131–138.

Taylor, R. D., – 1974 – A comparative study of Land-use under cattle and game on a Rhodesian lowveld ranch. Unpubl. M.sc. Thesis Univ. Rhod. Salisbury.

Van Wyk, P., & Fairall, N., – 1969 – The influence of the African elephant on the vegetation of the Kruger National Park. *Koedoe* 12: 57–89.

Van Zyl, J. H. M., – 1965 – The vegetation of the South African Lombard Nature Reserve and its utilisation by certain antelope. *Zool. Afr.* 1: 55–71.

Vesey-Fitzgerald, D. F., – 1960 – Grazing succession among East African game animals *J. Mammol.* 41: 161–172.

Vesey-Fitzgerald, D. F., – 1965 – The utilization of natural pastures by wild animals in the Rukwa valley, Tanganyika. *E. Afr. Wildl. J.* 3: 38–48.

Vesey-Fitzgerald D. F., – 1969 – Utilization of the habitat by buffalo in Lake Manyara National Park. *E. Afr. Wildl. J.* 7: 131–145.

Vesey-Fitzgerald, D. F., – 1972 – Fire and animal impact on vegetation in Tanzania National Parks. *Proc. Ann. Tall. Timbers Fire Ecol. Conf.* 11: 297–317.

Vesey-Fitzgerald, D. F., – 1973 – Animal impact on vegetation and plant succession in Lake Manyera National Park, Tanzania. *Oikos* 24: 314–325.

Volk, O. H., – 1966 – Einfluss von Mensch und Tier auf die natürliche Vegetation im tropischen Südwestafrika. *In:* Buchwald, K., Lendholt, E., & K. Meyer (ed.): Beiträge zur Landespflege 2: 108–131. Stuttgart, Ulmer.

Wager, V. A., – 1963 – Protecting protected areas. *Afr. Wildl.* 17: 11–16.

Walker, B. H., – 1974 – Ecological consideration in the managements of semi-arrid ecosystems in South-central Africa. Proc. First. Int. Congr. Ecol. Wageningen, Pudoc. pp. 124–129.

Walter, H., – 1939 – Grasland, Savanne und Busch der arideren Teile Afrikas in ihrer ökologischen Bedingtheit. *Jahrb. f. Wiss. Bot.* 87: 750–860.

Walter, H., – 1954 – Die Verbuschung, eine Erscheinung der subtropischen Savannen Gebiete, und ihre ökologischen Ursachen. *Vegetatio* 5–6: 6–10.

Walter, H., – 1962 – Die Vegetation der Erde in öko-physiologischer Betrachtung. I. Die tropischen und subtropischen Zonen. Jena, Fischer.

Walter, H., & Volk, O. H., – 1954 – Grundlagen der Weidewirtschaft in Südwestafrika, Stuttgart, Ulmer.

Werger, M. J. A., – 1973 – Phytosociology of the Upper Orange River valley, South Africa. A syntaxonomical and synecological study. Pretoria, V & R.

Werger, M. J. A., – 1976 – Environmental destruction in southern Africa: the role of overgrazing and trampling. *Proc. Symp. Environ. Cons. Tokyo.* 1974.

Werger, M. J. A., & Leistner, O. A., – 1975 – Vegetationsdynamik in der südlichen Kalahari. *In:* Schmidt, W. (ed.): Sukzessionsforschung. Ber. Int. Symp. Rinteln 1973. pp. 135–158.

West, O., – 1972 – Fire, man and wildlife as interacting factors limiting the development of vegetation in Rhodesia. *Proc. Ann. Tall Timbers Fire Ecol. Conf.* 11: 121–145.

Whittaker, R. H., Levin, S. A., & R. B. Root, – 1973 – Niche, habitat, and ecotope. *The Amer. Nat.* 107: 321–338.

4 FARMING REGIONS IN THE TROPICS: ENVIRONMENT, GEOGRAPHICAL COMPARISONS AND SOCIO-ECONOMIC DEVELOPMENT

B. ANDREAE

Contents

4 FARMING REGIONS IN THE TROPICS: ENVIRONMENT, GEOGRAPHICAL COMPARISONS AND SOCIO-ECONOMIC DEVELOPMENT

Throughout the world, agricultural zones are determined mainly by their climate and the stage of their economic development. However, an initial survey of farming regions in the tropics can use the climatic factor alone: since in these lower latitudes there are, so far, only developing countries whose economies do not differ greatly.

In the following, population density is considered as an additional factor in the causal relationship; though the primary concern is the connection between the farming regions and the climate. The interaction between natural and cultivated vegetation is particularly worthy of interest.

4.1 Classification of the Climatic Zones and Farming Systems

Firstly, in order to illustrate the climatically-determined differences between the farming regions of the tropics, it is necessary to identify types of climate on the one hand and types of farming on the other.

4.1.1 CLIMATIC ZONES OF THE TROPICS

This region, spanning the equator and extending between the two tropics, includes both the most arid and the most humid climatic belts of the earth. Figure 1 gives a rough indication of the place occupied by the tropical climatic zones in the climatic belts of the earth. A newer classification, based on climographs, is given by Walter (1971).

For the purpose of this study, however, a stricter classification of tropical climates is needed, optimally one based on vegetation formations. With increasing proximity to the equator and with rising humidity, six climatic zones can be distinguished. These are represented by the following six vegetation forms (cf. Fig. 2):

    1.   the *dry-hot desert*, with only slight and sporadic rainfall,

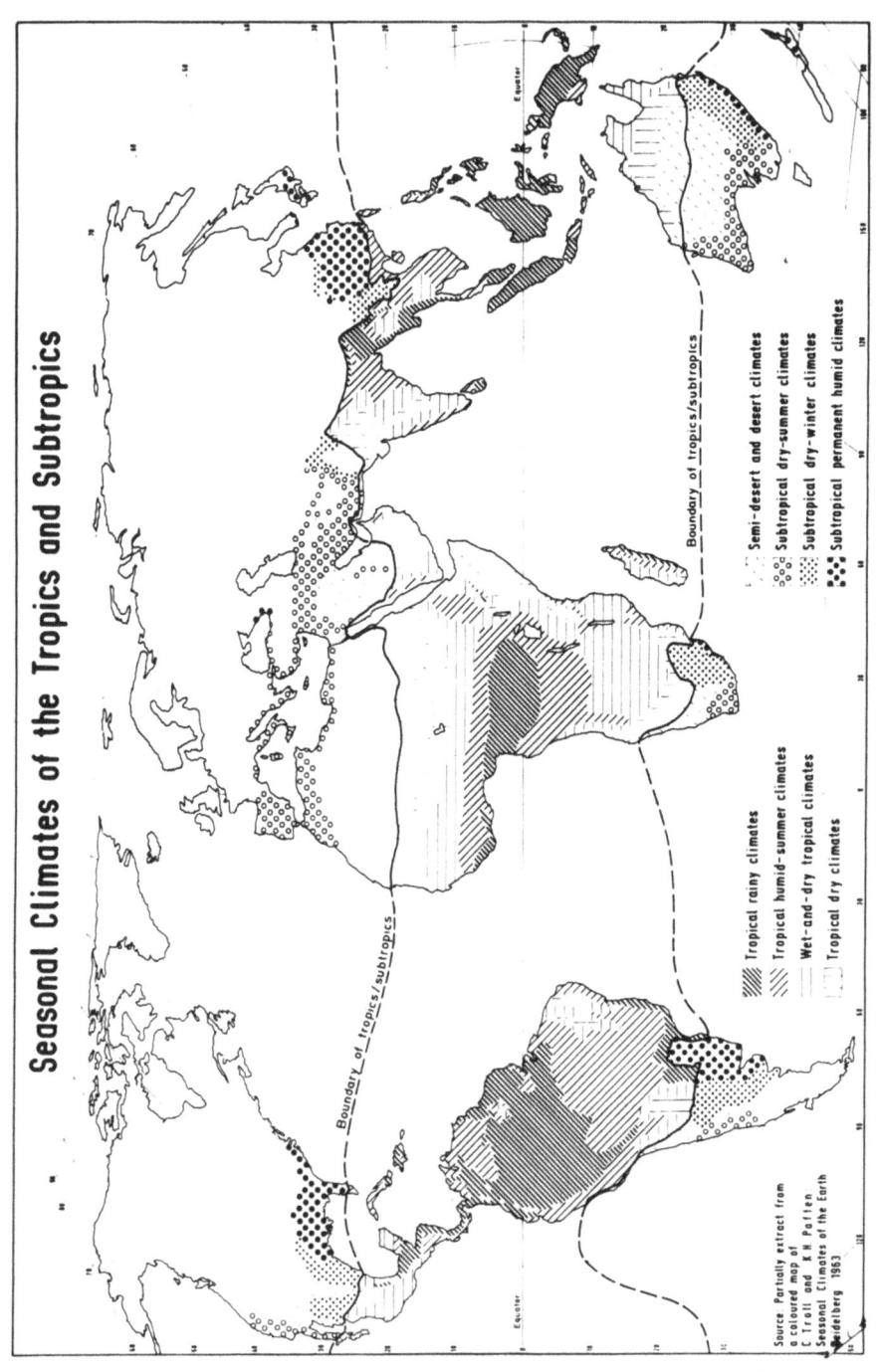

1. Climatic regions of the tropics and subtropics.

164

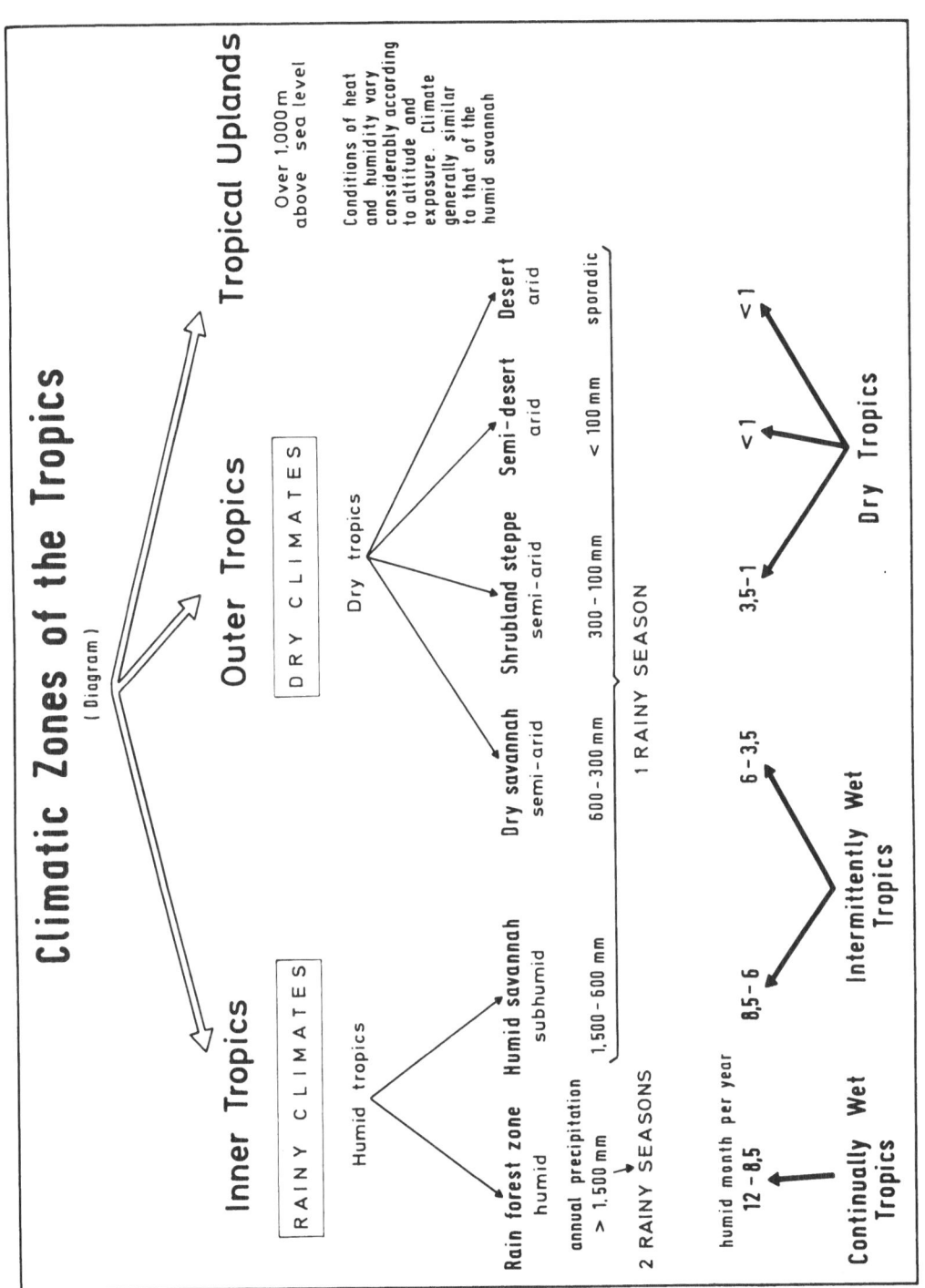

2. Climatic zones of the tropics (diagram)

which is mostly beyond the aridity limit for human habitation: (e.g., the Sahara, Kalahari and Central Australian Deserts).

2. the *arid semi-desert*, partly inside and partly outside the limit for animal farming, which can only be used occasionally by nomads, hunters and collectors (e.g., large parts of Egypt, Libya, Mauretania).

3. the *semi-arid shrubland-steppe* (shrub, salt steppe), with two to four humid months, which extends up to the agronomic aridity limit. Apart from a very little irrigated farming and some dry farming the steppe is used almost exclusively for extensive pasture farming – i.e. for ranching in the New World and often still by nomads in the Old World (e.g., large parts of Senegal, Upper Volta, South West Africa).

4. the *dry savannah*, with four to six humid months, situated between the agronomic and climatic aridity limits and thus capable of supporting rain-fed farming. The grassland which occurs is dry steppe, and the woodland dry forest which is green in the rainy season (e.g. Miombo). A short rainy season is followed by a long dry season (e.g., large parts of Chad, Tanzania, Somalia).

5. the *subhumid wet savannah*, with seven to nine humid months, situated between the climatic aridity limit and the humidity limit for pasture farming. The rainy season is now longer, the dry season shorter, the atmospheric humidity higher. The yearly rainfall is from approximately 800 to 1 500 mm. The grassland which occurs is high-grass savannah with tropical river forest, and the woodland is monsoon forest. This is the first climatic zone where even the small rivers contain water all the year round: in zones 1 to 4 they dry up soon after the rainy season ends (e.g., large parts of North Brazil, Thailand, Ghana).

6. the *tropical rain forest*, immediately on the equator and with the wettest climate. There are two rainy seasons with a total of at least 1 500 mm yearly rainfall, the mean temperature is 25 to 28°C and varies little throughout the year, and the humidity is rarely below 90%. This leads to a thoroughly humid, continually wet type of climate giving rise to evergreen tropical forest (e.g., large parts of Indonesia and of the Congo and Amazon basins).

7. in addition there are the *tropical uplands* (over 1 000 m above sea-level) with very different types of vegetation. However, they can be classified partially, or even wholly, with the above, especially with the dry and humid savannahs. Belonging to this group are large parts of East Africa, where agriculture has in some regions reached a remarkable stage of development because

166

of the influence of the upland climate, efficient sea-ports and foreign populations – particularly the white settlers.

In classifying the forms of agricultural economy in the tropics a progression from the dry to the rainy climates will be followed as far as possible. The most appropriate classification at present is the following, with only four parts:

1. *Pasture farming*, which in the tropics has so far been of a thoroughly extensive character, since there is no cultivation of any kind. The population simply appropriates whatever nature will grow without human intervention and does this by the use of undemanding grazing animals. Stabling is completely unknown, as is the conservation of fodder generally.

2. *Rain-fed farming*, also called *dryland farming*, which involves the cultivation of short-lived crops on the basis of the natural rainfall, i.e., without irrigation. The problem of soil fertility often makes it necessary to intersperse the cultivation of useful crops with bare fallow, grass, shrub or even forest fallow. In extreme cases (e.g. the forest-burning system of shifting cultivation) only 15 to 25% of the available land can produce a crop harvest at any one time.

3. *Irrigated farming*, which gives short-lived crops an artificially increased water supply, either throughout the year or simply in the dry season, so that crop production is often continued through the whole year. With some crops it is then possible to obtain two or even three harvests per year from the same land.

4. *Perennial crop farming*, which involves growth cycles of from several to a great many years, often several decades – as with the cultivation of coconuts, oil-palms, rubber, citrus fruit, coffee, cocoa and tea. The continuous land coverage and the technical processing of the harvested crop are important characteristics of this type of farming. Most farms practise intensive cultivation. There is a widespread tendency towards monoculture and to the plantation scale of farming.

Fig. 3 shows which types of arable farming and animal farming are native to the climatically-determined, natural vegetation belts of the tropics. To gain an overall impression of the vegetational characteristics and agricultural geography of the area, it is especially important to observe that in Figure 3:

— the *aridity limit for habitation* runs through the desert;
— the *aridity limit for animal farming* passes through the semi-desert;
— the *agronomic aridity limit* (limit for rain-fed farming) divides the shrubland steppe and the dry savannah;

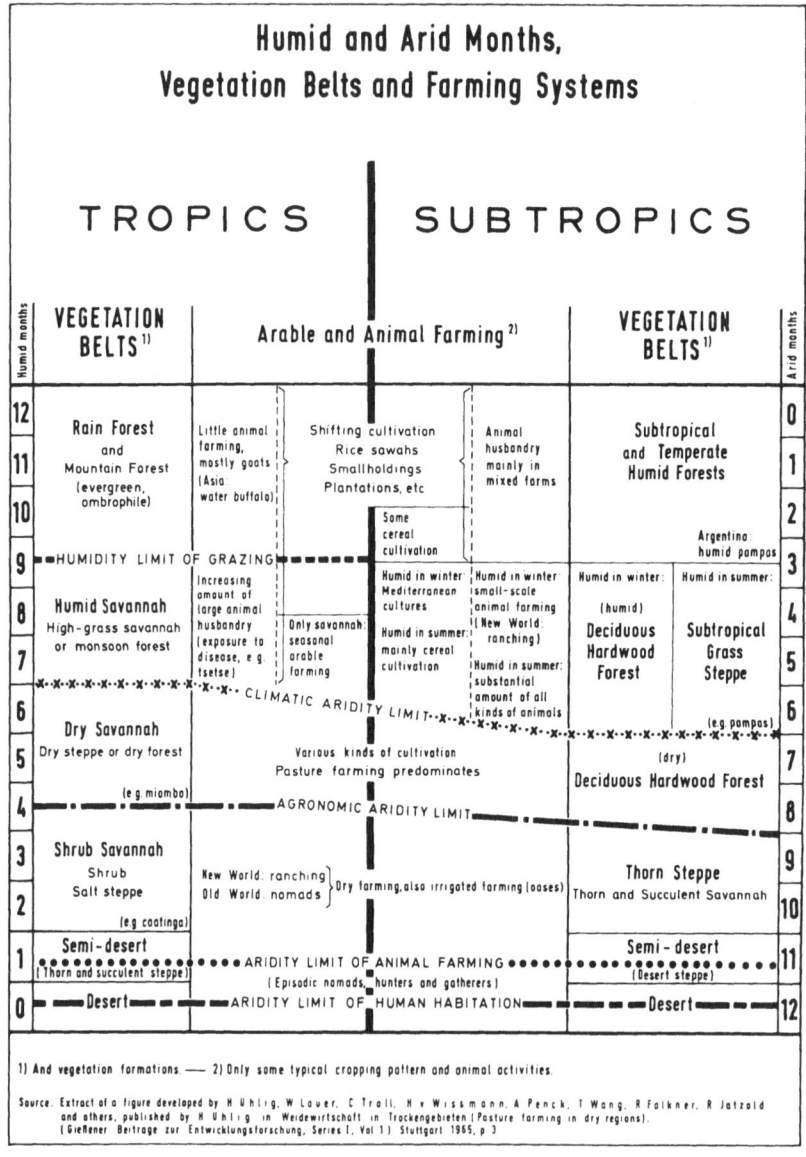

3. Humid and arid months and the forms of vegetation and agricultural economy in the tropics and subtropics.

— the *climatic aridity limit* runs between the dry and humid savannahs and

— the *humidity limit for pasture farming* is identical with the contact zone between the humid savannah and the rain forest.

Finally, in Fig. 4, an attempt has been made to give an over-all picture of the distribution of the four systems of agriculture within the six climatic zones.

4.2   Regions of Extensive Pasture Farming

Pasture farming can be found in all the tropical climatic zones except the rain-forest belt. In the rain forest, and also in parts of the humid savannah, any pasture-land rapidly reverts to bush or forest. A number of animal diseases also preclude pasture farming in the rainy tropics. The most dangerous of these in Africa is nagana, which is carried by the tsetse fly. The main centres of pasture farming are the shrubland-steppe and the dry savannah, i.e., the outer tropics. These regions are mostly semi-arid with about 200 to 400 mm of rainfall annually, though in most regions between 250 and 350 mm. The greater the differences

Farming Systems in the Climatic Zones of the Tropics
(Diagram)

Climax vegetation	Humid months	Ranching	Rain-fed farming	Irrigated farming	Tree and shrub cultivation
1. Deserts	—				
2. Semi-deserts	1	●			
3. Shrubland steppes	1 - 4	● ● ●	●	●	
4. Dry savannahs	4 - 6	● ● ●	● ●	● ●	●
5. Wet savannahs	7 - 9	●	● ● ●	● ● ●	● ●
6. Rain forests	10 - 12		● ●	● ●	● ● ●

● = infrequent   ● ● = moderately frequent   ● ● ● = widespread

4. Farming systems in the climatic zones of the tropics (diagram)

between the amounts of rainfall in individual years, the more the agronomic aridity limit is shifted into wetter zones. Thus in the area of Grootfontein (Fig. 5) in spite of a long-term mean annual rainfall in 534 mm, pasture farming is still predominant, because it is better able to withstand the drought years than dryland arable farming.

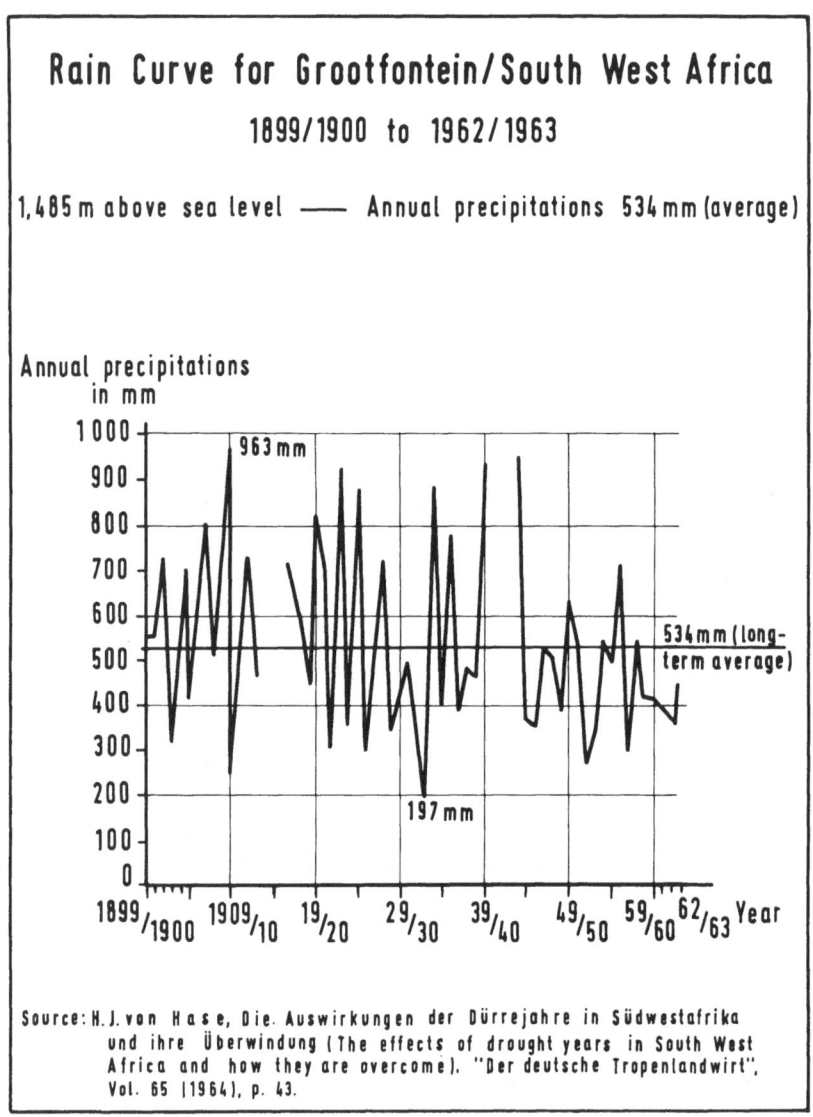

5. Rain curve for Grootfontein/S.W.Africa 1899/1900 to 1962/63

Since the variation of annual rainfall increases with the distance from the equator (Fig. 6), the outer tropics are especially predestined for extensive pasture farming. This is not only because of their general lack of rainfall but, also, because of their great flactuations of rainfall.

R. Schickele (1931) distinguished between three main categories of extensive pasture farming: nomadic herding, representative cattle farming and farming that find its way according to the requirements of the market.

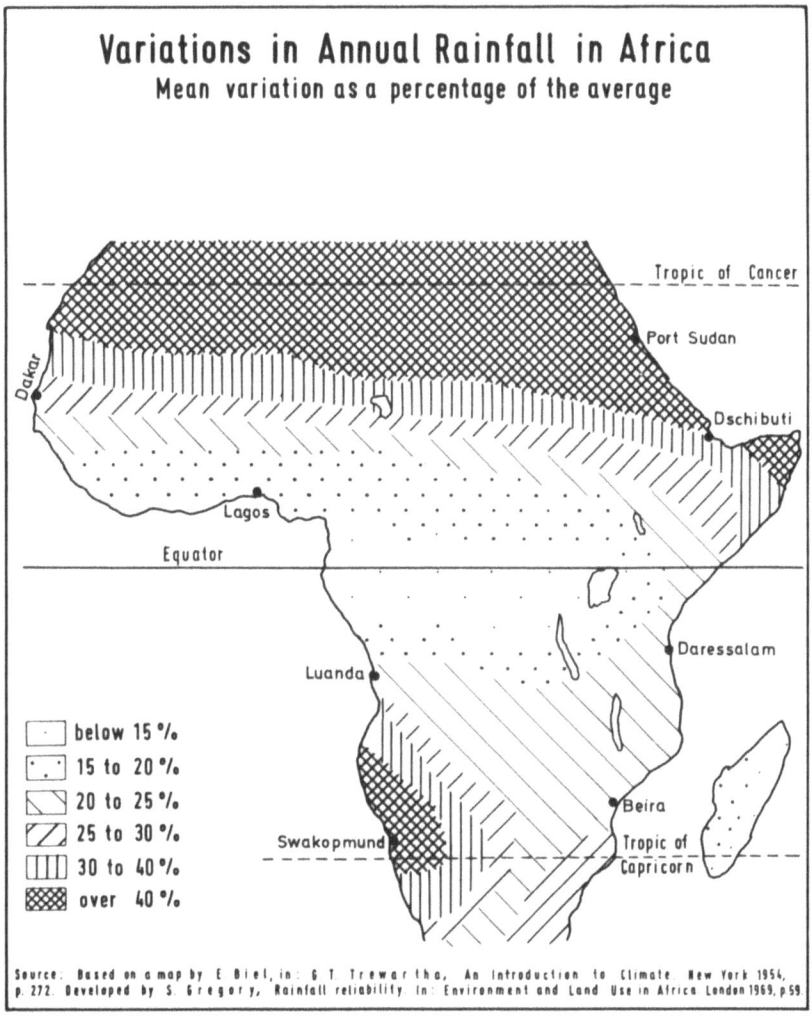

6. Variations in annual rainfall in Africa

171

In the nomadic economy there is a periodic migration of the whole population and herds. Production is predominantly for domestic consumption. Milk, as a basic foodstuff, is therefore the most important product. All kinds of domestic animals are used for milk production: camels, horses, asses, cows, goats and sheep.

The *desert nomads* are obliged to make especially long journeys. They therefore need animals which can walk long distances and at the same time withstand long periods of drought. They keep mainly camels (simultaneously for milk, as beasts of burden and for riding) but few smaller animals and no cattle. Camels, which need to be watered only every two to three days in the warm season and only every two to three weeks in the cool season, can be taken up to 80 km from their drinking place (Ruthenberg 1967). Cattle, on the other hand, have to go to their drinking place twice daily in the warm season in the Sahara, and every two to three days in the humid season. Cattle can therefore only graze within four kilometres of their watering place (Ruthenberg 1967). The marketable production of the desert nomads is very small but their contribution to transport (in the shape of camel caravans) is often considerable.

Higher marketable production is achieved by the *steppe nomads* who more often keep sheep and goats than camels. Horses, too, often play a large part (e.g. in many Arab countries). As the growth of fodder vegetation is somewhat better than in the desert, some cattle can also be kept. Often there are barter and trade relations with neighbouring peoples and thus they can receive grain, tea, tobacco, spirits etc. in exchange for their domestic products. The Mongolians, the Kirghiz and the Berbers of the Algerian plateau are steppe nomads, as were the hordes of Genghis Khan.

The *semi-nomads* are native to areas where fodder vegetation is somewhat more available. Their nomadic journeys are therefore much shorter. This means they are in a better position to keep cattle, which now predominate over sheep, camels and horses. The mountain nomads of North Africa and of Asia Minor are examples of this category. They are able to practise some arable farming at the foot of the mountains bordering the steppe or on the banks of rivers. However, the semi-nomads also include some purely pastoral peoples such as the Fula in Central Sudan, the Masai in Kenya and Tanzania, and to some extent the Heresos in South-West Africa.

As his second category of extensive pasture economy R. Schickele quotes savannah pasture farming with representative cattle herds. This is practised on a settled basis. Cattle predominate but have little economic importance as they are kept more for ritual, social, cult and prestige reasons.

The savannah peoples usually obtain their main food supplies from crop cultivation, and not from their cattle. Representative cattle herding is usual among many Bantu tribes in East Africa and among the Sudanese negroes of the Lake Chad area, for example.

Finally, the third and most important category of extensive pasture farming is the *commercial steppeland farming*. This is also settled farming, and production is mostly of a kind which is easy to transport, such as wool, hides and skins and lean-stock for the market. R. Schickele distinguishes between three degrees of intensity within this category:

Wild steppeland pasture farming is the most labour- and capital-extensive form, since protection of the herds is left to the stallions and bulls which lead them, keep them together and defend them against predators. Wild steppeland pasture farming is only possible with cattle and horses and today no longer plays an important part.

Free-range steppeland pasture farming has an increased labour demand, for the herds have to be under the constant supervision of a herdsman. This mostly involves wool sheep, which can walk for long periods and so cover long distances to reach their drinking places. This means that only a few wells are necessary. The strong herd instinct of the sheep simplifies the shepherd's task of keeping them together.

Finally, *enclosure or paddock farming* in which herdsman labour is partly replaced by capital investment, the pasture area being divided by wire fences, thus making numerous artificial drinking places necessary. Enclosure or paddock farming is therefore more labour *ex*tensive but also more capital *in*tensive than free-range steppe grazing. The result is a relatively high market productivity from sheep and cattle. This system is often combined with the free-range system on the same farm. In this case, the larger animals are kept in paddocks or enclosures near the farm, while the smaller are herded on the more distant and unfenced pastures.

The rainfall situation determines the type of pasture. For

each pasture type there is also a certain line of production from ruminants which is the most competitive.

The position of the aridity limits for the individual lines of production is dependent on the amount of rainfall, and naturally on its distribution also, on the warmth of the climate, the speed of the wind, the soil types prevailing, the geographical relief and other factors. For Central South-West Africa the following approximations can be quoted:

100 to 200 mm of rainfall means such sparse vegetation that only an undemanding breed of sheep, capable of sustained walking, can use it. Sheep also have an advantage over cattle here since they are able to graze more selectively with their pointed muzzles and pick out the most nourishing parts of the plants. Also, the sheep's own peculiar digestive tract makes it satisfied with a lower food concentration. Production is directed exclusively towards obtaining wool or skins.

With 250 to 300 mm of rainfall it is just possible to keep cattle, but only in the form of lean-stock rearing. Calf production is impossible as the sparse growth of fodder vegetation is insufficient for cows to be capable of developing a foetus or of raising a calf at the udder. Work-sharing lean-stock farms, for which calves are bought from zones with better fodder growth, are thus found.

350 to 450 mm rainfall is the typical location for the self-supplying fat-stock farm. The pasture quality is now so improved that home calf production as well as semi-fattening is possible. To avoid the difficulties of work-sharing and to minimize risks, both systems are often combined on the same farm. Thus a herd of cows for calving is kept, all the calves are reared and the semi-fattened stock are sold.

Finally, *from about 500 mm of rainfall* arable farming (and thus fodder-crop production) is possible. This is very important for fodder management, both with regard to the seasonal balance quantitatively and to improving the degree of nutrient concentration. Now high quality can be stressed: quality fat-stock and quality milk production.

4.2.3   SEASONAL FODDER BALANCE AS THE CENTRAL PROBLEM

In extensive pasture farming, not only may every kind of arable farming generally be precluded, but the farm may have to limit itself to one kind of animal and be dependent on a single

174

form of production from it. This means that monoproduction, with all its dangers, is widespread. The scope of production is narrow, its depth slight and the market risk accordingly high.

In addition, there is a high production risk related to the annual rainfall variations, to which an area where no fodder production or buying-in of concentrated fodder is possible, is particularly exposed. Here the farm is solely dependent (in its striking reliance on nature) on whatever plant-growth is produced by the natural rains. The severity of this problem can only be appreciated when one bears in mind that in the dry savannahs and shrubland-steppes, even in years of normal rainfall, there is a very long dry season following a very short and uncertain rainy season. To carry the animal herds through this dry season without great losses is the primary problem for every pasture farmer. This is also the key to any improvement in productivity.

In seasons when fodder is scarce, there are basically two different types of action the farmer can take to combat the situation: he can try to overcome it actively or he can adapt to it passively. Actively overcoming seasons of fodder shortage is the method commonly used in industrialized countries. It consists of collecting fodder in its natural season, conserving it as hay or silage etc. and thus providing a stockpile for the months of fodder shortage: or of buying-in already conserved fodder for these times of short supply. Such measures necessitate high expenditure and therefore presuppose a favourable price-cost ratio for the particular production.

If this condition is not fulfilled, (i.e., if the value of the animal products as against the cost of fodder conservation or purchase, is very low) then neither fodder conservation nor fodder purchase can be contemplated and passive adaptation to the season of fodder shortage must be considered. This consists of accepting the limits set by nature and attempting to avoid undue losses at the times of fodder shortage by a careful choice of the animal-husbandry system. This passive adaptation may take five forms:

1. *Species of animal* are chosen which are capable of withstanding times of fodder scarcity. Thus nomads keep numerous camels and sheep, which can cope with long periods of hunger and thirst. In the highlands of Ethiopia there are many goats, since they will graze and survive on bush foliage during times of drought.

2. Undemanding *animal breeds* are chosen. Fat-tailed sheep or zebus can withstand long periods of hunger because they accu-

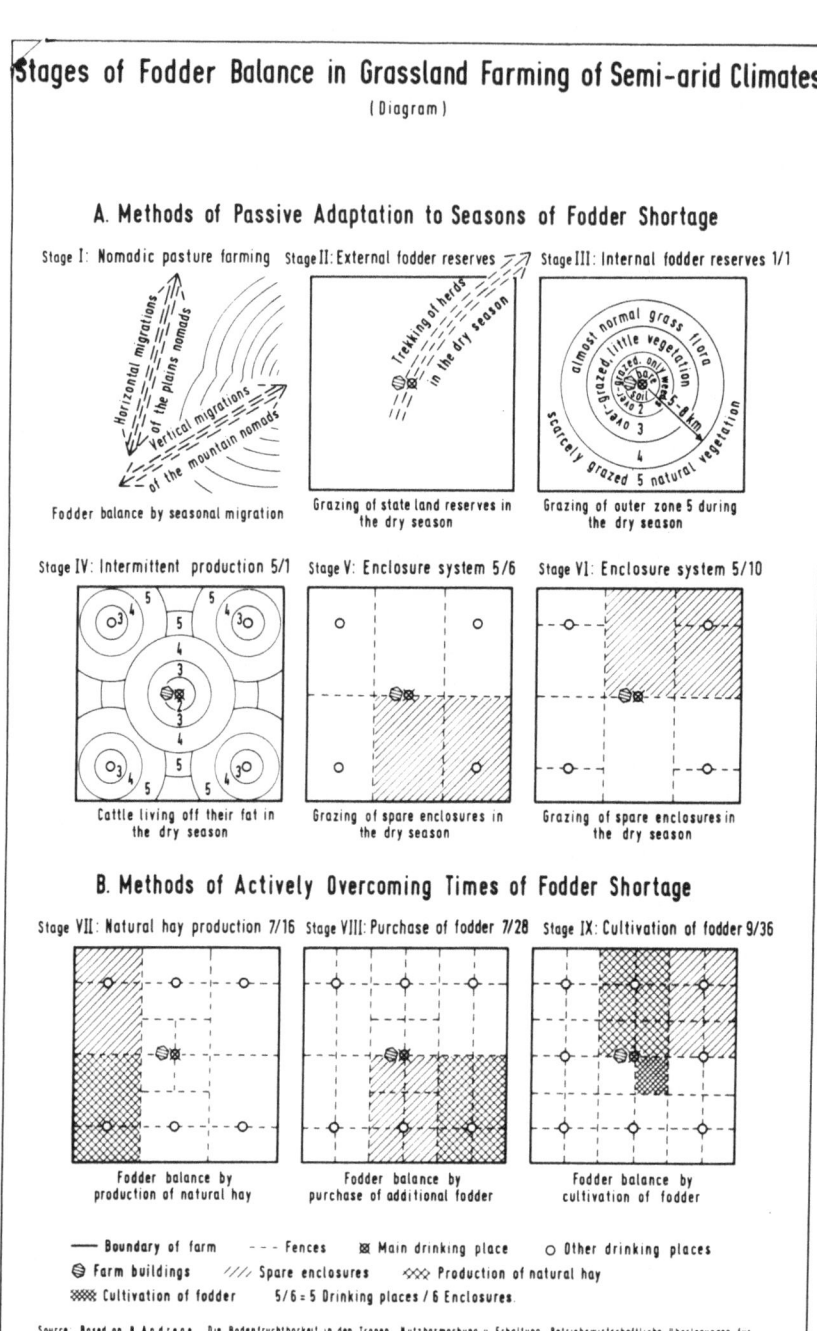

Stages of Fodder Balance in Grassland Farming of Semi-arid Climates
(Diagram)

A. Methods of Passive Adaptation to Seasons of Fodder Shortage

Stage I: Nomadic pasture farming

Horizontal migrations of the plains nomads
Vertical migrations of the mountain nomads

Fodder balance by seasonal migration

Stage II: External fodder reserves

Trekking of herds in the dry season

Grazing of state land reserves in the dry season

Stage III: Internal fodder reserves 1/1

almost normal grass flora
little vegetation
over-grazed, only 1 km – 8 km
scarcely grazed 5 natural vegetation

Grazing of outer zone 5 during the dry season

Stage IV: Intermittent production 5/1

Cattle living off their fat in the dry season

Stage V: Enclosure system 5/6

Grazing of spare enclosures in the dry season

Stage VI: Enclosure system 5/10

Grazing of spare enclosures in the dry season

B. Methods of Actively Overcoming Times of Fodder Shortage

Stage VII: Natural hay production 7/16

Fodder balance by production of natural hay

Stage VIII: Purchase of fodder 7/28

Fodder balance by purchase of additional fodder

Stage IX: Cultivation of fodder 9/36

Fodder balance by cultivation of fodder

—— Boundary of farm — — Fences ▨ Main drinking place ○ Other drinking places
Farm buildings ///, Spare enclosures ✕✕✕ Production of natural hay
Cultivation of fodder 5/6 = 5 Drinking places / 6 Enclosures.

Source: Based on B Andreae, Die Bodenfruchtbarkeit in den Tropen Nutzbarmachung u Erhaltung Betriebswirtschaftliche Überlegungen für die Arbeit in Entwicklungsländern (Problems of Soil Fertility in the Tropics Utilization and Maintenance Reflections on Farm-Management for Work in the Developing Countries) Hamburg and Berlin, 1965, p 92

7. Stages of fodder balance in pasture farming of semi-arid climates (diagram)

mulate fat in times of good fodder supply and thus operate for themselves a food storage system as between times of plenty and scarcity.

3. *Lines of production* are chosen which are least dependent on seasonal fodder balance. Thus with cattle, lean-stock rearing is preferable to milk production: and sheep are preferably used for their wool and skins rather than their meat.

4. *Animal output* is concentrated in the seasons when fodder is plentiful. Thus it is planned that young should be born just before the beginning of the rainy season, so that the mother animals do not go hungry during the months of suckling. The final fattening of the animals is also done during the rainy season.

5. The *size of the herds* is adjusted to the seasonal variations in fodder growth. The herd is stocked up, if possible, for the rainy season, and reduced for the dry season. All births and purchases are therefore planned for just before the beginning of the rainy season, and all slaughtering and sales take place at the beginning of the dry season.

In the diagram in Fig. 7, nine stages of extensive pasture farming are shown in relation to increasing economic development.

As long as labour and capital are scarce but land is plentiful, fodder balance must be achieved by means of using land reserves.

Stage I represents nomadic pasture farming. This system is found in extremely arid areas where communications are poor. For example, this is the type of economy of the steppe Bedouins, the Kurds or the Berbers. In Central Asia, nomadic herding is still widespread, e.g., among the Mongolians and the Kirghiz. In East Africa the Masai, the Fula or the Somalis could be termed at least semi-nomadic. A nomadic population is caused by the necessity of a fodder balance, which is attempted by long migrations. This type of economy is therefore particularly prevalent where the annual vegetation cycle varies from place to place, because of upland ranges, mountains and valleys. The migrations of the mountain nomads tend to be vertical, those of the nomads of the plains horizontal.

When the population later progresses, in *Stage I*, to a settled form of life, permanent ownership of the land occurs. However, as long as the population density is low, a land reserve remains in the hands of the tribe or the state. When fodder shortage occurs in dry seasons, these normally unoccupied areas of land are available as reserve pasture.

This situation changes with *Stage III*. Now all the land is privately owned. The farms are enclosed, but without subdivision into pastures. In the diagram of Stage III it is assumed that there is only one drinking place within the area belonging to the farm. Vegetation zones form around this centre. Near the drinking place there is only bare soil, since the herds kill off all plant growth by the concentration of their excrement and by the severity of their grazing and trampling.

In the *second zone* there is only a small amount of vegetation, of predominantly inferior weeds. Even the *third zone* is over-grazed, although the vegetation is more plentiful.

Only in *zone four* is there almost natural vegetation with many valuable grasses. However, this zone is as a rule already the outer ring of those areas grazed by the animals from the drinking place. The *fifth zone* is not used in the fodder-abundant rainy season, since the walking distance from the drinking place is too great. Natural vegetation predominates in this zone. Thus this fifth ring is available as a fodder reserve for the dry seasons when fodder is scarce.

However, in the course of further development land becomes scarcer and more expensive, so that productive use of it must be intensified. This is at first achieved by capital investment, which is by this stage cheaper. Additional wells are built for *Stage IV*. An increase in drinking places has a threefold effect on increasing animal productivity:

a. Walking distances to the animals' drinking places are reduced. The animals therefore use less energy for this task and can accordingly devote more energy to market output.

b. The central zones without vegetation now disappear completely since each drinking place is used less.

c. The little-used outer zone disappears too, since now all areas are grazed, even in the rainy season.

The farm's vegetation becomes more evenly distributed and more productive in the rainy season, so that herd numbers can be increased. However, the fodder situation in the dry season has become worse because now, not only are the regional land reserves and those outside the farms exhausted, but so also are those within the farms. The foodstuff reservoir for times of fodder shortage, which was previously available in the form of alternative pastures in undeveloped economies and which is stored in barns, stacks and silos in more developed agricultural economies, now has to be stored in the animals themselves in the form of fat deposits.

This stage IV is thus the one in which the animals suffer

most hunger during the dry season. However, this must be accepted here because of the prevailing price-cost ratio. Land is still cheap in comparison with labour and capital, so everything depends on managing production so as to save as much labour and capital as possible, even if this means using large areas of land. This is achieved by intermittent production. Not only is no meat produced during the dry season but weight losses are suffered, which have to be made good once the rainy season begins. This type of intermittent production means that the cattle are not ready for slaughter until they are four or five years old. This farming system is nevertheless economically appropriate since it is only in this way that a meat animal can be produced with a minimum of labour and capital, and thus at minimal cost.

Only when land becomes even more expensive and capital becomes cheaper, is it economic to make the larger capital investment which permits better land use. Only when *Stages V and VI* are reached, is pasture subdivision by fencing possible. Paddocks allow regulated grazing and this results in a more complete use of the land. Different animal species as well as animals of different ages can be assigned to different quality pasture. For the dry season, so-called reserve or spare enclosures are set up, rotated annually, and not grazed during the rainy season. The result of this method of bridging the times of fodder shortage is that the animals at least do not lose weight during the dry season, thus bringing down the age of an ox ready for slaughter to about three years. A three-year-old ox has consumed much less maintenance fodder than a five-year-old of the same weight. The percentage of productive fodder is therefore increased. There is also better land use since soil fertility suffers less than in stage IV, thus increasing the nutrient yield per hectare.

There are two main differences between stage VI and *Stage VII*: one of degree, in that the number of drinking places and paddocks is increased in the course of the intensification; and one of principle, in that a limited amount of natural hay is produced. Besides a further increase in capital investment, increased labour is thus now involved. The hay is stacked in small kraals. Hay in the stacks is richer in nutrients than as a standing crop because it is cut at a time when the nutrients from the shoots and blades of the grass have not yet passed to the seed or the root stock.

Natural hay production represents the first active step in the economic development process to overcome the seasonal fodder shortage. This can begin only when animal products have reached a higher level of purchasing power as regards labour and technical aids; for conservation costs a great deal in wages, machinery and draught animals.

At *Stage VIII* the number of paddocks has risen to 28, so the grazing technique is even further improved. The spare enclosures are now supported in their task of bridging the seasons of fodder shortage by the purchase of additional fodder, as well as by the production of natural hay. Where the degree of industrialisation is higher, the price of land tends to rise so that it is even more important for pasture farmers to increase their land productivity. In this situation, of course, the exchange value of meat animals against fodder concentrates is increased and it is now possible to supplement the feed of the animals with concentrates in their pastures. Maize from distant rain-fed farming zones is used for this purpose.

The transportation of maize has become possible, because at this stage of development communications in the country are considerably improved. Feeding concentrates to the animals has now become profitable for a further reason. The mass income of the population grows with increasing industrialisation, and with the rise in consumption levels, better-quality meat animals command higher prices. The extensive pastures themselves are unable to produce such high-quality meat animals, so final fattening with concentrates is undertaken in the dry season in order to attain a higher quality.

The difference in principle between *Stage IX* and stage VIII is that now a third and further method of actively overcoming the season of fodder shortage has been introduced: irrigated cultivation of fodder crops. This growing of fodder crops presupposes a higher stage of economic development: a high exchange value of animal products, not only against labour, but also against commercially-available farming aids such as machinery and mineral fertilizers.

4.3   Regions of Rain-fed Farming (Dryland Farming)

Of the six tropical climatic zones named at the beginning, only the last three, and especially the humid and dry savannahs plus various tropical uplands, can be considered for rain-fed farming. The basis of rain farming was in all cases the ancient system of shifting cultivation, in which only a few years of arable farming follow several or very many years of original vegetation. The further development of arable farming has followed and is still

180

following completely different courses, depending on the climate.

RAIN FARMING NEAR THE AGRONOMIC ARIDITY LIMIT (DRY SAVANNAH)

Rain farming in the dry savannah (also called short-grass savannah) is influenced by the atmospheric humidity, solar radiation, wind speed, type of crops, etc., but in general begins wherever the rainfall level is 250 to 400 mm (cf. Fig. 8) in Africa, 300 mm in Iran and 350 to 400 mm in Arabia and Central Asia. However, where the rainfall distribution is particularly disadvantageous, this agronomic aridity limit can be associated with as much as 600 mm of rain annually, as in parts of East Africa.

The origin of farming here is the shifting-cultivation system which also formed the basis of two thousand years of development in large parts of Europe. Referring to this cultivation system, C. Tacitus wrote in his "Germania": "Arva per annos mutant et super est ager". According to F. Aereboe, this is to be understood as follows: "The cornfields change each year and there is enough land left for arable use".

As long as the density of settlement is sparse and the land under consideration for arable use is predominantly natural grassland, the farmers break up a small piece of the grassland with a digging stick or hoe (or later with a plough) and then plant crops (mainly millet) on it for two to four years only – until yields become very low owing to the lack of thorough soil cultivation and an absence of any kind of manuring. Then the land is again left to revert to its natural grass cover for as many years as are needed for its soil fertility to be restored. Meanwhile another piece of land is taken instead and worked in the same way. Later on, the farmers may again grow grain on the pieces of land used previously, but only at irregular intervals of many years.

The further development of the steppe or savannah shifting-cultivation system towards more productive forms of agriculture
– is made necessary by population increase,
– is simplified by technical progress in soil cultivation, fertilisation, plant breeding etc., and
– is aimed primarily at increasing the proportion of land used productively.

If three years of cultivation follow fifteen years of grass-fallow, the proportion of land used is only one sixth. However, if

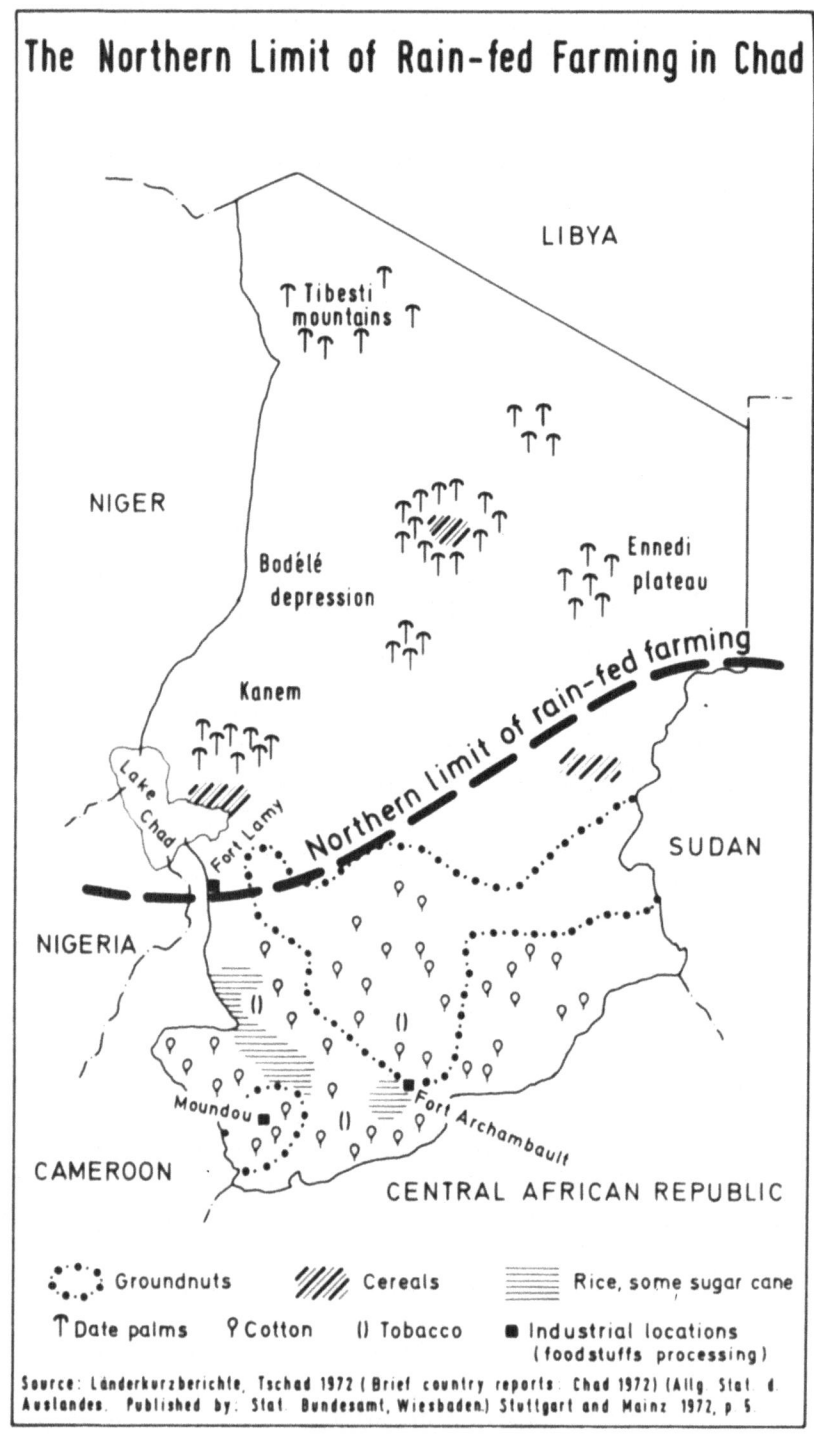

The Northern Limit of Rain-fed Farming in Chad

LIBYA

NIGER

⊤ Tibesti
mountains

Bodélé
depression

Ennedi
plateau

Kanem

Lake Chad

Northern limit of rain-fed farming

Fort Lamy

SUDAN

NIGERIA

Moundou

Fort Archambault

CAMEROON

CENTRAL AFRICAN REPUBLIC

⋰⋱ Groundnuts   ⫽⫽ Cereals   ☰ Rice, some sugar cane

⊤ Date palms   ♀ Cotton   () Tobacco   ■ Industrial locations
(foodstuffs processing)

Source: Länderkurzberichte, Tschad 1972 (Brief country reports: Chad 1972) (Allg. Stat. d.
Auslandes. Published by: Stat. Bundesamt, Wiesbaden) Stuttgart and Mainz 1972, p 5.

8. The northern limit of rain-fed farming in Chad

182

four years of arable alternate with twelve years of grassland, the proportion increases to one quarter. Finally, if eight years of grass are followed by four years of arable, a figure of one third is reached. A precondition for this sort of development is, however, that the fertility-increasing effect of the natural grassland is replaced by other measures encouraging fertility — but this is a difficult task in the dry savannah.

The more the grass-fallow is reduced, the more the water shortage is a limiting factor and determines the farm organisation. The farmer faces the alternative of

— being content with crops which need little water but whose yields are low, simply in order to be sure of obtaining an annual harvest,

— choosing, as in the dry farming system, crops needing more water but also yielding more, and at the same time accepting the need for fallow areas to store water.

For the *first method* crops needing little water are those such as millet, ground-nuts, Bechuana beans, chick peas, sesame and sisal. This explains the very basic importance of millet growing for all the arid regions of Africa and also the predominance of ground-nuts in the states bordering the Sahara, such as Niger, Chad, Upper Volta, Mali or Senegal. Sisal comprised 28% of Tanganyika's total exports in 1962, whilst ground-nuts and ground-nut products accounted for 69% of Niger's exports in 1967.

The *second method*, the dry farming system, generally demands a higher level of technology, and is therefore easier to practise in more industrialized states. Thus the main areas using the system are the arid regions of the U.S.A. and Canada, of Argentina and Australia, of the Republic of South Africa and of the U.S.S.R. This system is also found in the northern Sahara border states and in Iran and Turkey, as well as in many other countries. The main crops are wheat and barley; sometimes cotton. As shown in Fig. 9, in a very dry climate (about 300 mm of rain) with a crop rotation of a bare fallow followed by wheat (or with a shorter rainy season, a bare fallow followed by summer barley) half the arable area has to be devoted to fallow. With increasing rainfall, the proportion of fallow can be cut back, although soil conserving crops — apart from the ground-nut which needs very little water — are being introduced only slowly. The fallow serves to increase the yield of the following cereal crop, primarily by saving water and, in the second place, by releasing nutrients. This whole organisational principle of the dry farming system is, indeed, based on the fact that bare soil evaporates less water than a plant coverage, so that some of the rain falling in the fallow year can be of benefit to the following cereal crop.

Zones of Intensity of the Dry Farming System (Diagram)			
Rainfall mm / year	Long	Short	Fallow % arable land
	Rainy season		
ca. 300 mm	(A) 1. Bare fallow 2. Wheat	(B) 1. Bare fallow 2. Barley (four-row)	50 %
ca. 350 mm	(C) 1. Bare fallow 2. Wheat 3. Millet	(D) 1. Bare fallow 2. Groundnuts 3. Millet	33 %
ca. 400 mm	(E) 1. Bare fallow 2. Wheat 3. Wheat		33 %
450 – 500 mm	(F) 1. Bare fallow 2. Wheat 3. Maize, Blue lupins 4. Wheat	(G) 1. Bare fallow 2. Groundnuts 3. Sorghum, Millet 4. Millet	25 %
Transition to humid savannah	(H) 1. Sunflowers, Groundnuts 2. Maize	(J) 1. Dried peas 2. Wheat	──

9. Zones of intensity of the dry farming system (diagram)

The humid or high-grass savannah (also called the savannah forest zone) is the climatic area of the tropics where rain farming has the largest part to play. In more arid zones it is confined by grassland, and in more humid ones by tree and shrub cultivation.

The predominant short-life field crops in the humid savannah are maize, rice, phaseolus, ground-nuts, and others. Ecological conditions here are also acceptable, though not yet optimal, for cassava and yams. Table 1 shows some examples of crop rotations:

TABLE 1.

Crop rotations for rain farming in the humid savannah

A. West Africa	B. Malawi	C. India
1–15. forest fallow	1. ground-nuts	1. sugar cane
16. hill rice	2. cotton	2. vegetables
17. beans, yucca	3. rice	3. rice
18. cassava		

In Africa and India, mixed cropping is very important on small farms. Examples given by Könnecke (1967) are:

West Africa: yams and guinea corn; millet and guinea corn; maize and oil pumpkin; hill rice and cotton; maize and ground-nuts.

India: cotton and Italian millet; bush peas and Italian millet; cotton and coriander (spice plant).

Ethiopia has further important crops for its upland regions such as teff (*Eragrostis abyssinica*), from which the popular flat loaves (indjera) are made and whose straw is rich in nutrients and makes a welcome fodder in the dry season; or the oil-producing plant nug (*Guizotia olifera*) which is excellent for combating weeds and whose oil does not become rancid even under primitive storage conditions.

Where the climate of the humid savannah has some cool months, as in the tropical uplands, crops favouring moderate climates (such as wheat, barley, vegetables and also peas and potatoes) are grown. There is wheat growing in Kenya at 2 000 m above sea level, in Ethiopia (which is further from the equator) at 1 700 m above sea level. All African wheat lands also produce barley, but not vice versa. For where the rainy season is shorter and the climate drier (as in the countries between the Sahara and the Mediterranean) wheat is out-stripped by barley, because of the latter's special properties — a short life cycle and a resistance to drought. However, these areas can no longer be considered humid savannah.

In the humid savannah the ancient system of shifting cultivation again forms the basis of the field systems. According to the natural vegetation these are alternately forest and arable, or alternately grass and arable. Incentives and aims in overcoming this system are the same as in the dry savannah but the measures to be taken differ. The central problem here is not the economic management of scarce ground-water, but the maintenance of soil fertility without involving long years of natural vegetation. It is now necessary to combat humus losses without using traditional methods of years of forest or grass fallow to limit erosion damage, at least to moderate the erosion of nutrient content, to master weed growth and to prevent disasters caused by plant diseases and plant pests. All this is most succesfully achieved by plant cover throughout the year, but it is precisely this which is prevented by the dry season.

As the population grows, an increasingly large proportion of the usable land must be cultivated and the areas available for grass or forest fallow diminished. Thus in the humid savannah, the best replacement for fallow under original vegetation is a cultivated grass crop. Examples are shown in Fig. 10.

The percentage of grass cultivation necessary to conserve the soil is considerable and mostly far exceeds the amount usable for fodder. Ley farming is technically possible but, at the beginning, is an economically-expensive rotation system. As long as the market for milk and meat in developing countries remains poor and the communications systems incomplete, part of the fodder growth cannot be utilized. Therefore the total cost of the grass crop must be accounted as the cost of promoting soil fertility. But this situation will change. There is no doubt that time will operate in favour of the ley farming system of soil conservation in the humid savannah.

4.3.3   RAIN FARMING NEAR THE AGRONOMIC HUMIDITY LIMIT (TROPICAL RAIN FOREST CLIMATE)

There are not a large number of short-lived crops which prefer the hot and permanently wet climate of the tropical rain forests.

186

Root crops such as manioc (cassava), sweet potatoes and yams are grown, but these are typical crops for the farmer's household. As far as marketable grain crops are concerned, rice, with appropriate water supplies, and maize, because of its wide ecological distribution, are possibilities. Sugar cane grows well here. The

Ley Farming in African Humid Savannahs

Kenya Highlands (Kericho District)[1]

Large farms		Small farms	
A	B	C	D
1.-3. Cultivated grass	1.-3. Cultivated grass	1.-4. Natural grass	1.-5. Natural grass
4. Potatoes – Beans	4. Potatoes – Beans	5. English potatoes – Vegetables	6.-8. Maize
5.-7. Pyrethrum	5.-6. Wheat	6.-9. Maize	9. Sorghum, millet
8. Maize	7. Barley		10. Sweet potatoes – Vegetables
Grassland % of arable land			
38	43	44	50

South-East Africa

Rain-fed farming		Irrigated farming
E	F	G
Malawi[2]	North-Transvaal	Mid-Transvaal (Rustenburg)[3]
1.-3. Cultivated grass	1.-3. Cultivated grass	1.-3. Lucerne
4. Tobacco	4. Sorghum, millet	4. Maize silage
5. Cotton	5. Sorghum, millet	5. Summer: Grass silage; Winter: Green oats
6. Groundnuts	6. Bechuana beans	6. Summer: Bean hay; Winter: Wheat
7. Cotton		7. Summer: Tobacco; Winter: Green oats
8.-9. Maize		8. Summer: Tobacco; Winter: Wheat
		9. Summer: Maize silage; Winter: Wheat
Cultivated grass or lucerne % of arable land		
33	50	33

1) H. Niederstucke, Bodennutzungsformen in tropischen Höhenlagen (Types of land use in tropical highlands).„Landwirt im Ausland", Vol. 4 (1970), p. 76.— 2) 6ha useful agricultural land, 1300 mm rain, 4 workers per farm.— 3) 18 ha useful agricultural land, 600 mm rain, dairy cattle; 55 ar principal feedstuffs cultivation/cattle unit; gross return = 2,540 DM/ha, net return 1,000 DM/ha useful agricultural land.

10. Ley farming in the African humid savannahs

margins to be made from rain-fed farming for the market are not very wide, once the still widespread system of shifting cultivation by self-supporting farmers has been overcome. Above all, the soil fertility situation is even more precarious here than in the humid savannah, once the system of forest fallow in regenerating the soil can no longer be used.

The system of shifting cultivation (cf. Table 2) is the most important agricultural system of the rainy tropics and is still predominantly employed by almost 5.7% of the world population on over 25% of the land surface of the earth. The system is based on arable land being periodically given over to resting fallow under original vegetation. In the rainy tropics the annual rainfall is at least 1 200 mm, and the natural vegetation is forest. Thus the ancient shifting system really meant an alternation of

TABLE 2.

Examples of the forest-burning system (shifting cultivation) in the humid tropics.

Country or Region	Rainfall mm/yr	Crops	Original Vegetation	No. of Crop yrs	No. of Fallow yrs	Cropped area in % of land available
Zambia	approx. 1125		Miombo dry forest	2	up to 25	7 plus
Sarawak	approx. 3750	hill rice	rain forest	1	12 plus	at most 8
Liberia	2000–4500	rice, cassava	rain forest	1–2	8–15	11–12
Sumatra	approx. 2250	rice, tubers	rain forest	2	10–16	11–17
Assam	approx. 2500	rice, millet, maize	rain forest	2	10–12	14–17
Sierra Leone	2250–3250	rice, cassava	rain forest	1.5	8	16
Central Congo	1750 plus	rice, cassava, maize	rain forest	2–3	10–15	17
Guatemala	3075	maize	rain forest	1	4	20
Philippines	2500 plus	rice, maize, tubers	rain forest	2–4	8–10	20–28
Nigeria, Umuahia	approx. 2250	maize, yams, cassava	bush (Acioa barteri)	1.5	4–7	18–27
Abeokuta	approx. 1250	.	bush-thicket	2	6–7	22–25
West Africa	1500–2000	maize, cassava	semi-humid forest	2–4	6–12	25
Zambia	approx. 1250	.	bush-thicket	6–12	6–12	50

Source: Nye, P.H. & D.J. Greenland: The soil under shifting cultivation. (Commonwealth Bureau of Soils. Technical Communication. No. 51) Bucks 1960, p. 128.

field and forest, and – as forest can only be reclaimed for cultivation by burning – it is called the forest-burning system.

Thousands of years of dominance of the forest-burning system in the rainy tropics, and the tenacity of the inhabitants in retaining it, can be traced back in the first place to the fact that the clay-humus aggregates, which are so important in moderate climates, disintegrate in the soil of the rainy tropics. This is caused by rapid humus decomposition, once the natural vegetation has been removed. The clay by itself, unlike that found in the middle latitudes, possesses only a very limited power of sorption.

In Central Europe, retention, storage and subsequent releasing of plant nutrients are mainly functions of the clay-humus aggregates. In the rainy tropics, these functions have to be transferred, for the most part, to live or dead plant matter.

In the forest-burning system a large amount of raw material for humus is built up under forest fallow and a large supply of plant nutrients is stored. Forest clearance by burning mobilizes some of these nutrients in the form of the ash, so that the crops planted with hoes in the friable, loose, weed-free and nutrient-

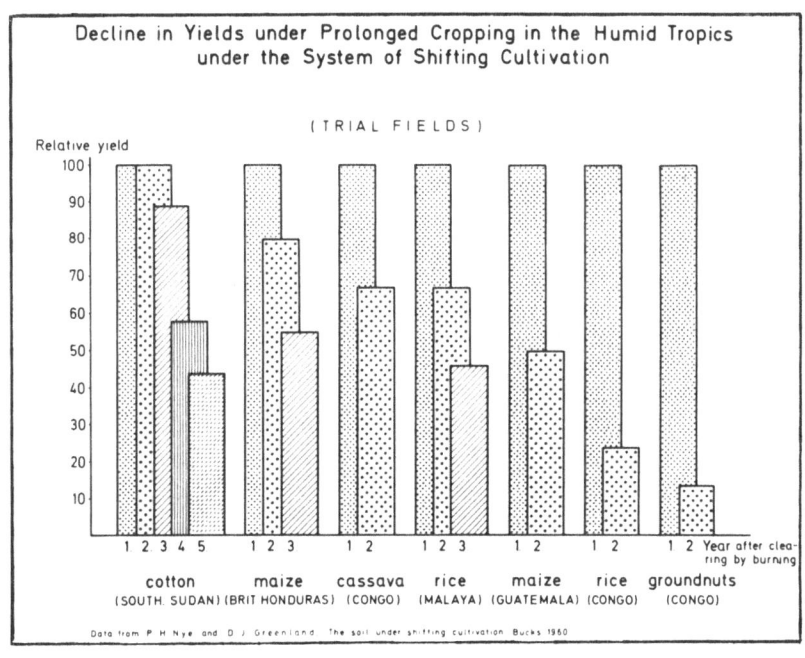

11. Decline in yields under prolonged cropping in the humid tropics

rich soil will produce relatively high and certain yields in the year of clearance. In the second and third years of cultivation there is still a supply of nutrients, because of mineralisation of the underground organic matter of the forest fallow. However, yields are already noticeably reduced while resistance to cultivation, structural disintegration, erosion and weed growth are all increased. The yield — expenditure ratio becomes progressively worse, and after three (or at the most four) years of cultivation is no longer satisfactory. The cleared area is then left to the forest with the aim of restoring the soil fertility, whilst another recovered patch of forest is burnt down and taken into cultivation instead (cf. Fig. 11).

Biologically speaking, this system of shifting cultivation must be regarded as an ideal solution for farming in the humid tropics. It is reasonable and economically acceptable in sparsely populated agricultural states only. Its great disadvantage is its very low capacity for foodstuff production, since only about 20% of the available land yields a harvest in any one year. Even on the fertile, young volcanic soils of the island of Java, the shifting-cultivation system is able to feed only 40 to 50 people per sq.km., and on poorer soils this figure is much lower. Today many developing countries have passed the critical limit and are faced with the problem (often still unsolved) of replacing the forest-burning system by more productive methods.

As this change causes serious difficulties, an attempt should first be made to make the system itself more efficient in food production. A first step can be a transfer from "haphazard" to "regulated" shifting cultivation (Fig. 12). The next step would be the introduction of artificial instead of natural afforestation, which with proper management could possibly allow the great technological stride from hoe to plough cultivation. Plough cultivation considerably increases both soil and labour productivity, since each worker can now cultivate a far larger area of land.

As the population increases further, however, even this farming system is no longer sufficiently productive and abandonment of the forest-burning system becomes inevitable. Forest fallow must now be replaced by a fallow under the kind of short-lived plants which can restore the soil fertility more quickly, thus resulting in a higher proportion of land use. At first legumes needing few fertilizers are chosen. Later, when an appreciable use of mineral fertilizers has become possible, deep-rooting grasses appropriate to the region can be planted. Admittedly, the surface growth cannot be used for cattle at first for several reasons — transport facilities in the tropics are insufficient, the population's purchasing power is low and also there is often widespread nagana disease. This means that the total cost of cultivating and planting the fallow areas must be regarded as a cost of conserving soil fertility. Nevertheless, the proportion of the available land which yields a harvest each year increases in the course of this development to between one and two-thirds. Only in a higher stage of development can all available land yield a harvest each year, for then quite different methods can be introduced. The increased earnings of the population (now predominantly urban) and the country's improved communications will now permit a change to ley farming in suitable locations. Other locations can start to concentrate on perennial crop cultivation (in

Yield Level of Crops under Shifting Cultivation
Depending on the Cropping Interval
(Examples)

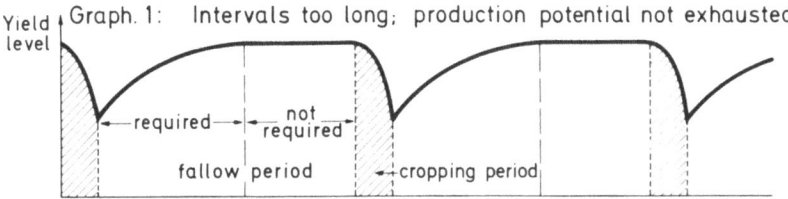

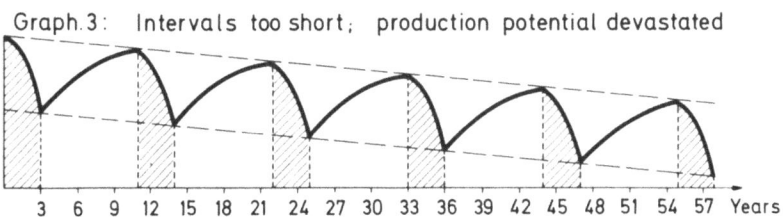

Adapted from van der Pool and H Ruthenberg, Agricultural Journal „Agrarwirtschaft", Hannover, vol 14 (1965),p. 28

12. Yield level of crops under shifting cultivation

line with increases in export and world trade). This can conserve soil fertility in the humid tropics far better than arable farming: and once export trade has been sufficiently developed, its products can be traded for basic foodstuffs from other climatic belts.

The succession of farming systems described here is directed towards an increasing intensification of cultivation, first by means of increased labour and later through more and more capital investment. As a development policy this evolution can be accelerated and directed by three different sets of measures:

— by an *infrastructure policy* aimed at developing the country's communications, lowering transport costs and reducing the difference between market prices and farm prices so that farmers in the humid tropics (who are mostly producing for very distant markets) can increase the exchange value of their products against commercially-manufactured equipment and labour;

— by an *industrialization policy* which raises general income levels and thus the level of agricultural incomes. Industrialization also reduces the price of the commercial goods needed as agricultural equipment; and finally,

— by an *export policy* which leads to world-wide trading, so that agriculture in the humid tropics can extend the cultivation of perennial crops much more than at present. Soil and climatic conditions in the humid tropics are most favourable to these crops, which are much better able to maintain soil fertility there than is arable farming generally.

4.4   Regions of Irrigated Farming

One basic function of irrigated farming in the inner tropics is the maintenance of soil fertility. The layer of surface water assumes to some extent the soil-protecting function of the forest. The water also prevents organic matter decomposing too rapidly and provides the soil with nutrients. Thus irrigated rice is the only crop that has been grown in the tropics for centuries on the same land without crop rotation or use of fertilizers and without causing soil disintegration. If the rain forest is the ultimate vegetation of the natural equatorial landscape, irrigated rice can be considered the ultimate vegetation of the agricultural landscape of the humid tropics (von Uexküll 1969) (Fig. 13). This function of the irrigation water, plus the fact that the inner tropics have a very plentiful water supply and temperatures are sufficient for

192

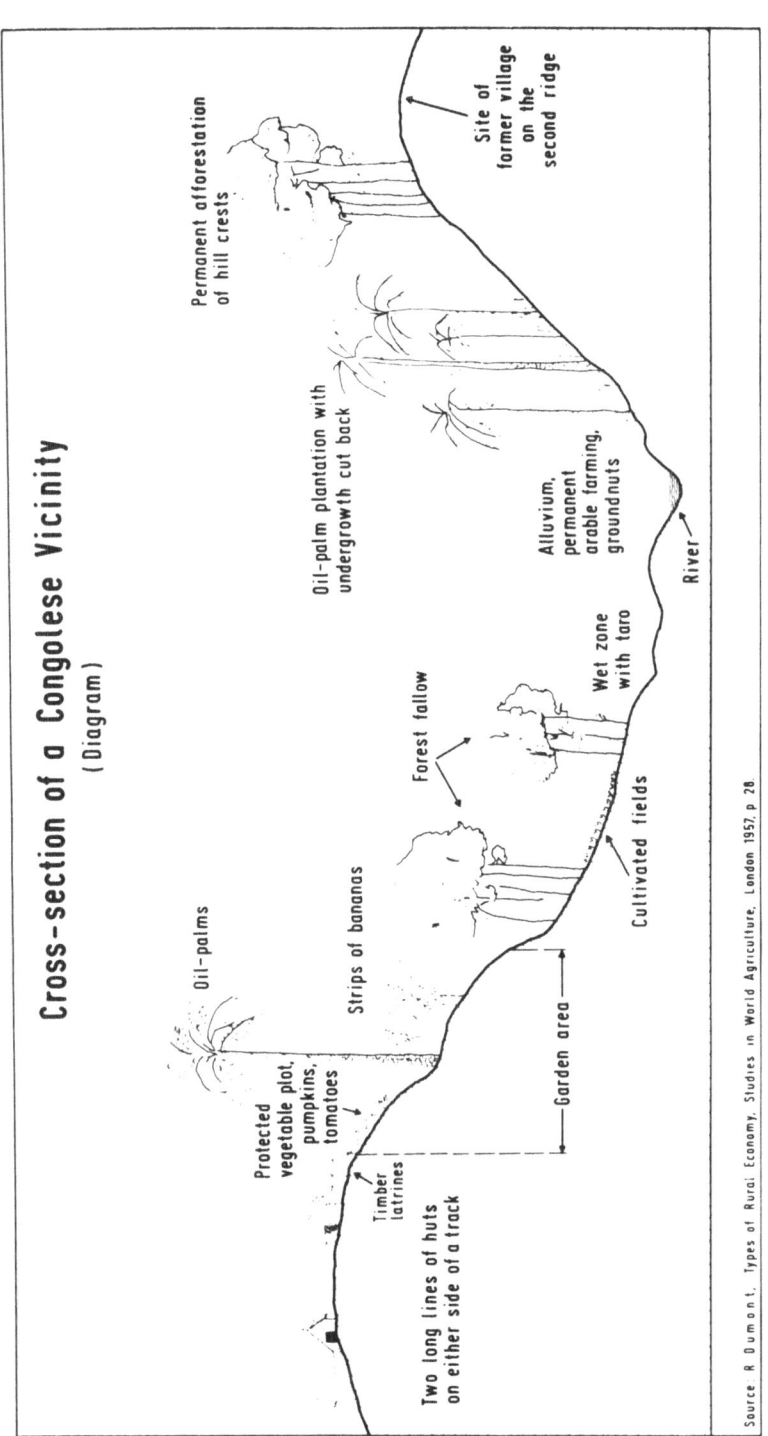

Cross-section of a Congolese Vicinity
(Diagram)

Permanent afforestation of hill crests

Site of former village on the second ridge

Oil-palm plantation with undergrowth cut back

Alluvium, permanent arable farming, groundnuts

River

Wet zone with taro

Forest fallow

Cultivated fields

Oil-palms

Strips of bananas

Garden area

Protected vegetable plot, pumpkins, tomatoes

Timber latrines

Two long lines of huts on either side of a track

Source R Dumont, Types of Rural Economy, Studies in World Agriculture, London 1957, p 28

13. Section of a Congolese communal field.

plant growth all the year round, explains why there is more irrigation in the tropical rainy climates than in the tropical dry climates. In South America 7.0% in Africa 10.7% and in Asia 21.8% of arable land is irrigated.

4.4.1 AGRICULTURAL AND ECONOMIC FUNCTIONS OF CROP IRRIGATION

Irrigated farming within and beyond this context has the following agricultural and economic functions (Ruthenberg 1967):

a. to increase gross yields per hectare by
— increasing the yield per hectare of the individual crop;
— allowing several crops per year;
— allowing cultivation of heavier-yielding crops;
— and allowing additional use of labour and production equipment.

b. to allow constant land use without interruption by forest, grass or bare fallow.

c. to balance out harvest fluctuations and thus provide a more even food supply for both man and animal.

d. to provide greater flexibility, as far as intensity and programme of production are concerned, than is the case in rain farming or with tree and shrub cultivation.

e. to increase the capacity for foodstuff production, i.e. to reduce the minimum size of farm that is necessary to a family. Using hoes, a family can hardly cultivate more than 2 to 3 hectares. In the rain-farming system this means a poor existence, but with irrigation it means an ample one.

f. to extend cultivation to dry areas not otherwise capable of producing crops.

These advantages, however, have to be acquired at considerable technological and material expense. Gross investment in water supply, canalization, levelling the land etc. amounts to 4 000 to 10 000 DM per hectare. Irrigated wheat growing in China needs 58% of its total labour for routine maintenance of the water supply (Ruthenberg 1967).

4.4.2 COMPARISON OF THE VARIOUS SYSTEMS

Table 3 contains some examples for irrigated farming. It shows, first, that this system is not limited to the rainy climates of the tropics, nor indeed to the tropics as a whole. It extends far

194

beyond these into the subtropics (Turkey, Egypt, Southern Japan). It is especially in these subtropics, where the summers are dry, that irrigation can increase land productivity immensely. The same applies to the tropics where humidity varies, for irrigation can

TABLE 3.

Examples of crop rotation for irrigated farming

A. Annual crop rotations

Egypt (200)		Southern China (300)	Formosa (400)
a) summer:	rice	a) rice	a) sweet potatoes
b) winter:	cereal	b) rice	b) rice
	legumes	c) wheat	c) vegetables
	fodder crops		d) rice
	vegetables		

B. Biennial crop rotations

Turkey (100)	India (150)	Southern Japan (200)
1. rice	1. sugar cane	1. a) rape
2. wheat	2. a) cotton	b) rice
	b) fodder crops	2. a) grass
		b) rice

C. Triennial crop rotations

Uttar Pradesh (100)	India (100)	Formosa (167)
1. sugar cane	1. vegetables	1. sugar cane
2. ratoon	2. rice	2. a) sweet potatoes
3. pearl millet	3. sugar cane	b) ground-nuts
		3. a) ground-nuts
		b) rice

Notes: (...) = degree of land use = annual harvest area in % of total arable area. – 1., 2. etc. = year, a), b) etc. = successive crops in the same year.

Sources: Ruthenberg, H.: Organisationsformen der Bodennutzung und Viehhaltung in den Tropen und Subtropen. In: Handbuch der Landwirtschaft und Ernährung in den Entwicklungsländern, edited by P.v. Blanckenburg & H.-D. Cremer, Vol. 1. Stuttgart 1967, pp. 159 & 167. – Tsuzuki, T.: Die Fruchtfolgen des japanischen Ackerbaues. "Berichte über Landwirtschaft", Hamburg and Berlin Vol. 41 (1963), p. 837. – Wang, Y., F. Nagel & H. Ruthenberg: Bodennutzung und technischer Fortschritt auf Taiwan. (Zeitschrift für ausländische Landwirtschaft, Sonderheft 7) Frankfurt/Main 1969, p. 18. – Franke, G., et al.: Nutzpflanzen der Tropen und Subtropen, Vol. II. Leipzig 1967, p. 189 ff.

make crop production possible throughout the year. Long-living perennial crops such as perennial fodder crops and sugar cane can then be succesfully cultivated, or, if short-lived crops (rice, vegetables, sweet potatoes) are grown, several successive harvests can be produced in the same year. Thus there are paddy fields in the tropics producing three, or even four harvests annually (cf. Fig. 13). In Table 3, the example of Southern China shows three harvests, and the example from Formosa even four harvests in the same year. This increase in cropping by irrigation is of great value, especially to overpopulated agricultural countries and to small farms, as well as under all conditions of production where the price-cost ratio makes increased land productivity the first economic goal.

Cotton is the most important cash crop to be irrigated. For example, in 1967 it represented the following percentages of the total export values: Chad 83%; Sudan 55%; Egypt 69%.

4.5 Regions with Perennial Crop Cultivation

Finally, the fourth major category of farming systems in the tropics (i.e. those cultivating tree and shrub crops) must be described. These mostly occur in the inner humid tropics, that is in the uplands near the equator, in the sub-humid savannah and especially in the humid and evergreen tropical rain forest. Peasant farmers in developing countries do not concentrate exclusively on tree and shrub crops, but operate rain-fed and/or irrigated farming at the same time, at least to the extent of supplying their own food (cf. Fig. 14).

4.5.1 AGRICULTURAL AND ECONOMIC CHARACTERISTICS OF PERENNIAL CROP CULTIVATION

In countries with tropical rain climates, considerable proportions of total exports consist of the products of perennial crop cultivation. For example:

Indonesia	40% Rubber,
Ghana	60% Cocoa,
Tanzania	28% Sisal and Coffee
Ethiopia	56% Coffee.

Moreover, the extreme importance of plantation crops such as coffee in Brazil, rubber in Malaya, tea in India and Ceylon, sisal in Kenya and sugar cane in Java or Cuba should be mentioned (though sisal and sugar cane are on the borderline between field crops and perennial crops).

Sisal, sugar cane and tea tend to be grown on a large-scale

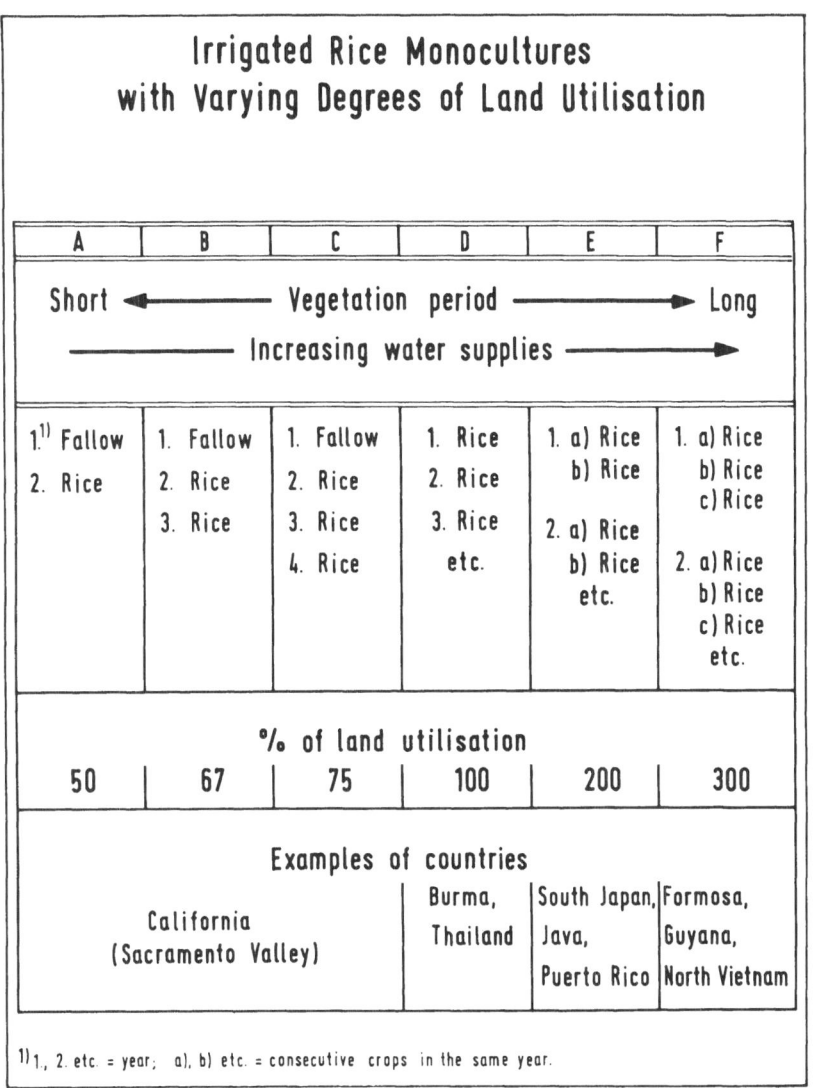

14. Irrigated rice monoculture with increasing degree of land use.

197

(plantation) level, so that the expensive factory plant can work at full capacity from a small collecting area. For sisal and sugar cane an additional problem is the transporting of a crop with so much bulky waste. This makes it imperative to grow the crop close to the factory. The fibre yield of sisal is only 3 to 4% but this processed product is well worth transporting all over the world. Almost the same applies to sugar.

On the other hand, oil palm, coconut palm and cocoa are true peasant crops. The widely varying quality of the large amounts that are being exported is a disadvantage and often requires help from marketing boards. Finally, rubber and coffee are possible crops for both the large-scale farmer and the peasant farmer. Coffee is predominantly a plantation crop in Angola, Kenya, Tanzania and Zaire, while production by peasant farmers has been developed in Uganda and Ethiopia. Thanks to the dry processing technique for coffee, production is possible even from the smallest area of land and just a few trees.

The particular characteristics of tree and shrub crop cultivation are mainly attributable to the fact that they are permanent crops. According to G. B. Masefield (1948), useful economic lives are as follows:

Pineapple	6 to 13 yrs	Oil palm	50 yrs
Bananas	20 „	Tea	over 50 „
Pepper	25 to 30 „	Coconut palm	80 „
Coffee (coffea arabica)	30 „	Date palm	75 to 100 „
Rubber	35 „	Coffee (coffea robusta)	over 100 „
Citrus trees	40 „		

The consequences of these lengthy productive cycles include:

a. the high initial investment needed for the planting, for the period when the young trees yield little or nothing, and until costs are finally recouped by the gradually increasing production. The costs for the oil palm are recouped about four years after planting, for rubber after about six years and for cocoa after about eight years (Ruthenberg 1967);

b. relatively stable yields subsequent to the initial period, because the deep root-systems of established trees and shrubs make them fairly insensitive to annual and seasonal variations in rainfall;

c. the constant ground cover, which is similar to the natural forest vegetation of the humid tropics and thus maintains soil fertility or at least conserves it to a great extent;

d. problems of harvesting, because many types of tree and

bush crop then require high inputs of labour and thus have a high total expenditure on labour. This can, of course, be absorbed only when there are correspondingly high gross proceeds or low wage levels. Most tree and shrub crops must therefore be regarded as intensive crops; and

e. the minimum size of farm sufficient for a peasant family of a given size is much less for perennial crops, and also for irrigation farming, than for rain-fed farming and for extensive pasture farming.

TABLE 4.
Examples of important tropical perennial crops

Tropical rain forest			Economic characteristics (approx. values)	Tropical uplands		
oil palm	rubber	cocoa		sisal	coffee	tea
50	35	over 20	useful economic life in yrs.	5–9	30	over 50
1,600	2,400	3,700	initial investment DM/hectare		1,350–3,000	4,000–12,000
350	960	300–2,000	labour in hours of work/hectare /yr	630	high	3,200
465	1,300	1,300	gross income in DM/hectare/ yr	1,200	1,500	1,300
(K), F	F	K	harvested crop processed in small plant (K), or factory (F)	F	K, (F)	F
B	B, P	B	specially suitable for peasant farms (B), or plantation (P)	P	B, P	(B), P

Sources: Franke, G., et al.: Nutzpflanzen der Tropen und Subtropen, Vol. I. and II. Leipzig 1967. – Ruhr-Stickstoff AG: Schriftenreihe über tropische und subtropische Kulturpflanzen. Bochum 1953–57. – Ruthenberg, H.: Organisationsformen der Bodennutzung und Viehhaltung in den Tropen und Subtropen. In: Handbuch der Landwirtschaft und Ernährung in den Entwicklungsländern, ed. P.v. Blanckenburg & H.-D. Cremer, Vol. 1. Stuttgart 1967.

It can be seen from the large number of branches of farming under consideration here that these stated agricultural and economic characteristics of perennial crops are valid in general principle, but hold less rigidly in individual instances. A selection of six tree and shrub crops in Table 4 illustrates the great difference between crops: differences which also give the general system its flexibility.

The amount of the initial investment for a particular crop is also governed by the law of minimal-cost combination. This makes imperative appropriate changes in the production procedure as economic development proceeds. However, the relationship between the initial investments for the six crops listed is probably essentially peculiar to them. The longer the expected productive period, the more the farmer can afford to spend on laying out his plantation carefully and thoroughly.

The differences in labour requirements vary almost from one to ten as between oil-palm cultivation and tea growing, while the other crops lie between these two. This amount of variation is basically attributable to the different methods of harvesting and different annual proportions of new planting. It shows that the cultivation of tree and shrub crops can be adapted by the choice of an appropriate form to very different farming situations.

However, the particular suitability of the individual crops for peasant farming on the one hand or for large-scale farming (plantations) on the other cannot simply be deduced from the amount of labour required. This is because of the very different technical processing requirements of the harvested crops. As long as tea-processing by hand was sufficient, in terms of the quality of the product, tea was grown as a garden crop on small peasant holdings — especially in Japan and China simply because of its high labour requirements. Since then the world market has demanded and been able to pay for a better and more standard quality tea, which can be processed only in the strict cleanliness of large tea factories. Thus tea has now become a definite plantation crop in India, Pakistan, Ceylon and Java, in spite of the amount of labour it requires.

Sisal needs a relatively small amount of labour but the expensive plants required for processing it must be worked to capacity from a small catchment area and the transport costs involved make it a subject for large-scale operation. The oil-palm is a typical peasant crop in West Africa, although it demands relatively little labour. The low initial investment makes it popular and the processing problem has been solved by a changeover from home oil production to the cooperative oil mills. The opposite is true of cocoa, which however is also a peasant crop in West Africa. Its high initial investment costs are accepted and a great deal of labour can be employed productively on a small area of cultivation.

4.6 The Dynamics of the Tropical Farming Regions in the Economic Development Process

In the course of economic development land becomes increasingly scarce so that farming must increasingly improve the productivity of the land it uses, i.e. use it more intensively. With the change from a sparsely to a densely-populated agricultural state, there is at first an increasingly large workforce available, whilst all capital goods continue to be very expensive. In this phase the necessary intensification of farming must therefore be achieved by increased labour use. Only when considerable industrialisation develops does the workforce decline and (like the land) become more expensive while commercially-produced capital goods become cheaper. In this latter phase of development, further productivity must be attempted by means of higher capital investment.

In the course of economic development, the productivity of the land, and subsequently that of labour also, must be increased. However, the three stages of production intensity which fulfil this demand are in sequential order:

extensive ⟶ labour-intensive ⟶ capital-intensive

4.6.1 CHANGES IN FARMING SYSTEMS IN THE INDIVIDUAL CLIMATIC ZONES

In the permanently-wet climate of the tropical rain forest the farming systems will always show a marked tendency to remain land-use systems and will in the course of economic development gradually change quantitively as follows:

Stage 1: haphazard shifting cultivation;
Stage 2: organised shifting cultivation with natural forest fallow;

Stage 3: organised shifting cultivation with artificial forest fallow;
Stage 4: organised shifting cultivation with plough cultivation;
Stage 5: system using legume fallow;
Stage 6: system using planted grass fallow;
Stage 7: ley farming;
Stage 8: systems cultivating perennial crops.

This progression of stages is governed by the necessity to raise land-productivity at first and labour-productivity later. It is therefore the general rule: but this does not exclude the possibility of some stages being skipped where there is a very rapid and dynamic economic development. In some cases, also, the sequence can differ somewhat. Haphazard shifting cultivation was replaced directly by oil-palm growing in parts of Nigeria, and by cocoa production in Southern Ghana, because the geographical position of these countries on an ocean allowed them to benefit from high prices in the world market. The combination of artificial forest fallow with plough cultivation presents problems and is skipped in many countries. Foreign trade often makes tree and bush cultivation possible at a time when the home market still cannot fully absorb the animal products of ley farming.

In general, however, the given sequence of farming systems will prevail during economic development. The order is realistic because an extension of the proportion of productive land with a simultaneous increase in yield per hectare is at first achieved by higher capital expenditure. The change from haphazard to organised shifting cultivation simply increases the proportion of cropped land, i.e., simply increases the amount of labour utilised. The change from natural to artificial forest fallow requires some capital in the form of plant stock, but this kind of capital can be created by labour on the farm itself. It is the changeover from hoe to plough cultivation that causes a considerable increase in capital expenditure — for implements and draught animals. When forest fallow is replaced by legume fallow, additional capital is required for seeds, and also for fertilizers when the legumes are replaced by grasses. The highest form of land use, ley farming or tree and shrub crops, necessitates considerable additional capital investment in the form of livestock, or permanent plant stock.

In the humid savannah the emphasis in farming will probably change as follows in the course of development:

Stage 1: forest fallow system;
Stage 2: bush fallow system;
Stage 3: green fallow system;

Stage 4: rain farming without fallow and without livestock;
Stage 5: rain farming without fallow but with livestock;
Stage 6: irrigated farming without fallow but with livestock.

Here, too, development is at first directed towards an increase in the proportion of productive land by a continuing decrease in the number of fallow years, changing the fallow vegetation from forest via bush to grass. Later an attempt is made to dispense with fallow altogether in the interests of land productivity, and it soon becomes obvious that livestock are a prerequisite of permanent rain farming in the interests of fertilizer production. The increased demand for animal products now makes this possible.

For its part animal husbandry, with its demand for a seasonal fodder balance, helps to bring about the last and decisive step to irrigated farming. The humid savannah is predestined for this, since it has a relatively small water requirement but abundant supplies of ground water. A rational system of reservoirs is also made possible by rivers which flow throughout the year. Besides the quantity, the quality of the water supplies in the tropical rain climates is favourable for irrigated farming; the low salt content of the river water means that damage from salt accumulation can be avoided.

As long as land productivity is the first concern and labour still cheap, various methods of irrigation can be used. From a certain wage level onwards, sprinkler irrigation will become more economic, since it requires much less labour and capital equipment becomes cheaper in the course of development. In conjunction with mineral fertilization, crop protection, etc., sprinkler irrigation ensures high labour-productivity whilst maintaining high land productivity.

In spite of the greater need for irrigation in the dry savannah, it will be less widespread owing to a shortage of water and/or danger of salt accumulation. In this case, the decisive sequence of changes in farming systems in the course of economic development can be described as:

Stage 1: extensive pasture farming;
Stage 2: savannah shifting cultivation;
Stage 3: cereal - fallow rotation besides extensive pasture farming;
Stage 4: crop rotation organised to include fodder crops and legumes in conjunction with extensive pasture farming;
Stage 5: extensive pasture farming intruding into the rain-farming area.

At first, the fodder-crop cultivation in stage 4 serves primarily the aims of soil biology rather than of fodder production. Planted grass fallow is at first introduced primarily in support of cereal growing, mostly millet or maize, and not as support for stock farming during times of fodder shortage. This does not, of course, exclude the possibility that the fodder crops can be used for grazing during the dry season or to produce hay. In this way they form the link between cattle fattening and maize, millet or barley growing, branches of farming which once existed side by side in isolation. Growing fodder crops enables crops to be rotated for cereal growing and the fodder basis for cattle fattening to be extended in the dry season. It provides arable crops with root humus and pasture animals with winter feed; the animals transform part of this feed into stall manure which in turn benefits the cereal crops. In this way the two agricultural branches of meat stock production and cereal growing, which once existed side by side in isolation, now give rise to an association, an integrated whole, a farming system.

In the course of further development, more productive crops are also cultivated, such as chick peas, ground-nuts, sunflowers, sesame and in some cases even cotton etc. This arable farming as a whole continues to encroach on the natural pasture areas, until it finally occupies all the farmland whose natural conditions make it suitable for arable use (gradient, depth of top soil, ground water table etc.).

However, in the thorn-shrub steppe, which lies outside the potential rain farming zone, the evolution can occur only within the extensive pasture-farming system. According to the principle
extensive system → labour-intensive system → capital-intensive system the sequence of changes in the course of economic development will be as follows: (cf. Fig. 7, p.   ):

Stage 1: nomadic herding;
Stage 2: settled pasture farming with seasonal treks;
Stage 3: bridging over the dry season with the farm's own pasture reserves and by drawing on the animals' fat deposits;
Stage 4: increasing the number of drinking places;
Stage 5: increasing pasture subdivision;
Stage 6: buying additional fodder;
Stage 7: farms producing own fodder for the dry season.

This gradual advance of capital investment is set in motion both by an improvement in the market exchange value of animal products for farming equipment and by an improvement in the

204

infrastructure, which reduces the farms' economic distance from the market. Depending on the natural conditions of production, a better fodder balance also allows changes in the systems of animal husbandry which mean increased productivity.

4.6.2   REORGANISATION OF THE TROPICAL AGRICULTURAL AREA
AS AN ASPECT OF THE FUTURE

Finally, we must concentrate once more on the four main farming systems in the tropics and look into the future. The subsequent rural picture for the lower latitudes, with its agricultural systems reshaped by the forces of competition and by changes of location, should unfold as follows:

A. *Expansion will occur in:*

1. *the perennial crop cultivation zones,* because a) world markets for most of these products are expanding; b) this is an intensive branch of agriculture and c) the inner tropics urgently need this system of farming simply because otherwise their soil fertility is at risk.

2. *The irrigated-farming zones,* because a) the economies are increasingly exploiting their water supplies by large-scale projects such as impounding dams etc.; b) irrigation increases both land and labour productivity and c) surface water protects the soil fertility in a similar way as natural vegetation.

B. *There will be a decline in:*

1. *the rain-farming zones* because they are being pushed back from three sides: a) in the rain forest belt by perennial crops; b) in the humid savannah by irrigated farming and c) in the vicinity of the agronomic aridity limit by the advance of extensive pasture farming;

2. *the extensive pasture-farming zones* because a) although they are gaining land in the present border areas of dryland farming (because this is losing its value as arable land owing to its low yields and to the rise in wage levels); b) this very gain is being more than offset by losses of land in the semi-desert and thorn-shrub areas. Extensive pasture farming will have to abandon these areas because of their extremely low productivity.

There will thus be encroachment on the arable-crop region from two sides: from the tropical border areas by extensive pasture-farming, and from the equator by tree and shrub cultivation. This means that increasingly large parts of the tropics will again have a vegetation cover approaching the natural state. The outer tropics will have grassland and the equatorial belt trees, even though the natural landscape has now been transformed into an agricultural landscape.

BIBLIOGRAPHY

Literature Consulted

Andreae, B., – 1965 – Die Bodenfruchtbarkeit in den Tropen. Nutzbarmachung und Erhaltung. Betriebswirtschaftliche Überlegungen für die Arbeit in Entwicklungsländern. Hamburg-Berlin.

Andreae, B., – 1972 – Landwirtschaftliche Betriebsformen in den Tropen. Bodennutzung und Viehhaltung im Spannungsfeld von Tradition und Fortschritt. Hamburg-Berlin.

Dehmel, R., – 1957 – Diercke Weltatlas, 92 nd ed. Brunswick.

Dumont, R., – 1957 – Types of Rural Economy. Studies in World Agriculture. London.

Franke, G., *et al.*, – 1967 – Nutzpflanzen der Tropen und Subtropen. Vols. I and II. Leipzig.

Gregory, S., – 1969 – Rainfall Reliability. In: Environment and Land Use in Africa. London.

Hase, H. J. von, – 1964 – Die Auswirkungen der Dürrejahre in Südwestafrika und ihre Überwindung. 'Der Deutsche Tropenlandwirt', Vol. 65.

Könnecke, G., – 1967 – Fruchtfolgen. Berlin. N.P. Länderkurzberichte, – 1972 – Tschad (Allg. Statistik d. Auslandes. (Ed.): Stat. Bundesamt Wiesbaden.). Stuttgart and Mainz.

Masefield, G. B., – 1948 – The Life of Perennial Crops. 'East African Agricultural Journal of Kenya'. Nairobi.

Niederstucke, H., – 1970 – Bodennutzungsformen in tropischen Höhenlagen. 'Landwirt im Ausland', Frankfurt/Main, Vol. 4, p. 74ff.

Nye, P. H., & D. J. Greenland – 1960 – The soil under shifting cultivation. (Commonwealth Bureau of Soils, Technical Communication No. 51) Bucks.

Ruthenberg, H., – 1965 – Probleme des Überganges vom Wanderfeldbau und semipermanenten Feldbau zum permanenten Trockenfeldbau in Afrika südlich der Sahara. 'Agrarwirtschaft'. Hannover, Vol. 14.

Ruthenberg, H., – 1967 – Organisationsformen der Bodennutzung und Viehhaltung in den Tropen und Subtropen. In: Handb. d. Landw. u. Ernährung in den Entwicklungsländern, Vol. 1. (Eds). P. v. Blanckenburg & H. D. Cremer, Stuttgart.

Schickele, R., – 1931 – Untersuchungen über die Formen der Weidewirtschaft in den Trockengebieten der Erde. Diss. Berlin.

Tsuzuki, T., – 1963 – Die Fruchtfolgen des Japanischen Ackerbaus. 'Ber. üb. Landw.', Hamburg and Berlin. N.F., Vol. 41, p. 833 ff.

Uexküll, H. R. von, – 1969 – Reis in Asien – Probleme und Möglichkeiten einer Produktionssteigerung. 'Zeitschrift f. ausländ. Landw.', Frankfurt/M., Vol. 8, p. 248 ff.

Uhlig, H., – 1965 – Die geographischen Grundlagen der Weidewirtschaft in den Trockengebieten der Tropen und Subtropen. In: Weide- Wirtschaft in Trockengebieten. Voraussetzungen, Grundlagen, Gegebenheiten und Möglichkeiten ihrer Intensivierung mit besonderer Berücksichtigung der Verhältnisse in den Tropen und Subtropen. 'Giessener Beiträge zur Entwicklungsforschung. Schriftenreihe des Tropen-Instituts des Justus Liebig-Universität Giessen'. Eds.: N. Atanasiu & H. D. Cremer *et al.*, Series I, Vol. I, Stuttgart.

Wang, Y., F. Nagel & H. Ruthenberg, – 1969 – Bodennutzung und technischer Fortschritt auf Taiwan. („Zeitschr. f. ausländ. Landw.", Sonderheft 7), Frankfurt/M.

Further Literature

Bähr, J., – 1970 – Strukturwandel der Farmwirtschaft in Südwestafrika. 'Zeitschr. f. ausländ. Landw.', Vol. 9, p. 147 ff. Frankfurt/M.

Baren, F. A. van, – 1964 – Fysisch-geografische aspecten van de aride en semi-aride gebieden. 'Tijdschrift van het Koninklijk Nederlandsch Aardrijkskundig Genootschap Amsterdam', Leiden, Vol. 81, p. 182 ff.

Bennett, M. K., – 1960 – A world map of food-crop climates. (Food Research Institute Studies, Vol. I.) Stanford (Calif.), No. 3.

Best, R., – 1962 – Production factors in the tropics. *Neth. J. Agri. Sci.*, Wageningen, Vol. 10, p. 347 ff.

Boughey, A. S., – 1957 – The physiognomic delimitation of West African vegetation. "Journal of the West African Science Association," Achimota (Gold Coast), Vol. 3. p. 148 ff.

Douglas, H. K. L., – 1957 – Climate and economic development in the tropics. New York.

Duvigneaud, P., – 1953 – Les savanes du Bas-Congo. Liège.

Edwards, D. C., – 1951 – The vegetation in relation to soil and water conservation in East Africa. Commonw. Bur. Pastures and Field Crops Bull. No. 41, p. 28 ff.

Eyre, S. R., – 1963 – Vegetation and Soil. A world picture. London.

Faucher, D., – 1949 – Géographie agraire. Types de cultures. Paris. N.P. Handbuch der Landwirtschaft und Ernährung in den Entwicklungsländern., – 1971 – (Eds) P. v. Blanckenburg & H. D. Cremer, Vol. 2, Stuttgart.

Hasselo, H. N., – 1961 – The soils of the lower eastern slopes of the Cameroon Mountain and their suitability for various perennial crops. Wageningen.

A History of Land Use in Arid Regions – 1965 – Arid Zone Research, (Ed.) L. D. Stamp, XVII, 2nd ed., Paris.

Holm, H. M., – 1965 – The agricultural economy of Ethiopia. Washington. Indices of Agricultural Production in Africa and the Near East 1962–1971, – 1972 – (Ed.) USDA, ERS-Foreign 265, Washington, D.C.

Irvine, F. R., – 1958 – A text-book of West-African agriculture, London.

Lauer, W., *et al* – 1952 – Studien zur Klima- und Vegetationskunde der Tropen. Bonner Geographische Abhandlungen, (Eds) C. Troll & F. Bartz, Heft 9. Bonn.

Lee, D. H. H., – 1957 – Climate and economic development in the tropics. New York.

McMaster, D. N., – 1962 – A subsistence crop geography of Uganda. Geographical Publications. The World Land Use Survey. Occasional Papers No. 2. Bude.

Mohr, E. C., & F. A. van Baren, – 1959 – Tropical soils. A critical study of soil genesis as related to climate, rock and vegetation. The Hague and Bandung, London, New York.

Mukerjee, H. N., – 1962 – Problems of soil of the paddy fields. Trans. International Soil Conference New Zealand. Sess. C. 13.

Ochse, J. J., M. J. Soule jr., M. J. Dijkman & C. Wehlburg, – 1961 – Tropical and subtropical agriculture. 2 Vols. New York.

Phillips, J., – 1959 – Agriculture and Ecology in Africa. London.

Possinger, H., – 1968 – Landwirtschaftliche Entwicklung in Angola und Mosambique. Afrika-Studien. (Ed.) Ifo-Institut für Wirtschaftforschung München. Nr. 31. Munich.

Production Yearbook, Vol. 23, 24, – 1969, 1970 – Publ. by the United Nations Food and Agriculture Organization, Rome.

Ruthenberg, H., – 1971 – Farming Systems in the Tropics. Oxford.

Schmithusen, J., – 1961 – Allgemeine Vegetationsgeographie. (Lehrbuch der allgemeinen Geographie. (Ed.) E. Obst, Vol. IV.) 2nd impr. ed. Berlin.

Statistical Yearbook, 1969. Publ. Statistical Office of the United Nations. New York 1970.

Tempany, Sir H., & D. H. Grist, – 1958 – An introduction to tropical agriculture. London, New York, Toronto.

Walter, H., – 1939 – Grasland, Savanne und Busch im ariden Teile Afrikas und ihre ökologische Bedingheit. Jahrbücher für Wissenschaftliche Botanik. Vol. 87. Berlin.

Walter, H., – 1971 – Klima und Vegetation. In: Handb. d. Landw. u. Ernährung in den Entwicklungsländern. (Eds) P. v. Blanckenburg & H.–D. Cremer, Vol. 2, Pflanzliche und tierische Produktion in den Tropen und Subtropen. pp. 1–25. Stuttgart.

Webster, C. C., & P. N. Wilson, – 1969 – Agriculture in the tropics. 3rd ed. London.

Whittlesey, D., – 1963 – Major agricultural regions of the earth. Annals of the Association of American Geographers. (Ed.) A. E. Parkins, Vol. 26, 1932, Repr. Vaduz. p. 199 ff.

Whyte, R. O., – 1967 – Milk Production in Developing Countries. London.

Wissmann, H. von, – 1956 – On the role of nature and man in changing the face of the dry belt of Asia. Univ. of Chicago Press. p. 278 ff.

Wrigley, G., – 1971 – Tropical Agriculture. The Development of Production. London.

Glossary of tropical crops

German	Botanical name	English

Plants producing stimulants

German	Botanical name	English
Kakaobaum	*Theobroma cacao* L.	cocoa, cacao
Kaffee	*Coffea* spec.	coffee
Tee	*Thea (Camellia) sinensis* L.	tea
Tabak	*Nicotiana* spec.	tobacco

Caoutchouc and rubber plants

German	Botanical name	English
Parakautschukbaum	*Hevea brasiliensis* Müll. Aarg.	Para rubber, caoutchouc tree
Guttaperchaliefernde Pflanzen	*Sapotaceae*	gutta-percha
Gummiakazien	*Acacia* spec.	gum arabic tree

Glossary of tropical crops

German	Botanical name	English

Oil and fat-producing plants

Ölpalme	*Elaeis guinensis* Jacq.	oil palm
Kokospalme	*Cocos nucifera* L.	coconut palm
Ölbaum	*Olea europaea* L.	olive
Soja	*Soja hispida Moench.*	soybean
Erdnuss	*Arachis hypogaea* L.	ground-nut, peanut
Sesam	*Sesamum indicum* DC.	sesame
Rizinus	*Ricinus communis* L.	castor bean, castor oil

Tuberous and root crops

Maniok	*Manihut utilissima* Pohl	cassava
Yams	*Dioscorea* spec.	yam
Batate or Süsskartoffel	*Ipomoea batatas* Poir.	sweet potato

Fruit crops

Banane	*Musa* spec.	banana, plantain
Ananas	*Ananas comosus* Merr.	pineapple
Zitrus	*Citrus* spec.	citrus plants
Dattelpalme	*Phoenix dactylifera* L.	date palm
Feigenbaum	*Ficus carica* L.	fig
Mangobaum	*Mangifera indica* L.	mango

Sugar-producing plants

Zuckerrohr	*Saccharum officinarum* L.	sugar cane

Cereals

Reis	*Oryza sativa* L.	rice
Mais	*Zea mays* L.	corn, maize
Hirsen, grosse	*Andropogoneae*	sorghum
Hirsen, kleine	*Paniceae*	millet

Fibre plants

Baumwolle	*Gossypium* spec.	cotton
Kapok	*Ceiba* spec.	kapok, silk-cotton tree
Jute	*Corchorus* spec.	jute
Kenaf	*Hibiscus cannabinus* L.	deccan hemp, kenaf
Roselle	*Hibiscus sabdariffa* L.	roselle hemp, sorrel
Ramie	*Boehmeria* spec.	ramie, Chinagrass, rhea
Sisal	*Agave sisalana Perrine*	sisal
Faserbanane	*Musa textilis* Nees	abaca, Manila hemp

PLANT SOCIOLOGY AND ECOLOGY APPLIED TO GRAZING LANDS RESEARCH, SURVEY AND MANAGEMENT IN THE MEDITERRANEAN BASIN

H. N. LE HOUEROU

Contents

212

PLANT SOCIOLOGY AND ECOLOGY APPLIED TO
GRAZING LANDS RESEARCH, SURVEY AND
MANAGEMENT IN THE MEDITERRANEAN BASIN

5.1   Introduction

Research and surveys on plant ecology and sociology in the
Mediterranean Basin have existed since the beginning of this century. Flahault produced an extensive report, with 5 coloured vegetation maps to the scale of 1/500,000, which was awarded the
Gay Prize of the French Academy of Science in 1897. This work
was published in a posthumous book by Gaussen in 1937. Flahault drew some 20 coloured vegetation maps to various scales
(1/80,000, 1/200,000, 1/500,000) between 1897 and 1934.

Later on, since about 1920, and for more than 50 years,
Braun-Blanquet and the numerous disciples (from all over the
world) he trained in his "Station Internationale de Géobotanique
Méditerranéenne et Alpine"[1] in Montpellier produced a great
number of studies, surveys and maps on the Mediterranean vegetation.

Applied phytosociology, however, appeared much later in
the Mediterranean, mostly after the Second World War (Br-Bl.,
1952). Although the use of phytosociology in grassland studies
appeared in Northern Europe as early as the late '30's (De Vries,
1937, 1938 etc. . . . .).

One of the first studies of agronomic and economic significance in applied phytosociology in the Mediterranean was devoted to grassland: "Les prairies de Crau" by Molinier & Tallon
in 1949.

Since the '50's, vegetation surveys for pasture development
have become extremely numerous.

Plant ecology and sociology are basic sciences for pasture
survey; they have, however, to be completed by other studies,
such as primary and secondary production, to become "operational".

Phytosociology and phytoecology belong to the daily routine in grazing land surveys and management schemes operated
by FAO field projects and by other agencies as well.

1. Which is often referred to as SIGMA. Braun-Blanquet's school is often called SIG-
MATIST and the work carried out with this methodology is referred to as SIGMATIST
PHYTOSOCIOLOGY.

In the developing countries of the Mediterranean Basin, almost all the pasture development and management schemes carried out during the last 20 years were based upon phytosociological studies, for instance by: Quézel, Nègre, Long, Le Houérou, Ionesco, Stefanesco, Thiault, Pabot, Floret, Le Floch, Claudin, Delhaye, Pouget, Loiseau, Sebillotte, Kassas, Froment, Van Swinderen, Rodin, Vinogradov, Mirotchnitchenko and many others.

The same is true to a lesser extent in the Northern Mediterranean, where applied grassland studies were less numerous. However, the names of Malato-Beliz & Teles, in Portugal, ought to be cited, as well as those of Montserrat Recorder, Bolos, Rivas Goday & Rivas-Martinez in Spain; Long, Jaccard, Daget & Poissonet (J. & P.) in France; Liaccos in Greece; Orshan, Tadmor, Seligman in Israel; Haussman, Giaccomini & Tomaselli in Italy.

5.2 Characteristics of the Mediterranean Environment

5.2.1 CLIMATE

5.2.1.1 *Precipitations*

The Mediterranean climate is characterized by precipitations concentrated on the cool season with short days whereas summers are hot and dry. However, there are roughly 3 precipitation regimes (Le Houérou, 1965, 1971; Akman & Daget, 1971).

— A western regime with maximum precipitations in the fall (Oct./Nov.)
— An eastern regime with maximum precipitations in winter (Dec./Jan.)
— An intermediate regime (continental) with maximum precipitations in spring (March/April).

The limit between eastern and western regimes is approximately the $16-18°$ of longitude East (Gulf of Sidra).

The amount of rains varies from about 2,000 mm in some mountain areas to less than 20 mm in the "True Desert". The length of the dry season is inversely correlated to the average amount of precipitations; it varies from 3 months in the more humid part (mid-June to mid-September) to 12 months where the rainfall is below $100-150$ mm. Following Bagnouls & Gaussen (1953), one considers as dry each month where $P \leqslant 2\,t$ (P being expressed in mm and t in degrees Celsius).

The upper limit of the Mediterranean climate was given by

Emberger (1955), with the formula $\dfrac{Pe}{Me} \leqslant 7$, where Pe represents summer rains in mm (June/July/August) and Me, the average daily maximum temperatures of the summer months. In practice, this means that the amount of summer rains is below 200 mm, since Me approximates 26–28°C in the transition zone. In the Northern Mediterranean, this corresponds almost exactly to the northern limit of the holm oak (*Quercus ilex*) and of the olive tree (*Olea europaea*); but this is not true in the south of the Mediterranean. According to Gaussen, the summer rains should not exceed 120–150 mm during the summer season, since $20 < t < 25$: ($20° \times 2 \times 3$ months = 120 mm; $25 \times 2 \times 3$ = 150 mm). The lower limit of the Mediterranean climate is the true desert where rains occur very rarely but in any season (Emberger, 1943). This corresponds roughly to the 20 mm isohyet.

5.2.1.2 *Temperatures*

The mean annual temperature varies from 18–20°C along the coasts to 10–15° in the higher elevations.

The average daily minimum temperature of the coldest months varies from 5–10°C on the shores and goes down to −2 and −5°C, above 2,000 m of elevation.

The average daily maximum of the hottest months varies from 30°C on the shores to 35–38°C in the hinterland and 40–42°C at the border of the Sahara.

5.2.1.3 *Potential Evapotranspiration (ETP)*

Potential evapotranspiration either measured or calculated through empirical formulas (Penman, Turc) varies. from 1,000–1,100 mm/year to the northern limit of the Mediterranean to 1,500–1,600 mm at the border of the Sahara or Near Eastern deserts. Potential evapotranspiration depends on solar radiation which, in its turn, depends on latitude and cloudiness. The ETP is thus inversely related to mean precipitation, and therefore, vegetation distribution and crop potential follow mean precipitation patterns.

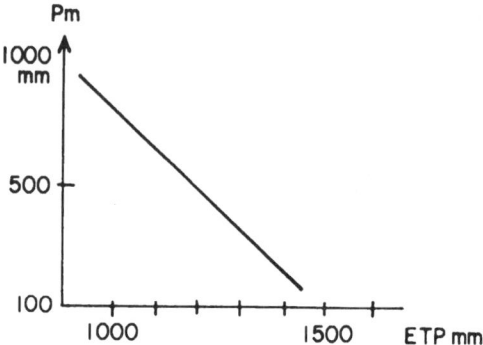

5.2.1.4 *Climatic Synthesis*

The classification increasingly used by phytoecologists in the Mediterranean is that of Emberger (1930 to 1970). This classification is based on 2 criteria:

— a *pluviothermic index* $Q2 = \dfrac{2,000\ P}{M^2 - m^2}$ where P = average annual precipitations in mm; M = average daily maximum of the hottest month and m = average daily minimum of the coldest month, both in degrees Kelvin.

The higher the Index Q2, the more humid the climate.

— the average daily minimum of the coldest month (usually January) "m". This figure is correlated with the length and intensity of the winter freezing period; in other words, with altitude and continentality.

However, I shall use here a method derived from that of Emberger, where the index "Q" is replaced by the average precipitations "P", which I consider a simplification and a progress since: 1. the method is simple, objective and easy to remember; 2. concrete, objective data are used, and not artificial abstractions; 3. it is excellently correlated with geobotanical, phytoecological and agronomic facts.

One could object that mean annual precipitation is not a very reliable parameter in Mediterranean climates, because of its uneven distribution. This is probably true, but nevertheless, it is a fact that mean annual precipitation is generally correlated inversely with three important factors:[1]

1. Le Houerou and Hoste, 1976, in press, found the following correlations in the mediterranean basin
— Rainfall and its variability: $\sigma = f\,(1/P)$, where $r = -0.73$; $n = 110$; $p < 0.001$
— Rainfall and number of rainy days: $np = f\,(P)$; $r = 0.79$; $n = 140$; $p < 0.001$
— Rainfall and potential evapotranspiration: $PET = f\,(1/P)$; $r = -0.75$; $n = 128$; $p < 0.001$.

216

- variability of precipitation
- potential evapotranspiration
- length of the dry period

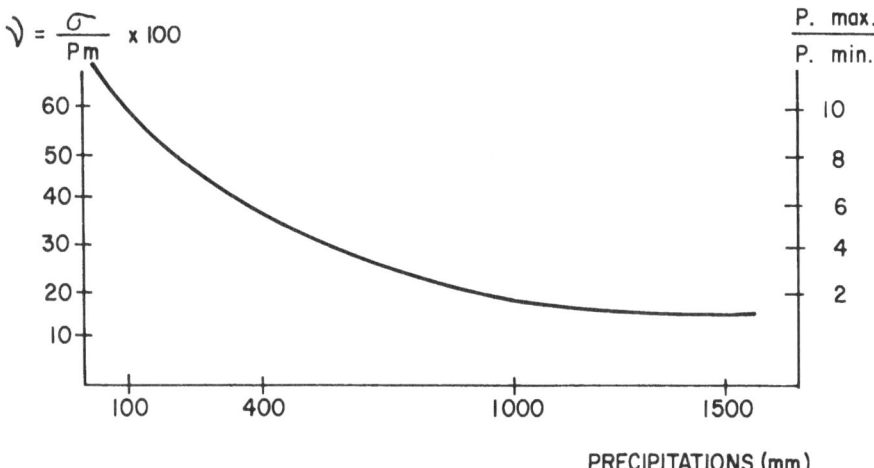

$$\mathcal{V} = \frac{\sigma}{P_m} \times 100$$

PRECIPITATIONS (mm)

There is no major inconvenience in replacing Q2 by P since:

a. Q2 claims to represent the quotient P/ETP (Emberger, 1930), which is perfectly true since I have found for Tunisia and Algeria a correlation of 0.95 between $\frac{100\,P}{ETP}$ and Q2 where $100\ \frac{P}{ETP} = 0.6\ Q_2 + 5$

b. ETP is roughly correlated inversely to P (see above).

c. ETP zonal variations are not of the same order of magnitude as P zonal variations: they are 200 times smaller since ETP varies from 1,000 to 1,600 mm whereas P varies from 2,000 to 20 mm in the Mediterranean Basin (60% against 10,000%). Therefore, I have set up a climagramme (1969) using P in ordinates and m in abscisses, which is almost similar to the classical one using Q and m. We have thus:

1,200 < P	Mediterranean	per humid climate
800 < P < 1,200	”	humid climate
600 < P < 800	”	sub-humid climate
400 < P < 600	”	semi-arid climate
300 < P < 400	”	arid climate upper
200 < P < 300	”	arid climate middle
100 < P < 200	”	arid climate lower
20 < P < 100	”	desert climate
P < 20	Eusaharan climate, non mediterranean	

217

Each one of these climate types is subdivided into eight sub-types according to average daily minimum temperature of January (coldest month; see Chart page 223):

9 < m	Very warm winters — no frost at ground level
7 < m < 9	Warm winters — no frost under shelter
5 < m < 7	Mild winters — 1-10 days frost
3 < m < 5	Temperate winters — 10-20 days frost
1 < m < 3	Cool winters — 20-30 days frost
−2 < m < 1	Cold winters — 30-60 days frost
−5 < m < −2	Very cold winters — 60-120 days frost
m < −5	High mountains — More than 120 days frost

We are thus left with theoretically 64 sub-types of Mediterranean climates. In fact, they are less as some combinations are impossible or, at least, practically never occur.

These criteria are not arbitrary but designed to match with phytogeographic and agronomic facts. They can be expressed more clearly with graphs showing the distribution of the climax types of vegetation in respect to these climatic criteria (see Figs. Pages 15, 16, 17, 18).

These thresholds have been defined by various authors in various ways and using either "Q" or "P" and "m".

Emberger (1930, 1955) distinguishes the following stages of vegetation and bioclimates in Morocco:

Mediterranean	Saharan	:	Q1 < 12 − 20
„	Arid	:	Q1 < 20 − 37
„	Semi-arid	:	Q1 < 30 − 70
„	Temperate	:	Q1 < 70 − 110
„	Humid	:	Q1 < 225

with the following varieties:

$$- \text{cool winters: } m < 0°C$$
$$- \text{medium } „ : m \simeq 0°C$$
$$- \text{warm } „ : m > 0°C$$

Long (1954) distinguishes two clear cut thresholds in Tunisia:

Q > 25: semi-arid
Q < 25: arid
m > 5.5°C: warm winters
m < 5.5°C: cool winters

Long (1957) recognized 4 bioclimates and 3 varieties in Eastern Jordan:

218

Q2 < 14	Saharan	30 < P < 120
25 < Q2 < 40	Arid	140 < P < 350
36 < Q2 < 70	Semi-arid	300 < P < 500
70 < Q2 < 90	Sub-humid	500 < P < 725

m < 3	cool winters variety
3 < m < 9	warm ,, ,,
9 < m	very warm ,,

Le Houérou (1958) distinguishes the following zones in Tunisia:

100 < Q2	humid	1,000 < P
60 < Q2 < 100	sub-humid	500 < P < 1,000
50 < Q2 < 60	semi-arid, upper	400 < P < 500
40 < Q2 < 50	semi-arid, middle	350 < P < 450
25 < Q2 < 40	semi-arid, lower	200 < P < 350
12 < Q2 < 25	arid	100 < P < 200
Q2 < 12	Saharan	P < 100

m < 0	cold winters
0 < m < 5.5	cool ,,
5.5 < m < 7.5	mild ,,
7.5 < m	warm ,,

Le Houérou (1959, May) has defined in Tunisia by studying the climatic distribution of some 1,200 species and 41 plant associations the following divisions and subdivisions:
a. according to "Q" and "P":

2 < Q2 < 5	Saharan, lower	20 < P < 50
5 < Q2 < 12	,, upper	50 < P < 100
12 < Q2 < 25	Arid, lower	100 < P < 200
25 < Q2 < 35	Arid, upper	200 < P < 300
35 < Q2 < 45	Semi-arid, lower	300 < P < 400
45 < Q2 < 70	Semi-arid, upper	400 < P < 600
70 < Q2 < 120	Sub-humid	600 < P < 800
120 < Q2 < 170	Humid	800 < P < 1200
170 < Q2	Per-humid	1200 < P

b. according to "m" the following varieties:

m < 0	cold winters
0 < m < 3	cool ,,
3 < m < 5	temperate winters
5 < m < 7	mild ,,
7 < m	warm ,,

This classification was followed by Gounot (1959, July) and Gounot & Le Houérou (1959, Oct.), and later on, with some slight modifications, by Bortoli, Gounot & Jacquinet (1969).

Sauvage (1963a and b) follows Emberger with some minor changes for the main divisions and recognizes most of Long & Le Houérou's varieties (although without quoting):

```
m < 0:    cold
0 < m < 3:   cool
3 < m < 7:   temperate
7 < m    :   warm
```

Le Houérou (1969) modifies his previous classification in the following way: the limit between cool and cold becomes + 1°C instead of 0°C.

The limit between Arid and Semi-arid becomes the isohyet of 400 mm or $Q2 = 45$ (hence lower semi-arid becomes upper Arid).

Le Houérou (1970, 1971, 1973, 1974) adds four subdivisions:

```
9 < m        very warm winters (Long, 1957)
−2 < m < +1   cold winters
−5 < m < −2   very cold winters
    m < −5   high mountains
```

Akman & Daget (1971) follow the classification defined by Le Houérou (1958, 1959) with some modifications. The following divisions and subdivisions are set forth in a survey of Turkey by these authors:

```
     Q2 < 30-40    Arid, upper
35 < Q2 < 45       Semi-arid, lower
45 < Q2 < 65-70    Semi-arid, upper
65 < Q2 < 95-140   Sub-humid
95 < Q2 < 200      Humid, lower
180 < Q2           Humid, upper

−12 < m < −7   glacial winters
−  7 < m < −3   very cold  „
−  3 < m <   0   cold       „
    0 < m <   3   cool       „
    3 < m <   7   temperate „
    7 < m <   9   warm       „
```

However, Akman & Daget do not give any geobotanical justification of the thresholds they have elected.

220

The classification used here (Le Houérou, 1970–1974) is justified by biogeographic and agronomic facts in the following way (see Le Houérou, 1969) based on the detailed study of some 1500 plant communities and the geographical and climatological distribution of over 5,000 plant species from Libya, Tunisia, Algeria, Morocco, Spain, Greece, Italy and Near East.

The 100 mm isohyet corresponds to the upper limit of the desert vegetation characterized by numerous plant communities (i.e. distributed on a contracted pattern). It is also the lower limit of *Argania spinosa* forest, of the dry farming cultivation and runoff agriculture, the lower limit of esparto steppes (*Stipa tenacissima*), of common sage (*Artemisia campestris*). The 200 mm line corresponds to the lower limit of dry farming without the use of runoff waters for instance for olive, almond or pistachio cultivation. It is the present lower limit of forest and degraded forests of Aleppo pine, Rosemary and Red Juniper. It is also the upper limit of many subdesertic plant communities dominated by *Hammada schmittiana* or *Hammada scoparia*, or *Laumea arborescens* or *Noaea mucronata* or *Traganum nudatum*, etc. . . .

The 300 mm isohyet is the lower limit of many forest species such as: *Erica multiflora, Tetraclinis articulata, Pistacia lentiscus, Phillyrea media, Lonicera implexa, Rhus pentaphyllum*, etc. as well as many common weeds such as: *Scolymus grandiflorus, Sc. maculatus, Arum italicum, Convolvulus cantabrica*, etc. . . . and also steppic species such as: *Sarcopoterium spinosum, Poa bulbosa, Tunica illirica, Minuartia geniculata* and many others.

The 300 mm isohyet also corresponds to the upper limit of many plant communities as for instance those dominated by the following species: *Gypsophila porrigens, Volutaria lippii, Thymelaea microphylla*, and so on.

The isohyet of 400 mm is a major limit: lower limit of Holm oak (*Q. ilex*), of Spiny oak (*Q. coccifera*) of *Juniperus oxycedrus* and many other species. It is also the upper limit of steppic vegetation: Esparto steppes (*Stipa tenacissima*), white sage steppes (*Artemisia herba alba*), *Peganum harmala* ermes and so on. It does also correspond to the lower limit of regular annual and profitable cereal cultivation in dry farming.

The 600 mm isohyet is the lower limit of the mediterranean Cedar (*Cedrus libanotica*), cork oak (*Q. suber*). This isohyet does also correspond to the lower limit of summer growing crops cultivation without irrigation such as Sorghum, Chick peas, melons, water melons, etc. . . . and also the upper limit of some types of forests: false Thuya (*Tetraclinis articulata*) for instance.

The 800 mm isohyet is the lower limit of the deciduous oaks (*Q. faginea, Q. afares, Q. pubescens, Q. toza*) of the Mediterranean Pine (*P. pinaster ssp. mesogeensis*) of the beech (*Fagus sylvatica*), of the firs (*Abies pinsapo*, etc. . . . .) of Laricio pine (*P. clusiana*). It is the upper limit of Aleppo pine and other species.

The 1200 mm isohyet is the lower limit of many species: *Taxus baccata, Ilex aquifolium, Acer obtusatum, Acer campestre, Acer opalus, Sorbus aria, Sorbus torminalis, Populus tremula* and many other trees, shrubs and forbs.

According to m: −5°C corresponds to the timber line on the high mountains (*Cedrus, Abies, Pinus, Quercus, Fagus, Junipertus*). −2°C corresponds to the lower altitudinal limit of *Cedrus* and *Abies* forest and of *Juniperus thurifera* woodlands and also on the cushion spiny Xerophytes (as a formation, not as individuals). (Le Houérou, Claudin & Haywood, 1975). It also correspond to the cold limit of numerous species such as Aleppo pine.

+1°C: This threshold was first defined by Le Houérou (1969). It is the lowest thermal limit of many mediterranean species and vegetation types such as: olive tree, *Pistacia lentiscus*, and also of an important naturalized species: *Opuntia ficus indica*. It is also the upper thermal limit of many cryophytic species such as: *Asphodelus acaulis, Cirsium echinatum, Erinacea anthyllis, Koeleria vallesiana, Bupleurum spinosum, Ormenis africana, Salvia argentea, Thymelaea tartonraira, Catananche caespitosa*, and many others.

This limit has very often, and arbitrarily, been considered as to be 0°C. A careful study of plant species and communities distribution has shown that the threshold is around m = + 1. The 0°C limit is purely anthropocentric (limit of melting ice at 2 m. above ground level, under standard shelter) and does not correspond to the reality of plant distribution.

m = 3: This threshold was first defined by Long (1957) and then by Le Houérou (1959) and, later on, by Sauvage (1963). It is the upper thermal limit of distribution of many continental species such as: *Alyssum alpestre, Alyssum montanum, Avena bromoides, Centaurea incana, Armeria plantaginea, Astragalus incanus subsp. numidicus, Launaea acanthoclada* and many others. It is also the cold limit of carob tree *Ceratonia siliqua, Periploca laevigata, Rhamnus alaternus, Rhamnus lycioides, Euphorbia dendroides* and so on. . . . It is also the lower thermal limit of citruses.

m = 5 (4.5 to 5.5 according to authors) corresponds to areas of a few days of light frost characterized by the cold limit of some cold sensitive species such as *Argania spinosa, Tetraclinis articulata, Chamaerops humilis, Quercus coccifera, Rhus pentaphyllum, Withania frutescens, Lavandula multifida, Sarcopoterium spinosum, Waronia saharae* and many other species. It is the lower thermal limit of productive orange growing.

m = 7 corresponds to the absence of frost under shelter. Several species are linked to these mild winters such as: *Ziziphus spina-christi, Solanum sodomaeum, Calotropis procera, Ononis vaginalis, Hegialophila pumila, Medicago helix, Trifolium nigrescens, Cichorium spinosum* and so on. Several tropical crops may be grown: suger cane, banana.

m = 9 corresponds to the absence of frost on ground level. It is the cold limit of several tropical species such as *Hyphaene thebaica, Salvadora persica, Ochradenus baccatus, Cassia obovata, Salenostemma argel, Aerva persica, Boerhavia repens, Acacia albida, Acacia nilotica, Acacia laeta, A. negevensis, Moringia aptera, Balanites aegyptiaca, Maerua crassifolia, Grewia villosa, Abutilon muticum* and many others. When m = 9, many tropical crops can be grown such as Sugar cane, Banana, Avocado, Papaya, Pacane, Mango and so on.

We are thus left with theoretically 8 x 8 = 64 combinations of Mediterranean climates as shown on the following table.

m \ P		A 1200	B 800–1200	C 600–800	D 400–600	E 300–400	F 200–300	G 100–200	H 100
a	9	+	+	+	+	+	+	+	+
b	7–9	+	+	+	+	+	+	+	+
c	5–7	+	+	+	+	+	+	+	+
d	3–5	+	+	+	+	+	+	+	+
e	+1–3	+	+	+	+	+	+	+	+
f	−2+1	+	+	+	+	+	+	+	+
g	−5−2	+	+	+	+	+	+	+	−
h	−5	?	?	+	+	+	?	?	−

+: Existing combination; − : Non existing combination.
We have, in fact, no more than 60 combinations at this level.

1. Cushion-like spiny xerophytes (high mountains)   (B h)
                                                     C h
                                                     D h
                                                     (E h)

2. *Fagus silvatica, Abies pinsapo*   A g
   A. maroccana, A. numidica, A. nebrodensis   B g
   A. cilicica, A. cephalonica etc.

3. *Pinus laricio*   C g (h)
   Juniperus thurifera   D g (h)

4. *Cedrus libanotica*   B g (f)
                                                     C g (f)
                                                     (D g f)

5. Deciduous Oaks: *Q. faginea*   Abc d e f
                                                     Bbc d e f

   Q. conferta   *Q. pubescens*   A   e f
   Q. cerris   *Q. toza* (= *Q. pyrenaica*) B   e f
   Q. macedonica   Bbc d e f
                                                     Abc d e f

6. Evergreen Oaks: *Q. suber*   Bbc d e f
                                                     Cbc d e f
                                                     (Dbc d e f)

                         Q. ilex   Bc d e f
                         Q. coccifera   C c d e f
                         Q. calliprinos   D c d e f

7. *Pinus halepensis*   C c d e f
                                                     D c d e f
                                                     E c d e f
                                                     F c d e f

8. *Tetraclinis articulata*   D a b c d (e f)
                                                     E a b-c d (e f)

9. *Argania spinosa*   D (a) b c d
                                                     E (a) b c d
                                                     F (a) b c d
                                                     G (a) b c d

10. Olive – Carob   B (a) b c d
    Ceratonia siliqua, Olea europea   C (a) b c d
                                                     D (a) b c d
                                                     E (a) b c d

11. *Poterium spinosum*   B a b c d
                                                     C a b c d
                                                     D a b c d
                                                     E a b c d

224

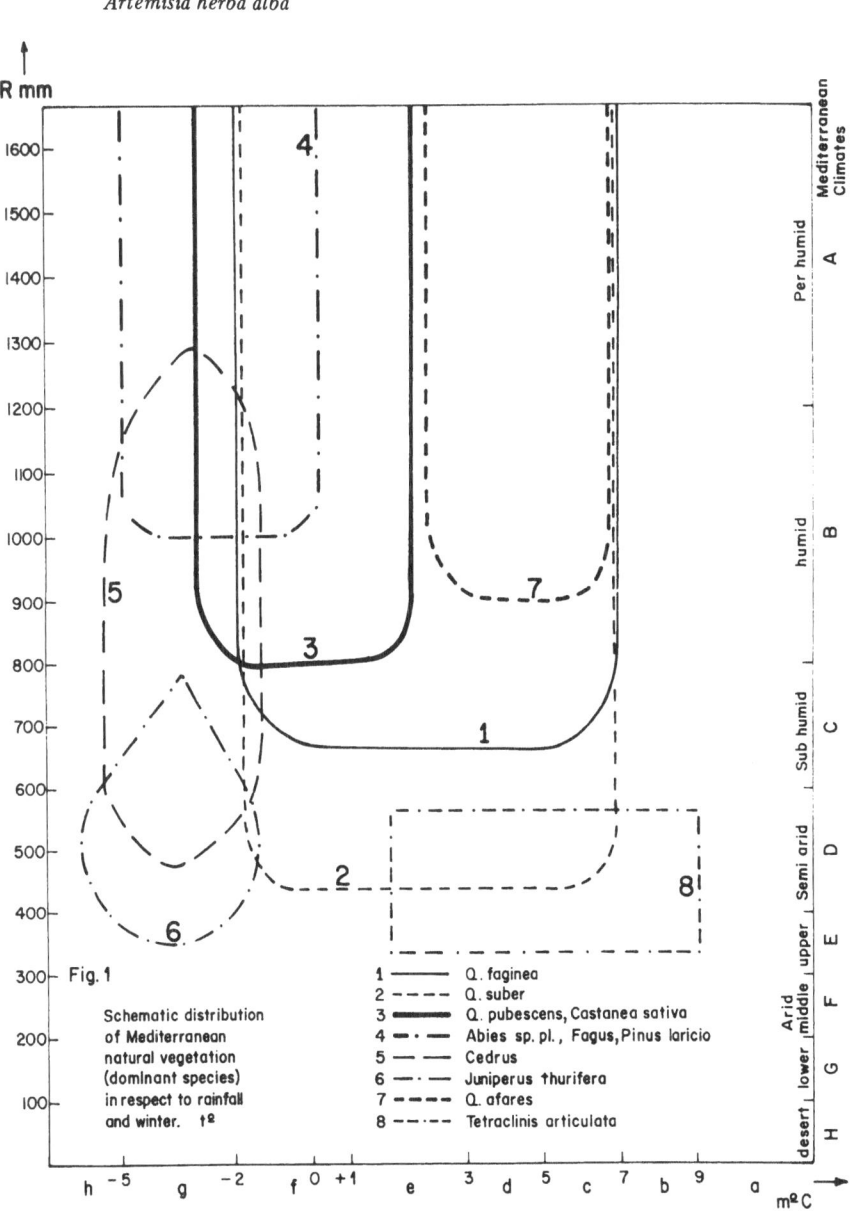

Fig. 1

Schematic distribution
of Mediterranean
natural vegetation
(dominant species)
in respect to rainfall
and winter. t°

1 ——— Q. faginea
2 - - - - Q. suber
3 ▬▬▬ Q. pubescens, Castanea sativa
4 —·—· Abies sp. pl., Fagus, Pinus laricio
5 ——— Cedrus
6 —··— Juniperus thurifera
7 ═ ═ ═ Q. afares
8 —···— Tetraclinis articulata

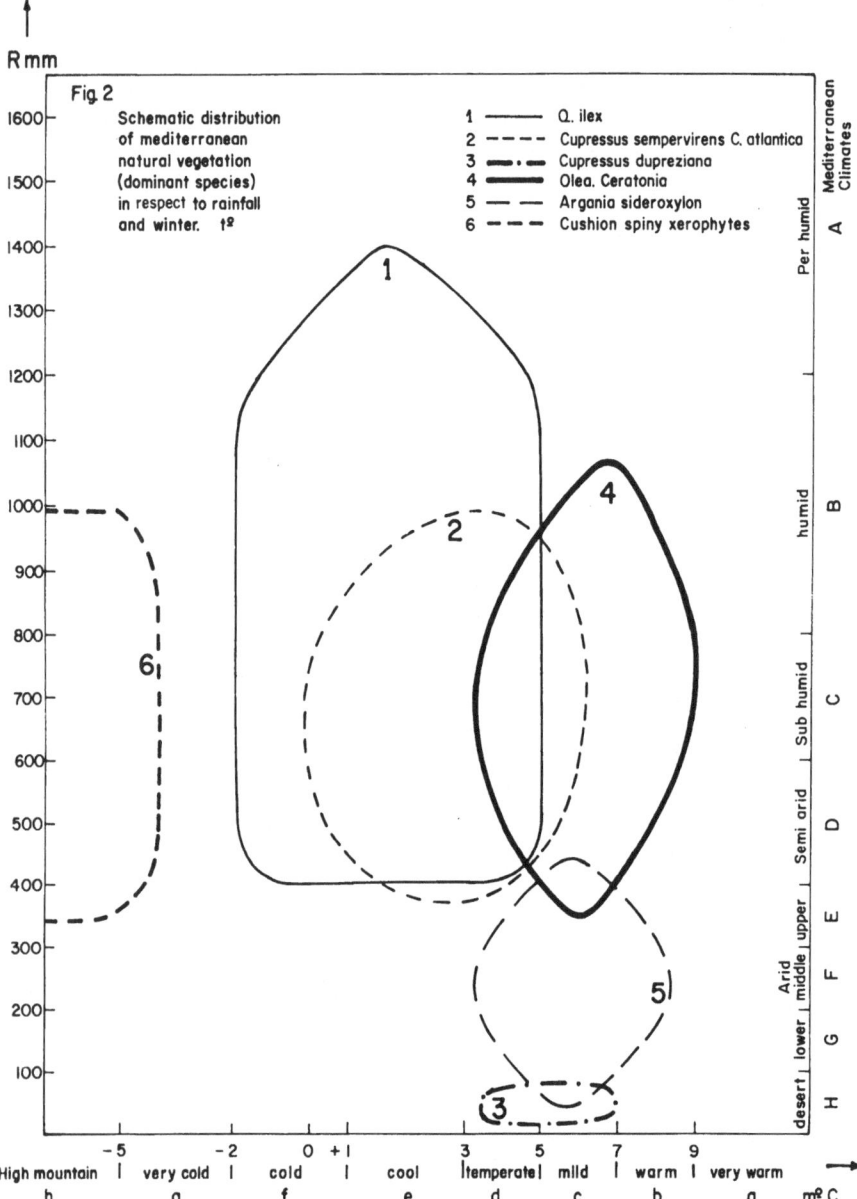

Fig. 2

Schematic distribution
of mediterranean
natural vegetation
(dominant species)
in respect to rainfall
and winter. t²

1 ——— Q. ilex
2 - - - - Cupressus sempervirens C. atlantica
3 ━·━·━ Cupressus dupreziana
4 ━━━━ Olea. Ceratonia
5 — — Argania sideroxylon
6 ━ ━ ━ Cushion spiny xerophytes

226

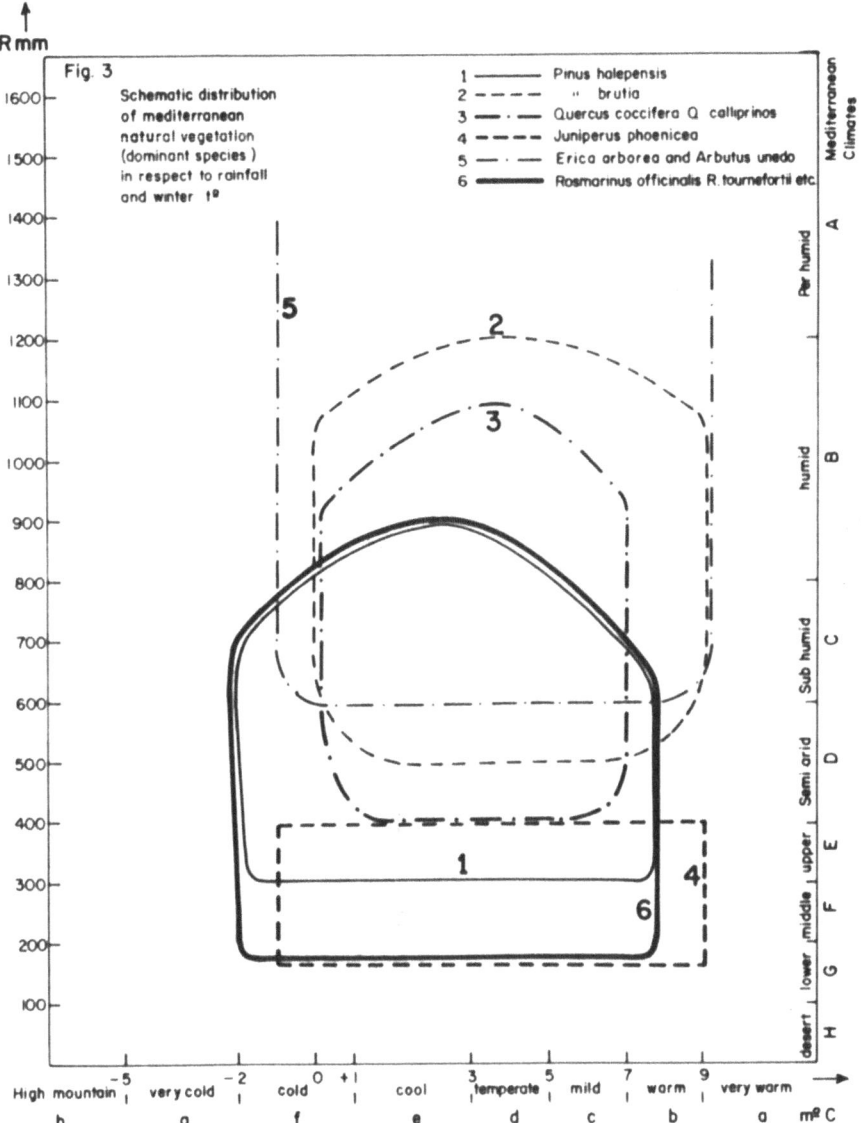

Fig. 3

Schematic distribution
of mediterranean
natural vegetation
(dominant species)
in respect to rainfall
and winter t°

1 ———	Pinus halepensis
2 – – –	" brutia
3 –·–·–	Quercus coccifera Q. calliprinos
4 – – –	Juniperus phoenicea
5 –··–··–	Erica arborea and Arbutus unedo
6 ▬▬▬	Rosmarinus officinalis R. tournefortii etc.

227

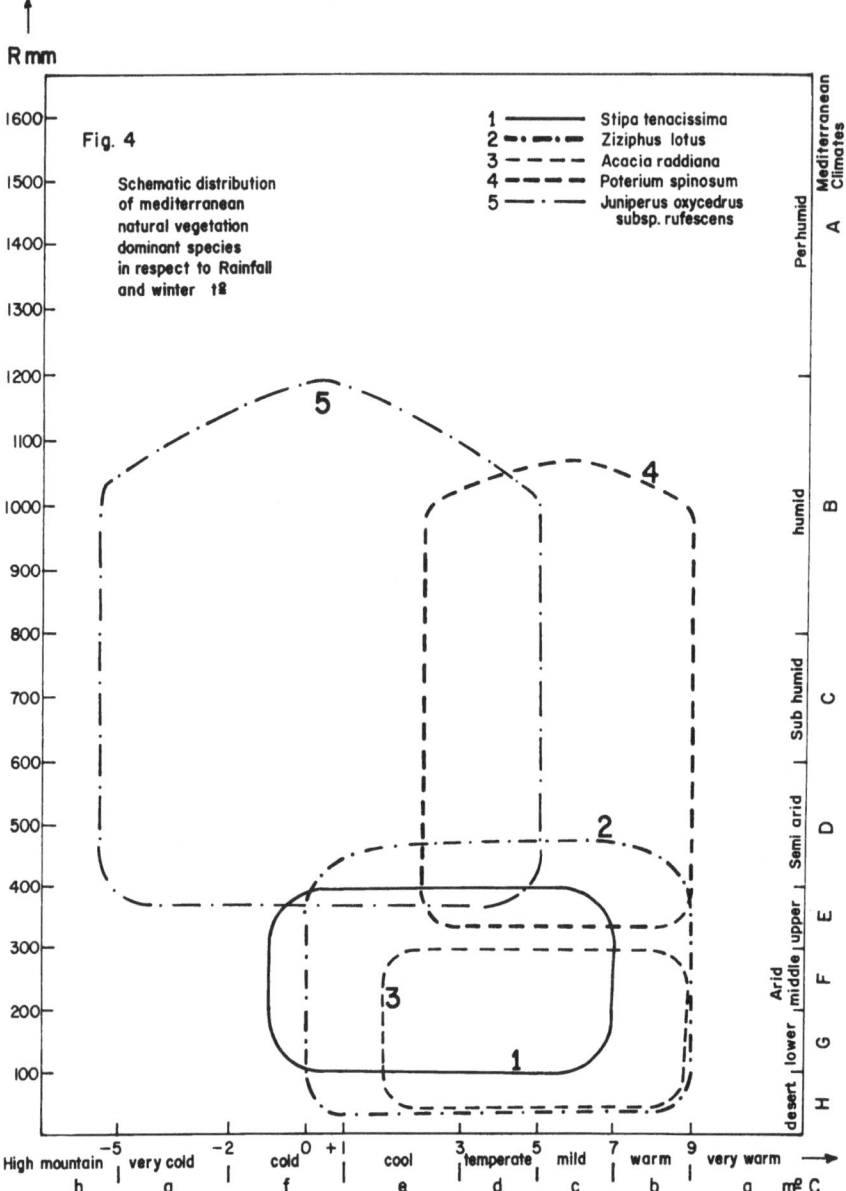

Fig. 4

Schematic distribution
of mediterranean
natural vegetation
dominant species
in respect to Rainfall
and winter t°

1 ———————— Stipa tenacissima
2 —··—··—·· Ziziphus lotus
3 ——————— Acacia raddiana
4 ————————— Poterium spinosum
5 ——·——·— Juniperus oxycedrus
                subsp. rufescens

If we make a table showing the distribution of climates in the different Mediterranean countries, we shall obtain the following results:

Climates / Countries	Per Humid	Humid	Semi Humid	Semi Arid	Arid	Desert
Portugal	+	+	+	(+)	−	−
Spain	+	+	+	+	+	−
France	+	+	+	(+)	−	−
Italy	+	+	+	+	−	−
Greece	+	+	+	+	−	−
Turkey	+	+	+	+	+	−
Syria	−	−	+	+	+	−
Lebanon	−	+	+	+	(+)	−
Israel	−	−	+	+	+	+
Jordon	−	−	+	+	+	+
Egypt	−	−	−	−	+	+
Libya	−	−	(+)	+	+	+
Tunisia	+	+	+	+	+	+
Algeria	+	+	+	+	+	+
Morocco	+	+	+	+	+	+
Cyprus	+	+	+	+	−	−
Malta	−	−	−	+	−	−
Iraq	+	+	+	+	+	(+)
Iran	+	+	+	+	+	+
Saudi Arabia	−	−	−	(+)	+	+
Yugoslavia	+	+	+	(+)	−	−

+ : Bioclimate existing in the country
(+) : Bioclimate dubious or existing on very limited surfaces
− : Bioclimate not represented in the country

5.2.1.5 *Bioclimates and Forage Production*

The bioclimatic and vegetation limits given above are also highly significant for pastures (natural or sown) and for fodder crops. For instance, the demarcation line between arid and semi-arid climates corresponds to the lower limit for reseeded pastures within the conventional pasture species. To my knowledge, all attempts to establish reseeded pastures below the 350—400 mm isohyet have failed, even in trials where establishment had been assured. This is due to the variability of precipitation which increases greatly below the 400 mm isohyet. The 100 mm isohyet corresponds to the lower limit of the sedentary type of pasture exploitation (with less than 100 mm, pasture utilization must be of the nomadic type) and of fodder shrubs which utilize run-off water.

The 600 mm isohyet (limit between semi-arid and sub-humid mediterranean bioclimates) is also the lower limit of several fodder crops in rainfed agriculture, e.g. Sudan grass, Maize, Italian Rye grass, Elephant grass, Rhodes grass and others. . . .

The sub-clovers (native or sown) develop only in the cool, temperate, mild and warm varieties of the semi-arid to humid bioclimates, i.e. above mild and warm varieties of the semi arid to humid bioclimate, i.e. above 350–400 mm of precipitation and above +1°C of daily minimum temperature in January. Similarly, the short cycle varieties of alfalfa (Flamande, Du Puits, etc. . . .) are the best in the cool to very cold varieties $(-2 < m < 3)$ but the long cycle varieties (Moapa, African, Sonora, Indian, San Joaquin, etc. . . .) yield much more in the temperate to very warm bioclimate varieties $(m < 3°C)$. The lower limit of rainfed Alfalfa crops is, again, the 400 mm isohyet, i.e. the border between the semi-arid and arid bioclimates.

The 5°C m is the cold limit for irrigated tropical pastures, *Setaria*, Elephant grass, *Paspalum, Cenchrus ciliaris, Digitaria, Pennisetum, Chloris gayana*, etc. The 3°C m is the cold limit for some forages, such as *Hedysarum coronarium*, long cycle alfalfa varieties, African, Hairy Peruvian, Sonora, Indian and *Trifolium alexandrinum*. The 1°C m is the cold limit for subterranean clover and *Trifolium resupinatum*.

I could give numerous examples of the application of this method to pasture and fodder crops studies and development (See tables* on the criteria of pasture and fodder crops selection, in fine).

The bioclimate affects forage potential in two different ways: 1. Water availability (rainfall); 2. Duration of the growing season.

The first factor greatly affects production, while the second affects the date of utilization.

The following figure represents the theoretical and practical curves of the number of tons of dry matter per hectare and year, in relation to rainfall.

* Tables 1 and 2.

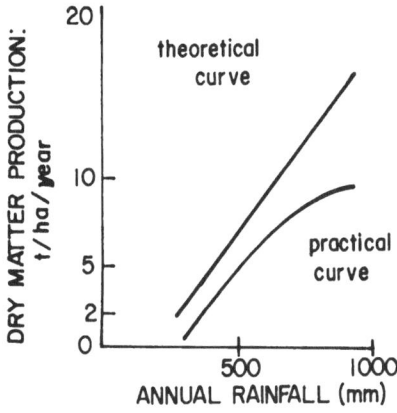

The upper biological limit is about 20 kg of dry matter per mm of mean annual rainfall, although more than 10 kg/DM/ha/ mm is rarely obtained. The highest production figure reached in trials is of 30,000 kg/DM/ha with *Chloris gayana*, under irrigation in Israel, and 33,000 kg/DM/ha with *Pennisetum purpureum* under irrigation in Tunisia. With irrigation, 1 kg of dry matter per 200 kg of water has been reached, but in practice, 500–600 kg of water is needed for 1 kg of dry matter.

The following figure indicates the increase of dry matter obtained with fertilized and unfertilized natural pastures:

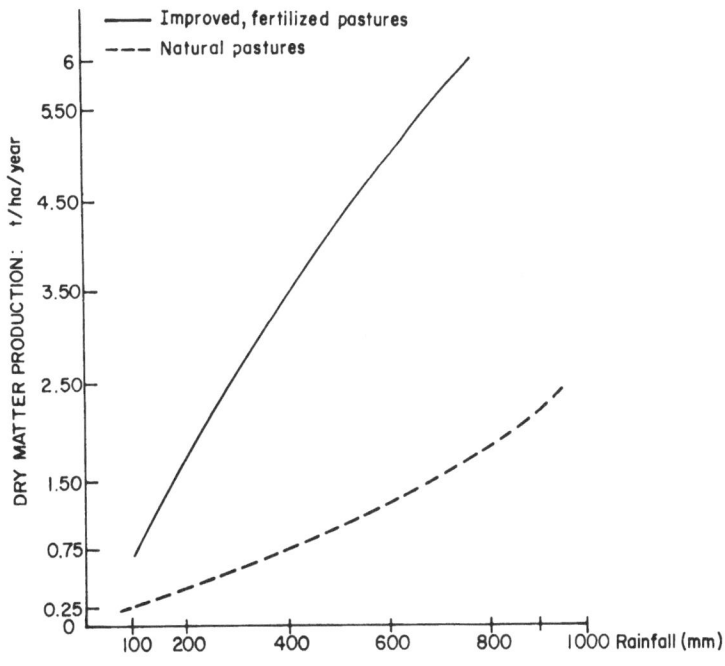

As the season of production is that of plant growth, it is important to determine the length of this season for each type of climate.

The length of the production season depends on: 1. the distribution and abundance of rainfall and correlatively the length of the dry season (vegetative rest); 2. winter temperatures (most species grow only at above 5°C); 3. the capacity of the soil to store water and then give it back.

Humid and Prehumid stages

warm winters	8–9 months growth	1 rest period:	summer		
temperate ,,	6–7 ,, ,,	1 ,, ,,	,,		
cool ,,	5–6 ,, ,,	2 ,, ,,	,, and winter		
cold ,,	4–5 ,, ,,	2 ,, ,,	,, ,, ,,		

Sub-humid stage

warm winters	8 months growth	1 rest period:	summer
mild ,,	7 ,, ,,	1 ,, ,,	,,
temperate ,,	6 ,, ,,	2 ,, ,,	summer and winter
cool ,,	5 ,, ,,	2 ,, ,,	,, ,, ,,
cold ,,	4 ,, ,,	2 ,, ,,	,, ,, ,,

Semi-arid stage

warm winters	6 months growth	1 rest period:	summer
mild ,,	6 ,, ,,	1 ,, ,,	,,
temperate ,,	5 ,, ,,	2 ,, ,,	summer and winter
cool ,,	4 ,, ,,	2 ,, ,,	,, ,, ,,
cold ,,	3 ,, ,,	2 ,, ,,	,, ,, ,,

Arid stage

warm winters	5 months growth	rest periods depend mainly
mild ,,	5 ,, ,,	on precipitations
temperate ,,	4 ,, ,,	,, ,, ,,
cool ,,	3 ,, ,,	,, ,, ,,
cold ,,	2 ,, ,,	,, ,, ,,

In the arid stage, the production period is often theoretical because of the irregularity of climate. In certain years the production period is non-existent (insufficient rain). The above figures show that the production of pasture lands is on the average at least as long as in temperate climates, where it rarely reaches 6 months per year.

1. These data are taken partly from M. Thiault, 1963 Rapp. FAO, ETAP, No. 1689.

We learn from paleogeography that all through the sedimentary ages, the Mediterranean (or Tethys, or Mesogaea for the geologist) was an immense and deep depression all around which deposited and emerged the sediments issued by the shields of the neighbouring continents: African, Scandinavian, Asiatic shields. The alternating transgressions and regressions of the seas left, all around the present day Mediterranean, important sedimentary series which developed particularly from the Trias to the present Era, with considerable extensions of secondary and tertiary formations. The nature of those formations depends on the depth where sedimentation took place: limestone and sandstone on the continental plateau, marls in deep synclines and subsidence zones. A lagune type sedimentation characterized by evaporites (gypsum, various salts) occured in shallow waters in contact with continents. These lagunary deposits can be found all along the sedimentary series with a particular development at Upper Trias, Lias, and Lower and Middle Eocene.

These sediments were folded again and again toward the end of the Secondary and the middle of the Tertiary (Oligocene, Miocene). Hence the Mediterranean mountains are young ranges with vigorous reliefs, and consequently they are very sensitive to erosion. Cristalline formations cover relatively small surfaces as well as volcanic and metamorphic formations.

Erosion has moulded those sedimentary formations and has given them their present forms.

Land form depends on two groups of factors: 1. structure and lithology and: 2. paleoclimates and present day climates.

5.2.2.1   *Structure of Surface and Lithology*

According to structure, we make a distinction between tabular, monoclinal and folded morphology.

Tabular formations are slightly folded or not folded at all. This is the case of Libya, Apulia, the Southern belt of Massif Central in France, Dalmatia, most of the Sahara, etc. . . . .

This morphology often lies at the origin of Karstic zones whenever the structural surface is formed by limestone, marly limestone (Upper Jurassic, Middle Cretaceous, Lutetian, etc. . . . .). Some typical forms are due to the dissolution of limestone by rainwater and snow: "Lappiaz", "Polje", "Dolines", "Avens",

"Gouffres", caves, etc. . . . A red coloured soil called red Mediterranean soil or terra rossa or "decalcification clay" develops in the anfractuosities of limestone. These soils are often devoid of lime but possess a calcium saturated absorptive complex. When thick enough, they are good soils for fruit crops. From a hydrological point of view, these regions are highly permeable and consequently, they are poor in water sources and rivers; they sometimes totally lack a hydrographical network which may be a very serious handicap for agricultural development (irrigation) as well as for urban or industrial development.

Monoclinal formations are typified by seams sloping all in the same direction and generally having slight inclination (less than 45° as a rule). This structure sometimes lies side by side with the preceding one and like that one, develops in the zones of relative tectonic stability.

This type of morphology can be found along the periphery of large sedimentary basins and causes a "cuesta" relief typified by series of ridges of tender rocks (marls) and hard rocks (limestone, sandstone, marly limestone). The hard rocks forming the top side of cuestas are generally covered with spontaneous vegetation (as they very often have skeleton or not very thick soils — rendzinas): garrigue, maquis, forest, grazing land; while the valleys that developed in looser materials give birth to deeper and generally cultivated soils: brown calcareous soils, vertisols, hydromorphic soils, terra rossa, etc. . . .

Folded Morphology. In the Mediterranean region, the anticlinal axes are very often folded in a Jurassic style, i.e., with steep sides and wide flat ridge tops whose centre is often hollowed in anticlinal "cluses" running parallel to the fold axis. Cluses and combes, cutting the anticlinal flancs in V shaped fashion, develop perpendicularly to the folds.

The types of soil and the subdivision of the vegetation are about the same as in the "cuesta" countryside. Marlo-calcareous alternances are covered with forests, garrigues or maquis, while marly cluses and valleys are cultivated, often for cereals.

The Morphology of Arid Regions. The characteristic features of arid regions are flat or slightly undulated large extensions of land that are interrupted now and again by hills or folded chains, monoclines or tablelands.

These arid plains are made up of "Erosion glacis" or "Pediments" that developed during the Quaternary and generally overlaid by calcareous crusts, or calcareous or gypseous incrustations that fossilized a loose substratum (marls, clays). There are generally four of these pediments; they are embedded and correspond to different periods of the Quaternary.

5.2.2.2 *Effects of Climate and Palaeoclimate on Soil Formation*

The action of climates over the various types of structure and lithology is shown either by a slowing down or, on the contrary, by an acceleration of glyptogenesis and sedimentation.

In highly rainy climates, denuded marls tend to slide in muddy flows that can be compared to those caused by cryoturbation phenomena at thaw time in subpolar regions. Those flows form a typical pattern with dented declivities that are alternatively concave and convex along the same line of incline, for instance, in the Appenines chains of Italy, the Rif range of Morocco or Kabylia mountains of Algeria.

Under less rainy climates, the same shales and marly (or clayey) substrata give way to a badland ravine (gully) for the more evolved forms, or to a "rill" or "sheet" eroded pattern as to less developed forms. Karst formations have generally developed in rainy and cold climates, or in rainy and warm climates giving way to sesquioxide earth.

Water erosion decreases as mean rainfall decreases. Towards the isohyets of 200-100 mm, aeolian erosion becomes increasingly important and causes a pattern characterized by regs, hammadas, barkhans, ergs and takirs, i.e., it causes typically desert land forms.

5.2.3 RUNOFF, EROSION, DESERTISATION

Run off and erosion are much more important in the Mediterranean zone than they are in most other climates. Run off depends on the violence of precipitation, on the pluviometric regime, on the relief and on the vegetal cover. It is of a 2% rate on catchment basins covered with forests but it may be over 10% on bare mountains. In some limited districts, it may be much greater, particularly at the end of summer (September-November).

Whenever the thunderstorms are very violent and the ground is bare, run off may arrive at 40% at the level of the catchment basin (the rate is inversely proportional to the catchment basin surface); as to very small catchment basins (a few hectares) it may arrive at 80% during very violent thunderstorms. In Tunisia, Tixeront & Berkaloff have set forth the following empirical formula:

$$R = \frac{H^3}{3}$$

where R = runoff in mm; H = precipitation in metres.

Erosion depends on numerous factors; rains intensity, the slope, plant cover, nature of the ground and its geological substratum, system and nature of cultures, etc.

Given a certain slope and substratum and certain climatic conditions, the rate of abrasion on bare ground can be 50 times as great as the one observed under forest (for example, from 25 to 0.5 metric tons/hectare/year). Erosion takes place especially in autumns and at the end of summer, from September to November. The erosion that takes place during this period of time is the 60–70% of the yearly total.

In the arid zone, desertisation affects tens of thousands hectares every year with sand removal of 0,1 to 2 cm per year (10–60 tons/year). (Le Houérou: 1962, 1968, 1973, 1974; Floret & Le Floch, 1972).

5.2.4   THE MEDITERRANEAN SOILS FROM THE ECOLOGICAL
POINT OF VIEW

From the pedogenetic point of view, the main soil categories of the Mediterranean region are:

scarcely developed soils:	aeolian alluvia
	river alluvia
calcimorphous soils:	Rendzines
	brown calcareous soil
sesquioxide soils:	Mediterranean terra rossa
	Mediterranean brown soil
vertisols:	modal on marls
	hydromorphic on alluvia
hydromorphic soils:	gleisoil
	pseudogleisoil
	with aquifer nodules and incrustations
halomorphic soils:	saline soils
	saline soils with tendency to alkalinity
	alcaline soils

236

inorganic humus soils:	leached brown	
	podzolic	} on high altitudes
	ranker	

soils with calcareous or gypseous crusts

inorganic mineral soils:	regosols on soft rocks
	lithosols on hard rocks

From the ecological point of view, the important characters are:

— soil chemism (acid, basic, toxic, poor, fertile, etc. . . .)
— morphology: texture, structure, depth
— hydrodynamic characters: porosity, permeability, water storage capacity.

The drier the climate, the more important the last point is. As to ecology and agriculture, the best soils possess good permeability and good water storage capacity.

The greater the amount of water stored in the ground, the greater the chances for the plants to withstand climate dryness. In this connection therefore, the role of a good Mediterranean soil is to compensate the irregularity of rainfalls and the consequences of the dry season.

In highly rainy areas (over 800 mm) the best soils are deep and of medium texture. Sandy soils are too dry and clayey soils too wet in winter (radicular asphyxia).

However, in arid areas, the best soils are deep sandy soils; in fact, fine textured soils are too dry because the little water that rainfalls may yield is retained by the absorption of colloidal particles with energies that are often greater than 16 atmospheres (pF 4.2) and consequently, the water is not at the disposal of plants even though the total water percentage is rather high (15 to 20% of weight for argillous soils at wilting point).

5.2.5   VEGETATION AND MAN

The Mediterranean region has been the cradle of numerous civilizations for 6,000 years: Sumerian, Hittite, Egyptian, Assyrian, Etruscan, Persian, Semitic, Phoenician, Greco-roman, Arabian civilizations, etc. . . .

This intense human activity finds its explanation not only in the geographical position that places this area in touch with three continents, but also in the ecological conditions prevailing in it. The mildness of the Mediterranean climate looks like being particularly favourable to man. It has always attracted barbarians from everywhere and even nowadays, it has increasingly been exercising an undoubtful attraction which manifests itself in that modern and peaceful invasion that we call tourism. These summertime invasions attract over 80 million tourists onto the shores of the Mediterranean. Whatever the reasons, the human concentration around the Mediterranean has inevitably led to the over-exploitation of the vegetal resources and to soil erosion. Deforestation, disorderly grazing and cultures on marly slopes have generally reduced natural vegetation to poor garrigues untiredly trodden by herds and flocks and periodically burned down.

As opposed to what happens in the temperate zones, the ground is not snowcovered in winter and the animals are left grazing all through the year because the nature of the climate does not force man to collect fodder for his livestock during the good season. This is the basic reason for the general overgrazing of Mediterranean vegetation.

The goat is bitterly blamed by all forest managers in the Mediterranean region, and literature clearly speaks on its crimes. Some countries have put it out of bound entirely, like Yugoslavia, or partially only. It must be admitted that differently from most other domestic ruminants (cattle, sheep), the goat, (like deer) likes to eat the young sprouts and the leaves of ligneous plants rather than herbaceous plants.

When goats graze in a disorderly way and in too large numbers, they become a real danger because they destroy all the young plants and consequently hinder forest regeneration. If a fire follows, the forest is lost for ever, because it will not be able to be born to new life from seeds. It is therefore necessary to eliminate goats from the patches to be regenerated or from the areas to be reafforested. Yet, the goat is the only domestic animal that can draw the best from the degraded vegetation of Mediterranean garrigues and maquis, and from this point of view, it is really "the cow of the poor". Hence, it must not be brutally eliminated if substitutive resources have not been found for the people who live on the goat.

Deforestation and cultivation are ancient practices: the Pharaohs of the 12th dinasty got their timber from the mountains of Lebanon as early as 4,000 years ago. In not so remote times, Massinissa, the king of Numidia (Eastern Algeria and Western Tunisia) delivered more than 200,000 quintals of wheat to the Romans between the years 200 and 170 B.C. while Carthago traded similar amounts of grains during the same period of time. So we know that the Mediterranean region was cultivated over large surfaces as early as in 150–200 B.C. At that time, Western Europe was still covered with dense forests inhabited by Neolithic men or their direct offspring who did not know agriculture yet.

On the other hand, the Mediterranean peoples had been constantly fighting one another since the remotest times. Numberless and short-lived empires and kingdoms have fought and defeated one another for 6,000 years. The policy of "burned down land" was often the policy of both the winners and the defeated of the moment, beginning with the conquests of the Pharaohs and of Alexander down to the Barbarian, Mongolian and Arabian invasions of old times and the Middle Ages.

At present, the demographic explosion and the use of archaic agricultural methods are rapidly reducing the little spontaneous vegetation that still survives and are speeding up the erosion process which is aggressive by nature. In my opinion, every year, the Mediterranean Basin is losing from 1 to 2% of its agricultural land. The effects of fires come in addition to land clearing; phytogeographers estimate that almost all the forests and matorrals in the Mediterranean Basin have suffered a fire at least once in a century and sometimes even every 20 or 30 years. In fact, every summer, the press and radio report the ruins caused by forest and shrubland fires in the South of France, in Italy and elsewhere. The total surface burnt each year is close to 200,000 hectares, in average, in the Mediterranean Basin (Le Houérou, 1973).

5.3   Biogeographical Definition of Mediterranean Grazing Lands

There is a Mediterranean flora well characterized by:

5.3.1   SCLEROPHYLLE PERSISTENT LEAF TREES

Quercus ilex	*Quercus suber*	*Quercus coccifera*
Quercus calliprinos	*Pistacia lentiscus*	*Ceratonia siliqua*
Juniperus oxycedrus	*J. phoenicea*	*Arbutus unedo*
Olive tree (*Olea europea*)	*Myrtus communis*	*Rhamnus alaternus*
R. lycioides	*Phillyrea media*	*Viburnum tinus*
Laurus nobilis	*Pinus halepensis*	*P. brutia*
P. laricio	*Cedrus libanotica* (Turkey, Cyprus, Lebanon,
Abies pinsapo (Spain)	Algeria, Morocco)
A. maroccana (Morocco)	*A. numidica* (Algeria)	*A. equitrojani* (Turkey)
A. cephalonica (Greece)	*A. borisiiregis* (Turkey)	*A. nebrodensis* (Sicily)

5.3.2 PERSISTENT LEAF AND AROMATIC SHRUBS (LABIATAE FAMILY MOSTLY)

Rosmarinus officinalis	*R. tournefortii*	*Thymus capitatus*
Thymus vulgaris	*T. coloratus*	*T. syriacus*
T. algeriensis	*T. hirtus*, etc.	*Satureja graeca*
Satureja nervosa	*S. thymbra*, etc.	*Lavandula stoechas*
Lavandula multifida	*L. dentata*	*L. officinalis*, etc.
Cistus villosus	*C. monspeliensis*	*C. crispus*
C. salviaefolius	*C. albidus*, etc.	*Erica multiflora*
Globularia nana	*Globularia alypum*	*Erica arborea*
Jasminum fruticans	*Daphne gnidium*	*Cneorum tricoccum*
Genista cinerea, etc. . . .		

5.3.3 NUMEROUS ANNUAL PLANTS (THEROPHYTES) AND GEOPHYTES

Colchicum, Scilla, Urginea, Asphodels, etc. . . . .

"The three summarized conditions are those that distinguish the Mediterranean region, and when one of them is missing, the other two disappear at once" (Durand & Flahault).

The olive-tree is often considered as the best representative of the Mediterranean flora so that some authors (Flahault & Gaussen) consider the Mediterranean region as extending as far as the olive tree area does. However, this is neither exact nor precise because it is necessary to distinguish the cultivated olive-tree — hence in artificial conditions — from the wild olive-tree and as a cultivated plant, its limits vary according to economic conditions: the olive-tree limit has lost ground everywhere, especially in France and Italy, as compared to the past centuries when it was cultivated up to its extreme limits. Thus, this limiting condition cannot be taken into consideration as it tends to vary according to the socio-economic conditions of each region.

The Aleppo pine and the holm oak have been employed in the same way (Flahault & Drude). But, though rare, the holm oak can be found also outside the Mediterranean region (Atlantic coasts) and so it is part of different, not mediterranean, plant communities.

Many botanists agree in having the limits of the Mediterranean region coincide with those of the holm oak. These limits roughly correspond to those of the olive-tree, at least in Europe; but the situation is quite different in North Africa (olive tree: $P > 200$ mm; holm oak: $P > 400$ mm). Neither is it the same in the Near East where the holm oak does not exist, e.g. it cannot be found in Lebanon, Syria, Palestine, Israel, or Jordan. In its details, the problem is not a simple one and the Mediterranean region as a whole cannot be defined with the aid of only one species.

5.4   Vegetation Types of the Mediterranean

Forest is often confused with shrubland in the Mediterranean. I call "forest" formations having over 100 trees over 5 m high per hectare.

There are many types of forest: dense, sparse, with glades, coppice, etc....

Matorral is a ligneous formation of less than 5 meters high. Ligneous species are either natural, i.e. specific in shape and dimensions, or artificial as a result of various mutilating treatments (fire, felling, browsing, etc....). It is almost always the case of degraded forests, comprising the notion of "garrigue", "maquis" and "Lande" where rosemary, inula, retam, etc. are growing.

Garrigue — edaphically dry calcareous soil (Kermes oak) sparse, open, short vegetation.

Maquis — edaphically moist non calcareous soil. Dense, high, thick vegetation. Calycotomes, Lentiscus, Myrtle, arborescent heath.

According to their hights, matorrals are sub-divided into low (H < 0.60 m), middle size (0.6 < H < 2.0 m) and high 2 m < H) matorrals. According to cover, they are subdivided into dense (75% < C), sparse (50% < C < 75) and thin (25 < C < 50) matorrals. According to structure, there are arbored matorrals with sociologically isolated trees (i.e. whose roots are not in competition): fewer than 100 trees/ha.; shrub matorrals: low and dense (chamaerops): and matorrals with thorny xerophytes in mountainous areas (*Erinacea anthyllis, Bupleurum spinosum, Acantholimon, Astragalus* (tragacanthes, etc....).

The terms "Lande" (moor), "scrub", etc.. must be eliminated from the Mediterranean region. Garrigues often have particular names such as "Tomillares" in Spain, "Batha" in the Near East and "Phrigana" in Greece; these three names correspond to very short degraded garrigues.

"Erme" is a low discontinuous herbaceous formation with a marked seasonal rhythm. Plants left by livestock, such as bulbous liliaceae (Asphodels, Urgineae, etc.), big umbelliferous plants

(*Ferula, Thapsia, Eryngium, Daucus, Foeniculum*, etc.) and this-
tles (*Carlina, Carduus, Scolymus, Silybum, Cirsium*, etc.).

Steppe is a vast arid expanse without reliefs. No trees, bare
soil showing between the tufts of perennial low vegetation (graminaceous or ligneous).

Graminaceous steppe where perennial ± coespitose grasses
are dominant, e.g. Alfa (*Stipa tenacissima*) Esparto, (*Lygeum
spartum*) *Aristida, Stipa* spp.

Chamaephytic steppe or ligneous steppe: a degradation
from graminaceous steppes where small shrubs grow: *Artemisia
herba alba, A. campestris, Rhanterium suaveolens, Hammada
scoparia*, etc.

Crassulescent steppes: e.g. halophytic chenopodiaceae,
*Atriplex, Suaeda, Salsola, Arthrocnemum, Salicornia, Haloc-
nemum*.

These crassulescent steppes correspond to saline depressions: solontchack soils.

Pseudo-steppes in which nanophanerophytes are domina-
ting. They occur on terraces, in non-saline depressions overfed
with water, and dunes of arid zones: Jujuby trees, *Retama rae-
tam, Haloxylon persicum, Tamarix* spp. *Calligonum, Ephedra
alata*, etc.

Meadow — low continuous formation, herbaceous plants
with little seasonal rhythm, made up of mesophytic or hygro-
philous species: strawberry clover, fescue-grass and alsike clover
grassland; they do not cover very large areas: near springs,
oozing, floodable areas of phreatic and aquifers near the ground
surface and often slightly saline water tables (5—10 mmhos).

Lawn — low continuous herbaceous formation with marked
seasonal rhythm (dry in summer) graminaceae such as *Poa bulbosa* and *Lolium perenne* and very numerous rosette compositae.

5.5   The Mediterranean Grazing Lands

5.5.1   GENERAL

It may seem an impossible task to deal with the pasture
lands of the Mediterranean Basin in such a limited space in view
of the fact that they are so varied, so numerous, and often,
unfortunately, little known.

This diversity reflects the extreme variation of ecological,
historical and agricultural conditions around the Mediterranean.

242

To speak only of the best known ecological factor, i.e., climate, at least 60 to 80 climatic sub-types all belonging to the Mediterranean type, may be found in the Mediterranean Basin. (See above: Par. 2).

Thus, we find at first estimate some sixty sub-types among the general climates. If, in addition, we introduce differential factors such as continentalism $(M - m)$, length and intentisity of dry season, water balance $(R - PTE)$, we arrive at several hundred climatic gradations, all of which are also based on phytogeographic criteria.

Thus we see that by introducing microclimatic and regional climatic gradations, geological, topographical-edaphic, historical and social conditions and variations of flora (western, eastern, African, etc...). we arrive at a multitude of types of Mediterranean vegetation and hence of pasture lands.

Each of these types has its own characteristics, i.e. periodicity, yield, potential and dynamism.

To cite the now well-known case of Tunisia, 1000 types of pasture lands have been referenced and mapped on a surface of approximately $160,000 \text{ km}^2$.

Similarly, the "Prodrome des groupements végétaux de la France méditerranéenne" includes no less than 150 plant associations, apart from sub-associations and variants: (Braun-Blanquet, Roussine & Nègre, 1952). Also in Spain, at the systematic level of phytosociological association nearly 140 types of pasture lands are distinguished (S. Rivas Goday & S. Rivas Martinez, 1963). In Algeria, in the Hodna Basin, I have distinguished 170 pasture types over an area of $25,000 \text{ km}^2$ (Le Houérou, Claudin & Haywood, 1975).

Applying this coefficient of variability to all Mediterranean vegetation would undoubtedly result in thousands of pasture types.

The foregoing indicates that the problems are very complex and can be treated here only in a very general way and solely on a bioclimatic basis.

5.5.2 TYPES OF PASTURE LANDS OF THE MEDITERRANEAN REGION

5.5.2.1 *General Information*

In accordance with what has just been said, we shall treat this subject on a bioclimatic basis, using the classification described above.

This classification reflects the climatic zonality, a function of latitude, altitude and continentalism. Certain types of pasture land, more or less azonal at the scale which concerns us, such as saltflats (plains) coastal vegetation, etc. will be examined separately.

The Roman writers classified plant landscapes into three categories (which we would now refer to as ecosystems), i.e. Silva, Saltus and Ager. This classification is so pertinent that it was taken up again by scholars such as Kulnholtz Lordat (1945) and Duvigneaud (1963). Silva is the forest, Saltus comprises degraded forests (low sparse vegetation, scrub) and pasture lands of various types, (meadows, prairies, steppes) whereas Ager refers to cultivated lands. This classification is all together static and dynamic as Saltus usually derives from forest or abandoned crop lands and Ager from the cultivation of Silva or Saltus. As Duvigneaud (1963) emphasizes, the agricultural prosperity of a given region derives from a harmonious balance of proportions and the localisation of Silva, Saltus and Ager.

Silva has, unfortunately, become vestigial in the Mediterranean Basin; hence the loss of balance and the gradual degradation into immense and unproductive Saltus, which we are now witnessing.

5.5.2.2   *Pasture Lands of the Saharan Mediterranean Zone*

The Saharan Mediterranean zone is hardly represented, except in North Africa and the desert zones of the Near East (Jordan, Arabia, Iraq, Syria, Palestine, Egypt). Two sub-zones may be distinguished, i.e. upper and lower.

5.5.2.2.1 The Upper Sub-Zone

The upper sub-zone is characterized on the whole by an average annual rainfall between 50 and 100 mm. From the point of view of grazing, this sub-zone is characterized by woody chamaephyte steppe or pseudo-steppe vegetation diffusely distributed that is, the perennial species are distributed more or less regularly along the regs and elevated grounds. The soil covering is always very slight: 1% to 30%. The thalwegs and wadis are covered with denser perennial vegetation and often arborescent and at times, even arbored (*Acacia raddiana, Maerua*, etc.). The vegetation of regs is often on the basis of salsolaceae:

Hammada scoparia	*Anabasis aphylla*
Hammada schmittiana	*Anabasis aretiodes*
Salsola spp.	*Anabasis articulata*

with some graminaceae, mainly of the *Stipa* and Aristida genuses (*A. plumosa, A. obtusa, A. pungens*, etc.) and numerous annuals, chiefly, cruciferae, leguminous plants, graminaceae and compositeae.

Non-salty depressions are often covered with a steppe of white sagebrush (*Artemisia herba alba, Artemisia judaica* or *Artemisia fragrans*, in Turkey). The thalwegs witness the development of nanophanerophyte vegetation:

Retama retam	*Rhus tripartitum*
Ziziphus lotus	*Acacia raddiana*
Lycium arabicum	*Maerua crassifolia*
Nitraria retusa	

with tall perennial grasses:

Aristida pungens	*Cymbopogon schoenanthus*
Hyparrhenia hirta	*Pennisetum elatum*
Pennisetum dichotomum	*Rottboellia hirsuta*, etc.

5.5.2.2.2 The Lower Sub-Zone

The lower sub-zone is distinguished by rainfall varying from 20 to 50 mm. Below 20 mm is a typically Saharan or Central Saharan climate, which no longer comes within the range of Mediterranean climates as the rare rainfall which is observed may occur in any season (Emberger, 1930).

The lower Saharan sub-zone is characterized by vegetation in narrow distribution (Monod, 1937); that is, the regs and interfluves are barren of perennial vegetation; the latter is localised in the wadis, thalwegs, low points, stony hamadas and at the foot of dune zones, hence in those places which are especially favoured in terms of water balance either through accumulation of run off waters or because they are below the level of particularly permeable zones. The vegetation of the lower sub-zone differs very little from that of the upper sub-zone from the standpoint of floral composition, but there is an increase in the proportion of tropical affinity species: Zygophyllaceae, Aristideae, Penicoideae, etc.

Zygophyllum spp. *Fagonia* spp. *Tribulus* spp. etc.

245

Sandy-bottomed dunes and thalwegs have ephedroid vegetation with *Calligonum*, *Ephedra alata*, *Haloxylon persicum* (Near East) *Ochradenus baccatus* (Near East), *Randonia africana*, *Retama retam*, *Genista saharae*, etc.

Salsolaceae are numerous: *Noaea*, *Holgeton*, *Haloxylon*, *Anabasis*, *Atriplex*, *Salsola*, *Choenolea*, *Suaeda*, *Cornulaca*.

The "pasture lands" of the Saharan Mediterranean zone are naturally characterized by a type the yield of which is both low and very irregular, i.e. "Aacheb", formed of annual eremophytes which develop several days after the rainfall. In the lower sub-zone, only certain thalwegs can be considered as pasture lands. The animals using them are mainly camelides, goats and certain breeds of highly resistant sheep, such as the "Mezragui" or "targui sheep" of Algeria, Libya, Hoggar, and Tibesti.

The upper sub-zone is mainly frequented by camelides, caprines and fat-tail Barbary sheep. Capacity is very small and may be estimated at approximately one sheep per 20—30 hectare for four to six months in the upper sub-zone. This concept no longer has any meaning in the lower sub-zone.

5.5.2.3 *Pasture Lands of the Arid Mediterranean Zone*

This stage is found in zones receiving 100 to 400 mm of rain per year. It is represented in the Near East: in Turkey, Iraq, Afghanistan, Saudi Arabia, Syria, Lebanon, Palestine, Jordan; in North Africa: Libya, Tunisia, Algeria and Morocco. There are some small enclaves in Europe, particularly in Spain (the region of Saragossa in the North-East, of Almeria and Murcia, in the South-East), in France (between Marseille and Cassis) and perhaps also in Southern Italy and Greece. It may now be stated that primitive vegetation is nearly everywhere a woodland or a steppe forest or an arborous steppe. The forest species are mainly red juniper (*Juniperus phoenicea*) and *Pistacia atlantica*, *Acacia raddiana*, False Thuya (*Callitris articulata*) and *Argania sideroxylon*; the last two named exist in North Africa only. Some enclaves of Aleppo pine may also be found towards the upper limit of the zone. The depressions, often more or less salty, are found by Tamarix woodlands. On the whole, these woodlands have now disappeared, giving way to the steppe. Alfa steppes (*Stipa tenacissima*) cover enormous surfaces in North Africa, i.e. Morocco, Algeria, Tunisia and Tripolitania. They are localized almost exclusively in the arid Mediterranean zone in various soils, which

246

usually include a crust or chalky deposit near the surface. Esparto steppes (*Lygeum spartum*) are also connected with this zone (North Africa, Cyrenaica, Near East); they cover much smaller surfaces and are localized in soils rich in sulphates. The Sage steppes of North Africa and the Near East may also be regarded as characteristics of the arid Mediterranean zone. These are the white Sage steppes (*Artemisia herba alba*) of Spain, Morocco, Algeria, Tunisia, Cyrenaica, Palestine, Jordan, Syria, Iran and Iraq. These steppes develop in more or less alluvial or clayey soils, sometimes rich in sulphates or even lightly chlori-

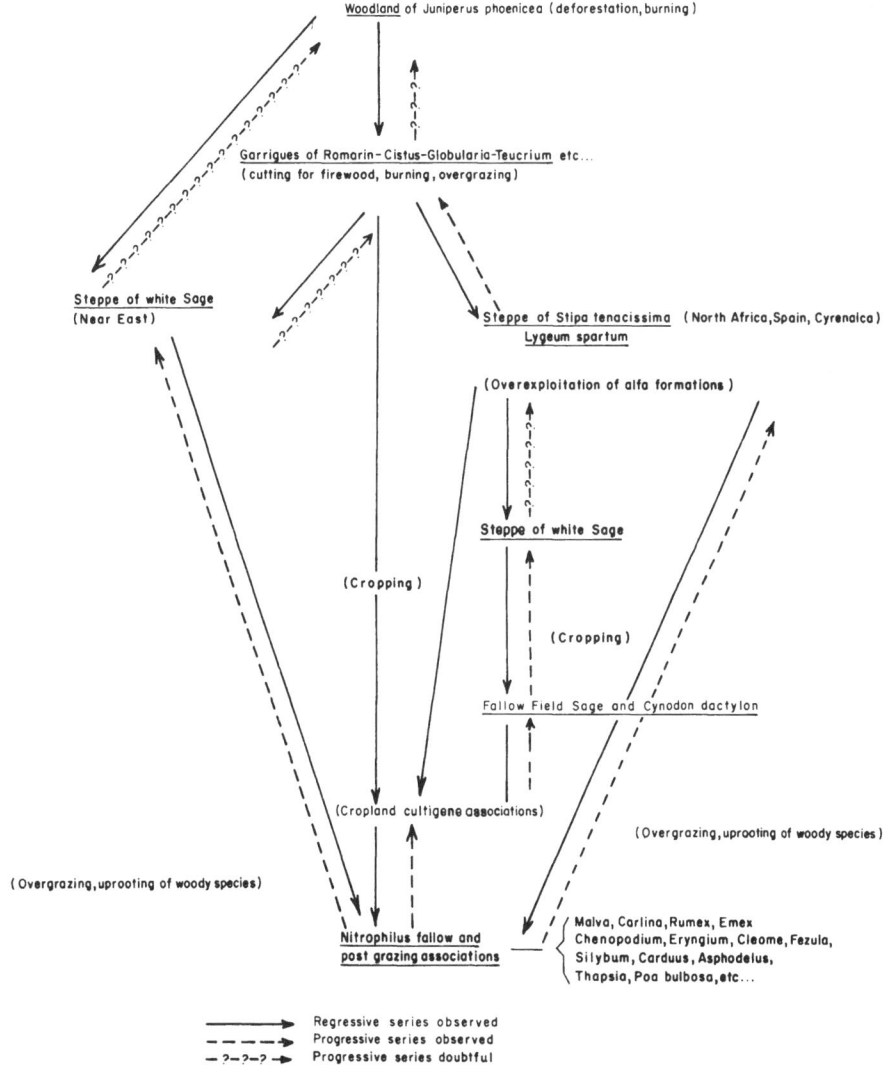

nated in depth. Their general characteristic is the presence at the surface of the soil of a sealed alluvial clayey film, which reduces infiltration of rain water. White Artemisia steppes are numerous and varied. In Tunisia, for example, where they are well known, at least thirty types are distinguished in accordance with their floral composition, reflecting climatic gradations and variations of the edaphic substratum. In Libya, some fifteen types are distinguished by the same factors.

The white Artemisia steppes are a dynamic stage from primitive arboreal vegetation (usually *Juniperus phoenicea*) according to the general pattern given on page 247.

It seems that at present certain stages constitute a paraclimax (specific steppes of white Artemisia of the lower arid sub-stage and of Alfa and Esparto of the same sub-stage).

The major partner plants of white Artemisia are:

Poterium spinosum (Near East, Cyrencia)	
Hammada scoparia	*Anabasis articulata*
Asteriscus pygmaeus	,, *oropediorum* (A.F.N.)
Plantago ovata	,, *aphylla* (Near East)
Eryngium ilicifolium	*Salsola rigida* (,, ,,)
Stipa retorta	*Achillea fragrantissima* (Near East)

Non-dune sandy soils are covered with a steppe of field Artemisia (*Artemisia campestris* ssp. *glutinosa*) and couch grass (*Cynodon dactylon*) (fallow), in the upper sub-stage, and *Rhantherium* in the lower sub-stage:

(*Rh. adpressum*: Morocco, Algeria)
(*Rh. suaveolens*: Algeria, Tunisia, Libya)
(*Rh. garcini*: Near East)
(*Rh. epapposum*: Near East)

To these dominant species must be added the following, which are good fodder plants:

Plantago albicans	*Echiochilon fruticosum*
Argyrolobium uniflorum	*Helianthemum lippii*
Aristida plumosa	*Pituranthos tortuosus*
,, *obtusa*	*Noaea spinosissima*
,, *adscentionis*	*Eragrostis papposa*
Stipa lagascae	*Cenchrus ciliaris*
,, *barbata*, etc.	

There is a large plant community of the following type to be found on sandy dune soils:

Aristida pungens	*Retama retam*
Danthonia forskhallei	*Rumex tingitanus*
Polygonum equisetiforme	

"Garrigue" skeletal soils are covered with chamephytic vegetation of:

Rosmarinus officinalis.	*Globularia alypum*
Thymus capitatus,	*Th. syriacus*
,, *hirtus,*	*Cistus libanotis,*
Phagnalon rupestre,	*C. monspeliensis,* etc
„ *saxatile,*	*Varthemia iphionoides*
Satureja spp.	,, *candicans*
Lavandula multifida,	*Ballota hirsuta*
Sideritis spp.	,, *undulata*
Teucrium spp.	*Dactylis hispanica*
Cistus salviaefolius,	*Rhamnus lycioides*
Erica multiflora,	*Calycotome villosa*
Cistus parviflorus,	*Fumana glutinosa,*
Helianthemum spp.	*F. arabica, F. laevipes,* etc.
	C. villosus,

5.5.2.4 *Pasture Lands of the Semi-Arid Mediterranean Zone*

The semi-arid zone may be regarded as especially representative of the Mediterranean climate (Madrid, Sevilla, Majorca, Constantine, Oran, Fez, Mostaganem, Tunis, Tripoli, Athens, Salonica, etc..) It is characterized climatically by an average annual rainfall of between 400 and 600 mm. The Emberger index is between 40 (cold winters) and 80 (hot winters). This zone covers immense surface areas all around the Mediterranean, i.e.: Spain, France, Italy, Yugoslavia, Albania, Greece, Turkey, Syria, Lebanon, Jordan, Palestine, Cyrenaica and northern regions of Tunisia, Algeria and Morocco.

The primitive vegetation type is forest. The dominant varieties are: *Tetraclinis articulata* (False Thuya), *Juniperus phoenicea* (red juniper), *Juniperus thurifera* (thurifer juniper), *Olea europea, Ceratonia siliqua, Pistacia lentiscus, Chamaerops humilis, Pinus halepensis, Cupressus sempervirens, Argania sideroxylon* (Morocco), *Quercus suber* (cork oak), *Quercus ilex* (Holm oak), *Quercus calliprinos, Quercus ithaborensis.* Most of these species are the leaders of dynamic series and the dominant species of as many types of semi-arid forests, except for olive tree-lentisk and olive-tree Caroub brush, which are already degraded stages of primitive forest. Many of these species are found again in the sub-humid stage but in different associations and belonging to a different dynamic series. There are, for example, notable differences between oak forests of cork oak of semi-arid, sub-humid and humid stages; few species are confined to a single stage of vegetation (this is the case for *Tetraclinis articulata*, which extends very little beyond the semi-arid zone). It is not possible to give here the countless types of pasture lands which derive from these "Silva". In general, they are "garrigue" hills or even scrubs (e.g. scrub of *Poterium spinosum* in the Near East and Cyrenaica), dry swards or meadows.

The majority of the dominant plants in the pasture lands of the "garrigues", i.e. rosemary, thyme and other xerophile labiatae and some perennial graminaceae, are unpalatable to animals, other than goats; however, good pasture species, if not abundant, are often present:

Avena bromoïdes	*Hordeum bulbosum*	*Hyparrhenia hirta*
Stipa barbata	*Stipa gigantea*	*Oryzopsis miliacea*
Dactylis hispanica	*Lolium perenne*	*Oryzopsis coerulescens*
Oryzopsis holciformis (Near East)		*Melica ciliata*
Phalaris truncata	*Phalaris tuberosa*	

Included amongst the perennial or biennial leguminous plants are:

Lotus creticus	*Lotus corniculatus*	*Hedysarum coronarium*
Hedysarum flexuosum	*Onobrychis spp.*	*Coronilla minima*
Medicago tunetana	*Medicago arborea* (Greece)	*Cytisus spp.*
Argyrolobium linneanum	*Hippocrepis spp.*	

Among other palatable species, *Sanguisorba minor*, should be mentioned.

These species are usually not very abundant due to overgrazing and often they owe their survival to the protection afforded by spiny shrubs such as: *Calycotome, Crataegus, Rhamnus, Rhus, Ziziphus*, etc.

Scrub of *Poterium spinosum* develops when rainfall exceeds 300 mm. It includes a bushy stratum where *Poterium spinosum* is the dominant species, covering 20% to 80% of the surface of the soil. There is a mat-like greensward of the ephemeroïd *Poa bulbosa* between the *Poterium* bushes where the animals walk.

The palatable species are numerous but not abundant:

Dactylis hispanica	*Stipa lagascae*	*Avena bromoïdes*
Phalaris tuberosa	*Stipa parviflora*	*Oryzopsis holciformis*
Phalaris truncata	*Stipa capensis*	*Oryzopsis miliacea*

Hordeum bulbosum	*Stipa bromoïdes*	*Stipa barbata*
Hordeum murinum	*Medicago spp.*	*Elymus delileanus*
Avena barbata	*Lotus creticus*	*Cynosurus coloratus*
Phalaris canariensis	*Aristida adscentionis*	*Festuca eliator* (humid stations)

The dry swards "Pelouses" of the semi-arid zone are formed mainly of annuals where *Stipa capensis* and *Plantago albicans* dominate. Among the perennials are:

Oryzopsis miliacea	*Dactylis hispanica*
Oryzopsis coerulescens	*Lolium perenne*
Avena bromoïdes	*Stipa parviflora*
Phalaris truncata (on marl)	*Lotus creticus*
Hedysarum coronarium (on marl)	*Onobrychis argentea*
Hedysarum pallidum	*Onobrychis viciaefolia*
Psoralea bituminosa	*Ebenus pinnata*

Among the annuals:

Avena sterilis	*Avena barbata*
Hordeum murinum	*Medicago spp.*
Melilotus spp.	*Vicia spp.*
Trifolium spp.	*Trigonella arabica* (Near East)

and mainly: *Stipa capensis*

Meadows are localized in flooded low points on soils which are often "tirsified" (vertisols) and more or less hydromorphic or salty, or on permanently wet soils with a temporary or permanent water table.

The most characteristic species are:

Juncus maritimus	*Juncus acutus*	*Festuca elatior* sup. *arundinacea*
Trifolium fragiferum	*Agropyrum elongatum*	*Medicago ciliaris*
Trifolium resupinatum	*Hordeum maritimum*	*Medicago intertexta*
Phalaris truncata	*Pucinellia distans*	*Agropyropsis lolium*
Phalaris coerulescens	*Poa trivialis*	*Alopecurus pratensis*
Phalaris tuberosa		

5.5.2.5 *Pasture Lands of the Sub-Humid Mediterranean Zone*

The sub-humid zone is very extensive mainly in the Mediterranean zone of Europe: Portugal, Spain, France, Yugoslavia, Italy, Greece and Turkey. It also occupies large surfaces in the mountain regions of the Near East: Lebanon, Syria, Jordan, Palestine, Iraq, Iran.

In the North of Africa, it is represented by a small zone in Cyrenaica and large surfaces in the northern regions of North Africa from Bizerta to Tangier and around the high Algerian-Moroccan Atlas Mountains.

The zone is defined climatically by an average annual rainfall between 600 and 800 mm and an Emberger index of between 70 (cold winters) and 120 (warm winters). The soils are usually chalky, but sometimes washed, because of the high rainfalls on permeable substrata, unsaturated and even oligotrophic, particularly in sandstone and sandy geological formations such as the "sandstone of Numidia" (oligocene) or acid crystalline rocks (granite).

The primitive vegetation is always forest, which is usually degraded into a "garrigue" (calcareous soil) or "maquis" (acid soil). The key species are:

Quercus ilex (North Africa, Europe)
Quercus coccifera (N. Afr. Europe)
Quercus calliprinos (Near East)
Quercus infectoria (Near East)
Quercus aegilops (Near East)
Quercus suber (N. Afr. Europe)
Cupressus sempervirens (N. Afr. optimum semi-arid)

Ceratonia siliqua (N. Afr. Europe, Near East. mild or warm) winters; optimum semi-arid).
Olea europea (N. Afr., Europe, Near East. mild or warm winters. optimum semi-arid).

Pinus maritimus (North Africa, Europe)
Pinus laricio (Europe)
Pistacia lentiscus (N. Afr. Europe)
Pistacia palaestina (Near East)
Pistacia terebinthus (Europe)

Pinus halepensis (N. Afr. Near East, Europe, optimum semi-arid)
Juniperus phoenicea (N. Afr. Cyrenaica, Near East)

The low, sparse vegetation deriving from these forest plants is very similar to that of the semi-arid zone. They often develop on more or less karstic chalky massifs, or often decalcarated terra-rossa soils or marl (Aleppo pine).

The plant associations of heath and degraded heath are numerous:

Quercetalia ilicis — Quercetum gallo provinciale
Quercetum mediterrano montanum
Calycotomo-Myrtetum
Cocciferetum
Quercetum calliprini
Oleo-Ceratonion

Rosmarinetalia   – Aphyllanthion
                  – Rosmarino-Ericion

Cisto Lavanduletea

Thero-Brachypodetalia— Brachypodion phoenicoides
                       Thero -Brochypodion
                       Brachypodietum ramosi
                       Asphodeletum fistulosi, etc.

From the standpoint of pasture land, we note:
Brachypodietum ramosi and Brachypodie-
tum phoenicoides of Europe,
Greenswards of *Hyparrhenia hirta* of the Near East,
Scrub of *Poterium spinosum*, sub-humid, of the Near East,
Matorral of Cistes, of Europe, Africa and the Middle East,
Matorral of Kermes oak of Europe and the Near East (ass. of
Quercus coccifera, of *Quercus calliprinos* and *Pistacia palaestina*).

Scrub develops on unsaturated soils of Europe and North
Africa. It is a regressive series characterised by acidiphilous species:

Myrtle (*Myrtus communis*)
Halimium halimifolium (N. Afr.)
Heath: (*Erica arborea, E. scoparia*)
Strawberry tree (*Arbutus Unedo, A. andrachne* (Near East)
            A. pavarii (Cyrenaica)

Among the pastoral species of the maquis and garrigue, the
following may be found:

Brachypodium ramosum	*Oryzopsis holciformis* (Near East)
Brachypodium phoenicoides	*Oryzopsis thomasii* (acid soils)
Agropyrum glaucum	*Trifolium subterraneum* (acid soils)
Dactylis hispanica	*Medicago orbicularis*
Dactylis maritima	*Hordeum bulbosum*
Lolium perenne	*Medicago falcata*
Festuca elatior	*Lotus corniculatus*
Festuca ovina ssp. duriuscula	*Lotus creticus*
Trigonella monspeliaca	*Trifolium resupinatum*
Melilotus sulcatus	*Trifolium angustifolium*
Anthyllis vulneraria	*Bonjeania hirsuta*
Medicago minima	*Sanguisorba minor*
Medicago tribuloides	*Ampelodesmos mauretanica* (N. Afr.)
Avena bromoides, etc.	

This zone is defined climatically by a rainfall of between 800 and 1200 mm and an Emberger index of between 90 (cold winters) and 150 (warm winters). Examples: Draguignan, Grasse, Coimbra, Pamplona, Tabarka, Djidjelli, Bougie, Ifrane. It covers limited surfaces in the mountainous regions of North Africa, Spain, France, Italy, Yugoslavia, Greece and the Near East. The characteristic vegetation comprises forests of:

Cedars: *Cedrus libanotica, C. atlantica, C. brevifolia* (Lebanon, Algeria, Morocco, Cyprus, Turkey)

Firs: *Abies pinsapo* (Spain)
- „   *cilicia* (Near East)
- „   *numidica* (Morocco, Algeria)
- „   *cephalonica* (Near East, Greece)
- „   *regis Borisii* (Greece, Yugoslavia)
- „   *equi trojani* (Turkey, Greece)

Pines: Maritime Pine
Pinus laricio (Corsica, Calabria, N. Africa)
P. pinaster ssp mesogeensis
P. clusiana
P. mauritanica
P. hispanica
P. italica, etc ...

Deciduous Oaks:
Quercus toza (Morocco, Spain, France)
- „   *faginea* (Portugal, Morocco, Algeria, Tunisia, Spain)
- „   *cerris* (Near East, Greece, Italy)
- „   *pubescens* (France, Italy)

Non deciduous oak forests and shrublands:
Quercus ilex (Europe, N. Afr.)
- „   *calliprinos* (Near East)
- „   *suber* (Europe, N. Afr.)

Because of high rainfall and frequent hard winters (often high mountains are involved), the soils are decalcarated and often leached (podzols) and usually very rich in humus under the forest (oxyhumic soils of H. Del Villar).

Forest degradation produces: Cistus scrub (*C. salviaefolius, C. villosus*)

Heath (*Erica arborea, E. scoparia*)
Myrtle, Eagle fern (*Pteridium aquilinum*)
Halimium spp.
Arbutus unedo, A. andrachne

which finally give way to greensward with:

Dactylis glomerata	Festuca arundinacea
Phalaris tuberosa	Brachypodium phoenicoides
Phalaris coerulescens	B. sylvaticum
Digitalis purpurea	Helianthemum tuberaria
Oryzopsis paradoxa	Anthoxanthum odoratum
Vicia sp. plur. etc.	Cynosurus cristatus
	Holcus lanatus

5.5.2.7 *Pasture Lands of the Prehumid Mediterranean Zone*

This zone is defined climatically by a rainfall over 1200 mm. The Emberger index is between 150 (cold winters) and 200 (temperate winters). It is localised on the coastal mountains of North Africa, Spain, Portugal, Italy, Yugoslavia, Greece, Turkey and the Near East. Examples: Aïn Draham (Tunisia), Ketama (Morocco), Montalegra (Spain), Collo in Algeria, etc. . . . .

The forest vegetation comprises conifers (*Abies pinsapo, A. cephalonica,* etc. . . .) and deciduous oaks (*Quercus faginea, Q. pubescens*) or sometimes, non-decidious (*Quercus suber*), the latter arriving here at its humid ecological limit and being a degradation stage of *Q. faginea* forest (Quezel, 1954).

The degradation of these forests produces "maquis" of briar, eagle fern, juniper oxycedrus and *Rubus* sp. plur. which give rise to greenswards of *Dactylis glomerata Trifolium pratense, Trifolium camprestre, Trifolium repens, Festuca elatior, Brachypodium sylvaticum, B. pinnatum,* etc.

This zone covers limited surfaces only and therefore, we shall not emphasize it.

5.5.2.8 *Pasture Lands of the High Mountain Mediterranean Zone*

This Mediterranean zone corresponds to the Alpine and sub-Alpine zones of Europe. The lower altitudinal limit varies according to latitude and continentality: 1500–2000 m. in Europe to 2000–2800 m. in Morocco (P. Quezel, 1957). The climate remains Mediterranean type (as defined by Emberger) i.e. relatively cold, dry and bright with a rainy winter season and a dry summer. This type of climate, none the less, has great affinities with the high mountain climates of temperate regions, to the extent that for H. Gaussen, this stage does not belong to Mediterranean vegetation. For this author, the Mediterranean region is limited to the area of extension of olive tree, that is, to zones with relatively mild winters.

For Emberger, on the contrary, the Mediterranean region is characterized above all by a rainy winter and a dry summer, the low winter temperatures being only a secondary factor. This point of view makes it possible to include the high mountain stage in the Mediterranean region. In my opinion, Gaussen's point of view is mainly valid for the mountains of Europe (Pyrenées, Cévennes, etc....) whereas Emberger's is perfectly adequate for North Africa and also for the Near East, as P. Quezel has shown.

The vegetation of the high mountain stage is asylvatic and includes 60 to 90% chamephytes and hemicryptophytes (the latter represent at least 50% of the species).

There are the following bushes: *Juniperus oxycedrus, J. thurifera, J. excelsa, J. communis*, Shadberry, *Sorbus aria, Rhamnus alpina, Bupleurum spinosum, Lonicera pyrenaica*, etc....

Thorny xerophytic bushes, also called "tragacant" vegetation, dominate physiognomically in the lower part of the stage:

Alyssum spinosum	*Pseudocytisus spinosus*
Arenaria pungens	*Pseudocytisus Mairei*
Erinacea anthyllis	*Bupleurum spinosum*
Acantholimon spp.	*Astragalus spp.*
Acanthophyllum spp.	*Cousinia spp.*

The highest zones are covered solely by herbaceous species:

Nardus stricta	*Avena montana*
Festuca mairei	*Poa* sp. plur.
„ *rubra*	*Carex fusca*
„ *indigesta*	*Trifolium repens*
Cardamine pratensis	

5.5.2.9 *Types of Azonal Vegetation*

5.5.2.9.1 Halophile Vegetation

Salty pasture lands cover large surfaces in the Mediterranean region. Their relative size increases with the aridity of the climate. Their optimum development seems to be the arid and upper saharan bioclimates.

256

The types of salty pasture lands are highly varied and depend on the water balance in the soil (height of level, periodicity, submersion, drainage, nature and quantity of salinity, etc.). The salsolaceae dominate physiognomically:

Salsolaceae:

Atriplex halimus
  ,,   glauca
  ,,   malvana
Salsola tetandra
  ,,   tetragona
  ,,   vermiculata
  ,,   sieberi
  ,,   baryosma
Suaeda fruticosa
  ,,   pruinosa
Salicornia fruticosa
  ,,   herbacea
Arthrocnemum indicum

Halocnemum strobilaceum

Halopeplis amplexicaulis

Plumbaginaceae:

Limoniastrum spp.
Limonium spp.

Zygophyllaceae:

Zygophyllum album
Nitraria retusa

Frankeniaceae:

Frankenia thymifolia
  ,,   loevis

Graminaceae:

Pucciniella distans
Hordeum maritimum
Polypogon maritimum
Lepturus incurvatus
Sphenopus divaricatus

5.5.2.9.2 Swamps and Humid Depressions

They are obviously more frequent as the climate is more humid. The characteristic vegetation consists of rushes:

Juncus bufonius
J. maritimus
J. spuarrosus
J. acutus
Scirpus holoschoenus
Carex sp. plur.
Schoenus nigricans
Phragmites communis
Gyceria spp.
Lythrum spp.
Alhaghi maurorum (desert zones)

Potamegeton spp.
Ranunculus flammula
  ,,   aquatilis
Eleochoris palustris
Isoetes spp.
Mentha pulegium
Triglochin spp.
Pulicaria spp.
Imperata cylindrica (arid and desert zones)

Good pastures species sometimes dominate in between the rushes:

Festuca elatior ssp. *arundinacea*	*Bonjeania recta*
Agropyrum elongatum	*Tetragonolobus siliquosus*
Trifolium fragiferum	*Melilotus spp.*
„ *resupinatum*	*Phleum spp.*
Medicago ciliaris	*Phalaris coerulescens*
„ *intertexta*	*Lotus corniculatus*
Phalaris arundinacea	*Lotus uliginosus*

5.5.2.9.3 Coastal Vegetation

is generally well-known. Association of:

Agropyrum junceum	*Ammophila arenaria*
Cakile maritima	*Euphorbia paralias*
Pancratium maritimum	*Medicago marina*, etc.

It is often of limited interest from the standpoint of pasture.

5.5.2.9.4 Fallow Pasture Lands

Crop lands comprise a large and usually underestimated part of Mediterranean pasture lands. These are plant communities of weeds of cereal fallow as well as stubble left after harvesting.

Pasturing fallow is a pernicious practice which is widespread in the Mediterranean region, especially where the level of agricultural technology has remained traditional. Fallow is bad pasture land with nearly unpalatable annual species and weeds used for 7 to 8 months of the year (November–June), whereas stubble is used for 2 to 3 months (July–August–September). In this system of grazing, the intervals between harvests are always difficult (summer interval, June–July, and autumn interval, October–November). These intervals occur: a. between the period of the end of production of fallow (June) and the beginning of the use of stubble (July). b. between the end of the use of stubble (September–November) and the beginning of the use of fallow after the first autumn rains (October–November).

In the arid zone and on sandy soils, certain couch-grass fallows (*Cynodon dactylon*) are highly advantageous because of their yield and particularly because of their production period (May–October).

The production of fallow varies naturally with climate; it may be estimated at:

258

200 to	400 Kg/DM/ha/year	arid zone
400 to	800	„ „	semi-arid zone
800 to	1200	„ „	sub-humid zone
1200 to	1400	„ „	humid zone[1]

The production of stubble and straw is about the same (the grains and ears which fall to the ground are eaten by cattle and comprise a large part of the production of this type of grazing.).

5.6 Production and Output of the Mediterranean Pasture Lands

The yield of Mediterranean pasture lands is generally very low because of their degradation. They are typically underdeveloped, particularly as the bio-climate is more humid. The lowness of this yield is emphasized again by its irregularity, which increases as average precipitation decreases.

We shall estimate production in fodder units, calculating that the normal ration of one female unit of goat or sheep is around 300 to 400 feed units per year or on the average approximately 1 F.U. per day, as against 5 F.U./day for an average female bovine unit. Average yield is estimated from references which made it possible to construct the following interpolative curves (See Table No. III). The solid line curves represent the yield of natural, non-developed pasture lands under continuous grazing. The dotted line curves represent natural pasture lands of good quality rationally used with fodder reserves. The broken line curves represent sown pasture lands and meadows which are rationally exploited, as well as fodder crops.

These curves were constructed on the basis of results obtained in North Africa, Near East, France, Spain and Italy. Beyond 1,000 mm the curves were extrapolated in the absence of data. They show that the disproportion between pasture lands of good and bad quality grows when rainfall increases. They also show the influence of rational management on yield and the great advantages of sown pasture lands and fodder crops when average annual rainfall exceeds 350–400 mm. Sometimes the yield of one hectare of pasture land can thus be multiplied by ten (a figure realized at Sedjenane, in Tunisia, with a rainfall of

1. I reckon that for grasses, legumes and herbs, 1 kg of dry matter equals 0.5 scandinavian feed units, whereas for shrubs and trees, 1 kg of dry matter equals 0.33 feed units, and 1 kg of cereal grains and pulses is worth 1 F.U., 1 kg of dry matter of straw equals 0.25 F.U.

800 mm by permanent prairies and fodder crops on clearing of a Myrtle heath).

The figures given in the curves are, of course, "average" figures. It is certain, for example, that far higher yields can be obtained on sown pasture lands or with fodder crops, for example, 3,000 to 4,000 F.U. with 450 mm of rain and 6,000 to 8,000 F.U. with 800–900 mm of rain, but such high yields are still exceptional. It may be said that on good land and with a high level of agricultural technology, we may obtain in fodder crops and prairies between 5 and 10 F.U. per hectare and per mm of rain, in the 400–1,000 mm interval.

Conclusions

Phytosociology and Phytoecology provide an indispensable tool in the study and mapping of grazing lands in allowing a scientific and sound classification of range and grassland types, their ecology and their dynamics.

However, phytosociological studies and mapping do not provide sufficient information to permit management planning.

For practical purposes, the knowledge of many other parameters is needed; but such parameters can only be meaningfully studied once the phytosociological surveys have given a rational basis on where and what to measure. These additional parameters are:

Biomass
Primary production
Secondary production
Carrying capacity
Seasonal and annual variability of primary and secondary productions as linked to rainfall distribution
Functioning of the ecosystems
Reaction to grazing and browsing by various livestock species
Reaction to fertilization
Reaction to bush and weed control
Reaction to reseeding
Water availability

It is only when in possession of all this information that sound and rational pasture management scheme can be established.

Management schemes are usually achieved through empiric know how at farm level by professional stockmen, but such empiric knowledge does not permit generalization and hence planning at regional or national level.

Therefore range and grassland surveys are more numerous and necessary in modern economy.

Phytosociological studies provide an indispensable and preliminary starting point for these detailed studies: in a word, they are necessary but insufficient.

TABLE I. Choice of Annual Species in Dry Farming (depending on rainfall and nature of soil)

Rainfall (in mm)	Saline soils	Wet soils (+ or − saline)	Heavy, well Soils − marl	Calcareous tufa (+ or − skeletal)	Sandy soils
300–400 Arid Arid (upper)	Green barley *Hedysarum carnosum* *Melilotus alba*	*Melilotus alba* *Hedysarum album* *Hedysarum carnosum*	*Phalaris minor* *Lolium rigidum* Medics	Subterranean clover "Clare" Field peas *Lolium rigidum* Medics	Subclovers Medics *Lolium rigidum*
400–600 Semi-arid	idem Beets	Persian clover Beets (spring)	Field peas Bersim Vetch-oats mixture *Medicago rugosa* *Medicago hispida*	Field peas *Medicago truncatula* Subterranean clover "Clare"	*Lolium rigidum* Subclovers
600–800 Sub-humid	idem Beets	Persian clover Beets (spring) Sudangrass Italian ryegrass	Italian ryegrass Persian clover Bersim Winter beets Vetch-oats	Bersim Subterranean clover "Clare" Persian clover Field peas Vetch-oats	Italian ryegrass Subclovers
800–1200	idem	Sudangrass Persian clover	Sudangrass Italian ryegrass Persian clover Bersim Winter beets Vetch-oats	Bersim Subterranean clover "Clare" Persian clover Field peas Vetch-oats	Italian ryegrass Subclovers

TABLE II. Choice of Perennials in Dry Farming (depending on rainfall and nature of the soil)

Rainfall (in mm)	Saline Soils	Flood Plains (+ or − saline)	Heavy, Well Drained Soils — Tell Marl	Hamri Soils (Red Soils with Calcareous Crust)	Sandy Soils
300–400 Arid (upper)	Atriplex species	Tall fescue Agropyrum elongatum Phalaris truncata Lotus corniculatus Atriplex sp. Trifolium fragiferum	Atriplex Hedysarum carnosum Melilotus alba Phalaris truncata	Lotus creticus Burnet Ehrharta calycina Brachypodium phoenicoides Oryzopsis miliacea Agropyrum elongatum Medicago tunetana	Ehrharta calycina Lotus cytisoides Cenchrus ciliaris (on the coast) Digitaria nodosa (on the coast) Cactus (prickly pear)
400–600 Semi-arid	idem	idem	Atriplex Phalaris truncata Hedysarum coronarium Bromus catharticus	Ehrharta calycina Lotus creticus Medicago sativa Onobrychis viciifolia "Belvédère" ryegrass Oryzopsis coerulescens „ holciformis „ miliacea	Ehrharta calycina Lotus cytisicides Medicago sativa Onobrychis viciifolia Cactus (prickly pear)
600–800 Sub-humid	idem	idem	Beets Phalaris truncata „ tuberosa Hedysarum coronarium Bromus catharticus	Onobrychis viciaefolia Medicago sativa Oryzopsis miliacea Ehrharta calycina Brachypodium phoenicoides "Belvédère" ryegrass Oryzopsis coerulescens „ holciformis Medicago arborea	Onobrychis viciaefolia Ehrharta calycina Lotus creticus Oryzopsis holciformis Oryzopsis thomasii Medicago sativa Cactus (prickly pear)
800–1200 Humid	idem	Phalaris coerulescens	Phalaris coerulescens „ tuberosa Hedysarum coronarium Purple clover Bromus catharticus Beets	idem	Ehrharta calycina Lotus creticus Oryzopsis Thomasii Oryzopsis holciformis Medicago sativa Onobrychis

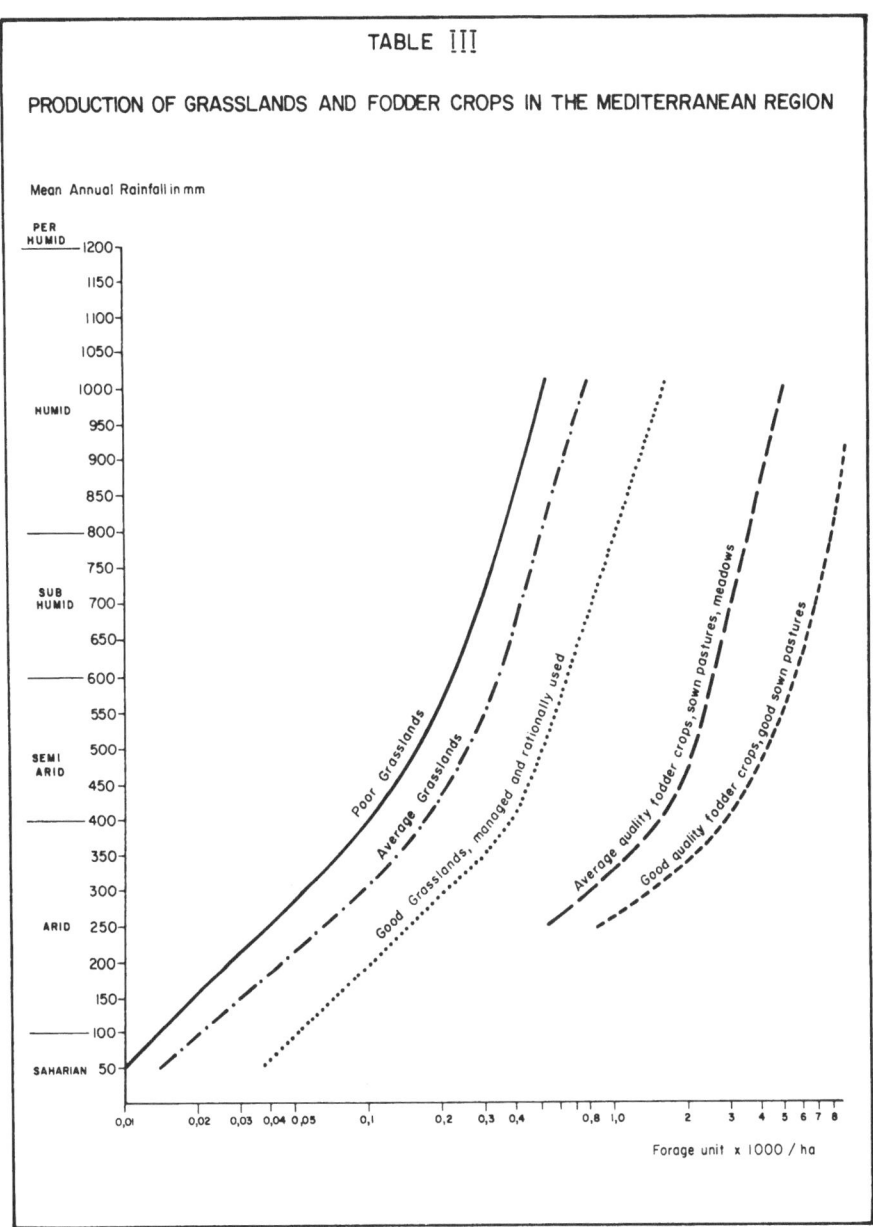

TABLE III

PRODUCTION OF GRASSLANDS AND FODDER CROPS IN THE MEDITERRANEAN REGION

Mean Annual Rainfall in mm

PER HUMID — 1200

1150

1100

1050

HUMID — 1000

950

900

850

—— 800

750

SUB HUMID — 700

650

—— 600

550

SEMI ARID — 500

450

—— 400

350

300

ARID — 250

200

150

—— 100

SAHARIAN — 50

Poor Grasslands

Average Grasslands

Good Grasslands, managed and rationally used

Average quality fodder crops, sown pastures, meadows

Good quality fodder crops, good sown pastures

0,01 0,02 0,03 0,04 0,05 0,1 0,2 0,3 0,4 0,8 1,0 2 3 4 5 6 7 8

Forage unit x 1000 / ha

263

Bibliography

Arènes, J., – 1929 – Les associations végétales de la Basse-Provence, Thèse, Paris.

Akman, Y. & Ph. Daget, – 1971 – Quelques aspects synoptiques des climats de la Turquie. *Bull. Soc. Lang. Géogr.* 5, 3: 269–300.

Aschmann, H., –1973 – Distribution and peculiarity of Mediterranean ecosystems. In Di Castri & Mooney (eds.), 1973: 11–19.

Auriau, Ph., – 1957 – Quelques données sur le climat du blé dans le Proche-Orient. *Ann. Amél. Pl.,* 1: 5–19.

Bagnouls, F. & H. Gaussen, – 1953 – Saison sèche et indice xérothermique. *Bull. Soc. Hist. Nat. Toulouse.* 88: 193–239.

Bagnouls, F. & H. Gaussen, – 1957 – Les climats biologiques et leur classification. *Ann. Géogr.* 66, 355: 193–220.

Bagnouls, F. & P. Legris, – 1970 – La notion d'aridité en Afrique du Nord et au Sahara. 12 p., 4 pl., 1 carte, Trav. Lab. Forest; Fac. Sces. 1, V, III, Toulouse.

Bagnouls, F., H. Gaussen & P. Lalande *et al.,* – Carte de la végétation de la région Méditerranéenne, 1/5.000.000, Notice 90 p. Bible. UNESCO-FAO.

Baldy, Ch., – 1965 – Climatologie de la Tunisie Centrale. 84 p., 20 cartes coul., 38 tabl. FAO, Rome.

Barry, J.P., J.C. Celles *et al.,* – 1967–1975 – Carte Internationale du tapis végétal; feuilles de Chardaia: 1/500.000 et Notice 1967; Biskra: 1/500.000 et Notice sous presse; Ouargla: 1/1.000.000 et Notice sous presse; In Salah: 1/1.000.000 et Notice sous presse.

Baumer, M., – 1964 – Rapport au Gouvernement d'Algérie sur les pâturages et l'élevage sur les hauts plateaux Algériens. Rapport FAO/PEAT No. 1784. FAO, Rome 94 p.

Bernard, A. & N. Lacroix, – 1906 – L.évolution du nomadisme en Algérie. 341 p., 1 carte H.T. Challamel, Edit., Paris.

Bharucha, F. R., – 1932 – Etude écologique et phytosociologique de l'Association à Brachypodium ramosum et Phlomis lychnitis des garrigues languedociennes. SIG-MA, comm. No. 18,11, 50. Montpellier.

Birot, P., – 1964 – La Méditerranée et le Moyen-Orient, P.U.F., Paris, 550 p.

Bortoli, L., M. Gounot & J. C. Jacquinet, – 1969 – Climatologie et bioclimatologie de la Tunisie septentrionale. *Ann. Inst. Nat. Rech. Agron. de Tunisie.* 42, 1: 1–235, (+ annexes).

Boudy, P., – 1948-1958 – Economie forestière Nord Africaine. 4 Vol., Larose, Edit., Paris.

Boulos, L., – 1972 – Our present knowledge on the flora and vegetation of Libya. A bibliography. *Webbia* 26: 365–400, Bibl. 791, Firenze.

Bourges, J., Ch. Floret & R. Pontanier, – 1973 – Etude d'une toposéquence typique du Sud Tunisien – Djebel Dissa, 42 p., 16 fig., 13 tabl., ORSTOM, Tunis.

Braun-Blanquet, J. & O. de Bolos, – 1957 – Les groupements végétaux du bassin moyen de l'Ebre et leur dynamisme. An. de la Est. Exp. de Aula Dei 5 (1–5), 266 p.

Braun-Blanquet, J., N. Roussine & R. Nègre, – 1952 – Les groupements végétaux de la France Méditerranéenne, 297 p., XVI Pl. Phot. Bibl. 220 – CNRS, Paris.

Burollet, P.A., – 1927 – Le Sahel de Sousse. Monographie Phytogeographique. Ann. Serv. Bot. Tunisie. 270 p.

Caees, – 1964 – Perspectives d'amélioration des productions fourragères et animales en Algérie. 2 tomes. Roneot. Alger.

Calvet, C., – 1964 – Le quotient pluviothermique de L. Emberger et l'évaporation. *C.R. Soc. Nat. et Physiques du Maroc.* : 55–61.

Calvet, C., – 1965 – Le quotient pluviothermique d'Emberger et l'évaporation, *La Météo*: 53–57.

C.E.P.E., – 1967 – Carte phytoécologique de la Tunisie septentrionale. *Ann Inst. Nat. Rech. Agron. de Tunisie.* 1966, 39, 5: 1– 213; 1967, 40, 1: 1–340; 1967, 40, 2: 1–426 (+ annexes).

Cordier, G., – 1947 – De la composition de quelques produite fourragers Tunisiens et de leur valeur pour l'alimentation du mouton. *Ann Serv. Bot. Agron Tun.* 20: 22–108.

Daget, Ph., & J. Poissonnet, – 1969 – Une méthode d'analyse phytologique des prairies. Critères d'application. *Ann Agron.*, 22, 1: 5–41, Paris.

Daget, Ph., – 1971 – Le quotient pluviothermique d'Emberger et l'évapotranspiration globale. *Bull. Rech. Agron Gembloux. H.S.*: 87–94.

Daget, Ph., – 1975 – Bibliographie classée sur le climat méditerranéen. C.N.R.S.-C.E.P.E., Montpellier.

Delhaye, R., H. N. Le Houérou & M. Sarson, – 1974 – L'amélioration des pâturages et de l'élevage dans le Hodna (Algerie). 115 p., 5 phot., 17 fig., AGS: DP/ALG/66509. Rapp. Techn. No. 2, FAO, Rome.

De Vries, D.M., – 1937 – Methods of determining the botanical composition of hay fields and pastures. Rep. fourth. Int. Grassl. Congr., Aberystwyth.: 474–480.

De Vries, D. M., – 1938 – The plant sociological combined specific frequency and other method. *Chronica botanica.* 4: 115–117.

De Vries, D. M., – 1950 – Grasland typen en hun oecologie. *Ned. Kruidk. Arch.*, 57: 28–31.

De Vries, D. M. & J. Koopmans, – 1949 – La relation entre l'indice de qualité des herbages et les conditions de milieu (en Néerlandais). *Landbouw Tijdschrift. CILO*, 61, 1: 21–37.

De Vries, D. M., A. A. Kruijne & H. Mooi, – 1957 – Fréquence des plantes prairales et leurs valeurs indicatrices des conditions de milieu. (en Néerlandais), Jaarboek IBS – 183-191, Wageningen.

Di Castri, F., – 1973 – Climatological comparisons between Chile and the Western coast of North America. In Di Castri & Mooney, 1973, 21–36.

Di Castri, F. & H. Mooney, – 1973 – Mediterranean type ecosystems, Origine and structure. Springer Verlag, Berlin, 405 p.

Dubuis, A. & P. Simonneau, – 1954 – Contribution à l'étude de la végétation de la region d'Ain Skrouna (Chott Ghergui Oriental). 124 p., 16 pl., 2 cartes H.T. Serv. Et Scient. Appl. à Hydraul., Alger.

Durand, E. & Ch. Flahault, – 1886 – Les limites de la région méditerranéene en France. *Bull. Soc. Bot. Fr.* 35: 24–34.

Eig, A., – 1946 – Synopsis of the Phytosociological Units of Palestine. *Pal. Journ. of Bot. Jerus.* Ser., III, 4. Jerusalem.

El Hamrouni, A & M. Sarson, – 1974 – Exploitation de parcours forestiers en Tunisie Centrale. Cpte-rendu Groupe FAO Pâtur. Médit. Acc. Géogofili Firenze 9 p.

Emberger, L., – 1930a – La végétation de la région méditerranéenne, essai d'une classification des groupements végétaux. *Rev. Gen Bot.* 42: 641–662, 705–721, Paris.

Emberger, L., – 1930b – Sur une formule climatique applicable à la géographie botanique. *C.R. Acad. Sc.* 191: 389–391, Paris.

Emberger, L., – 1938 – La définition phytogéographique du climat desertique. In "La vie dans la région desertique nord-tropicale de l'ancien monde'', Soc. Biogéogr., Paris; 9–14.

Emberger, L., – 1939 – Aperçu général de la végétation du Maroc. 175 p. 1 carte 1/750.000. Bull. Soc. Nat. et Phys. du Maroc. Mem. H.S., Rabat.

Emberger, L., – 1955 – Une classification biogéographique des climats. *Trav. Inst. Bot.* 7: 3–43, Montpellier.

Emberger, L., H. Gaussen, M. Kassas & De Philippis, – 1963 – Carte bioclimatique de la région méditerranéenne. 2 feuilles 1/5.000.000. UNESCO/FAO. Paris-Rome.

Emberger, L., – 1971a – Considérations complémentaires au sujet des recherches bioclimatologiques et phytogéographiques-écologiques. In Emberger, 1971b, 291-301.

Emberger, L., – 1971b – Traveaux de botanique et d'écologie. Masson, Paris, 520 p.

Evenari, M., L. Shanan & N. H. Tadmor, – 1971 – The Negev – The challenge of a desert. Harvard Univ. Press, Cambridge, Mass. 345 p., 203 fig.

Fallon, L. E., – 1972 – Rapport préliminaire sur le projet d'amélioration de l'élevage et des terrains de parcours au Maroc. US Aid, Rabat, 36 p. miméo.

Flahault, Ch., – 1937 – La distribution géographique des végétaux dans la région méditerranéenne française. 180 p., IV pl. H.T., Le Chevalier, Paris.

Flahault, Ch., – 1893 – La distribution géographique des végétaux dans un coin de Languedoc (Hérault). Montpellier.

Floret, C., – 1971 – Recherches phytoécologiques entreprises par le CNRS sur le biome "zône aride" en Tunisie. CNRS, CEPE, Doc. 57, Montpellier. 26 p.

Floret, C., – 1973 – Notice de la carte phyto-écologique des zônes pastorales d'Hadj Gacem. Echelle 1/50.000. In "Améliorations pastorales", Projet Parcours Sud, Inst. Nat. Rech. Agron. Tun. Etude Techn. 3, 13 p. ron. 1 car. h. t.

Floret, Ch., & E. Le Floch, – 1972 – Désertisation et ressources naturelles dans la Tunisie présaharienne. Journées d'Etude sur la desertation. 25-27 Dec. 1972. Cabes. Min. Agric. Tunis. 12 p.

Floret, Ch., M. Gounot, C. Rossetti & D. Schwaar, – 1968 – Conception générale des travaux de cartographie phytoécologique réalisés par le CNRS en Tunisie septentrionale. Ann. Inst. Nat. Rech. Agr. Tun. 41, 1, 142 p.

Floret, C., & E. Le Floch, – 1972 – Désertisation et ressources pastorales dans la Tunisie présaharienne. Min. Agric. INRAT, Tunis 12 p.

Floret, C., & E. Le Floch, – 1973 – Production, sensibilité et évolution de la végétation et du milieu en Tunisie présaharienne. 45 p., 13 fig. dont 9 cartes noir et coul., 4 phot. Inst. Nat. Rech. Agron. Tunis et CEPE (Dec. No. 71). Montpellier.

Floret, C. & R. Pontanier, – 1973 – Etude de trois formations végétales naturelles du Sud tunisien: production, bilan hydrique des sols. Inst. Nat. Rech. Agron., Tunis. 80 p.

Floret, Ch. & R. Pontanier, – 1974 – Etude de trois formations végétales naturelles du Sud tunisien. Production, bilanhydrique des sols. 44 p. mimeo., table. Inst. Nat. Rech. Agron. Tunis.

Floret, Ch. & T. Talahique, – 1974 – Evaluation des ressources pastorales, Domaine du Chahal. 17 p. mimeo. 1 carte phyto-écol. 1/25.000, 1 carte amenagement 1/25.000. FAO/PNUD-TUN/69/ 001-INRAT, Tunis.

Foret, D., – 1958 – Le cheptel tunisien et les problèmes d'ensemble de son alimentation. *Terre Tunisie* 4: 11–28, S.E. à l'Agric. Tunis.

Foret, D., – 1958 – L'élevage ovin tunisien. *Terre Tunisie* 5:5–30, S.E. à l'Agric. Tunis.

Franclet, A. & H. N. Le Houérou, – 1971 – Les Atriplex en Afrique du Nord et en Tunisie. 249 p., 27 fig., 50 phot. Bibl. 375. Div. For., FAO-Rome.

Froment, D., – 1966 – L'exploitation pastorale en Tunisie centrale. 1 carte. FAO/ FSNU, Proj. plan rur. integr. Tun. Centr. Tunis. 39 p. roneot.

Froment, D., – 1970 – Aménagement des parcours et leurs relations avec les cultures fourragères en Tunisie centisle. AGS 17 Rapp. Tech. No. 3, FAO, Rome. 117 p.

Froment, D. & D. Schwaar, – 1964 – La cartographie du couvert végétal en région méditerranéenne aride à l'aide de l'interprétation des photographies aériennes. *Ann. Inst. Agron. Gembloux* 70: 31–44.

Froment, D., H. van Swinderen & D. Schwaar, – 1964–1966 – Cartes physionomiques au 1/100.000ème de la végétation de la Tunisie centrale. R'Gueb (1964) – Sbeitla-Djilma-Hadjeb El Afoun (1965) – Thala Foussana (1965) – Kasserine Shiba (1966) – Feriana-Maadjen Bel Abbes (1966) – Gafsa-Maknassy (1966). Doc. ined. Proje. plan rur. intégr. Tun. centr. FAO/FSNU, Tunis-Rome.

Froment, D., – 1971 – Steppisation du couvert végétal en Tunisie centrale occidentale. Bull. pédi. Agron. Gemloux, Sem. et probl. Médit. 117-133.

Gaussen, H., A. Vernet, *et al.*, – 1958 – Carte internationale du tapis végétal 1/1.000.000. Feuille de Tunis-Sfax et Notice. Trav. Lab. For., Fac. Sces, Toulouse. 34 p.

Giacobbe, A., – 1958 – Ricerche ecologiche sull'aridita nei paesi del Mediterraneo occidentale, *Webbia 14*,1: 81–159.

Giacobbe, A., – 1959 – Nouvelles recherches écologiques sur l'aridité dans les pays de la méditerranée occidentale. *Nat. Monsp.* 11: 7–28.

Giacobbe, A., – 1962 – I caratteri mediterranei della flora montana appennica. Italia Forest. e. Mont.

Gosselin, M., – 1963 – Les possibilités d'aménagement hydraulique et pastoral du Sud tunisien. S/SE à l'Agric., Div. Hydraul. et équip. rural. Tunis. 25 p. roneot., 3 cartes.

Gounot, M. & H. N. Le Houérou, – 1959 – Carte des bioclimates de Tunisie. In Gounot, 1959, et Le Houérou, 1959b.

Gounot, M., – 1959 – Contribution à l'étude des groupements végétaux messicoles et ruderaux de la Tunisie. (Thèse Doct. Sc.), Ann. INRAT, 31. 282 p.

Gounot, M., – 1960 – Méthodes d'étude et d'inventaire de la végétation pastorale et prairiale. *Fourrages* 4: 46–52. Paris.

Guillerm, J. L., – 1969 – Relations entre la végétation spontanée et le milieu dans les terres cultivées du bas Languedoc. 165 p., 33 tabl., 16 fig. 2 cartes. CEPE, Montpellier.

Guinet, Ph., – 1953 – Carte internationale du tapis végétal. Feuille de Beni Abbes 1/200.000. Trav. Lab. For. Fac. Sces, Toulouse.

Guinet, Ph., – 1958 – Notice détaillée de la feuille de Beni Abbes. Bull. Serv. Carte Phytogéogr., Ser. A, III. CNRS, Paris. 96 p.

Guinochet, M., – 1951 – Contribution à l'étude phytosociologique du Sud tunisien. *Bull. Soc. Hist. Nat. Afr. Nord* 42: 131–153.

Hausman, G., – 1965 – L'esperienza forragera nel bacino mediterraneo. Acc. Econ. Agrar. dei Georgofile XII, 7, 141. Firenze. 27 p.

Hedin, L. & E. Duval, – 1967 – Intérêt de l'étude de la production primaire en écologie prairiale. Problèmes de productivité biologique, Masson, Paris: 93–112.

Ionesco, T., – 1966 – Considérations bioclimatiques et phytoécologiques sur les zônes arides du Maroc. Cah. Rech. Agron. 19: 1–130. 10 cartes HT 1/1.000.000. Bibl. 542, Rabat.

Ionesco, T., – 1968 – Remarques méthodologiques concernant l'étude des ressources pastorales du Maroc. *Al Awamia* 29: 35–67.

Ionesco, T., – 1972 – Pastoralisme et desertisation. Symposium sur la désertisation. Gabes, PRDIRPT, Doc. No. 7/Dir/FAO.

267

Ionesco, T. (Coll. M. Cabee), – 1972 – Démonstration d'aménagement de pâturage dans la région de Sbeitla. Etude consultant FAO auprès du projet FAO/TUN/ 71/ 525.

Ionesco, T., – 1974 – Principales données integrées dans la méthodologie appliquée en matière d'améliorations pastorales dans la steppe tunisienne. FAO/PNUD – TUN/69/001. INRAT, Tunis. 32 p. mimeo.

Ionesco, T. & M. Cabee, – 1972 – Améliorations pastorales – Démonstration d'aménagement de pâturages dans la région de Sbeitla. 35 p. mimeo. 8 Tabl., 1 carte, 1 graph. FAO/PNUD – TUN/69/001. INRAT, Tunis.

Ionesco, T., M. Cabee & D. Schweisguth, – 1974 – Améliorations pastorales – Périmètre de Sbeitla. 26 p. mimeo, 6 Tabl. Et. Techn. No. 5. FAO/PNUD – TUN/69/001. INRAT, Tunis.

Ionesco, T., M. Cabee, F. Foglino & D. Schweisguth, – 1974 – Améliorations pastorales – Périmètre de Hadj Gacem. Et. Techn. No. 3. FAO/PNUD – TUN/69/001. INRAT, Tunis. 28 p. mimeo.

Ionesco, T., M. Cabee & D. Schweisguth, – 1974 – Améliorations pastorales – Périmètre d'El Adala. 20 p. mimeo. 9 Tabl. Et. techn. No. 2. FAO/PNUD – TUN/69/001. INRAT, Tunis.

Ionesco, T., M. Cabee, M. Hadjej & T. Telahique, – 1974 – Améliorations pastorales – Périmètre du Chahal. 31 p. 7 Tabl. Et. Techn. No. FAO/PNUD – TUN/69/001. INRAT, Tunis.

Ionesco, T., M. Cabee, M. Hadjej & T. Telahique, – 1974 – Améliorations pastorales – Périmètre du Chahal. 31 p. 7 Tabl. Et. Techn. No. FAO/PNUD – TUN/69/001. INRAT, Tunis.

Ionesco, T., M. Cabee, Ch. Floret, M. Hadjej & E. Le Floch, – 1974 – Améliorations pastorales – Périmètre d'Oglat Merteba. 73 p. mimeo. 1 carte phyto-écologique et pastorale 1/25.000 et Notice. Et. Techn. No. 4 FAO/PNUD – TUN/69/001. INRAT, Tunis.

Ionesco, T. & Ch. Sauvage, – 1962 – Les types de végétation du Maroc. Essai de nomenclature et de définition. *Rev. Géogr. Maroc* 1–2: 75–86, Rabat.

Jacquart, P., Ph. Daget, J. Poissonet & G. Laroche, – 1968 – Expression de l'évolution du potentiel de production et de la composition botanique d'une formation herbacée dense. CNRS – CEPE Doc. No. 47. 22 p.

Kassas, M., – 1952–1953 – Habitat and plant communities in the Egyptian Desert. *Journ. of Ecol.* 40: 242–251; 41: 248–256.

Kassas, M. & M. S. El Abyad, – 1962 – On the phytosociology of the desert vegetation of Egypt. *Ann. of Arid Lands* 1–2: 54–83. Jodhpur.

Kassas, M., – 1968 – Dynamics of Desert Vegetation. Intern. Biol. Progr., ÇT. Techn. Meet. Hammamet. 25 p. mimeo.

Kassas, M., – 1970 – Desertification versus potential recovery in circum-Saharan territories. In "Arid Lands in Transition", Amer. Assoc. Adv. Science, Washington, D.C., 123–139.

Kerguelen, M., – 1960 – Quelles indications peut-on retirer de l'analyse botanique des herbages? *Fourrages* 4: 70–82.

Killian, Ch., – 1948 – Conditions édaphiques et réaction des plantes indicatives de la région alfatière algérienne. *Ann. Agron.* 18: 4–27, Paris.

Killian, Ch., – 1949 – Observations sur la biologie des végétaux des pâturages mis en défense en Algérie. Ann. Inst. Agric. Algérie 4 (9). 27 p.

Killian, Ch., – 1950 – Nouvelles observations sur les conditions édaphiques et la réaction des plantes indicatrices dans les reserves de pâturage de la région alfatière algérienne. Ann. Inst. Agric. Algérie 4 (10).

Killian, Ch., – 1953 – La végétation autour du Chott Hodna, indicatrice des possibilités culturales et son milieu édaphique. Ann. Inst. Agric. Algérie 3 (5).

Killian, Ch., – 1954 – Plantes fourragères types des hautes plaines algériennes, leur rôle particulier en période sèche. Ann. Amél. Plantes 4: 505–528. INRA, Paris.

Killian, Ch., – 1961 – Amélioration naturelle et artificielle d'un pâturage dans une reserve algérienne, le Mergueb. Mém. Soc. Hist. Nat. Afr. du Nord (6). 62 p.

Klapp, E., – 1954 – Wiesen und Weiden, Berlin.

Kuhnholtz-Lordat, G., – 1938 – La terre incendiée. Essai d'agronomie comparée. 361 p. Nîmes.

Kuhnholtz-Lordat, G., – 1949 – Evolution des pacages en costière nimoise et sa cartographie. Ann. agron. 5, Paris.

Kuhnholtz-Lordat, G., – 1946 – La silva, le saltus et l'ager. Ann. Ec. Nat. Agric. XXVI, L, Montpellier, 82 p.

Kuhnholtz-Lordat, G., – 1958 – L'écran vert. Mém. Mus. Hist. Nat. B, 9, Nîmes. 265 p.

Kuhnholtz-Lordat, G., – 1959 – La cartographie de la végétation et les services qu'elle peut rendre à l'Agronomie. Cah. Ing. Agron. Fr. 132: 25–28, Paris.

Laumont, P., – 1960 – Les prairies naturelles en Algérie. Doc. Rens. Agric. Bull. (5). Nlle série. Algér. 23 p.

Laumont, P., & M. Gueit, – 1953a – Les principes de l'étude agronomique du "Range". Ann. Inst. Agric. Algérie 7 (10). Algér. 31 p.

Le Floch, E., – 1974 – Notice de la carte phyto-écologique et des ressources pastorales du périmètre d'Oglat Merteba. Projet Parcours Sud. 12 p. roneo. 1 carte.

Le Floch, E., – 1973 – Notice et carte phyto-écologique et des ressources pastorales d'El Adala. 11 p. roneo. 1 carte. Projet Parcours Sud. Gabes.

Le Floch, E., & C. Floret, – 1972 – Désertisation, dégradation et regénération de la végétation pastorale dans la Tunisie présaharienne. Symposium sur la désertisation, Gabes. PRDIRPT. Doc. No. 6, 11 p.

Le Floch, E., – 1973 – Carte phyto-écologique des régions présahariennes de la Tunisie. Feuille de Mareth 1/100.000. Inst. Nat. Rech. Agron. Tunis. CEPE, Montpellier.

Le Houérou, H. N., – 1955 – Contribution à l'étude de la végétation de la région de Gabes. Ann. Serv. Bot. Agron. Tunisie 28: 141–180. 1 carte couleur 1/200 000, Tunis.

Le Houérou, H. N., – 1958b – Vue d'ensemble sur les pâturages du Sud tunisien. Terre de Tunisie 4: 40–49. SE à l'Agric., Tunis.

le Houérou, H. N., – 1958c – L'élevage dans le Sud tunisien et ses perspectives en fonction de la valeur des parcours. Terre Tunisie 4: 50–51. SE à l'Agric., Tunis.

Le Houérou, H. N., & M. Gounot, – 1959 – Carte bioclimatique de la Tunisie. In Le Houérou, 1959.

Le Houérou, H. N., – 1959a – Recherches écologiques et floristiques sur la végétation de la Tunisie mériodionale (thèse). Mem. Inst. Rech. Sah. Université Algér. 2 vol. 229 et 281 p.

Le Houérou, H. N., – 1962 – Les pâturages naturels de la Tunisie aride et désertique. 106 p. 12 pl. 4 cartes, RT. Inst. Sc. Econ. Appl. Tunis/Paris.

Le Houérou, H. N., – 1963 – Méthodes d'inventaire de la végétation naturelle et leur relation avec la production et l'utilisation des herbages. 7ème réun. Groupe Médit. Pat. Cult. Fourr. Madrid. 12 p. roneot.

Le Houérou, H. N., – 1964 – Les pâturages du bassin méditerranéen et leur amélioration. FAO Goat Raising Sem. Rome. 26 p. roneot.

Le Houérou, H. N., – 1965 – Improvement of natural pastures and fodder resources. Report to the Government of Lybia. 46 p. 5 graphs, 3 maps. EPTA Rep. No. 1979. FAO, Rome.

Le Houérou, H. N. & D. Froment, – 1966 – Définition d'une doctrine pastorale pour la Tunisie steppique. *Bull. Ec. Nat. Sup. Agron.* 10–11: 72–152, Tunis.

Le Houérou, H. N., – 1968 – Rapport au Gouvernement de la Tunisie sur le développement de la production fourragère et pastorale. Rapport PNUD/AT 2386. FAO, Rome 14 p.

Le Houérou, H. N., – 1968 – La désertisation du Sahara Septentrional et des steppes limitrophes (Libye, Tunisie, Algérie). Progr. Biol. Intern. C.T. Colloque Hammamet. 40 p., et *Ann. Algér. de Géogr.* 3, 6: 2–27, Algér.

Le Houérou, H. N., – 1969a – North Africa: Past, present, future. In "Arid Lands in Transition", 227–278. Amer, Ass. for Adv. of Science, Washington D.C.

Le Houérou, H. N., – 1969b – Principes, méthodes et techniques d'amélioration fourragère et pastorale en Tunisie. 291 p. 9 graph. 4 cartes, 64 phot. Bill. 250. FAO, Rome.

Le Houérou, H. N., – 1969c – La végétation de la Tunisie steppique (avec références aux végétations analogues d'Algérie, de Libye et du Maroc). 624 p., bibl. 225, 40 phot. 39 fig. 1 carte coul. H.T. 1/500.000 (128.000 km^2). 21 tabl. H.T. Ann. Inst. Nat. Rech. Agron. 42 (5), Tunis.

Le Houérou, H. N., – 1971 – Les bases écologiques de la production pastorale et fourragère en Algérie. Div. Prod. Prot. Pltes, FAO, Rome. 60 p.

Le Houérou, H. N., – 1971 – An assessment of the primary and secondary production of the arid grazing lands: ecosystems of North Africa. FAO, Rome, 25 p; Proceed. Intern Symp. on ecophysiological foundations of ecosystems productivity in Arid Zones, ed. Nauka, Leningrad: 170–172.

Le Houérou, H. N., – 1972 – The useful shrubs of the Mediterranean basin and of the Sahelian belt of Africa. In "Wildland Shrubs: their Biology and Uses". Intern. Symp. Utah State, Univ. Logan and Intermountain Forest and Range Experiment Station, Ogden, Utah. 26–36.

Le Houérou, H. N., – 1973a – Ecological foundations of agricultural and range development in Western Libya. Rep. to the Govern. of the Lib. Arab Rep., FAO, Rome, 20 p.

Le Houérou, H. N., – 1973c – Peut-on lutter contre la désertation? Coll. Intern. Désertific. Nouakchott, 17–19 déc. 1973. Bull. IFAN Dakar. 13 p. (in press)

Le Houérou, H. N., – 1973 – Contribution à une bibliographie des phénomènes de désertisation et de l'écologie végétale. Coll. Nouakchott desertis. Bill. IFAN. 120 p. 1500 réf. (sous presse).

Le Houérou, H. N., – 1974 – Deterioration of the ecological equilibrium in the Arid Zones of North Africa. *Journ. of Agron. Rehovot.* 11p. (in press)

Le Houérou, H. N., – 1974 – Fire and vegetation in the Mediterranean basin. Tall timbers fire ecology conference 13: 237–277, Tallahassee, Fla.

Le Houérou, H. N., J. Bigot, D. Froment & D. Van Swinderen, – 1966 – Carte phytoécologique de la Tunisie steppique au 1/500.000ème. CEPE/ISEA, Montpellier/ Tunis. 1969.

Le Houérou, H. N. & T. Ionesco, – 1973 – L'appétabilité des espèces de la Tunisie steppique. Projet parcours Sud-INRAT, Tunis, 68 p.

Le Houérou, H. N., J. Claudin & M. Haywood, – 1975 – La végétation du Hodna (Algérie). 145 p. Graph. Tabl. 2 cartes coul. 1/200.000 (23.000 km^2). AGS; DP/ ALG/66/509. Rapp. Techn. No. 6, FAO, Rome.

Liacos, G. L. & Ch. Moulopoulos, – 1967 – Contribution to the identification of some range types of *Quercus coccifera*. University of Thessaloniki. 54 p.

Loiseau, P. & M. Sebillotte, – 1972 – Etude et cartographie des pâturages du Maroc oriental. 540 p. 12 cartes coul. 1/100.000 Min. Agric. Rabat, ERES-SCET/Coop., Paris.

Long, G., — 1950 — Contribution à l'étude des pâturages tunisiens. Les méthodes et leur application dans la Tunisie centrale. Possibilités d'amélioration pastorale dans la zône prédesertique de Ben Gardane et de Bordj Sidi Toui. *Ann. Serv. Bot. Agron. Tunis.* 23: 139—160.

Long, G., — 1953 — Aspects agronomiques de l'amélioration des pâturages du centre et du sud Tunisien. *Tunisie Agricole*: 95—110, mai.

Long, G. A., — 1954 — Contribution à l'étude de la végétation de la Tunisie centrale. *Ann. Serv. Bot. Agron.* 27: 1—388, 1 carte coul. 1/200.000 H.T., 21 Tabl. Tunis.

Long, G. A., — 1955 — The study of natural vegetation as a basin for pasture improvement in the Western desert of Egypt. *Bull. Inst. Desert of Egypt* 5: 18—48, Cairo.

Long, G. A., — 1955 — Grazing problems in the Western Desert of Egypt, 55/4/2497, FAO, Rome, 37 p. 1 map.

Long, G., — 1960 — Cartographie de la végétation prairiale et pastorale. *Fourrages* 4: 53—59, Paris.

Long, G., — 1960 — Les terrains de parcours de plaine, de plateau et de basse montagne dans la région méditerranéenne. *Fourrages* 4: 99—127, Paris.

Long, G. A. & M. A. Ayyad, — 1956 — Phytoecological map of Ras el Hekma, prepared for planning range improvement and determining carrying capacity. 56/9/7130, FAO, Rome, 38 p.

Long, G., F. Fay & M. Thiault, — 1964 — Possibilité d'utilisation de la Garrigue par le mouton. FNCETA, Et. No. 982, CEPE, Montpellier. 6 p.

Long, F., F. Fay, M. Thiault & L. Trabaud, — 1967 — Essais de détermination expérimentale de la productivité d'une garrigue à Quercus coccifera. Doc. No. 39, CNRS. Distr. Limit. 28 p. mimeo.

Long, G., — 1974 — Le diagnose phyto-écologique et l'aménagement du territoire. 2 tomes. Masson edit. Paris.

Maignan, F., — 1973 — Cours d'aménagement des parcours. Bibl. 424 FAO/PNUD. Ecole Nation. Forest. d'Ingénieurs, Sale-Rabat, Maroc. 277 p. mimeo.

Malato-Beliz, J., — 1953 — Sur les pâturages du Portugal méditerranéen et leur amélioration. FAO/53/4/2654, Alger/Rome. 9 p. ron.

Malato-Beliz, J. & J. P. Abreu, — 1951 — Enasio fitosociologico numa pastagem espontanea da leziria do Rio Guadiana. *Melhoram ento* IV: 75—122, ELVAS, Portugal.

Molinier, R. & G. Tallon, — 1949 — Les prairies de Crau. Ann. Agron. 3, Paris.

Molinier, R. & G. Tallon, — 1950 — La végétation de la Crau (Basse Provence). *Rev. gén. Bot.* 56: 525—636, Paris.

Monchicourt, Ch., — 1906 — La steppe tunisienne chez les Fraichiches et les Majeurs. Bull. Dir. Agric. Comm., Col. 38—76, 159—199. 4 fig. 6 pl.

Monjauze, A., — 1960a — Essai sur l'utilisation rationnelle des terres en zônes arides et semi-arides. Dir. Gen. Agric. Alger. 97 p.

Monjauze, A., — 1960b — Solutions doctrinales du problème pastoral dans les régions de climat xerothérique. Coll. cons. rest. sols. UNESCO, Teheran, 307-317.

Monjauze, A. & H. N. Le Houérou, — 1966 — Le rôle des Opuntia dans l'économie agricole Nord-Africaine. *Bull. Ec. Nat. Sup. Agric. Tun.* 7, 8: 85—164.

Montserrat, P., — 1961 — Plant ecology and pasture problems in the Mediterranean provinces of Spain. Proc. 8th Intern. Grassl. Congr., 336—339, Oxford.

Montserrat, P., — 1956 — Las pastizales aragoneses. 190 p. Minist. Agric. Madrid.

Nègre, R., — 1956 — Recherches phytosociologiques sur le Sedd-el-Messjoun. 190 p. 45 fig. 36 tabl., 2 pl. 1 carte coul. 1/50.000. Trav. Inst. Scient. Cherif. Ser. Bot. No. 10, Rabat.

Nègre, R., — 1959 — Recherches phytogéographiques sur l'étage méditerranéen aride (sous-étage chaud) au Maroc occidental. 385 p. 1 carte coul. 1/750.000, 4 pl. 60 fig. Trav. Inst. Scient. Cherif., Ser. Bot. No. 13. Rabat.

271

Nègre, R., – 1974 – Les pâturages de la région de Syrte (Libye), projet de régénération. Feddes répertorium 85, 3: 185–243. 5 fig. 1 carte, Berlin.

Ozenda, P., – 1958 – La végétation ligneuse du Sahara. Rivières et fôrets, 9–10.

Ozenda, P., & J. Kéraudren, – 1960 – Carte de la végétation de l'Algérie au 1/200.000. Feuille de Guelt Stell-Djelfa. Publ. Gouv. Gén. Alger.

Pabot, H., – 1957 – Rapport au Gouvernement de la Syrie sur l'écologie végétale et ses applications. Rapport EPTA No. 663. FAO, Rome.

Pabot, H., – 1960 – Comment briser le cercle vicieux de la désertification dans les régions sèches d'Orient. FAO, Inst. Franç. Coop. Techn., Coll. Téhéran, 120–125.

Pabot, H., – undated (1960?) – L'action de l'homme sur la végétation naturelle et ses conséquences dans les pays secs ou arides du Proche et du Moyen Orient. Plant Div. FAO, Rome, 14 p. mimeo.

Pearse, C. K., – 1970 – Range deterioration in the Middle East. XI Int. Grassland Congress, Proc. Surfers Paradise, Queensland, Australia, 13–23 April, 26–30.

Pearse, C. K., – 1971 – Grazing in the Middle East: Past, Present, Future. *Journ. Range Managt.* 24: 13–16.

Perrin de Brichambaut, G. & C. Wallen, – 1964 – Une étude d'agroclimatologie dans les zônes arides et semi-arides du proche-orient. OMM, No. 141.TP.66, Note techn. No. 56, Genève, 69 p.

Planhol, X., de & P. Rognon, – 1970 – Les zones tropicales arides et subtropicales, A. Colin, Paris, 487 p.

Poissonnet, P. & J., – 1969 – Etude comparée de diverses méthodes d'analyse de la végétation des formations herbacées denses et permanentes. Conséquences pour les applications agronomiques. 120 pp. 32 fig. 11 Tabl. Doc. No. 50. CEPE, Montpellier.

Pons, A., – 1953 – Le repeuplement végétal sur les anciennes cultures de la région du Grand Luberon. *Rec. Trav. Lab. Bot. Géol. Zool. Fac. Sces. Ser. Bot.* 6: 135–147. Montpellier.

Pouget, M., – 1974 – Etude écologique et pédologique de la région de Messad. 1 carte phyto-écologique 1/100.000. DERMH, Alger.

Pouget, M. & H. N. le Houérou, – 1971 – Etude agropédologique du bassin du Zahrez Gharbi (Feuille Rocher de sel). 160 p. 9 fig. 30 tabl. H.T., 4 cartes coul. 1/100.000. Groupements Végétaux, Aptitude à la mise en valeur, Pédologie, Nappe phréatique. DERMH, Alger.

Poupon, J., – 1965 – La mise en oeuvre de la mise en défense pastorale et de l'aménagement des parcours par rotation. S/SE à l'Agric. Dir. Hydr. Equip. Rural, Tunis. 44 p. roneot.

Pujos, A., – 1958 – Etude des érosions dans le bassin de la Moulouya. 4 cartes phyto-écologiques coul. 1/200.000. 500 p. mimeo. nombr. graph., fig. et photos. Div. Forest, Min. Agric. Rabat.

Quezel, P., – 1954 – Contribution à l'étude de la flore et de la végétation du Hoggar. Trav. Inst. Rech. Sah. Mon. Reg. No. 2. 164 p. 10 pl. Alger.

Quezel, P., – 1958 – Mission botanique au Tibesti. 357 p. 15 pl. Mém. No. 4. Inst. Rech. Sah., Alger.

Quezel, P., – 1958 – Quelques aspects de la dégradation du paysage végétal au Sahara et en Afrique du Nord. UICN, 7e heun. Tech., Athènes. 7 p.

Quezel, P., – 1965 – La végétation du Sahara, du Tchad à la Mauritanie. 333 p., 72 fig. 15 cartes, 93 tabl. Fischer, Stuttgart.

Quezel, P., – 1971 – Flora und Vegetation der Sahara. In H. Schiffers "Die Sahara. . . .". Band I: 429–497. Weltforum Verlag, München.

Quezel, P. & S. Santa, – 1962–1963 – Nouvelle flore de l'Algerie et des régions désertiques méridionales. I, 565 p. 1 carte H.T., 20 phot. 24 cartes, 51 pl. dess. (1962); II, 494 p. 20 phot. 18 cartes, 61 pl. dess. (1963). CNRS edit., Paris.

Regazzola, T., − 1968 − Premiers résultats de l'enquête sur le nomadisme et le pastoralisme en Algérie. 62 p. mimeo. 4 cartes. Div. Statist. Minist. Finances et Plan. Alger.

Rivas Goday, S. et al., −1959 − Contribution al estudio de la Quercetea ilicis hispanica. An. Inst. Bot. cavanilles, XVIII, 11: 285−406.

Rivas Goday, S., − 1955 − Los grados de vegetacion de la peninsula iberica. An. Inst. Bot. Cavanilles, XIII: 269−331, Madrid.

Rivas Goday, S. & S, Rivas Martinez, − 1963 − Estudio y classification de los pastizales espanoles. Minist. de Agricultura, Madrid. 269 p.

Rodin, L., B. Vinogradov, et al., − 1970 − Etudes géobotaniques des pâturages du secteur ouest du département de Medea (Algérie). 124 p. 2 cartes coul. 1/200.000 (8.500 km^2). Nauka Edit., Leningrad.

Sabeti, H., − 1969 − Les études bioclimatiques de l'Iran. Publ. Univ. Téhéran, No. 1231, 253 p.

Sagne, J., − 1950 − Algérie pastorale. 267 p., Alger.

Sarfatti, G., − 1954 − Ricerche sui pascoli della Sila (Calabria). Webbia, 10 (1): 319−439.

Sarson, M., − 1970 − Résultats d'un essai sur l'alimentation du mouton en période de disette fourragère au Centre d'Ousseltia. Note Techn. No. 6. FAO, FSNU/TUN 17, Tunis. 6 p. mimeo.

Sarson, M., & A. El Hamrouni, − 1974 − Valeur alimentaire de certaines plantes spontanées ou introduites en Tunisie. 3 p. 17 tabl. Note de recherche No. 4. Inst. Nat. Rech. Forest. Projet UNDP/FAO − TUN/71/540, Tunis.

Sauvage, Ch., − 1963 − Etages bioclimatiques du Maroc. 44 p. 1 graph H.T., 1 carte coul. 1/2.000.000. In "Atlas du Maroc". Com. Nat. Géogr., Rabat.

Sauvage, Ch., − 1963 − Etages bioclimatiques. Notices explicatives de l'Atlas du Maroc, Rabat. 44 p.

Sauvage, Ch., − 1963 − Le coefficient pluviothermique d'Emberger, son utilisation et la représentation graphique de ses variations au Maroc. Ann. Ser. Phys. du Globe et de la Météo. Inst. Sc. Chérifien 20: 11−23

Seltzer, P., − 1946 − Le climat de l'Algérie. 219 p. cartes H.T. Inst. Météor. Phys. Globe. Fac. Sces. Alger.

Simonneau, P., − 1952 − La végétation halophile de la plaine de Perregaux. Serv. Et. Scient. Appl. Hydraul. Min. Trav. Publ., Alger. 279 p.

Squalli, O., − 1974 − Le périmètre d'amélioration pastorale de l'Arid (Boumia, Midelt). Semin. Pastoral. Ec. Nat. Forest. Ing. et Direct. Elevage, Min, Agric. Ref. Agr. Rabat. 5 p.

Stewart, Ph., − 1968 − Quotient pluviothermique et degradation biosphérique. Bull Soc. Hist. Nat. Afr. du Nord 59, 1−4: 23−36. 1 carte H.T. 1/3.500.000. 1 tabl. H.T. Alger.

Soroceanu, E., − 1936 − Recherches phytosociologiques sur les pelouses mesoxérophiles de la plaine languedocienne (Brachypodietum phoenicoidis). SIGMA. Comm. No. 40. Montpellier.

Teles, A. M., − 1953 − As ervagens de anafe dos arredores de Lisboa. Agronom. Lusitanica, 15: 259−313. Lisboa.

Teles, A. N., − 1970 − Os Lameiros de Montanha do Norte de Portugal. 130 p. 5 pl. Agron. Lusitania 31: 1−2.

Thiault, M., − 1955 − L'évolution des pâturages en Tunisie en fonction du mode d'exploitation. Ann. Serv. Bot. Agron. 28: 181−208. Tunis.

Thiault, M., – 1958 – Les perspectives ouvertes à l'élévage par les nouvelles méthodes de culture de l'herbe. Tunisie agricole, avril-mai. 22 p.

Thiault, M., – 1960 – Valeur pastorales des plantes fourragères spontanées. *Fourrages*: 63–80.

Thiault, M., – 1963 – Rapport au Gouvernement de la Tunisie sur l'amélioration des pâturages et de la production fourragère. Rapport FAO/PEAT No. 1689. FAO, Rome. 62 p. roneot.

Thiault, M., F. Fay, G. Long & L. Trabaud, – 1967 – Essai de détermination expérimentale de la productivité d'une garrigue à *Quercus coccifera*. Doc. 39. CEPE, Montpellier. 29 p. mimeo.

Thrower, N., & D. Bradbury, – 1973 – The physiography of méditerranean lands with special emphasis on California and Chile. In Di Castri & Mooney, 1973: 37–52.

Tisseron, J., –1958 – L'élévage ovin dans ses rapports avec la production végétale en Tunisie. *Terre Tunisie* (4): 31–49. SE à l'Agric. Tunis.

Tomaselli, R., – 1948 – La pelouse à Aphyllanthes (Aphyllanthion) de la garrigue montpellieraine. SIGMA, comm. No. 99. Att. Inst. Bot. Univ. Ser. 5, Vol. VII (2). Pavia.

Turc, L., – 1961 – Evaluation des besoins en eau d'irrigation, évapotranspiration potentielle. *Ann. Agron.* 12, 1: 13–49.

Van Zwinderen, H., – 1973 – Développement pastoral et ménagement forestier en Algérie (Aurès et Belezma). 51 p., 10 cartes. FO:SF/ALG/68/515. FAO, Rome.

6 REINDEER PASTURES IN THE SUBARCTIC TERRITORIES OF THE U.S.S.R

V. N. ANDREYEV

Contents

6 REINDEER PASTURES IN THE SUBARCTIC TERRITORIES OF THE U.S.S.R.

6.1 Introduction

The Subarctic includes the whole tundra zone and the forest-tundra zone, which is the area transitional between tundra and taiga zones. The subzone of northern taiga is also added here by some authors. In such a wide interpretation Subarctic coincides roughly with the term "Far North", broadly used in literature, especially in the economic one. In the north the Subarctic borders on the Arctic desert zone of Arctic region, in the south – on the taiga zone of Boreal region.

The Far North covers about a half of the territory of the USSR, being the least populated and insufficiently utilized with respect to agriculture part of the Soviet Union. The diverse biological resources of this area are being gradually drawn into the economic turnover. Among them the reindeer pastures occupying nearly a half of the Far North territory deserve much attention. It is the basis of one of the perspective branches of the northern economy – the reindeer-breeding industry.

The reindeer-breeding was one of the ancient occupations of native peoples; Nenetzs, Evenks, Evens, Chukchi, Koryaks and others. Presently the reindeer-breeding has become the most profitable branch of Subarctic Agriculture, one of an important economic value. One pays a great attention to its development. About 2.5 million head of reindeer are concentrated in the big state farms, most of which have 10 thousand head and more. Reindeer-breeding gives 32 thousand tons of important dietary product – meat and 650 thousand of hides – a valuable raw material for clothes and footwear preparation, for chamois and other articles of manufacture.

The expansion of reindeer-breeding industry and increase of its productivity considerably depends upon its fodder base. Reindeer graze on their natural pastures throughout the year and that is why they, as no other domestic animal, depend on the vegetation cover, its feeding value and accessibility for usage. We will consider these problems here.

The primary information on the reindeer natural fodder and pastures can be found already in the works of naturalists and travellers of 19th century. The importance of forage lichens (so called "yagels") in reindeer-breeding was marked first by the academician of Lomonosov's school — I. I. Lepekhin, who wrote that reindeer eat in winter "the bitter white moss, that grows on bogs and is called 'yagol' " (Lepekhin, 1805).

A. Shrenk, who has accomplished a tour through the Bolshezemelskaya tundra, gives more detailed data on reindeer pastures. Academician A. F. Middendorf (1869) has brought from his travellings into the Far North of Siberia, lower reaches of Yenissei river and Taimyr peninsula the conviction that the well-being of reindeer-breeders depends on the "yagel", its distribution and accessibility. G. I. Tanfilyev (1894) has first collected some data on the rate of yagel accretion. However the whole problem of reindeer pastures did not concentrate anybody's special attention at the prerevolutionary times.

The Soviet period in the investigations of the Far North was begun in a short time after the Great October socialist Revolution.

The scientific-economical expedition, organized 4-th of March 1920 according to Lenin's instruction, started its investigations of the biological resources of the East-European North. One of its tasks was the inspection and study of reindeer pastures.

The special study of reindeer pastures has been carried out first by B. N. Gorodkov in North Urals and at the lower reaches of Ob river. The investigations of reindeer pastures based on the modern geobotanical methods were started in 1928 in the tundras of Arkhangelsk region (Andreyev, 1930, in the symposium "Reindeer pastures of the North").

A broad programm of Siberian reindeer pastures study was outlined by I. A. Perfilyev (1931) and V. B. Soczava (1932). But it was not before the organization of Institute of Reindeer-Breeding Industry in 1931 that the study of reindeer pastures acquired the systematic and balanced character. This Institute included the Section of forages and pastures, which directed a number of experimental stations, where under guidance of the author of present paper worked many geobotanists specialized in reindeer pastures study (A. S. Salazkin, V. A. Kovakina, P. A. Mashistova — in Murmansk; L. E. Arens, A. A. Dedov, D. M. Glinka, A. V. Vazinger, G. K. Karev, V. V. Utkin — in

Naryan-Mar; K. N. Igoshina, M. M. Antonova, I. D. Kildjushev-sky, M. N. Avramchil — at the Yamal station; V. A. Sheludya-kova, B. N. Ovchinnikov — at Bulun station; M. N. Avramchik — at the Anadyr station; O. N. Mironenko — at the Turin station etc.) Under guidance of B. N. Gorodkov the Institute has accomplished the reindeer pastures inventory all over the Far North, and this inventory gave the preliminary tentative data on the distribution, typology and fodder reserves of reindeer pastures (Soviet Reindeer Industry, 1933, 1934). The inventory was used both for practical and theoretical purposes and followed by more detailed studies in Murmansk region (Salazkin, Sambuk, Polyans-kaya & Pryakhin, 1936), in the Ural part of Yamal-Nenetsky district (Andreyev, Igoshina & Leskov, 1935) and on the Kol-guyev island (Bogdanovskaya-Gienef, 1938). Later the work on pastures inventory was handed over to the special expeditions on the organization of reindeer farms' territory proceeded by "Roszemproekt" in the frames of Ministry of Agriculture of R.S.F.S.R.

As time passed on, the Institute of Reindeer-Breeding Industry had been incorporated into the Institute of Far North Agriculture. On the experimental stations of this institution a number of studies on the forage base of reindeer-breeding had been accomplished, such as: a. the study of yagel increment and the rate of its restoration (Gorodkov, 1936; Salazkin, 1937; Igoshina, 1939, Andreyev, 1954); b. the elaboration of the scientific basis of pasture rotation and introduction of this rotation in a number of districts (Andreyev, 1940b, 1954; Karev, 1956; Antonova, 1957); c. the discovery of seasonal changeability of pastures and the elaboration of the pastures classification according to this changeability (Gorodkov, 1934; Igoshina, 1937; Glinka, 1939; Avramchik, 1939; Igoshina & Florovskaya 1939); d. the finding out of methods of study and economic evaluation of reindeer pastures (Sambuk, 1931; Sochava, 1934a; Andreyev Panfilovsky, 1938; Andreyev, 1940a, 1952); e. work on the problem of reindeer pastures improvement (Barashkova, 1961, 1963b; Kovakina, 1967); f. the finding out and study of plants eaten by the reindeer (Alexandrova, 1940; Alexandrova a.oth., 1964).

The problems of pastures distribution, their productivity and economic value have been studied almost on the whole territory of the Far North. Being the members of the expeditions on the organization of land exploitation, many geobotanists have collected the valuable materials on the reindeer pastures on the territories hitherto unexplored geobotanically. The scientists who scientifically contributed to the study of reindeer pastures during that period were: V. D. Alexandrova (1937), M. K. Baryshnikov & V. N. Vasiljev (1936), V. S. Govorukhin (1933, 1949), V. I. Dushechkin (1937), A. I. Zubkov (1932), N. N. Prakhov (1957), T. A. Rabotnov (1940), Z. N. Smirnova (1938), V. B. Sochava (1933, 1934b), A. A. Shakhov & V. A. Sheludyakova (1938, 1948), L. V. Shumilova (1933), M. I. Yarovoi (1939) and many others.

The organization of land-and-water exploitation, conducted in the Far North from 1930 to 1940, along with the better knowledge of pastures, helped us to distribute them among separate farms. However, the development and consolidation of reindeer-breeding kolkhozes and sovkhozes required much more detailed and exact data on the distribution and productivity of pastures. So the new stage of work on the land organization in reindeer-breeding farms on the base of topographic maps of larger scale, was begun in the 1940.

Much attention was paid on the geobotanical works. The precise demands of accuracy in the forage production estimations and minute details of geobotanic maps were established along with the requirement of the single geobotanical methods. The organization of large expeditions (up to 100 members and even more) took place. These expeditions covered vast territories and began to use the modern airphotography methods. Their activity did not stop during the Great Patriotic war and gained in scope significantly in the post-war period. Then were put into practice such great expeditions as the Northern inter-regional expedition (in Komi ASSR, Arkhangelsk and Murmansk regions), Yakutian expedition (north Yakutian districts, Taimyr and Evenk native territories), Magadan expedition (Magadan and Kamchatka regions) and others. Changes in pastures together with the development of reindeer-breeding farms call for the reorganization of land exploitation and such reorganization is carried out every ten years (the 4-th round of such work is being accomplished now in the majority of regions). The constant branches of "Roszemproekt" and protracted expeditions have presently replaced the temporary expeditions in the regions of Far North.

6.3   **Methods of Reindeer Pastures Inventory and their Grazing Capacity**

The land organization in the reindeer-breeding farms is based on the natural and economic pre-conditions of reindeer-breeding industry — the forage resources calculations first of all. The methods of such works was first elaborated by the author of this paper in 1940. In the process of its application these methods have been tested and improved by large teams of specialists in different districts of Far North, and are used up to present time.

The most important characteristics to have been gained in the study of forage resources are as follows: a. the stock of fodder mass; b. the composition of fodder and feeding value of its constituents; c. the distribution of forage resources over the territory under investigation; d. the practical accessibility of fodder plants in given conditions; e. the seasonal changes in the quality and quantity of fodder and the rates of fodder stock restoration; f. the optimal quantity of reindeer grazing on every pasture given; g. the optimal and the possible times of pasture exploitation. Only the farm, which has such data in its disposal, can use its forage resources on the scientific ground.

The geobotanical mapping of reindeer pastures land is carried out as follows: on the basis of all materials of previous investigations and aerial photographs one draws on the topographic map of the territory under investigation the contours of geobotanical units. Then all the territory is covered with the parallel aerial routes at the distance of 4–7 km from each other. During the flight on a helicopter or on a small plane at the height 300–400 m one makes the contours of the units more accurate and also obtains some data on the visual developments of the units. These data consist not only in determination of plant communities inside every unit, but also of the rough calculation of their percentage and thus the evaluation of the units and their classification according to the accepted classification system. The well trained specialist can fulfil this task to within ± 10%. The mozaic character of the tundra and forest-tundra vegetation is well known, and it manifests itself even inside very small contours. That is why the calculation of communities percentage as it is seen from above constitutes the first important feature of this method.

The second feature is the estimation of forage plants standing crop mass in every contour separately. The usual method of standing crop evaluation on the grounds of average forage stock in certain types of pastures is not applicable to the reindeer pastures. The range of fodder stock fluctuation in the limits of the same type of pasture is rather great. In the Far North conditions even insignificant changes in water-supply, steepness or exposition of slopes, microrelief, depth and time of frozen ground melting and snow cover distribution as well as pastures utilization provoke the considerable response in plant development and significantly tell on the standing crop mass of the plot. Therefore, if to be guided by the average values of fodder stock in the pasture types, one can be badly mistaken while estimating the forage stock of certain contours. That is why this standing crop should be evaluated on the spot in every contour separately. Herefrom follows the third feature of our method.

To carry out the mass estimation (in every contour, for every pasture type) one needs such an index, that can be easily and quickly gained. The method of harvested plots is too labour-consuming and is not a suitable for every scale of work. For instance, the territory under pasturage in the average reindeer-breeding farm makes up about 1 million ha, the mean area of a contour on the map of used scale averages 650 ha and there are more than 1.500 contours on this territory. Every contour contains not less than 3—4 pasture types, consequently not less than 4.5—6.0 thousand of measurements should have been carried out. Such a work cannot be done quickly, at a fixed date. In this case it is much more profitable to use the projection method. It was L. G. Ramensky (1938), who had ascertained the correlative dependance between the projective cover and the weight of herbaceous plants. The similar dependance was found for the tundra shrubs and fruticosae lichens (Andreyev, 1954).

For several regions of the Far North separately we elaborated a series of correlation tables, where the weight of forage plants is confronted with the gradients of their hight and projective cover. If geobotanists possess such tables, their field work may be reduced to the estimation of projective cover of different groups of forage plants (lichens, willows, sedges etc.). The geobotanist with a correct eye can make every estimation in several seconds and all this work can be done looking from a helicopter or a plane simultaneously with the geobotanical mapping and the calculation of intra-contour percentage of pasture types.

If we obtain some mass data on the projective cover of the most important species or groups of forage plants — this makes the reliable ground for the standing crop estimation. And the application of the projective cover estimation method is the third essential feature of our methods. If we dispose of data on the standing crop mass on different parts of every contour of our geobotanical map, we can calculate the permissible grazing capacity of pastures for reindeer. But we should have in addition the data on the accessibility of pastures in different seasons, as well as the knowledge about the necessary amount of food for the reindeer of different age and sex (in different seasons too) and about the maximal and minimal areas of pasturage. The last value depends on the system of reindeer maintenance adopted by certain farm (the pasturage by herds or inside fences, free or half-free maintenance). The areas of pasturage also depend on the strain of reindeer and their biological peculiarities. For example, the reindeer of Chukotka tundras keep together and select and use the plots rich in fodder. The Evenk reindeer graze in more dispersed manner and therefore they successfully use the poorer pastures. The maximal pasture area by Evenk reindeer is 1.5–2 times as big as one by Chukotka reindeer. In winter this area averages 3,3 are per one Chukotka reindeer per day and 6,5 are – by Evenk reindeer correspondingly. In the case when the stock of fodder on such an area unit is lower than the diurnal need of a reindeer – the pasture is not suitable for reindeer pasturage. If the pasture is too rich in fodder, it cannot be completely utilized by reindeer because these animals need some area for their movements; the minimal size of this area per one reindeer (of the majority of strains) per day is in winter 1 are, in summer 8 are. The feeding ration of one reindeer is 4–6 kgr. of airdry fodder. The accessibility of pastures ranges usually from 30% to 80%.

The study of pastures is carried out by geobotanists and zootechnicians in one complex. All the data requested are gained for every plot of pasture separately.

On the basis of the methodics described above we have accomplished the inventory of reindeer pastures of the USSR on the area about 8 million square kilometers. 489 million ha of pastures with the grazing capacity of 3.412 thousand head of reindeer have been investigated (Andreyev, 1968). 300 million of them (grazing capacity of 2.500 thousand head) are situated in the tundra zone and the adjacent regions of forest-tundra and parklands. But no more then 2.200 thousand head are pastured here practically (136 ha per one reindeer). Thus we have in reserve the area of pastures with the grazing capacity more than 900 thousand reindeer. This reserve includes mainly the almost inaccessible and poorly investigated pastures, which are used now by wild reindeer. But the main part of pastures area in such reindeer-breeding regions as Nenetzky region and Komi ASSR, Yamalo-Nenetzky region, Chukotsky and Koryaksky region is close to the full saturation of grazing capacity of pastures.

With the increase in general culture of reindeer-breeding and more rational usage of pastures the grazing capacity of pastures may become much bigger. Probably, in the near future the total capacity of tundra and forest-tundra pastures will increase from 2.5 to 3.0—3.5 million head. Such increase in pasture capacity and hence in the productivity of reindeer-breeding industry will be possible.

6.4   Natural Features of Reindeer Pastures

The climate of tundra and forest-tundra regions is characterised by the short period of vegetation (80—120 days) first of all. The mean summer temperature does not exceed 14°C, in majority of forest-tundra regions it is 11—12°C, in tundra — 8—9°C.

In the western (Murmansk region and the coastal part of Nenetzky national territory) and in the extreme eastern (coastal districts of Magadansky and Kamchatsky regions) regions the climat has the continental-marine features, the annual amplitude of average monthly temperatures does not exceed 25—28°, and the annual precipitation amount reaches 400—500 mm. In the western regions with rather strong influence of Atlantic cyclones and in the eastern ones — under influence of Aleutain cyclones — the wind velocity averages 6—8 m/sec, snow storms are infrequent. The snow cover is very uneven and averages 50 cm and more.

Eastward from Pechora river and westward from Chukotsk peninsula and the city of Magadan the climate both of tundra and forest-tundra assumes the sharp continental character, the amplitude of monthly temperatures averages 30—40°C. The most continental climate characterizes the region between Yana and Indigirka rivers. Here the mean temperature of the coldest month reaches —45—50°C, the precipitation does not exceed 150—200 mm, the mean wind velocity is reduced to 1—2 m/sec, the snow cover is even and friable, does not exceed 15—20 cm. The permanently frozen grounds (permafrost) prevail everywhere; up to the end of summer the soil thaws out not more than 25—30 cm (peaty one) or 100—200 cm (sandy one).

The soils on watersheds are shallow, mozaic and often disturbed by solifluction, with various forms of micro- and nano-relief (frost cracks, spots (medallions), peat hillocks, hummocks etc.). Biochemical and microbiological processes are slowed

284

down, the acid gley and gleyish soils prevail, which are poor in macro- and microelements necessary for plant development. Although in tundra and forest-tundra the plants grow in conditions of high relative air humidity and soil paludification, at the beginning of vegetation they undergo water deficiency.

The region of northern reindeer-breeding industry is rather various in plant cover. In USSR more than 80% of this industry is concentrated in tundra and forest-tundra zones, where one can trace from north to south the following distinctly separated subzones.

6.4.1 LATITUDINAL ZONALITY

6.4.1.1 *The Subzone of North Arctic Tundra*

The subzone of North Arctic tundra covers the northernmost part of the Taimyr peninsula and the northern parts of the southern island of Novaya Zemlya and Novosibirsk islands. It is characterized by the open vegetation cover occupying 40–60% of surface, with one layer, as a rule, and with prevalence of mosses and lichens (the species of crustaceous lichens are abundant). Not numerous higher plants belong mainly to the genera *Draba, Papaver, Senecio, Cardamine, Ranunculus, Carex, Eriophorum, Poa, Calamagrostis, Luzula.*

6.4.1.2 *The Subzone of South Arctic Tundra*

The subzone of South Arctic tundra lays on the southern island of Novaya Zemlya, Vaigach island, northern parts of Yamal, Gydan and Taimyr peninsulas, at the delta of Lena river, in the coastal districts between Yana and Kolyma rivers. It is characterized by the continuous vegetation cover, usually of two layers and with considerable quantity of herbaceous perennials (*Carex stans, Eriophorum scheuchzeri, Alopecurus alpinus, genera Draba, Papaver, Pedicularis* etc.) mosses and lichens.

Further to the south three subzones of Subarctic tundra stretch as nearly uninterrupted stripes. Here the hypoarctic and boreal floristic elements appear usually as edificators or dominants of phytocenoses.

6.4.1.3 *The Subzone of North Subarctic Tundra*

The subzone of North Subarctic tundra reaches from the Pechora river up to Chukotka and is characterized by the tussock communities of *Eriophorum vaginatum,* small-hummock tundras with prostrate willows, shrubby and mossy tundras with *Vaccinium vitis idaea, V.. uliginosum, Empetrum nigrum, Ledum palustre,* lichen tundras of fruticosae lichen species, such as *Cetraria cucullata, Alectoria ochroleuca, Cladonia sylvatica, Cl. rangiferina.*

6.4.1.4 *The Subzone of Middle Subarctic Tundra*

The subzone of Middle Subarctic tundra represents a continuous belt with wide distribution of shrubby willows (*Salix lanata, S. glauca, S. phylicifolia, S. pulchra*) and dwarf birches (*Betula nana, B. exilis*), which either form here vast shrubby communities or present an essential component in the phytocenoses similar to ones mentioned above for the north Subarctic tundra. As in the later subzone, considerable plane areas (river terraces, lake and sea terraces, plane watersheds) are covered here with tundra bog complexes, as a rule. Eastward from Yenisei river there prevail crack-polygonal and ridge-polygonal complexes, westward from this river — the different variants of hummocky wet tundras.

6.4.1.5 *The Subzone of South Subarctic Tundra*

The subzone of South Subarctic tundra is the broadest one. It skirts the southern border of tundra zone from Kola peninsula up to the Chukotka peninsula. The thick active layer of soil together with the development of "frost boiling" and soil-distending processes are responsible here for various bumps, hummocks, medallions and other forms of micro- and nano-relief There are even more shrubs, dwarf birches especially, and also *Alnus fruticosa* appears. The thickness of lichen-moss layer is bigger either. In the vegetation prevail: dwarf-shrub tundras, willow and dwarf birch thickets, lichen-moss (with *Cladonia alpestris, Cl. sylvatica, Cetraria cucullata*) tundras, hummocky and polygonal mires. In the south this South Subarctic tundra is contiguous to the polar limits of woody vegetation, which is represented here by open woodlands and light forests.

286

6.4.1.6 *The Subzone of Open Woodlands*

The subzone of Open Woodlands ("redkolesya") is more or less expressed on the territory from the White sea to Anabar river in the east. Woodlands occupy here only 25–40% of area, the rest being covered by tundras and swamps characteristic for the south Subarctic tundra subzone. Trees grow here only on the areas with good drainage and deep thawing of permafrost. On the plain areas without permafrost ("taliki") one can meet the bogs with the hummock ridges (aapa complex).

6.4.1.7 *The Subzone of Light Forests*

Southward from the Open Woodlands-subzone we find the sub zone of Light Forests (parklands), where forests have 0.3–0.4 density of canopy and cover more than 50% of the territory. Here the remnants of tundra still manifest themselves as separate tundra phytocenoses along the creeks and small rivers, as well as the tundra plants in the forests. In the north this subzone borders with the subzone of open woodlands only up to the Anabar river, and further to the east it adjoins independantly to the subzone of southern Subarctic tundras.

We unite this subzone with one of open woodlands in the zone of Forest-Tundra, but some authors include it into the northern taiga zone, which borders it from the south.

6.4.2 LONGITUDINAL ZONALITY

Side by side with latitudinal zonality the longitudinal changes in vegetation are rather well expressed, that is closely connected with the climatic factors (the eastward increse of continentality up to Indigirka river and subsequent abatment of it beyond this river) and the history of floras formation. In this direction the following provinces can be distinguished if we move from the west to the east.

6.4.2.1 *Kola province*

The forest-tundra with wide distribution of open thickets of *Betula tortuosa*. At the forest limits we meet *Pinus sylvestris*.

287

Here the winter reindeer pastures with lichens prevail along with the deficiency in summer pasturess.

6.4.2.2 *East-European province (from the White sea up to Urals)*

The moss, lichen-moss and lichen tundras prevail, as well as the open woodlands with dwarf shrubs. *Picea obovata* makes the forest limit. The summer pastures prevail, winter ones are scanty, and on them the lichens are heavily trampled down.

6.4.2.3 *Ural-Novozemelsk province*

Mountain tundras, where there are only summer pastures for reindeer.

6.4.2.4 *Yamal province*

Peaty shrub tundras, vast areas of the secondary shrub-moss and lichen-moss tundras on the places of the trampled-down lichen tundras, willow thickets, grassy polygonal bogs. The forest limit is formed by *Larix sibirica*. The summer pastures prevail.

6.4.2.5 *Gydan-Pyasina province*

Mainly moss tundras with *Betula nana, B. exilis,* willows of shrubby form and dwarf shrubs. Also *Alnus fruticosa* is fairy abundant. Pure lichen tundras absent, but in mossy tundras lichens are frequent. *Larix sibirica* grows at the forest limit. The summer pastures somewhat prevail over the winter ones.

6.4.2.6 *Taimyr-Anabar province (From Pyasina river up to Olenek river)*

The strong deficiency in snow cover tells on vegetation: the areas covered by shrubs are restricted and the shrubs are lower. The dwarf shrub-moss and lichen-moss tundras, as well as cotton-grass (*Eriophorum vaginatum*) tussock tundras prevail. Here the forest limit is most moved northward and is formed by *Larix gmelini*. In the Olenek river basin the subzone of the tundra open woodlands comes to naught. Surplus of summer pastures.

6.4.2.7 Lower-Lena province (From Olenek river to Omoloi river)

Mainly mountain tundras of *Alectoria-Cetraria* dwarf shrub-moss and tussocky phytocenoses. The area of summer pastures is equal to that of winter pastures.

6.4.2.8 Yana-Indigirka province

Strongly continental climate. The least precipitation (120–150 mm) and the shallowest snow cover in the whole tundra zone. Dwarf shrubs are mostly prostrate, moss tundras absent. The tundras on frost-cracked grounds – small-hummocky, dwarf-shrub and lichen-dwarf-shrub ones – prevail. The tussocky cotton grass tundras have their maximal development in this province. The forest limits here, as in the next province, formed by *Larix cajanderi*. The summer pastures prevail.

6.4.2.9 Lower-Kolyma province (From Alazeya river up to Anyui Mountains)

The climate is less continental. More shrubby and mossy tundras. Small-hummocky dwarf-shrub and cotton grass tussocky tundras and polygonal bog complexes are most typical. The summer pastures prevail.

6.4.2.10 Chukotka province

The alternation of mountain and plain shrub, dwarf-shrub and cotton grass-sedge (*Eriophorum vaginatum* – *Carex lugens*) tundras. An original subzone of *Pinus pumila* (prostrate pine) stretches here between the south Subarctic tundra and open woodland tundra subzones. Forest limit is formed by *Larix cajanderi* with admixture of *Populus suaveolens* and *Chosenia macrolepis*. The summer and winter pastures are equal in quantity.

Reindeer pastures are present in all the abovementioned provinces. Only in the north Arctic tundra subzone there are no domesticated reindeer and all pastures are used by wild reindeer. The subzones of south Arctic and north Subarctic tundras are used by reindeer-breeders mainly in summer time. The farms on Kolguev and Vrangel islands are an exception: there they use their reindeer pastures the whole year round. The pastures of the middle and south Subarctic subzones are used in spring and in autumn, when the reindeer herds are driven from their winter living places in the forest regions to the summer pastures in the regions of Arctic sea coast and vice versa. Only part of reindeer herds, chiefly in Chukotka, stay for the winter in the south Subarctic zone. The subzone of open woodlands also is used in the period of calving in early spring. Winter pastures of the majority of farms are concentrated in the subzone of tundra light forests and in the northern taiga subzone. Some reindeer farms are situated in the middle taiga of Siberia and Far East constantly (taiga reindeer-breeding).

6.5   Forage Plants and Reindeer Pasturage Seasons

6.5.1   FORAGE PLANTS

The peculiar ecological conditions of tundra and forest-tundra determine the features of their floras. With the exception of several species, here grow the perennial plants, which are able to use the brief Arctic vegetation period most productively. About 400 of these plants are of feeding value, mainly the following: fruticose forage lichens (58 species) known under the combined name "yagels", shrubby willows and birches (44 sp.), sedges (34), grasses (52 sp.) legumes (24 sp.), and of lesser value – plants of C o m p o s i t a e , P o l y g o n a c e a e and E q u i s e t a c e a e families. Of the forage lichens the following ones have the most importance for reindeer both in tundra and forest-tundra: *Cladonia alpestris, Cl. sylvatica, Cl. rangiferina, Cetraria cucullata, C. nivalis, C. islandica, Stereo-caulon paschale.* Especially great mass they form in the forest-tundra. The prevailing forage grasses are *Arctophila fulva, Cala-magrostis langsdorfii, C. neglecta, Festuca ovina, Deschampsia flexuosa, Poa pratensis, P. alpigena, Alopecurus alpinus.* Of the sedge family the main feeding value have *Carex aquatilis, C. stans, C. lugens, Eriophorum vaginatum, E. angustifolium.* Among the forbs one should mention *Polygonum bistorta, Coma-rum palustre, Menyanthes trifoliata, Polemonium acutifolium, Pedicularis sudetica, Cirsium heterophyllum, Senecio arcticus, Nardosmia frigida, Galium boreale.* The most valuable forage shrubs are: *Salix glauca, S. lanata, S. phylicifolia, S. pulchra, S. alaxensis, Betula nana, B. exilis.* Of willingly eaten by reindeer legumes *Astragalus alpinus, Hedysarum arcticum, Oxytropis sordida* and *Vicia cracca* are most widely distributed.

The total stock of forage lichens (yagels) reaches its maximum in forest-tundra, where 60—70% of pastures are distinguished by predominance of yagels. Farther northward both the areas under lichen communities and the total lichen mass decreases. The coastal tundras are usually deprived of yagel pastures. But the stock of perennials increases to the north. So, in the forest-tundra and south Subarctic tundra of East-Europaean province the total stock of perennials makes 2.2—2.5 dezitons/ha, in the middle Subarctic tundra — 5.1 dt/ha, in the north Subarctic tundra — 6.4 dt/ha. It is the sedge family that has the major weight in the total mass of perennials: up to 90% in the forest-tundra and southern Subarctic, about 55% in the middle and northern Subarctic tundras. The role of forbs, especially of those fairy eatable by reindeer, grows to the north, their relative weight figure increases from 11—12% in the forest-tundra and south Subarctic tundra to 31—34% in the middle and northern Subarctic tundra. The mass of grasses on the watersheds of tundra and forest-tundra is scanty (0.05—0.1 dt/ha), but it is somewhat bigger in the northern part of tundra zone (up to 0,7 dt/ha), where they grow on the trampled out parts of the summer pastures. Forage shrubs are more widely distributed in the southern and middle Subarctic tundras while in forest-tundra their mass is less. The shrubby willows most valuable as a fodder comprise as much as about 30% of the whole forage mass in forest-tundra, 50% in south tundra and 90% in the middle Subarctic tundra. On table 1 one can see the forage masses in different subzones of the East-Europaean province. Of these data one can be convinced of the fact that the northern part of tundra zone is most favorable for pasturage in summer and forest-tundra — in winter. The same is valid for a number of other provinces.

In mountain regions, especially in Eastern Siberia and Far East, with their vertical (altitudinal) zonality of vegetation, the winter pasturage has to take place on the forested mountain slopes, and in summer reindeer should be driven to the treeless highlands rich in grass and shrubs.

The reindeer pastures are characterized by the sharply expressed seasonal changes in the composition, mass, feeding value and degree of accessibility of fodder plants. As for their accessibility and seasonal changes the fodder plants of reindeer can be subdivided in two main groups: 1) plants nearly unchanged throughout the year, 2) those utilized by reindeer mainly or ex-

TABLE 1.

The distribution of reindeer pasture forage resourses over different subzones (East-European province)

Forage	Zones and subzones	Tundra			Forest-Tundra	
		Northern	Middle	Southern	Open woodland	Light forests
Total stock of green fodder dt/ha[1]		6.4	10.5	6.9	6.4	6.8
Including (in %): forbs		35	15	5	4	4
Willow foliage . .		—	47	32	22	15
Birch foliage . . .		—	5	33	44	49
Sedges, grasses . .		65	33	30	30	32
Forage lichens (yagels), Total dt/ha[1]		0.1	0.9	2.7	4.2	3.0
Yagel pastures (% of the total area)		18	32	54	64	25

[1] air-dry weight

clusively at the vegetation period. The yagels belong to the first group. Although lichens have the maximal photosynthesis activity and the thallus increment during the vegetation period (Barashkova, 1963), actually their feeding value and their stock of biomass remain unchanged throughout the year (Andreyev, 1940), that appears as the main reason for their being the most important fodder of the reindeer in winter. The second group of plants may be subdivided into four subgroups: a. one of greatest value at the beginning of vegetation, and thereafter getting coarse and nearly uneatable for reindeer (most species of C y p e r a -c e a e and G r a m i n e a e Families, the foliage of dwarf birches); b. ones useful during the whole vegetation period (most forbs and the foliage of willows); c. ones used practically at the end of vegetation period, when they develop abundantly (mushrooms); d. ones especially valuable in winter time because their green shoots or dying but still eatable parts — "vetosh" remain unchanged under the snow cover (so called "winter-green plants" — some species of *Carex*, *Eriophorum* and *Gramineae*).

The accessibility of pastures at the period of vegetation depends mainly on the possibility of reindeer movement through them. The depth of swampy, frozen ground thawing, the height and density of shrub thickets, the steepness of slopes in mountains — all this can be sometimes of decisive importance. In winter the accessibility of pastures depends on the depth and firmness of snow cover first of all. Besides, the quickness of swampy bogs freezing has some importance at the beginning of winter, and the degree of pasture protection against winds — at the end of winter, in the calving period.

In connection with the feeding value of pastures and the pasturage conditions on them, we distinguish in our northern reindeer-breeding industry six pasturage seasons and accordingly six seasonal types of pastures.

6.5.2 PASTURAGE SEASONS

6.5.2.1 *Winter season*

From November up to April inclusive. The main fodder consists of yagels (lichens), an important role (and the decisive one in regions poor or lacking of yagels) plays the winter-green fodder. The accessibility of pastures decreases as snow cover gets thicker, denser and firmer. The tundra reindeer is able to dig out the snow cover 75—80 cm deep if the latter has no more than 0.35 density. The forest reindeer digs out snow cover up to 1.500 cm deep and 0.28—0.300 dense. In the region of scanty snow-falls between Khatanga and Kolyma rivers the snow cover does not hinder reindeer's grazing, however in other regions, especially in those of continental-marine climate, the snow cover results in the strong reduction of grazing area. The valleys of creeks and small rivers, which are just the areas rich in winter green fodder, become inaccessible first of all; then snow drifts develop in more dense forests (of dark conifers especially) and at last the lower slopes, shrub thickets become inaccessible too. In March the frozen snowcrusts appear on the plain interriver divides, which remain inaccessible for pasturage. The snow gets as dense as 0.3—0.4 and even more. At the beginning of winter, while the snow is loose (the fresh snow is of 0.12—0.15 density) and shallow, all the pasture area is accessible for use. 25—30% of area becomes inaccessible in december, 40% in February, 50—60% — in March-April and to the end of winter the accessible pasture area often reduces up to 10%. The degree of pastures accessibility fluctuates considerably on years.

Among the tundra and forest-tundra winter pastures of most importance are the lichen ones, then follow moss-lichen,

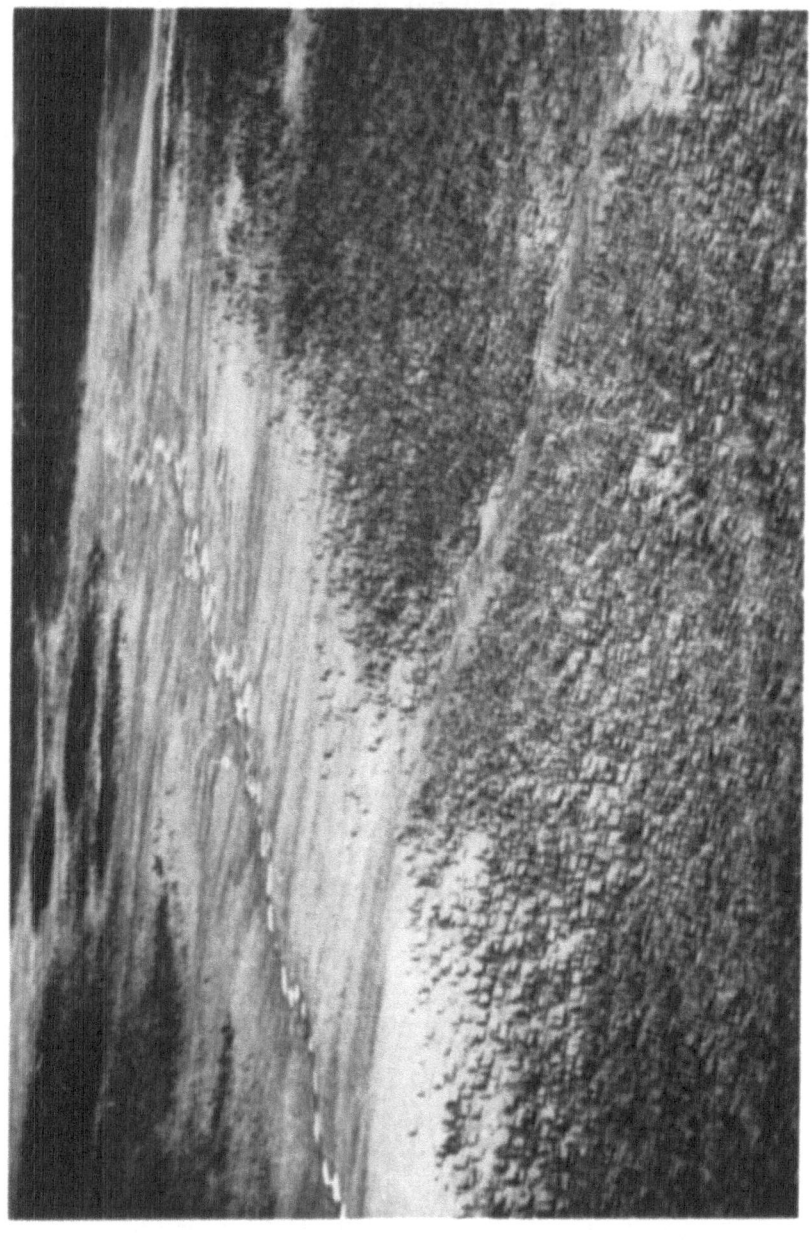

Fig. 1. Reindeer winter pasture within the forest-tundra. Clear larch-forest with undergrowth of lichens, appropriate to pasture. Dominant *Cetraria cucullata*. Alongside the watercourse tundra with *Betula exilis* and *Eriophorum vaginatum*.
Aerial photo. (See p. 294).

Fig. 2. Reindeer summer pasture. Mountain tundra with *Betula exilis* and shrubs. (See p. 295).

lichen-dwarf-birch, tussocky tundras, lichen frost-cracked and hummocky bogs, all kinds of lichen open woodlands (larch, birch, spruce ones), lichen light forests (both larch and pine ones), lichen alder- and pinus pumila thickets, sedge bogs, grass and sedgegrass meadows.

In early winter those kinds of pastures are used, which later become inaccessible: lichen spruce open woodlands, lichen alder- and pinus pumila thickets, lichen dwarf-birch tundras, meadows along small rivers and creeks. But the more accessible and fodder-rich pastures are left for the late winter, that is profitable in the view of compensation of energy expenditure of animals during the digging out of snow. It has been found that this expenditure increase 1.5 times as compared with one in early winter (Gulchak, 1950). Besides, the need in food of reindeer increases towards the end of winter in connection with fetus development in the pregnant females (and they make as much as 50% of the whole cattle stock). In this period of most importance are lichen open woodlands (larch and pine ones), lichen tundras and all kinds of the lichen bogs, hummocky ones first of all.

6.5.2.2 *Early spring season*

Month of May — the time of reindeer calving. Yagels continue to remain the main reindeer fodder as a rule, but with addition of winter-green plants. The inaccessibility of a great part of pasture area and the necessity of neglecting wind-exposed areas in this period result in significant narrowing of the total pasture area. At the same time the decreased mobility of herds containing newly born calves leads to the more intense utilization of pastures. The demands of this early spring season are met best by the fodder-rich lichen and hummocky tundras situated in mountain valleys or the plain pastures alternating with open forest islands which can give protection against the wind. Also the peat bogs rich in yagels are widely used, in forest-tundra especially.

6.5.2.3 *Late-spring season*

In the majority of tundra and forest-tundra regions it falls on June. It is the time of intense snow thawing and beginning of vegetation. On pastures the first green fodder — young shoots of sedges, grasses and cotton grasses is added to the yagels. The ratio between yagels and green plants ranges broadly depending on the stock of these fodders on the pasture, but its average value is 1 : 1. In this period the reindeer use all the pastures except of water-logged places and ones still covered by snow, which reindeer cannot dig at this time. They disperse widely in search of green fodder and collect yagels on moss-lichen and hummocky tundras; as the main sources of green fodder they use fens and shrub thickets rich in sedges and grasses. It is of special importance to use during this period those pastures inaccessible in the summer season.

6.5.2.4 *Summer season*

Lasts from early July to the middle August. The main fodder of this time are various green plants, chiefly forbs, and shrub foliage. The summer pasturage is possible in all subzones of tundra and forest-tundra, but it is most profitable to use during this period the regions of north Subarctic and south Arctic tundras, where there are the greatest masses of forbs and willow foliage − the most profitable summer fodder. Besides of that, these subzones are distinguished by cool summer, strong winds and consequently by lesser quantity of reindeer disturbing insects. The same advantage have the mountain pastures of Urals, Eastern Siberia and Far East, situated above the altitudinal forest limits.

The most important summer pastures are: forb-dwarf-shrub, willow and sedge-cotton-grass tussocky tundras, forb-sedge-birch and forb-willow-shrubs, tundra and mountain meadows, sedge cotton-grass fens, Arctophila and forb-grass lowland and lake-shore meadows, sedge-grass sea cost marshes. When the weather is very hot, one uses the shrubless pastures situated on more elevated relief forms and high banks of big lakes; in the cool weather it is possible to graze in shrubby and lowland pastures.

6.5.2.5 *Early autumn season*

Lasts from middle August to early October. This is the time of green plants weathering and snow cover establishment, of reindeer fattening and intense growth of young reindeer. The pasture fodder is of mixtured kind − green plants and yagel. One tries to use as much of green plants — the sedge-, grass- and cotton-grass shoots — as possible, while most forb species are dead and shrub foliage fallen off. Mushrooms are eaten willingly in this period and contribute much to the reindeer fattening due to the high content of proteins, fats and vitamins in them. With the deficiency in green fodder the reindeer begin to eat yagels. The utilization of the maximal mass of green fodder and mushrooms appears as an important task of reindeer-breeders in this period. Having this in mind, they select for this season such pastures provided with long preserved green fodder plants and with such areas and relief that permit reindeer to disperse in search of valuable fodder plants. They stay on such places as long as possible and not before the snowfall they move to the places where the yagel pastures prevail.

The most valuable of early-autumn pastures are the lake-side meadows (khasyrei) with predominance of *Arctophila fulva*, *Senecio arcticus*, *Eriophorum* species and remnants of several forb species; also the sea-side grass-sedge meadows, tundra- and mountain meadows, sedge-cotton-grass fens, sedge-grass-willow thickets, as well as shrubs and meadows along the banks of small rivers and creeks. As the source of yagel fodder in this season can serve the same moss-lichen and hummocky (poor-in-yagel) tundras, as in the previous season.

6.5.2.6 *Late-autumn season*

It lasts two months — October and November. The mixture of fodder with yagel prevalence is typical. Also in this season the problem of green fodder utilization (as long as possible) is not removed. This kind of fodder is most valuable for reindeer fattening; autumn fattening may continue up to the middle of winter period (December, January). Another reason for green fodder utilization is the general deficiency in yagels, the necessity of economizing in their usage.

The green fodder is found in this season on the same pastures, as in previous one. Of yagel pastures the *Alnus*- and *Pinus pumila* thickets, spruce- and open woodlands, moss-lichen and hummocky tundras. lichen-dwarf birch tundras are to be used in this season, all of them comparatively poor in yagels.

The correct elaboration of annual roaming route of herd offers a very difficult and complicated problem, which may be successfully solved only if one has a geobotanical map, pastures inventory materials and a good knowledge of the whole pasture territory. The rational alternation of pastures according to the day-time and weather conditions represents the addition to this route. The long-term, periodical and diurnal weather forecasts are of great importance for the reindeer-breeders helping them to plan the order in pastures usage beforehead.

6.6   The Yagel Problem

Yagels (forage fruticosae lichens) are one of the most important fodder resources in the reindeer-breeding industry. Their stock in the Far North regions reaches now as much as 50 million tons of air-dry weight. Usually yagels make 2/3 of the annual quantity of fodder consumed by reindeer on pastures. In spite of their feeding inferiority (low content of proteins and minerals), yagels represent the important and available source of carbohydrates, which are almost completely assimilated by reindeer. The reindeer digest yagels much better than other cattle, sheep, swines and horses (Aksenova, 1937). The capability of yagel nutrition throughout the year is an indisputable advantage of reindeer over other northern ungulates in the process of evolution. Apparently, in connection with the broadening of fodder base (at the expense of yagels) stays the numerical superiority of reindeer over the other species of wild ungulates. Owing to the yagel nutrition the herd maintenance of domestic reindeer is possible. If the reindeer sought for the green fodder, scanty in winter, the herd would disintegrate.

If the previous opinion that the yagels are the one and only irreplaceable reindeer fodder is wrong, the underestimation of the role of yagels in reindeer-breeding is also undesirable. We should keep in mind that in a number of regions the under-snow stock of green fodder is so little that by no means it can make up a deficiency of yagels. For instance, in Kola peninsula and Pechora river territory, where the pine lichen forest is used as winter pasture, the standing crop of yagels in this forest makes 90—95% and green plants 5—10%. Nearly the same figures are valid for the vaste Siberian lichen forests (of larch or pine). In the Eastern Siberia and in the East-North of the Europaean part of USSR the territories with the prevalence of yagels compose more than 50% of the whole winter pastures stock. The reduction of yagels on this territory cannot be compensated by green fodder and would lead to the total reduction of the fodder base.

The problem of yagel stock protection and their quick regeneration after being grazed now acquires the more and more importance. It is necessary to take some precautions against reduction of yagels on winter pastures.

The biology of yagel development is peculiar. At present we know some features of growth and regeneration of yagel. There are three growth stages in the development of fruticosae lichens: 1. Podetium (*fruticosae thallus*) formation and growth stage takes 8—14 years; 2. Podetium renovation, i.g. the simultaneous growth and dying off (from below) of podetium. This two processes roughly counterpoise each other and the size of living part of thallus remains of the same size; 3. Podetium dying off, when the rate of dying of exceed the accreation (Andreyev, 1954). As was shown by E. A. Barashkova (1963a), in winter time the activity of cell division in algal component of lichens falls down and the growth of podetium ceases. The growth of yagels during vegetation period can be shown as a curve with two peaks. The first rise of curve concerns the early summer, when podetia are water-saturated as a consequence of snow melting. In the middle of summer there is a deceleration of growth in connection with the dryness of air and of soil surface. In the conditions of continental climate (Taimyr, Yakutia) this curve depression is deeper and more prolonged. The second rise begins in autumn and it lasts up to the snow cover establishment.

Economically it is most profitable to use yagels at the point of first stage of their growth completion. The second stage can be compared with a larder in which the useful fodder stock is kept for long, but is permanently renewed. However, in well organized economy the presence of such larders is not profitable, for the yearly dying off fodder mass does not produce any economical effect. The accumulation of dead parts of podetia with their slow decay rate leads to the creation of acid litter. When the thickness of this litter exceeds 5 cm, it begins to influence negatively the further development of yagels. Thus, it is desirable to aim at the complete utilization of yagels, before they finish their first growth stage, and not letting them stay for long at the second growth stage. But we should stress that the utilization of yagels, which have not yet completed the stage of podetia formation threatens with more serious loss of fodder mass. As our investigations have shown, the fodder mass accumulation is much more intense at the second half of podetia formation process. This stage is 10—12 years long by most species of lichens. If we take the standing crop mass of developed yagels, which have fully completed their first growth stage towards the age of 12 years, for 100%, we will find that at the age of 10 years their mass reaches 76% of that value, at the age of 8 years — 50%, at the age of 6 years — 33%. Thus, if we utilize these yagels within 2 years before they complete their growth, we will have a shortage of 24% of fodder mass possible, within 4 years — 50%, within 6 years — 67% (Andreyev, 1954).

The utilization of yagels at the optimal time is possible only on the basis of pastures rotation. It has been found that under normal grazing process the reindeer do not eat yagels entirely, but only 25—35% of area, selecting the plots with more developed lichen cover (Igoshina 1939; Florovskaya 1939; Andreyev, 1954; Makhaeva, 1966). Consequently, during three or four years they will use the whole area. Under normal conditions of pasturage the time of yagel standing crop renewal is equal roughly to the time of podetium formation, i.g. 10—12 years. Of this follows that in the 10—12 year period this pasture can be used 3—4 times, i.g. 3—4 years of pasture utilization should correspond with 7—8 "blank" years, when the pasture "takes rest". Of all this we may conclude that the three-year pasture rotation period, when 1 year of utilization is followed by 2 years of rest, will do. The practice of such rotation periods in a number of reindeer-breeding farms testifies its high efficiency (Besumov & Preobrazhensky, 1952; Antonova, 1957; Karev, 1956; Bagaev & Shatalov, 1968). The introduction of pasture rotation results in the even utilization of forage resources and the complete restoration of yagel forage stock. It is one of the most important measures on the enhancing of forage base of reindeer-breeding industry and the most important conditions for the increase of the culture of reindeer-breeding practice.

According to composition and fodder standing crop, the yagel pastures can be divided into three categories:

1. Pastures with vigorous development of lichen cover, where *Cladonia alpestris* predominates and gives as much as 5—10 tons of air-dry forage mass per 1 ha of area. Such pastures are poorly or irregularly used or not used at all.

2. Pastures with less developed lichen cover and with predominance of quick-growing *Cladonia sylvatica* and *Cl. rangiferina*. They give 2—5 tons of forage mass and are normally and systematically used.

3. Pastures with strongly disturbed lichen cover in which, besides of above mentioned cladonias, the weedy lichens are developed, such as *Cladonia coccifera, Cl. bellidiflora* and others.

These lichen pastures appear as a result of incorrect pasture usage. Under the influence of permanent utilization the yagel pastures of first category are transferred into ones of second category. The stock of lichen phytomass decreases. However, the yagel pastures of second category are of greater economic value because they regenerate more quickly and on them there are some green plants, which are practically absent on the pastures of first category.

The peculiarities of different categories of yagel pastures must be taken into consideration when one elaborates the regime of their utilization. If on the pasture of first category the pasture rotation will be one of two years, this will hasten its transformation into one of the second category. On the yagel pastures of the third category the intensity of usage should be decreased with the aim of restoration of lichen cover and their transformation into ones of second category. On the pastures of second category the pasture rotation should be one of three years. The pasture rotation is a fairy effective means of pasture preservation, increase of their productivity, but also its effectiveness can be considerably rised if we organize the "superficial bitting off".

The mass of annual accretion of yagels depends on the total living mass of yagels (Andreyev, 1954). The lower the reindeer cut the podetia, the more time is necessary for their full regeneration. If we take the podetia with living part of 4.5 cm high and remove 1 cm of their upper part, this removed top will regenerate in 4 years, removed 3 cm − in 10 years, 4 cm − in 15 years. The plants die, as a rule, if nearly all living parts of podetia are removed. The fodder mass of the upper, fruticosee parts of yagel podetia is bigger than one of their lower parts. If we divide the 4.5 cm thick yagel cover into horizontal strata 1 cm thick each, we will find that their weight from the upper stratus downward will make correspondingly 28%, 28%, 24% and 15% of the total fodder stock and the lowest 0.5 cm will comprise 5% of it. Thus, if we remove only the upper parts of podetia comprising 20% of the total mass, they will regenerate in 2 years and the annual accretion will be 14% of the total mass. If we remove the upper strata, i.g. 56% of the total mass, they will regenerate in 4 years, 14% yearly. However, if we remove 3 strata, i.g. 80% of mass, the regeneration period will be as long as 10 years and annual accretion will not exceed 8%. In the case of 4 strata removal (95% of mass) we will need 15 years for regeneration with annual increment 6%. As our practice shows, under conditions of correct organization of pasturage and normal accessibility of fodder the reindeer bite off approximately 1/3 upper part of living podetia, i.g. 42% of their mass and this part can regenerate in 3 years, 14% of mass yearly. Such way of yagel pastures utilization we call "superficial bitting off". Economically it is most effective; it secures the highest rate of yagel increment and the quickest restoration of the removed fodder mass. The failures in maintainance of regular pasturage can result from the growth and condensation of snow cover in late winter, but also from the wrong ways of pasturage, lack of active control over the herd. Such failures lead to the lower bitting off the yagels and the utilization of 65–70% of entire standing crop; the annual incremant mass may fall in these cases up to 8%. In especially difficult cases the reindeer bite off the yagels close to the ground (more than 80% of mass), the increment falls to 6% and the pastures undergo a great damage.

Unfortunately, such a low bitting off is not rare in practice. Besides of bad pasturage, it may be a consequence of low capability of reindeer for snow digging in the case they are wasted or sick. Then the reduction of reindeer quantity on the pastures becomes necessary in order to avoid the pastures degradation. The quantity of reindeer in this cases should be two or more times less.

With the aim of securing the superficial bitting off, we also control the movement of grazing herds, do not let reindeer to linger on the pastures beyond the certain time.

In perspective we see the broad use of different ways of yagel growth stimulation by means of air-chemical methods, we started the search for them still in 1957. The possibilities of chemical stimulation of yagel growth have been successfully tested in a series of experiments carried out in field conditions at the Kola and Taimyr peninsulas and also in the laboratories of the Institute of Far North Agriculture (Norilsk) and Komarov Botanical Institute of Academy of Sciences (Barashkova, 1963b). In despite of the opinion of previous investigators, who had thought that it is "beyond man's power" to hasten the growth of yagels (Tanfilyev, 1894, p. 30) it proved possible to hasten the increment of fruticosae lichens 1.5–1.7 times with the influence of a very weak solution of gibberellic acid, thiamine and sodium salt 2.4 – D. The last solution, of 0.0015 concentration in dozage 400 gr of salt (In 600 liter of water) per 1 ha is of greatest economic value.

The "yagel problem", which had arosen as the consequence of trampling and destruction of yagel pastures and worried much

the progressive-minded tundra investigators from the beginning
of the last century (Boguslav, 1848; Tanfilyev, 1894) presently
can be solved by means of scientifically-based ways of rational
utilization and improvement of pastures.

6.7   The Green Fodder Problem

The green fodder plants constitute from 30% up to 100%
(in certain cases) of the annual mass of forage plants eaten by
reindeer. This is the feeding stuff of full value, especially in the
vegetation period, when their feeding properties are the highest
ones. At this period the reindeer has its most intensive growth
and development and the accumulation in his organism of those
store matters, which will be spent later in winter time. In this
view the rational utilization of green fodder appears as one of
most important problems in our task of further reindeer-breeding
development.

The value of different groups of plants is various. Sedges
and grasses are most valuable at the very beginning of vegetation
period, when they are most rich in nutrients and most palatable,
and also in winter period, when another green plants are nearly
absent, Forbs appear as the most valuable and diverse group of
summer fodders. Also the foliage of shrubby willows is of great
importance in summer. The shrubby birches are less valuable,
willingly eaten only at the first half of summer. Thus the forbs
and the willows give the summer pastures their high fodder qualities.

As was noted above, the pastures of highest feeding value
are situated in the north Subarctic and south Arctic tundra subzones. Proceeding from the requirements of zoohygiene too, a
number of investigators came to the conclusion on the highest
efficiency of summer grazing in these very subzones (Nikolaevsky, 1951).

Also the practice of reindeer farming confirms the high
efficiency of summer pasturage of reindeer in the north of tundra
zone, especially in coastal districts. It is here that the best fattening and lowest degree of reindeer morbidity have been registered.

Groundless is the shortening of routes of reindeer-breeders'
teams at the expence of summer grazing in more southern tundra
subzones. It can be permissible in particular cases only, for instance, if there are hills or mountains in southern tundra. To
disregard the inner connections between the reindeer and the

environment is inadmissibly. In the southern tundra there are concentrated enormous reserves of valuable green fodder for reindeer. However, they are nearly not utilized in reindeer-breeding industry (Vakhtina, 1963). Their utilization might increase the summer pastures grazing capacity in several times.

The green fodder, mainly grasses, plays a considerable role in winter ration of reindeer either. In the standing crop of forage plants, which remain on pastures in winter time, the "vetosh" (the dry brownish shoots and leaves, just died after vegetation) prevail, only 5—10% falling on the share of living green shoots, which hibernate beneath the snow. Of the biochemical data we know that 50% of the summer quantity of proteins remain in the green parts of hibernating plants in winter, while in the "vetosh" there remain 35—40% (Florovskaya, 1939; Fedorova, 1958; Kovakina, 1968, Andreyev, and others, 1974). In winter the "vetosh" of most sedges and grasses, composing the main mass of forage plants stock, contains 5—6% of proteins (in absolutely dry matter), while in yagels there is 2—3% of proteins. Thus, "vetosh" and green under-snow parts of grasses appear as an important source of proteins. It is known, that the reindeer left on pure yagel ration experience the ash-protein exhaustion, in late winter especially. But if they are supplied with green fodder, they keep their nourishing up to the end of winter period.

The domesticated reindeer have accomodated themselves to the considerable prevalence of lichens in their feeding ration because reindeer-breeders prefer to graze them on lichen-rich pastures, where reindeer are more concentrated and it is easier to control them (Andreyev, 1971). Such a share of yagels is never reached in the ration of wild reindeer even in the case they have unlimited possibilities of yagels utilization. The studies of reindeer paunches in winters have shown that the share of yagels does not reach more than 30—40% by wild reindeer (Michurin & Makarova, 1962; Egorov, 1965) whereas by domesticated reindeer there is usually 70% of lichens or more (Alexandrova et al., 1964). The reason for this is the prevalence of yagel pastures, which are more convenient for herdsman, but not at all the best ones if to look from the physiological point of view.

The reduction in yagel resources in a number of regions causes gradually some changes in the reindeer winter ration. This can be seen, for example, at the reindeer-farms situated in the Pechora river basin, where in 1940, 1952 and 1963 the stock-taking of yagel resources has been carried out. The data of this inquiry has shown that on the winter and early-spring pastures of most farms the stock of yagel was reduced during 23 years for 50–70%. But this has not provoked a catastrophe in reindeer-breeding. As we found from the paunch content analyses, the winter ration of reindeer at these farms had changed profoundly: the share of yagels had reduced from 85–90% in 1940 to 30–45% in 1963, while the share of herbaceous fodder increased from 51% to 55–70%. Such ration with high content of green fodder is closer to the physiological norm. In spite of this, the decrease in yagel stock in the Pechora basin cannot be considered as a positive fact. The deficiency in yagels can be compensated at the expense of green fodder (the under-snow one), but we should remember that the standing crop of green fodder in different years is very inconstant and there can be underfeeding and even starvation and death.

The reindeer can do without yagel, if they are sufficiently provided with under-snow green fodder. We have a good experience concerning this problem. The recent investigations have shown that reindeer cannot be adopted to the yagelless ration quickly. In the process of natural and artificial selection the special strains of reindeer adopted to the yagelless nutrition were created. Among them of special value is the Chukotka tundra reindeer called "khargin" in Yakutia. This animal is distinguished by its exterier, biological features and behavior. It is a short-legged, stumpy creature with massive barrel-like body. It is capable of quick graziery and fattening and high dressing percentage (up to 60%). Khargin is capable too to live in winter in tundra on yagelless pastures (Kurilyuk, 1969; Yakovlev & Kurilyuk, 1969; Pomishin & Kachan, 1972; Rumyantsev, 1975). The khargin herds are closely grouped on pastures and this helps them to break up the firm tundra snow and extract the fodder hidden beneath that snow. Khargin is adapted to the utilization of tundra pastures with under-snow fodder grasses stock (tussocky and sedge-forb tundras of the North-East of the European part of USSR).

Most farms, which use the tundra zone and the adjacent forested territories, have plenty of pastures with green fodder. They use them in summer, leaving yagel pastures for autumn, winter and spring. According to their grazing capacity, the summer pastures can secure at an average in the Far North more then twice bigger reindeer herd than the yagel ones. As a result of unbalanced utilization of summer and winter pastures, more then half of green fodder pastures are underused at present. These extra pastures appear as an enormous reserve for the further development of reindeer-breeding industry, which has already fully assimilated the yagel pastures in a number of regions, but the further growth of reindeer head is restrained there by the lack of winter pastures. A certain part of ungrazed tundra pastures undoubtedly can be used as winter pastures. For this purpose one must properly select the pastures rich in wintering forage grasses. Such pastures, represented by tussocky and sedge-forb tundras, are widely distributed eastward from Yenisei river. But they can be utilized only by those strains of reindeer adopted to the yagelless ration, Chutkotka reindeer — khargin first of all. The full utilization of winter forage resources would result in the reindeer stock capita considerable increase.

The following types of ration may be distinguished correspondingly to the different ratio between the green and yagel fodder by two reindeer strains (Table 2).

When ascertaining the grazing capacity of pastures one must take into consideration the ration permissible in given conditions. In the regions well provided both with yagels and green fodder, the first ("usual") type of ration is permissible; at the places with surplus of yagel stock and shortage in green fodder — "high-yagel" ration; at the places with the rich green fodder stock and lack or deficiency of yagel — "poor-in-yagel" ration. Probably, the ration of wild reindeer corresponds best to the physiological standard.

In spite of the under-snow green fodder surplus and under-utilization of it in the North as a whole, the need in its increase takes place in some regions, especially in those with abundance of yagel pastures. The problem of reindeer pastures enrichment in green fodder might be solved by the following ways.

First way — the draining of tundra lakes with following regulation of water regime in lake depressions. A number of scientists noted the appearance of rich grass stands on the self-

TABLE 2.
Reindeer grazing rations (in %)

| | Domesticated reindeer | | | | | | Wild reindeer | |
| | Usual ration | | High-yagel ration | | Poor-in-yagel ration | | | |
	winter	summer	winter	summer	winter	summer	winter	summer
Yagels	70	5	95	20	5	—	30	5
Green fodder	30	95	5	80	95	100	70	95

drained lakes (called "alas" by Yakuts, "khasyrei" by Nenetz, "pnoutki" by Chukchi). We tried to drain artificially several tundra lakes by means of drains and then recommended this method as a way of pastures enrichment in green fodder (Andreyev, 1948). The methods of tundra grassing and the possibilities of meadows creation on the areas arising as a result of lake drainage have been specially studied by A. V. Pryanishnikov in Khatanga tundra (Pryanishnikov, 1954, 1955). Presently the analogous studies are carried out at Chukotka in connection with the problem of thermokarst and having in mind certain practical aims (Tomirdiaro, 1969). When draining lakes one must keep in mind the problems of water supply of population as well as the interests of fishery and hunting.

Second way — the undersowing of winter-green grasses on the trampled off pastures and slash fires. As the result of several experiments we have recommended an assortiment of grasses to be sown and urged on the necessity of nitrogenous fertilizers, in minimal dosage at least (Andreyev, 1946; Kovakina, 1958; Mashistova, 1967).

Third way — the effect of chemicals on plants, of such substances, which are able to make the advacement in age of leaves slower and stimulate the autumn increment in vegetative mass of plants. There are only data of rather tentative but encouraging experiments in this respect (Kovakina, 1967).

The problem of rational utilization of green forage plants stock and of its increase has, together with the yagel problem, a great importance in the strengthening of fodder base of reindeer-breeding industry. At the expense of green forage plants full utilization the further increase in reindeer stock capita in all regions of the Far North becomes possible.

6.8   Reindeer Pastures Protection and Improvement

The grazing of reindeer influences the vegetation cover, and partly the soils, of tundra and forest-tundra very much. Under this influence the plant cover becomes more homogeneous with predominance of grazing-resistant grasses, such as *Festuca ovina, F. brevifolia, Poa pratensis, P. alpigena, Calamagrostis neglecta, C. lapponica, Deschampsia borealis, D. arctica, Arctagrostis latifolia* and others. Yagels are replaced by uneatable lichens (the goblet like cladonias and others), grazing resistant green mosses and dwarf shrubs; the shrub thickets are thinned out or destroyed, the forest regeneration gets poor and then die off.

Besides of the independant effect (bittening off, trampling) the grazing of great quantity of animals renders a great indirect influence on vegetation through the thickening of soil, intensifying of ground movement (solifluction) and formation of micro-relief (hummocks, bare spots etc.). Nearly all the territory of tundra and forest-tundra bears the traces of reindeer grazing.

There are two categories of reindeer's influence on vegetation: 1. inevitable influences connected with the normal reindeer grazing, 2. influences connected with the breach of normal pasturage, such as pasture overgrazing, unnecessary movements of reindeer or their too dense arrangement etc. These kinds of influence can be avoided. The overgrazing effect is the most harmful one. We know that in a number of tundra and forest-tundra regions the quality of pastures has been lowered sufficiently.

The normal pasturage does not lead to such undesirable after-effect. There are many farms where the pastures used completely during many years and nevertheless they keep their good state. Even the correctly organized pasturage improves the conditions of pastures: it prevents the overmature yagel pastures formation and superfluous development of mosses (that oppress the growth of the grasses), makes the shrub thickets not so dense and favours the development of forbs and other forage plants.

In order to avoid the negative consequences of grazing and strengthen its positive effect, one should use the abovementioned methods of rational use of pastures, keep to the right grazing capacity of pastures, without overgrazing, to the pasture rotation, superficial use of yagels. The recommended measures are elaborated separately for different reindeer-breeding regions and both specialists and reindeer-breeding farms' administration, as well as land organization service keep an eye to their maintenance.

The problems of pasture melioration are not worked out enough, and we will not discuss them here. Better we arrest on the endeavours of increase in winter accessibility of pastures and their protection of fires.

The control of frozen snow-crust and general snow thickening on winter and early spring pastures is of great importance. The proposed tenon-rollers has shown good results. The self-propelled roller is under construction at present. The expences on the mechanized snow-crust control are worth while, since the snow crust covering enormous areas of pastures and bearing the features of natural calamity (especially under sea-continental climate) leads to the loss of many thousand reindeer.

At present we use the strewing of snow surface with ash or with other dark powder not harmful for animals to provoke the accelerated snow melting at the period of pre-spring and early-spring snow crust. We make this from aeroplane and the results of such measure in the conditions of the North are rather satisfactory (Antrushin, 1956).

Fires make a considerable damage for reindeer pastures, especially in lichen forests and woodlands. But the careless handling with fire and thunder storms lightning may provoke them.

At present all the territory of reindeer pastures in forest-tundra and northern taiga is under air-patrol supervision at the time of possible fires. Special aeroplanes fly over the reindeer pastures along the previously elaborated routes. The measures for fire liquidation are taken immediately. These flights are paid mainly by the state. As the result of these patrol service the burned areas are considerable reduced recently. The restoration of vegetation on slashes is observed. The utilization of grasses, of *Calamagrostis lapponica* especially, for reindeer grazing is possible in 5–6 years after fire, the utilization of lichens – in 20–25 years.

Our observations on the reindeer (and caribou) pastures in Finland, Sweden, Norway, Canada and Alaska have shown that in their natural conditions they are very similar to the reindeer pastures of the Soviet Union. The problems of rational utilization, protection and improvement of reindeer pastures are the international problems and they are to be solved by joint efforts of scientists of many countries. These problems have the considerable importance not only for reindeer-breeding industry, which has now excellent perspectives in northern countries, but also for protection and improvement of the whole natural complex of Subarctic.

6.9   Summary

After an introduction into the commercial importance of reindeer breeding in subarctic U.S.S.R., the author gives a review about the hitherto done investigations on reindeer pastures. Then a description follows of the methods which are employed to investigate the feeding value of vegetation in its dependence of seasonal influences. Importance is attached to estimation of standing crop mass in the wide territory with its highly complicated plant cover. After this, the typical tundra flora in its dependence upon climate and soil is described. This chapter includes a geobotanical survey of the tundra region between Murmansk and the territories eastward the Lena river. The following chapter deals with the single plants which are eaten by reindeer, including phanerogamous plants, mosses, lichens, and mushrooms. In this connexion, the seasonal changes in the feeding value of pastures are explicated.

Among the fodder plants of reindeer, the lichens (*Cladonia* spec. div.) are of particular importance. They are eaten in great quantity and therefore menaced by excessive live stock pressure. Therefor it is necessary to give them occasion to regenerate without beeing pastured within some years. Additionally there are experiments with a solution of a phytohormone which is spread from an aeroplane. In a similar way the author deals with the green fodder plants of reindeer. Finally, mention is made of pasture protection.

References

Alexandrova V. D., − 1937 − Winter food of domesticated reindeer in Novaya Zemlya. Sovetskoye Olenevodstvo, 9, Leningrad.

Alexandrova V. D., − 1940 − Feeding characteristics of plants in the Far North. Trudy Instituta Polarnogo Zemledelia, Zhivotnovodstva i Promyslovogo Khozyaistva. Ser. "Olenevodstvo", 11, Leningrad.

Alexandrova V. D., Andreev V. N. et al., − 1964 − Feeding characteristics of the Far North plants. Moscow-Leningrad.

Aksenova M. Y., − 1937 − Problems of reindeer nourishing. Sovetskoye Olenevodstvo, 10, Leningrad.

Andreyev V. N., − 1930 − Geobotanical investigation of tundra reindeer pastures. Sovetskyi Sever. 5, Moscow.

Andreyev V. N., Igoshina K. N. & Leskov A. I. Reindeer pastures and vegetation cover of the territories adjacent to Polar Urals. Sovetskoye Olenevodstvo, 5, Leningrad.

Andreyev V. N. & Panfilovsky A. L., − 1938 − Inspection of tundra reindeer pastures with help of aeroplane. Trudy Instituta Polarnogo Zemledelia, Zhivotnovodstva i Promyslovogo Khozyaistva, Ser. "Olenevodstvo", 1, Leningrad.

Andreyev V. N., − 1940a − Aerial-by-sight inspection method in reindeer pastures inspection. Trudy Instituta Polarnogo Zemledelia, Zhivotnovodstva i Promyslovogo Khozyaistva, Ser. "Olenevodstvo", 12.

Andreyev V. N., − 1940b − Pastures and pasture rotation in reindeer breeding Trudy rasshirennogo plenuma Komissii Krainego Severa Vsesoyuznoi Akademii Selskokhozyaistvennykh Nauk imeni Lenina "Voprosy olenevodstva Krainego Severa" (Problems of Far North reindeer breeding).

Andreyev V. N., − 1946 − Brief survey of the Naryan-Mar reindeer-breeding station activities in 1940−1945. Naryan-Mar.

Andreyev V. N., − 1948 − Fodder and pastures of reindeer. In book: "Severnoye Olenevodstvo" (Northern reindeer-breeding), Moscow.

Andreyev V. N., – 1952 – Application of aerial methods in the geobotanical mapping and fodder areas inventory. *Botanichesky Zhurnal,* 37, 6.

Andreyev V. N., – 1954a – Forage lichens increment and the ways of its regulation. *Trudy Botanicheskogo Instituta Akademii Nauk SSSR, Ser.* 3, *Geobotanika,* 9, Leningrad.

Andreyev V. N., –1954b – Vegetation cover of the East-Europaean tundra and some arrangements aimed to its utilization and transformation. Leningrad.

Andreyev V. N., – 1966 – Features of zonal distribution of aboveground phytomass in East-Europaen North. *Botanichesky Zhurnal,* 51, 10.

Andreyev V. N., – 1968 – Problems of the rational utilization of reindeer pastures. In symp.: "Problemy Severa" (North Problems), 13, Moscow.

Andreyev V. N., – 1971a – Methods of forage phytomass stock estimation and mapping in Subarctic. *Rastitelnye Resursy,* 7, 3, Leningrad.

Andreyev V. N., – 1971b – Ecological factors of present evolution of the reindeer. In symp.: "O fisiologo-biokhimicheskikh i geneticheskikh problemakh Severa" (On the physiologie-biochemical and genetic problems of the North), Yakutsk.

Andreyev V. N. et al., – 1974 – Tebenyevochnye pastures in the North-Eastern Yakutia. Yakutsk.

Andreyev V. N., – 1975 – Main problems of the reindeer pastures investigation In symp.: "Magadansky olenevod" (Magadan reindeer-breeder), 27, Magadan.

Antonova M. M., – 1957 – Experiments on reindeer pastures rotation introduction into the kolkhozes of Yamal-Nenetz territory. Byulleten nauchno-tekhnicheskoi informatsii Instituta Selskogo Khozyaistva Krainego Severa, 2, Leningrad.

Avramchik M. N., – 1939a – Winter food of reindeer in Yamal North. Trudy Instituta Polagnogo Zemledelia, Zhivotnovodstva i Promyslovogo Khozyaistva. Ser. "Olenevodstvo", 4, Leningrad.

Avramchik M. N., – 1939b – Aftergrowth in some forage plants on tundra pastures. Trudy Instituta Polarnogo Zemledelia, Zhivotnovodstva i Promyslovogo Khosyaistva, Ser. "Olenevodstvo", 4, Leningrad.

Bagaev I. I., Shatalov V. S., – 1968 – Introduction of pasture rotation as the most important condition of the further development of the reindeer-breeding. Problemy Severa, 13, Moscow.

Barashkova E. A., – 1963a – Seasonal features in the development of forage lichen – reindeer cladonia. Trudy Instituta Selskogo Khozyaistva Krainego Severa. 12, Norilsk.

Barashkova E. A., – 1963b – Speeded up restoration of forage lichen stock as influenced by the growth stimulants. In symp.: Problemy Severa" (North Problems), 7, Moscow.

Besumov F. A. & Preobrazhensky B. V., – 1952 – Pastures use in the reindeer-breeding kolkhoz. Sotsialisticheskoye zhivotnovodstvo. 12, Moscow.

Bogdanovskaya-Gienef I. D., – 1938 – Reindeer pastures and natural conditions of reindeer-breeding on Kolguev island. Trudy Instituta Polarnogo Zemledelia, Zhivotnovodstva i Promyslovogo Khozyaistva. Ser. "Olenevodstvo", 2, Leningrad.

Boguslav I., – 1848 – Bolshaya Zemlya and its economic enterprises. Trudy Volnogo Economicheskogo obtshestva. 2, Petersbourgh.

Dedov A. A., – 1954 – Pastures of the forest reindeer-breeding in the Pomozdinsk region of Komi ASSR. Trudy Komi filiala Akademii Nauk SSSR. 2, Syktyvkar.

Druri I. V., – 1955 – Reindeer-breeding. Moscow-Leningrad.

Dushechkin V. L., – 1937 – Reindeerr pastures in Kharaulakh mountains (Yakutia). Trudy Arcticheskogo Instituta, 63, Leningrad.

Egorov O. V. – 1965 – Wild ungulates of Yakutia. Moscow.

Fedorova O. A., – 1958 – Chemical composition of forage plants, hay and succulent fodder of the Far North. Trudy Instituta Selskogo Khozyaistva Krainego Severa, 6, Tyumen.

Florovskaya E. F., – 1939 – Chemism of the undersnow fodder of reindeer on the winter pastures of Saranpaul reindeer-breeding state farm (Sovkhoz). *Botanicheskyi Zhurnal*, 24, 4.

Glinka D. M., – 1939 – Pasture seasons in reindeer-breeding and the conditions of winter feeding of reindeer in Nenetz territory. Trudy Instituta Polarnogo Zemledelia, Zhivotnovodstva i Promyslovogo Khozyaistva, Ser. "Olenevodstvo", 4, Leningrad.

Govorukhin V. S., – 1933 – Essaz on the reindeer summer pastures vegetation in the tundras if Ob-Tas peninsula. *Zemlevedenie*, 35, I, Moscow.

Govorukhin V. S., – 1949 – Seasonal pastures of reindeer. In the book: "Kalendar prirody SSSR" (Calendar of nature of USSR), 2, Moscow.

Gorodkov B. N., – 1926 – Reindeer pastures in the north of Ural region. Symposium: "Ural" (Urals), 8. Sverdlovsk.

Gorodkov B. N., – 1934 – On the grounds and methods of economic classification and valuation of reindeer pastures. Sovetskaya Botanika, 1, Leningrad.

Gorodkov B. N., – 1935 – Vegetation of the tundra zone of USSR, Moscow-Leningrad.

Gorodkov B. N., – 1936 – Review of lichen increment studies. Sovetskoye Olenevodstvo, 8, Leningrad.

Gorodkov B. N., – 1938 – Vegetation of Arctic and Mountain tundras of the USSR. Rastitelnost SSSR, 1, Moscow.-Leningrad.

Gorodkov B. N., – 1946 – Botanic-geographical essay on the Far North and Arctic regions of the USSR. Uchenye Zapiski Leningradskogo Pedagogicheskogo Instituta imeni Gertsena, 46, Leningrad.

Grigoryev A. A., – 1956 – Subarctic. Moscow.

Gulchak F. N., – 1950 – Influence of stabling and nourishing conditions on the development of reindeer. Sovetskaya Zootekhnia, 6, Moscow.

Igoshina K. N., – 1939 – Growth of forage yagels in the Ural North. Trudy Instituta Polarnogo Zemledelia, Zhivotnovodstva i Promyslovogo Khozyaistva. Ser. "Olenevodstvo", 8, Leningrad.

Karev G. I., – 1956 – Experiment on the introduction of pasture rotation in the reindeer-breeding kolkhozes of Nenetz native territory Bulleten nauchno-teknicheskoi informatsii Instituta Polarnogo Zemledelia, Zhivotnovodstva i Promyslovogo Khozyaistva, 1 Leningrad.

Kovakina V. A., – 1958 – Biological peculiarities of some winter-green plants of the Far North. Botanicheskyi Zhurnal, 43, 9.

Kovakina V. A., – 1967 – Effect of some growth stimulants and microelements on the development of vegetative mass and on winterhardiness of *Eriophorum vaginatum* L. Trudy Instituta Selskogo Khozyaistva Krainego Severa, 14, Norilsk.

Kovakina V. A., – 1968 – Hibernating of higher plants in the forest-tundra on the right bank of Yenisei river. Trudy Instituta Selskogo Khozyaistva Krainego Severa, 15, Krasnoyarsk.

Kurilyuk A. D., – 1969 – Reindeer-breeding of Yakutia. Yakutsk.

Kurilyuk A. D. Rumyantsev V. V., – 1974 – Biological peculiarities of the khargin reindeer. In symposium: "Fisiologicheskie adaptatii k prirodnym faktoram vysokikh shirot ' (Physiology of the adaptation to the nature conditions of high latitudes), Vladivostok.

Lepekhin I. I., – 1805 – Daily notes by the academician Ivan Lepekhin during his journey over different provinces of the Russian state 4, Petersbourg.

Makhaeva L. V., – 1959 – Use of winter pasture fodder plants in the reindeer-breeding of Murmansk region. Problemy Severa, 3, Moscow.

Makhaeva L. V., – 1963 – Sizes of grazing areas necessary for reindeer. Trudy Instityta Selskogo Khozyaistva Krainego Severs, 11, Norilsk.

Mashistova P. A., – 1967 – Experiments of undersowing of winter-green grasses on the slash fired and strongly trampled lichen pastures. Sbornik nauchnykh rabot Murmanskoi olenevodcheskoi opytnoi stantsii, 1, Murmansk.

Middendorf A. F., – 1869 – Journey over the northern and eastern Siberia 2, 5 Siberian fauna. Petersbourg.

Michurin L. N. & Makhaeva L. V., – 1962 – Food of wild reindeer on the Taimyr peninsula. Zoologicheskyi Zhurnal, 41, 12.

Nikolaevsky L. D., – 1951 – Necrobacillosis of reindeer and its control. Moscow.

Perfilyev I. A., – 1931 – Scheme of organization of reindeer pastures inspection in Siberia. Sovetskyi Sever, 1, Moscow.

Pomishin S. B. & Kachan, A. P., – 1972 – Reindeer-khargin, Yakutsk.

Pomishin S. B., – 1975 – From the reindeer-breeding history. In symposium: "Olenevodstvo Vostochnoi Sibiri" (Reindeer-breeding of Eastern Siberia), Yakutsk.

Prakhov H. H., – 1957 – Main elements of vegetation on the Verkhoyan range. Trudy Instituta Biologii Yakutskogo filiala Sibirskogo otdeleniya Akademii Nauk SSSR, 3, Yakutsk.

Pryanishnikov A. V., – 1954 – Tundra grassing. Botanicheskyi Zhurnal, 39, 1.

Pryanishnikov A. V., – 1955 – Meadow formation on the bottom of drained lakes in tundra. *Botanicheskyi Zhurnal*, 40, 3.

Rabotnov T. A., – 1940 – Physico-geographical essay of Aldan territory of the Yakutian ASSR. Materials on the land-water organization of Far North territories., 6, Moscow.

Reindeer pastures of the Northern Territorics. 1, Arkhangelsk, 1931, 2, Leningrad 1933.

Ramensky L. G., – 1938 – Introduction into the complex soil-geographical exploration of lands. Moscow.

Rumyantsev V. V., – 1975 – Comparative data on the winter pastures utilization by different reindeer strains. In symposium: "Olenevodstvo Vostochnoi Sibiri" (Reindeer-breeding of Eastern Siberia, Yakutsk.

Salazkin A. S., – 1937 – Growth rate of forage lichens. Sovetskoye Olenevodstvo, 11, Leningrad.

Salazkin A. S., Sambuk F. V., Polyanskaya O. S. & Pryakhin M. I., – 1936 – Reindeer pastures and vegetation cover of Murmansk territory Trudy Arkticheskogo Instituta, 72, Leningrad.

Sambuk F. V., – 1931 – Methodics of route investigations in tundra pastures. Leningrad.

Smirnova Z. N., – 1938 – Plant associations of Kolguev island. Botanicheskyi Zhurnal, 23, 5–6.

Sochava V. B., – 1932 – Tundrology and yagel pastures exploitation. Symposium on reindeer-breeding, tundra veterinary and zootechny. Moscow.

Sochava V. B., – 1933 – Feeding value of plants in the Far North. *Sovetskaya Botanika*, 3–4, Leningrad.

Sochava V. B., – 1934a – Geobotanical studies in the process of economic arrangements of reindeer-breeding sovkhoses and kolkhoses. Leningrad.

Sochava V. B., – 1934b – Plant associations of the Anabar tundra. *Botanicheskyi Zhurnal*, 19, 3.

Sochava V. B., – 1936 – Green fodder of reindeer in the Yamal North as the source of reindeer mineral nutrition. Sovetskoe Olenevodstvo, 6, Leningrad.

312

Sovetskoe Olenevodstvo (Soviet Reindeer-Breeding). 1 – 1934, 2 – 1933.

Tanfilyev G. L., – 1894 – Along the tundras of Timan Samoyeds in the Summer of 1892. Isvestia Russkogo Geograficheskogo obtshestva, 30.

Temnoev N. I., – 1939 – Wintering of belowground organs in some herbaceousplants in the Far North. Trudy Instituta Polsrnogo Zemledelia, Zhivotnovodstva i Promyslovogo Khozyaistva, Ser. "Olenevodstvo, 4, Leningrad.

Tikhomirov B. A., – 1937 – *Arctophila fulvas* (L.) Trin., its ecological features and feeding value. Trudy Dalnevostochnogo filiala Akademii Nauk SSSR, 2, Moscow.

Tomirdiaro S. V., – 1969 – Problems of the economic utilization of thermokarst on permafrost. Symp.: "Problemy rasvitiya proisvoditelnykh sil Magadanskoi oblasti" (Problems of productive forces development in Magadan region), 1, Magadan.

Vakhtina T. V. .– 1963 – Tundra shrubs and their usage in reindeer-breeding. Trudy Instituta Selskogo Khozyeistva Krainego Severa, 11, Norilsk.

Vasilyev V. N., Reindeer pastures of the Anadyr territory. Trudy Arkticheskogo Instituta., 62, Leningrad.

Yakovlev M. R. & Kurilyuk A. D., – 1969 – Organization of breeding work in the reindeer-breeding of Yakutian ASSR. Symp.: "Plemennaya rabota v severnom olenevodstve" (Breeding work in the northern reindeer-breeding industry), Norilsk.

Yarovoi M. I., – 1939 – Vegetation of Yana river basin and Verkhoyan ridge. Sovetskaya Botanika, 1, Leningrad.

Zubkob A. I., – 1932 – Tundras of Gusinaya Zemlya. Trudy Botanicheskogo Museya Akademii Nauk SSSR, 25, Leningrad.

J. LOOMAN

Contents

316

7.1   Introduction

The North American Great Plains extends from 24°N to 52°N, for a distance of about 3300 km, and from 100°W to the Rocky Mountain range, a distance of about 450 km. The great Plains encompass an area of approximately 2.5 million km² which presently produce large amounts of cereal grains and support large herds of cattle.

Fig. 1. Approximate boundaries of the North American Great Plains; the Canadian Prairie Provinces outlined.

317

The Canadian prairies occupy the northernmost part of the Great Plains, from 49°N to the southern margin of the boreal forests (Fig. 1). This is an area of 525,000 km² about 60% of which is presently cultivated (census of Canada, 1961, 1971). The remaining 40% uncultivated land and 15% of the cultivated land is used for pasture.

This amounts to approximately 260,000 km² of native and cultivated pasture, wooded range and hayland, which support 7,800,000 head of cattle, 3,000,000 horses, and 750,000 sheep, or about 60% of Canada's livestock. In addition, conservative estimates place the numbers of wild ungulates at about 900,000 head grazing and browsing in this area.

7.2   The Area

The Canadian prairies and parklands are located in an area with a pronounced continental climate, but because of the distribution over 4 or 5° of latitude, 20° of longitude, and a general slope, from about 300 m altitude in eastern Manitoba to about

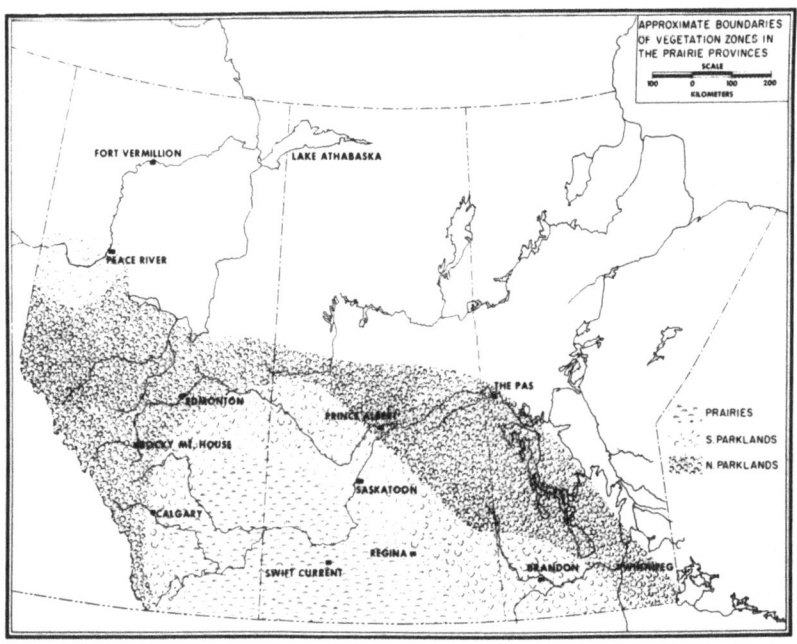

Fig. 2. Approximate boundaries of vegetation zones, based on climatic and vegetational data.

1500 m altitude in western Alberta, considerable variation in climatic conditions, soils and vegetation exist. It is thus possible to divide the area into three more or less distinct zones: the virtually treeless prairies, the southern parklands, where open grasslands predominates over tree groves, and the northern parklands, where grassland openings are relatively small and few (Fig. 2).

7.2.1 CLIMATE

Summers in the prairies and parklands are warm and dry, with temperatures of 35 to 40°C not uncommon. Winters are very cold, with temperatures to −40°C with great regularity during the part of the winter. The main differences in climate between the three zones are due to variations in precipitation, or more accurately, its effectiveness.

Climatic diagrams (Walter 1963) for the three zones are presented in Figs. 3 to 5. Each diagram is a composite of 25 stations, and illustrates conditions in the zone. For comparison a

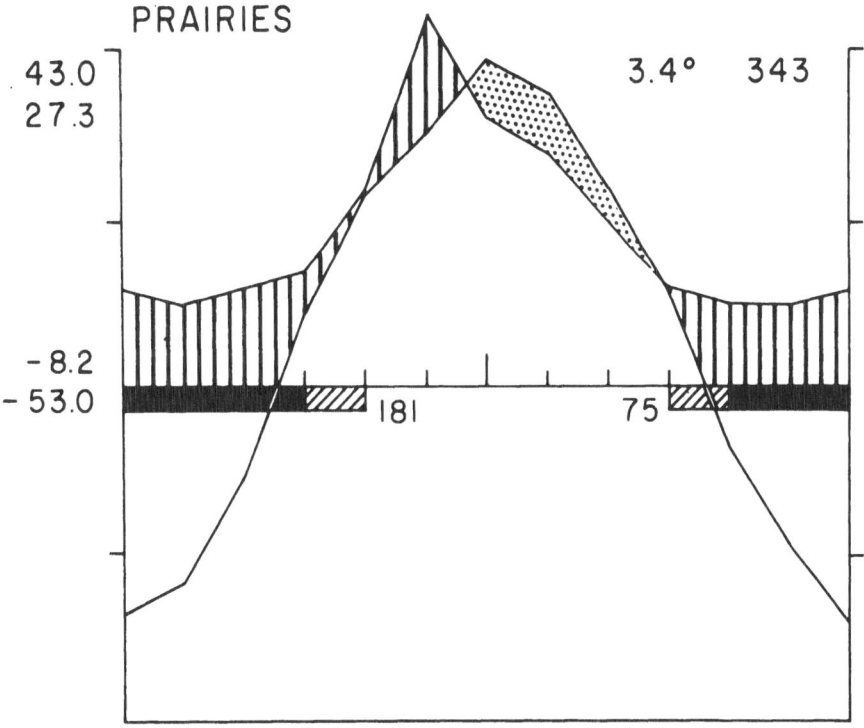

Fig. 3. Climatic diagram for the prairies. Values averaged for 25 stations.

319

climatic diagram for Swift Current, in the prairie zone, is also presented (Fig. 6), which shows that the average has the main characteristics of the heart of the prairies.

A comparison of the diagrams shows that the main differences between the zones are that the prairies have a "drought period" in July to September, which does not occur in either of the parkland zones in average years. In the southern parklands, however, the July to September period shows a strong tendency towards dryness, and in below average precipitation years drought is most likely to occur in this period.

In the northern parklands, the peak of precipitation evident in the other zones in June, is much less pronounced. In fact, of the 25 stations, 13 have more precipitation in June than in July, while at the remaining 12 stations the reverse is true. Also, at five stations the month of August has more precipitation than June, and at another five stations more than July.

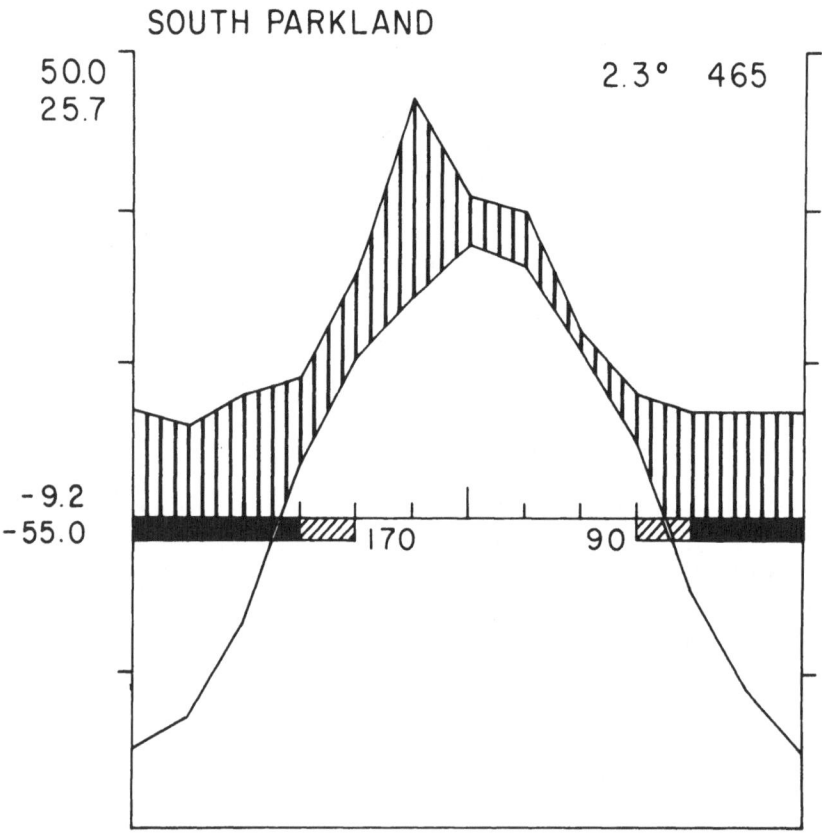

Fig. 4. Climatic diagram for the southern parklands. Values averaged for 25 stations.

In comparison June has more rainfall than July at 19 stations in the southern parklands, and at 23 stations in the prairies. The month of August averages more rainfall than July at seven stations in the prairies, but no station shows more rain in August than in June in this zone. In the southern parklands three stations record more rainfall in August than in June, and at eight other stations in August rainfall exceeds that in July.

In each of the zones temperature and precipitation for a given month vary considerably from station to station. For the months of June, July and August, for example, a temperature difference of 5°C between highest and lowest exists in the parklands, but only about 2.5°C in the prairies. Similarly, high and low precipitation in each of the summer months may be in a ratio of 2:1 in all zones. However, the averages for the 3-month period are very close for both temperature and precipitation in all zones.

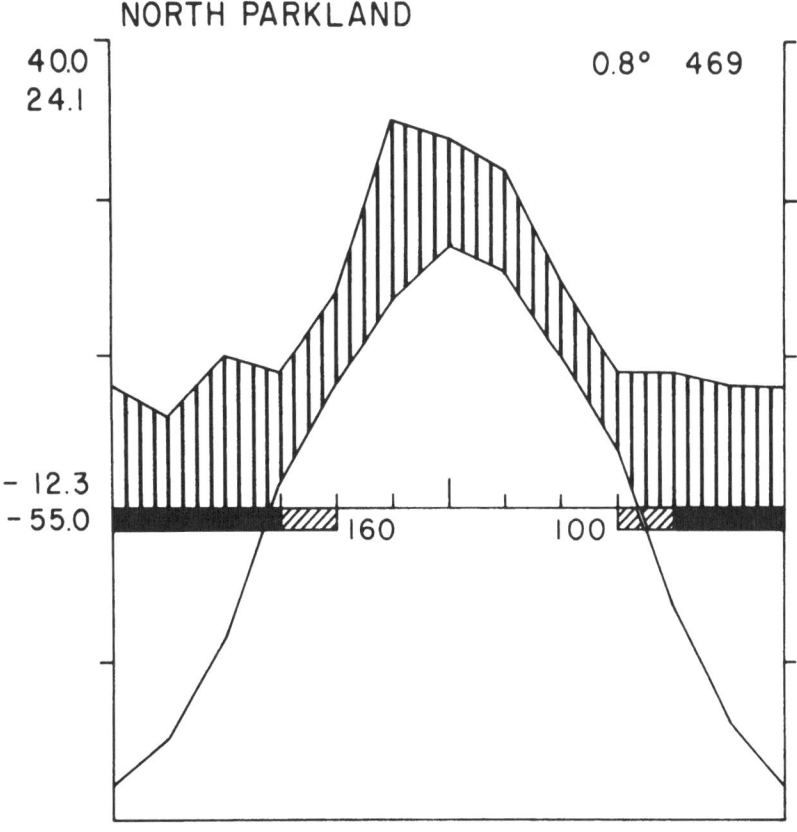

Fig. 5. Climatic diagram for the northern parklands. Values averaged for 25 stations.

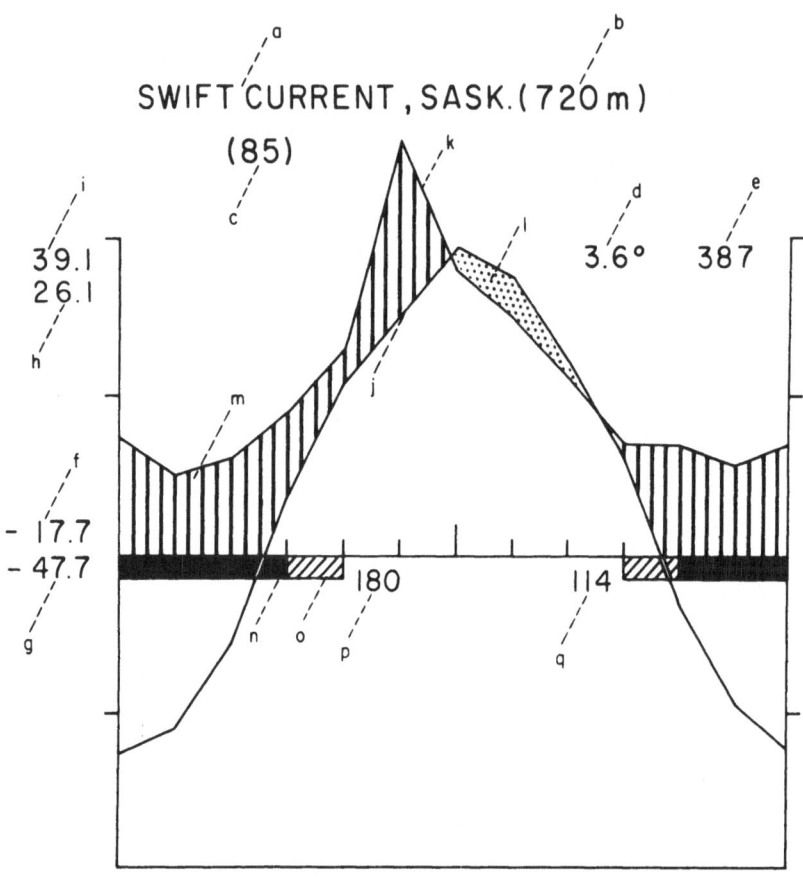

Fig. 6. Climatic diagram for Swift Current, Saskatchewan; a. Station name; b. Altitude in meters; c. Number of years of observation; d. Mean annual temperature in °C; e. Mean annual precipitation in mm; f. Mean daily minimum in coldest mo; g. Absolute minimum; h. Mean daily max. in warmest mo.; i. Absolute maximum; j. Monthly mean temperature in °C; k. Monthly mean precipitation in mm; l. Drought period; m. Humid period; n. Months with daily min. below 0°C; o. Months with absolute minimum below 0°C; p. Average length of growing season; q. Average number of days with measurable precipitation.

7.2.2   SOILS

Virtually the entire area of the Prairie Provinces was glaciated at one time or another during the Pleistocene. Consequently, soils in the area have developed mainly on glacial till, or on bedrock modified by glacial action.

Till deposits range in texture from fine sands to clay loams, and from stone-free to very stony. Many of the deposits are

322

highly calcareous, particularly the loam soils. Modified bedrock is mostly of Tertiary and Cretaceous origin, and includes sands, loams and heavy clays. The loam soils in this category are often very stony, while the clay soils are usually derived from marine shale and tend to be saline.

Postglacial deposits are present as extensive areas of undifferentiated sands, sand dunes, lacustrine and alluvial deposits. In the northern parklands, organic soils occur in the form of peat bogs and muskegs. In the prairies and southern parklands shallow postglacial deposits may overlie marine shales and salinity is common in these situations.

Soils in the Prairie Provinces have been surveyed, classified and mapped, and are described in the various Soil Survey Reports of Alberta, Manitoba and Saskatchewan for selected areas (e.g., Mitchell *et al.* 1944; Ehrlich *et al.* 1953; Odynsky *et al.* 1961).

The main soil types described in the area are chernozemic, with a more or less well developed calcium-rich horizon. These soils are classified in the Brown, Dark Brown, and Black soil zones, developed in the prairies, southern parklands and northern parklands, respectively. There is considerable overlap between the soil zones and the vegetation zones, but in general the correspondence indicated is valid, although much of the Dark Brown soil zone consists of open grassland.

Soils in the area are usually of good fertility. Some parameters for grassland soils are presented in Table I. Samples used for this table were taken in the course of phytosociological stu-

TABLE I.
Soil properties in the Prairie Provinces.

	Clay		Loam		Sand	
	till (24)	coll (17)	till (16)	coll (10)	till (12)	coll (10)
pH	7.64	7.24	7.30	7.20	6.61	6.50
Ec	.71	.65	.60	.67	.42	.36
P	8.2	20.0	10.0	18.2	5.5	22.3
K	100	110	132	153	41	45
Ca	3722	2900	4210	3900	2442	1230
Mg	542	400	421	372	187	54
N	1.6	9.5	4.4	3.5	1.3	1.8

Ec — millimhos/cm; elements = parts per million.

dies and represent grasslands in the three vegetation zones as well as in the soil zones.

In respect to soil fertility, Mitchell *et al* (1944) stated: "Our Prairie and Parkland soils are in general to be considered highly fertile, but their fertility is not inexhaustible . . .".

The data presented in Table I show that nitrate nitrogen is usually in low supply. Phosphorus also is mostly in low supply, except in colluvial deposits, Although potassium is usually plentiful, this element is limited in several localities in the northern parklands, particularly in light soils. Sulphur has also been found to be deficient in some parkland areas.

Continuous agricultural use, which subtracts nutrients from the soil without replenishing the supply, can therefore result in severely lowered fertility, as foreseen by Mitchell *et al.* (1944).

7.2.3   THE VEGETATION

7.2.3.1   *Grasslands*

The grasslands of the prairies and parklands have been described and classified by various authors. Most of these have followed the climatic climax concept (Clements 1928; Weaver & Clements 1938).

Grasslands in Saskatchewan were classified by Clarke *et al.* (1942), who recognized three main types within the grassland climax. The "short-grass prairie" was described as the B o u - t e l o u a - S t i p a, Association; "mixed prairie" as the S t i - p a - A g r o p y r o n - B o u t e l o u a Association, and "submontane mixed prairie" as the F e s t u c a - D a n t h o n i a Association.

Following this classification, Moss (1944) described the "mixed prairie" in Alberta as the S t i p a - B o u t e l o u a Association. Moss remarks that Weaver & Clements (1938) would classify this association as "merely a subdivision or faciation of the mixed prairie association of North America."

The vegetation of the Black Soil zone in Alberta, in particular that of the parklands, was described by Moss (1932) as the "northern prairie", dominated by *Festuca scabrella*.* In southwestern Alberta the "natural and climax prairie" on black soil was described as the F e s t u c a - D a n t h o n i a Association

* Species names according to Boivin (1967).

324

(Moss 1944). Later, the relationship of this association with the northern prairie was investigated in more detail, and the fescue grasslands were described as the F e s t u c a   s c a b r e l l a Association (Moss & Campbell 1947).

A classification of the grasslands of Saskatchewan as subdivision of the mixed prairie or S t i p a - B o u t e l o u a Association, of Weaver & Clements (1938), is given by Coupland (1950). In this classification the associations of Clark *et al.* (1942) and Moss (1944) are reduced to faciations.

The S t i p a - B o u t e l o u a ˙ faciation is considered the "characteristic climax type" on medium textured soils of the moisture brown and dry dark brown soils. On soils of medium texture in the drier parts of the Brown soil zone the B o u t e - l o u a - S t i p a faciation is the characteristic climax vegetation.

Soils of medium texture on undulating to gently rolling topography in the Dark Brown soil zone are occupied by the S t i p a - A g r o p y r o n faciation, while the A g r o p y - r o n - K o e l e r i a faciation occurs on uniform clay deposits, as the beds of former glacial lakes.

Solonetzic soils on clay loam support stands of the fifth faciation, B o u t e l o u a - A g r o p y r o n .

In addition to these five faciations, Coupland describes the A g r p y r o n - M u h l e n b e r g i a facies, and the A g r o p y r o   s m i t h i i consocies. The former occurs on eroded areas of rolling topography, the latter on slightly alkaline areas in clay flats, and is associated with the A g r o p y r o n - K o e l e r i a faciation.

The fescue grasslands of Saskatchewan are described as the F e s t u c a   s c a b r e l l a association (Moss & Campbell 1947) by Coupland & Brayshaw (1953). Thus, while the "mixed prairie" of the Canadian Plains is considered a subdivision of the Great Plains grasslands, the fescue grasslands are recognized as a separate climatic climax. The five faciations, and the fescue prairie association are confirmed by Coupland (1961).

Grasslands in the Peace River region of northwestern Alberta are considered to be a separate climax association by Moss (1952), who described the A g r o p y r o n - S t i p a - C a r - e x community with the A g r o p y r o n - S t i p a faciation as the most common type. However, Moss considers this grassland as an edaphic, rather than a climatic climax.

Classification of grasslands in the Prairie Provinces in accordance with phytosociological methods of the Zurich-Montpellier school (Braun-Blanquet 1951) has been presented by Looman

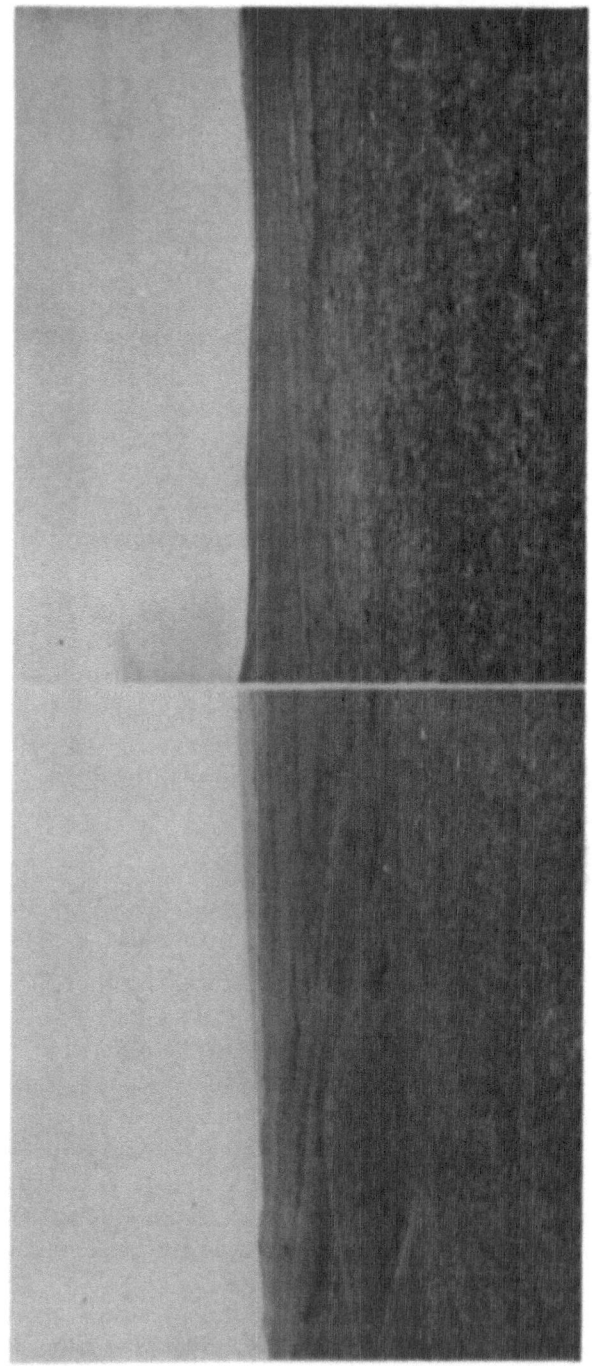

Fig. 7. Grassland in the B o u t e l o u i o n g r a c i l i s ; at left with e r i o g o g o - n e t o s u m phases on hills in background, at right typical A s t r a g a l e t u m p e c t i n a t i .

Fig. 8. Grassland in the Stipion curtisetae ; at left looking north from Cypress Hills plateau, strip-farming in background, at right looking south over ravine with Salix community.

(1963, 1969). In this classification the faciations in the *Stipa* dominated grasslands, recognized by Coupland (1950, 1961) and Moss (1944, 1952) are included in a class A s t r a g a l e t o - S t i p e t a . This class includes the mixed prairie in the prairie zone on brown soils in order A s t r a g a l e t o - S t i p e t a - l i a   c o m a t a e   (Fig. 7), and the mixed prairie of the dark brown soils in the prairie and southern parkland zones in order S t i p e t a l i a   c u r t i s e t a e   (Fig. 8).

Most of the "dry prairie" is included in the alliance   B o u - t e l o u i o n   g r a c i l i s   but a separate alliance is recognized for the communities occurring on eroded slopes and crests, the E r i o g o n i o n   f l a v i . This alliance is considered "pioneer vegetation."

The fescue grasslands were originally placed with the grass-lands in mesic to wet soils as the order D a n t h o n i e t a l i a i n t e r m e d i a e   in   the   class C a l a m a g r o s t e t o - D a n t h o n i e t e a . These grasslands are now considered to form the order F e s t u c o - S t i p e t a l i a   in the A s t r a - g a l e t o - S t i p e t e a   (Fig. 9).

While none of the workers in the Clementsian school separates the grasslands on sandy soils, Looman (1963) places these grasslands   in   the   class C a l a m o v i l f e t e a   l o n g i f o - l i a e ,   in which the association R u m e c e t u m   v e n o - s i   is described as pioneer vegetation on unconsolidated sand dunes.

Grasslands in Manitoba can be included in part in the   A s - t r a g a g a l e t o - S t i p e t a l i a   c o m a t a e , S t i p e t a l i a c u r t i s e t a e   and F e s t u c o - S t i p e t a l i a .   However, very little is left of the "true" and "tall grass" prairie, and clas-sification of the few remnants as the   " S t i p i o n   s p a r - t e a e "   is still provisional (Looman 1965).
1965).

A summation of the Manitoba grasslands, more or less in accord with the Clementsian classification, is given by Bird (1961). Remnants of the tall grass prairie are described as the A g r o p y r o n -   P o a - S p a r t i n a   grasslands   and   S t i - p a - A n d r o p o g o n   sand prairie. Fairly large stands of the latter type still exist in presently protected areas.

7.2.3.2   *Woodlands*

Woods of the parklands are described by Bird (1961), who recognizes six woodlands communities, which are named by the common names of the dominant trees, but will be designated by the latin names here. The most important woodland type is the Populus tremuloides community, which is common throughout the Prairie Provinces. Bird states that the "aspen poplar grows in pure stands except in poorly drained soil . . .", but in most northern parkland areas *Populus balsamifere*, as well as *Betula papyrifera* are more or less important admixtures. Bird considers small trees, as *Prunus* spp., in the ". . . well-marked shrub . . . stratum."

The flood-plain community Acer negundo - Ulmus americana - Fraxinus pennsylvanica occurs along the rivers in eastern Manitoba, but is lacking in the western parts. However, variants of this community occur in large ravines and deep coulees throughout western Manitoba, southern Saskatchewan and southeastern Alberta.

Unique to southern Manitoba and southeastern Saskatchewan is the Quercus macrocarpa community. The farthest western penetration of this community is along the As-

Fig. 9. In the foothills of the Rocky Mountains Festucetum scabrellae grassland is often converted to tame hayland where possible, and used for grazing on slopes too steep or stony for plowing.

siniboine-Qu'Appelle river system in Saskatchewan. Understories of these communities are listed in Table II.

Widespread throughout the parklands are the S a - l i x and L a r i x L a r i c i n a - P i c e a m a r i a n a communities. The S a l i x communities may have as few as two or three to as many as seven species of willows, as well as *Alnus* spp., *Cornus stolonifera*, and other shrubs. The L a r i x - P i c e a community usually has an admixture of shrubby growth with *Salix* spp., *Betula* spp., *Ledum groenlandicum*, and a thick layer of mosses in which *Sphagnum* spp. predominate. Several *Carex* spp., and in the drier muskegs, lichens, with the C l a - d o n i e t u m a l p e s t r i s (Looman 1964) commonly occurring in the ground layer.

TABLE II.
Understory in woodland communities.

	Aspen poplar	Maple ash	Bur oak
Shrubs			
Cornus stolonifera	x	x	x
Corylus americana	x	x	x
Corylus cornuta	x	x	x
Viburnum opulus	x	x	(x)
Prunus virginiana	x	(x)	(x)
Prunus pensylvanica	x	(x)	(x)
Amelanchier alnifolia	x	(x)	(x)
Shepherdia canadensis	x	(x)	(x)
Humulus lupulus	—	x	x
Herbs:			
Actaea rubra	x	—	—
Aralia nudicaulis	x	—	—
Aster ciliolatus	x	x	x
Aster conspicuus	x	x	x
Cornus canadensis	x	x	x
Maianthemum canadense	x	x	x
Pyrola asarifolia	x	x	x
Pyrola secunda	x	x	x
Arctostaphylos uva-ursi	—	—	x
Lathyrus ochroleucus	x	—	x
Vicia americana	x	—	x
Astragalus frigidus	x	—	—
Osmorhiza spp.	x	—	—
Sanicula marilandica	x	—	—

A local community is the P i c e a g l a u c a Sandhill community, an outlier of the northern spruce forests, occurring in the Spruce Woods Forest Reserve in south-central Manitoba. This community is probably edaphically determined and forms a "parkland" with areas of the S t i p a - A n d r o p o g o n prairie.

7.3 History of the Great Plains

7.3.1 ORIGIN

The vegetation of the Great Plains was considered to be a "climatic climax" by Clements (1928). Though the prairies were grazed by large herds of ungulates, and fires − both natural and man-made − were frequent, Clements & Chaney (1936) state that these influences were not causal in the origin of the North American prairies.

Uplift of the Rocky Mountains, and subsequent changes in climate as a result of the "rainshadow" east of the mountain range, are considered to be the major cause of the "climatic climax prairies" by Weaver & Clements (1938). This establishes an age of 20 to 30 million years for the Great Plains grasslands. Although the distribution of plant species is determined firstly by climate (Good 1964), the climate is not the only determinant of plant distribution, and hence vegetation. Soil conditions, use of the vegetation by herbivores, fires, floods and other often catastrophic events exert their influence.

7.3.2 HISTORICAL DEVELOPMENT

In the Great Plains these influences have played important roles in the shaping of the vegetation, but in its history three events stand out in particular, and make it unlikely that the present-day vegetation is a climatic climax, dating several millions of years back.

7.3.2.1 *The Pleistocene*

The first event was the Pleistocene, lasting some two million years, and covering the entire northern Great Plains under a great ice sheet (Flint 1955, 1971). The climatic changes brought about by this glaciation were felt far south, and at the end of the Pleistocene the vegetation of the southern Great Plains had a distinctly cold-temperate character. Thus, the vegetation of the Llano Estacado included coniferous species whose present distribution limits lie at much higher altitudes or much greater latitude (Wendorf 1961).

7.3.2.2 *The Buffalo*

The second event was the arrival of the buffalo, *Bison bison*, on the North American continent via land bridges existing during the Pleistocene. This arrival dates probably back to the Illinoian glaciation, the presumed counterpart of the Riss or Saale of Europe, about 300,000 B.P. (Flint 1971).

Remains of bison, possibly *B. antiquus* or *B. latifrons*, have been found in New Mexico, and date back to 20,000 B.P. These remains were contemporaneous with several other mammals: mammoth, mastodon, *Camelops* and *Equus* among them. In finds of later dates, 10,000 to 8,000 B.P., *Bison antiquus* was contemporaneous with *B. bison*. Remains of the latter in these finds were more numerous than those of the former (Wendorf 1961). This may indicate that the modern buffalo replaced its predecessors, including the mammoth and other mammals, towards the end of the Pleistocene.

7.3.2.3 *Man*

The third event is the arrival of man in North America. Like the buffalo, he probably crossed the "Beringia landbridge" in Pleistocene times, but more recently than the bison. Although there is no certainty about the date, it is considered possible that man may have arrived during the Wisconsin glaciation as early as 60,000 B.P. (Flint 1955).

Fossil finds are all much more recent, in the order of 15 000 to 10,000 B.P., as indicated by C^{14} dating (Wendorf 1961; Flint 1971). However, although artifacts of the Folsum culture of this age are the oldest known, there are indications of pre-Folsum cultures. These ancient cultures lived by hunting, with mastodon, mammoth, bison, and other now extinct mammals as their quarry (Flint 1955). Early man may thus have been instrumental in the disappearance of these animals.

332

7.3.2.4   *Changes in the Great Plains*

7.3.2.4.1   The Southern Plains

Far from having been a climax for millions of years, the prairies of the Great Plains have been subject to repeated changes. Perhaps the most drastic of these changes has occurred since the Pleistocene, and possibly in the last few hundred years. For, as the available evidence indicates, both Indian and buffalo became important influences only recently.

The climatic changes caused by glaciation of the northern Great Plains as well as in the Rocky Mountains, extended far south. The Llano Estacado, at 32°-35°N latitude, had a climate about 10°C cooler than today during the Tahoka Pluvial 22,500-15,000 B.P. The vegetation at that time was largery coniferous.

Extensive data are available on the more recent San Jon Pluvial, 13 000-6000 B.P. This pluvial has been subdivided into three subpluvials: Blackwater subpluvial, 13,000-12,000 B.P. The fauna included mammoth, horse, bison, *Capromeryx*; Lubbock subpluvial, 10,000 B.P., with a fauna including *Bison antiquus* and *Capromeryx*; Portales subpluvial 8,000 B.P., with *B. antiquus*, possibly *B. occidentalis*, and *B. bison*.

During the San Jon Pluvial and the preceding Monahan Interpluvial the coniferous woodlands of the Tahoka Pluvial were gradually replaced by parkland and grassland. However, the climate during this pluvial was still colder and wetter than that of the present (Wendorf 1961).

The faunal sequence from about 15,000 B.P. to the end of the San Jon Pluvial about 6,000 B.P. is one from a relative rich fauna in which the modern buffalo was either absent or very rare to one with few species in which the buffalo was common. Concomitantly, the vegetation changed from sub-boreal woods to open parkland.

This suggests a warming trend and recession of the glaciers during the Monahans Interpluvial, followed by a series of re-advances and recessions during the San Jon pluvial. During the final recession, following the Portales subpluvial, the ice retreated to approximately the present situation.

This sequence of advances and recessions is summarized by Wendorf (1961).

Lake Arkona recession	14,000-13,500 B.P.	
Mankota readvance	13,000	B.P.

Two creeks recession	12,000-11,000 B.P.	
Valders readvance	10,700	B.P.
Cochrane readvance	6,700	B.P.

7.3.2.4.2 The Northern Plains

During the early part of this sequence the Prairie Provinces were wholly or almost wholly covered by the ice sheet. In several areas of the southern Prairie Provinces buried remains of trees, including spruce, have been found together with mosses, insects and snails. Pollen analyses and C^{14} dating place the age of these finds at about 11,000 B.P. (Ritchie 1964; Ritchie & de Vries 1964; de Vries & Bird 1965).

The existence of spruce forests in the region at that time is therefore indicated, and these forests were buried by a readvance of the ice sheet correlative with the Valders in eastern North America. This places the front of the ice at about 50°N latitude, or little more, about 10,000 B.P. (Christiansen 1961, 1965).

Since the speed with which the ice retreated is estimated at 90 to 100 m per year (Christiansen 1956), a recession to about 53°N, and a re-advance correlative with the eastern Cochrane, to about 52°-30'N along the 108th meridian at 6,700 B.P. seems possible.

Assuming the vegetational changes after recession of the northern Great Plains were similar to those in the Llano Estacado, a change from coniferous forest to parkland would have occurred at some distance south of the ice front. The time required for this change — about 7000 yr in the Llano Estacado, would be influenced in the Prairie Provinces firstly by location, and secondly by further climatic changes after 6,700 B.P.

The distance of the Llano Estacado from the northern ice front was about 12 to 15° latitude, but the distance from probably glaciated mountains was only 150 to 400 km. Altitudes are comparable for the western part of the prairie Provinces, but are less elevated in the eastern parts. On an average, however, conditions in the western areas of the northern Great Plains might have shown a pattern similar to that in the Llano Estacado, while in the central and eastern parts the pattern would have differed in tempo of development.

On this basis, and in view of the fact that the climate of the northern Great Plains is rather similar to that assumed for about 7,000 B.P., it might be expected that the southern part of the

334

Prairie Provinces should be parkland with coniferous forest at higher elevations, if not entirely covered by spruce forest. That this is not so would indicate a strong influence on the postglacial development of the vegetation by factors other than climate. These factors may safely be assumed to have been man and wildlife.

7.3.2.5 *Indian Culture*

Early Indian cultures have been traced in many areas, and appear to have moved north with the gradual retreat of the ice. Most of these early cultures were sedentary, with primitive agriculture as the major source of subsistence. Amongst the crops raised by many tribes were corn, beans, squash and tobacco.

From New Mexico, through the midwestern USA to southern Manitoba evidence of several cultures has been found. This includes the so-called cliff dwellers and Pueblo Indians of the southwestern USA who were mainly farmers (Quimby 1968), the mound and Effigy mound builders (Stuart 1972; Thro 1973; Curtis 1959), and the Whiteshell in southern Manitoba (Bird 1961).

The movement northward was on the eastern flank of the Great Plains. Apparently the immigrants arriving via the Beringia land bridge moved along the coast to the southwestern USA leaving very little evidence of settlement along the way. Settlement of the Great Plains after retreat of the ice seems to have been entirely from the south and east. Moreover, there is considerable evidence that this movement was largely under pressure of settlement of the Atlantic regions by white immigrants (Roe 1951; Curtis 1959).

The early cultures span a period of about 7500 yr, from about 6000 B.C. to 1500 A.D. (Flint 1955) spelled the end of glaciation and the beginning of considerable human activity. During this span of 2,000 yr forest gave way to savannah, and finally to large expanses of grassland.

Since the Climatic Optimum, or Hypsithermal Interval (Flint 1971), the climate has become cooler again but human activities, grazing by herds of buffalo, and other ungulates, and fires exerted influence on the vegetation. The adverse effects of the warm interval, shown in the separation of Indian cultures into strata separated by windblown material (Malin 1957), were thus continued. There are indications that the Indian cultures had entered a period of decline before the Conquistadores arrived (Malin 1967, 1956; Roe 1951).

The nomadic and seminomadic tribes of the northern Great Plains are therefore in all likelihood a relatively recent phenomenon. Although the main means of subsistence of these Indians, as the Blackfoot, Assiniboine and Blood, was hunting, gathering of plants for food or other uses was also important.

The Blackfoot Indian utilized more than 200 plant species (Johnston 1970), several of which were used for food. Thus, *Amelanchier alnifolia* was gathered in large quantities, as were *Prunus virginiana, Shepherdia argentea* and *Cornus stolonifera*. These fruits were used to eat out-of-hand, as flavoring of soups, or in pemmican (Grinnell 1962).

Roots of several plants were also used as food, e.g., *Psoralea esculenta* was dug up in great quantities, as were the bulbs of several species of *Allium* and in particular of *Camassia quamash*. Although grasses were used for bedding, there is little evidence that seeds were used for food, despite the assertions of several writers that *Oryzopsis hymenoides* was gathered for this purpose (Johnston 1970). The exception is *Zizania aquatica* in the eastern fringe of the plains.

7.3.2.6   The Ungulates

Food gathering activities of the various Indian tribes undoubtedly exerted a considerable influence on the vegetation. But the effects of this influence must have been minor compared to those resulting from the grazing of the buffalo and other ungulates. This is particularly so if the possibility is considered that the decline of sedentary Indian cultures coincided with a rapid expansion of wildlife herds.

It is estimated that towards the end of the 18th century the Great Plains supported 60,000,000 buffalo, equal numbers of *Antilocapra americana** and *Odocoileus virginianus*, and at most 10,000,000 each of *Odocoileus hemionus* and *Cervus canadensis*. In terms of modern range management, these herds were equal to at least 100,000,000 animal units, with one A.U. equal to a 450-kg cow. This amounts to a stocking rate of about 2,5 ha/ A.U.

---

* Species names according to Bird 1961.

Allowing for the different grazing and browsing habits of the various species, and allowing for the greater productivity of the eastern fringes of the Great Plains, the grazing pressure must have resulted in serious damage to the vegetation in large areas. Together with the influence exerted by man, particularly through the practice of burning prairie and woodlands, the effects on the grassland have been very great.

If it is true that the change from sedentary culture to that of a nomadic or seminomadic way of life was forced upon the Indian and coincided with a period of expansion of the buffalo herds, the implication is that these herds were not very large in the 15th and 16th centuries. This further implies that rapid expansion of the herds occurred from the 18th to the late 19th century. As Roe (1951) states: "To assume that between 1540 and, say 1750 overcrowding for the first time reached an acute stage, would, in effect, be to postulate an incredibly recent date for their appearance . . . as the historic buffalo . . . in their historic habitat."

It is not necessary to postulate "an incredibly recent date" for the appearance of the buffalo. What is necessary is to reevaluate available information, to abandon the notion that Indian and buffalo have "always been a part of the environment", and to view the Great Plains grasslands not as a climatic climax, but as the result of the conjunction of the influences of climate, Indian and buffalo.

Early white travellers refer to herds seen as counting "hundreds" of animals (Roe 1951); the estimates of hundreds of thousands or even millions date from the late 18th and early 19th centuries. Indian folklore often refers to starvation, e.g., in the folklore of the nomadic Blackfoot Indians, because ". . . the buffalo was scarce at that time . . .". Several of the Indian stories also make it clear that herds were not counted in thousands, but hundred of animals or less (Grinnell 1962).

The rapid increase in numbers of buffalo and other wildlife is not difficult to explain. Compared to natural enemies, bear, cougar, and wolf, the Indian was an inefficient hunter until he acquired the horse. To make up for his relatively poor physical equipment as a hunter, the Indian used fire to create lush pasture and thus attract the herds and keep them in the vicinity, or to surround the herd and flush game out of woods.

Besides the destruction of woods, and enlarging the areas of grassland, burning also destroyed large numbers of birds and small game (Bird 1961). In addition, large numbers of predators

337

were destroyed or forced to migrate to other areas. The combined effect of increased availability of open grassland, easier escape from predators and decreased numbers of predators meant a greater rate of survival.

If it is assumed that the shift from a sedentary way of life to nomadism occurred in the southern Great Plains as a result of crop failures during the Climatic Optimum, the time may have been the interval corresponding with the Blytt-Sernander Subboreal phase, about 1000 B.C. The Indian population of the Great Plains has been estimated at about 750,000 prior to the coming of the white man (Roe 1951).

The following assumptions can be made to explain the rapid increase in wildlife numbers. Firstly, the Indian population remained stable at 750,000 persons. Secondly, the consumption of buffalo meat per person per day was 500 g, so that 500,000 animals per year were needed. Thirdly, these requirements were equal to one-half of deaths by all causes. Fourthly, the average calf crop was about 40% of herd numbers.

These assumptions then lead to the conclusion that at the end of the Climatic Optimum at least 2,500,000 buffalo were needed to sustain the annual losses. Failure of crops, and a greater need for meat may have led to the extermination of several species of game, and a reduction in buffalo numbers to 500,000 animals by 1000 B.C.

The decrease of the herds was halted through the expansion of the grasslands, and was gradually reversed during the cooler and moister sub-Atlantic. If at first the excess of births over deaths was 0.1%, with a gradual increase to 0.5%, the herds could have increased to 60 million animals by 1750 A.D. The number of buffalo in 1540 A.D. would have been about 20 million, and early Spanish travellers may well have encountered sizeable herds. Since the numbers of other large game would have increased in a similar manner, the abundance of game remarked upon by many of the early travellers (Roe 1951) can also be accounted for.

7.3.2.6.1 Decline of the Buffalo Herds

The assumed rate of increase is small enough that the balance could rapidly swing to a decrease when the Indian became a more efficient hunter with horse and firearms. And, when the white man's slaughter for profit began, the days of the buffalo herds were counted. In the southern area one to two thousand white men went shooting buffalo with an average kill of 10 animals per day per man (Gard 1972). About 4,500,000 animals were killed in this area in 1872–74 (Roe 1951).

If the assumptions regarding the increase are correct, the large herds which grazed in densities of up to 50 or so animals per hectare (Roe 1951) were actually short-lived, and existed for perhaps 50 to 100 years. However, this brief period would have been quite sufficient to thoroughly change the vegetation from, presumably, parkland to "bald prairie"

7.4 The Period of Settlement

Extermination of the buffalo, and greatly diminished numbers of other game, did not result in recovery of the vegetation to its original state. Even before the great slaughter began, domestic cattle were replacing the buffalo where settlements had been established. Ranching took over immediately wherever the buffalo had been exterminated.

In the last decades of the 19th century ranching had become established from the southern USA to southern Canada (Atherton 1972), and large areas of the prairies were being overgrazed by domestic cattle.

7.4.1 SETTLEMENT IN THE CANADIAN PRAIRIES

The Canadian Prairies occupy a unique position in the Great Plains. Firstly, the area is entirely postglacial, and secondly, there was a time lag between extermination of the buffalo and introduction of domestic cattle. As a result, large areas once considered "buffalo country" by early travellers, became wooded after the buffalo was gone and burning ceased.

Thus, while buffalo were numerous along Lake Manitoba and in the Swan River, Manitoba area and the region from Carlton, Saskatchewan to Fort Pitt, Saskatchewan was ". . . country proverbial then and later (1846) for its unfailing plenty (of buffalo) . . . ." (Roe 1951). These areas are now well wooded and in places densely so. The author has met early settlers in these areas who can still remember that much of this land was open grassland.

In addition to the somewhat slower tempo of settlement, the development of agriculture in the Prairie Provinces has been substained almost from the beginning by the establishment of agricultural Research Stations. The first Dominion Experimental Farm in the Prairie was established at Brandon, Manitoba, and Indian Head, Saskatchewan in 1886, and at Lethbridge, Alberta in 1906 (Anon 1939).

Although research in these establishments was primarily aimed at farming, some attention was given to animal husbandry. The growing of forage crops to supplement native pasture was encouraged. In support of this effort plant breeding to obtain improved strains of introduced forage crops, as *Melilotus* spp. and *Medicago* spp. was one of the main contributions of the Experimental Farms Service in the early period.

At the Lethbridge station research programs were initiated for the management of native range. As a centre for irrigation research, this station offered excellent opportunities to test various forage crops as supplemental fodder. In these tests native range was used for summer grazing, and the forage crops for winter feeding and finishing.

7.4.2.1   *Range Research*

In 1926 a Range Experimental Station was established at Manyberries, Alberta, in the middle of the "short-grass" plains. The main objective of the research program here was to find ways of regrassing abandoned farmland. Many farms, established in the dry areas of Alberta and Saskatchewan had failed as a result of drought and severe soil erosion.

Most of the abandoned farmland had been in short-grass prairie before being plowed, and detailed botanical surveys were made of the various prairie types. This survey was made to obtain information on the carrying capacity of the native range, forage quality and agricultural (Clarke & Tisdale 1936). Thus, it became possible to compare the gains of cattle on native range with those on "tame" pasture, and select the best varieties of forage and management practices.

In 1936 most of the range research work done at Manyberries shifted to Swift Current, Saskatchewan where an Experimental Farm had been established in 1920 (Campbell 1971). This was primarily as a result of the development of government owned and operated community pastures.

7.4.2.2   *Community Pastures*

Several community pastures had been established in Saskatchewan in 1921. When in 1935 the Government of Canada brought in the Prairie Farm Rehabilitation Act (P.F.R.A.) to alleviate the results of drought and farm abandonment, the community pasture program in this Act utilized the provincial experience. This was, to a large extent, possible through the research work done at Manyberries and Swift Current, and over the years the P.F.R.A. made extensive use of this research work as well as of the researchers.

This arrangement was of mutual benefit. Early research work was rather local in character (Clarke 1930; Clarke & Tisdale 1936), or aimed at specific aspects of the short-grass prairie (Clarke *et al.* 1943; Clarke & Tisdale 1945). Cooperation with the P.F.R.A. offered the opportunity to expand the scope as well as the area of investigation.

Thus, it became possible to apply regrassing techniques on a large scale and in a wider area than before (Clarke & Heinrichs 1941). Ecological and botanical studies led to the evaluation of the vegetation as well as individual plant species (Clarke *et al.* 1942; Clarke *et al.* 1950), and formed the basis for range management (Campbell *et al.* 1962).

7.5   The Present

7.5.1   SYNOPSIS

Range management research is presently aimed at increasing the carrying capacity of the area available for cattle production. This can be achieved only in one of two ways. The one way is to expand the area available by seeding submarginal land, now used for wheat production, to forage crops. The second way is to convert low yielding native range to tame forages.

Native grassland has a low production capability. The need for increased beef production, with a prognosis for an increase from the present 8,000,000 to 11,000,000 by 1980, must therefore be satisfied by applying one or both methods of expanding the available carrying capacity.

7.5.2 PRODUCTIVITY OF THE NATIVE RANGE

7.5.2.1 *Carrying Capacity*

Carrying capacity is usually expressed as the area of range needed to support one animal unit (A.U.)., i.e., one mature cow of 450 kg. Since the amount of feed required to support one A.U. for one month (A.U.M.) is 300 kg, the carrying capacity of range can be expressed as "unit area needed per A.U.M."

The productivity, and hence the carrying capacity, of grassland can be established in various ways (Brown 1954; Campbell 1969). In most studies on carrying capacity in the Prairie Provinces the "point method" (Levy & Madden 1933) has been used (Campbell *et al.* 1962; Lodge *et al.* 1971). In wooded areas this method is not applicable and quadrats have been used by this author to determine yields in the parklands in woods as well as in grasslands.

7.5.2.2 *Carry-over*

Early studies on carrying capacity have shown that it is desirable to leave a "carry-over" of unused production. This carry-over serves several purposes. Firstly, it allows a part of the plants to take part in the reproductive process; secondly, these plants remain in a healthy condition; thirdly, leaving carry-over forms a reserve of feed in years of low production.

The amount of carry-over which insures a healthy range condition over the years is generally taken as 45% of the current year's production, so that 55% can be used. Hence the average carrying capacity of range can be expressed:

$$\text{ha/AUM} = \frac{300}{.55 \times \text{avg. annual production}}.$$

This basic formula can be extended to determine the area needed during the grazing season, i.e., ha/A.U.S., by multiplying 300 by the number of months in the season. In the prairie zone this is about seven months, while in the parklands it is four or five months.

The carrying capacity thus determined is the long-term carrying capacity since it is based on average annual production. The actual production in any given year depends to a large extent on precipitation. For the short-grass prairie the period April-July determines the yield (Lodge *et al.* 1971).

342

7.5.2.3   *Variability of Forage Production*

The value of carry-over may be realized when the long-term precipitation trends are considered in relation to carrying capacity. Seasonal precipitation accounts for about 80% of the variation in annual yield, as shown by the correlation coefficient of the regression equation for annual production of mixed prairie at Swift Current (Lodge *et al.* 1971). Soil moisture reserves, and hence total precipitation in the previous year(s) probably account for much of the remaining variation.

Precipitation data for the Research Station at Swift Current, Saskatchewan show a long-term average annual precipitation of 365 mm. However, over a period of 85 years only 28% of these years had 330 to 400 mm precipitation; 29% of the years exceeded 400 mm, but 43% of the years brought less than 330 mm. Moreover, the average for the years with high rainfall was 475 mm or 110 mm above average, while the average for the dry years was 240 mm or 125 mm below average.

Data for seasonal precipitation show an average of 170 mm precipitation for May, June and July. During the 85-yr period 44% of the years had less than 150 mm, and 37% more than 190 mm seasonal precipitation. In the dry years the average was 105 mm for a deficit of 65 mm, while in the wet years the average was 240 mm for an excess of 70 mm.

Chances that the season will start with a soil moisture deficit are somewhat greater than those for favorable moisture conditions. Furthermore, in an average decade 45% of the years will be deficient in seasonal precipitation, 15% will be average, and 40% above average.

7.5.2.4   *Pasture Management*

Since good pasture management involves stocking of the range according to the carrying capacity in average years, carry-over will be available in dry years. An average year, grazed at 55% of its production will provide enough carry-over for two average dry years, and the range will recover in two average wet years.

In balance, therefore, the adjustments to be made in the stocking rates of native range are minimal if the average carrying capacity is the basis of range management. However, it is evident that the low productive capability of the native range severely limits the livestock numbers which it can support. Some gener-

TABLE III.
Production and carrying capacity of native range.

Prairie type		Alliances	Yield kg/ha	ha/A.U.M.
–	1.	Rumicion venosi	200	–
–	2.	Eriogonion flavi	100	–
Stipa-Bouteloua	3.	Boutelouion gracilis	430	1.5
S-B-Agropyron	4.	Stipion curtisetae	600	1.0
Festuca scabrella	5.	Danthonion parryi	800	0.7
	6.	Festucion scabrellae	600	1.0
Aspen grove	7.	Populus tremuloides all.	200	2.2
„ „		„ „	250	2.0
„ „		„ „	450	1.1
„ „	8.	P.t.-P. balsamifera all.	900	0.6
„ „		„ „	850	0.7
True tallgrass prairie	9.	Stipion sparteae	1200	0.6
–	10.	Muhlenbergion squarrosae	600	1.0

alized data in carrying capacity of native range in good to excellent condition are presented in Table III. Overgrazing can rapidly reduce the carrying capacity and on range in poor overgrazed condition more than twice the area given in the table may be required.

Carrying capacity is not only determined by vegetation type and range condition, but also by such factors as topography and soil type. Thus, although the Rumicion venosi and Eriogonion flavi communities have a nominal carrying capacity, grazing at a lighter rate than this capacity soon results in severe erosion. Hence, areas occupied by these vegetation types are not considered in the calculation of carrying capacity in large pastures.

7.5.2.4.1 Overgrazing

Overgrazing is not always the result of stocking above carrying capacity, and good range management involves more than leaving the recommended carry-over. One of the main factors to be considered is the time at which to commence spring grazing.

In most plant communities only six or seven grasses, one or two sedges, and a few palatable herbs form 90 to 95% of the total production. The most important of these species, their yield based on 100% density, yield rating in relation to *Stipa comata*,

344

TABLE IV.

Ratings of species in various communities; numbering as in Table III alliances.

Name	Yield 100%	Yield factor	Value rating	Communities
Agropyron dasystachyum	7,200	1.28	6	1,2,3
Agropyron smithii	7,200	1.28	6	2,3,4,10
Agropyron subsecundum	7,200	1.28	6	4,5,6,7,8,9
Agropyron trachycaulum	7,200	1.28	6	7,8,10
Andropogon gerardi	7,900	1.58	4	9
Andropogon scoparius	5,700	1.00	4	9,10
Bouteloua gracilis	1,900	0.33	7	2,3
Bouteloua curtipendula	3,500	0.60	7	9
Bromus ciliatus	5,700	1.00	8	7,8
Calamagrostis canadensis	6,800	1.20	1	7,8
Calamagrostis inexpensa	5,700	1.00	2	10
Calamovilfa longifolia	8,00	1.40	2	1
Danthonia intermedia	4,500	0.80	3	5,6
Danthonia parryi	5,700	1.00	7	5
Elymus canadensis	8,000	1.40	3	1
Elymus innovatus	5,700	1.00	3	7,8
Festuca idahoensis	4,500	0.80	7	5,6
Festuca ovina	4,500	0.80	7	4,5,6
Festuca scabrella	11,000	1.94	7	5,6
Koeleria cristata	5,500	0.97	7	1,2,3,4,5,6,9,10
Muhlenbergia cuspidata	4,500	0.70	5	2
Muhlenbergia richardsonis	4,000	0.70	5	6,9,10
Poa canbyi	4,500	0.80	7	4,10
Poa cusickii	3,400	0.60	7	4,5,6
Poa pratensis	5,700	1.00	7	5,6,7,8,9,10
Poa secunda	3,400	0.60	7	2,3,4
Schizachne purpurascens	3,400	0.60	3	7,8
Spartina gracilis	6,800	1.20	2	1
Spartina pectinata	8,000	1.40	1	9
Stipa columbiana	6,300	1.10	7	5
Stipa comata	5,700	1.00	7	1,2,3,4
Stipa curtiseta	6,800	1.20	7	4,5,6,9,10
Stipa spartea	6,800	1.20	7	9
Stipa viridula	8,000	1.40	7	4,5,6,9
Carex filifolia	4,600	0.80	6	2,3
Carex pensylvanica	2,500	0.45	1	4,5,6
Carex stenophylla	2,500	0.45	1	2,3
Astragalus danicus	8,000	1.40	7	4,5,6
Astragalus striatus	10,000	1.76	7	3,4,5,6
Lathyrus ochroleucus	8,000	1.40	10	7,8
Lathyrus venosus	10,000	1.76	10	7,8
Vicia americana	8,000	1.40	10	7,8
Vicia minor	5,000	0.88	10	3,4,5,6
Agropyron cristatum	14,000	2.54	8	—
Bromus inermis	10,000	1.76	9	—
Festuca rubra	8,000	1.40	10	—
Phleum pratense	8,000	1.40	9	—
Medicago sativa	14,000	2.54	10	—
Elymus junceus	10,000	1.76	10	—
Poa pratensis	8,000	1.40	9	—

palatability rating, and the major communities in which they occur, are listed in Table IV. For comparison, the main introduced forage species are also listed.

7.5.2.4.2 Early Grazing

Grazing at too early a date can result in considerable changes in the composition of the sward. Species as *Stipa comata*, *S. curtiseta*, *Festuca scabrella* and *Agropyron* spp., may be prevented from reproduction. Replacement by sedges and low yielding grasses commonly follows, and results in a greatly reduced yield. For example, replacement of *Festuca scabrella* by *Danthonia intermedia* reduces the yield, often by as much as 40%, as well as the utilization, since *D. intermedia* is not very palatable.

Similarly, too early grazing of range dominated by *Stipa* spp. can result in a take-over by *Bouteloua gracilis*, which also results in up to 50 or over 60% reduction in yield. In both instances, further reductions in yield occur when overgrazing is continued, mainly by increases in unpalatable herbs.

7.5.2.4.3 Poisonous Plants

Another concern of range managers is stock poisoning. Several constituents of the vegetation are highly or potentially poisonous to livestock. However, good management can reduce the dangers of poisoning to a minimum, and sometimes prevent losses altogether.

One of the most poisonous species, *Delphinium bicolor*, occurs primarily in the southern fescue prairie and aspen groves. This species commences growth early in spring before all snow is gone, and usually completes its life cycle before the end of June. Stock poisoning in pastures where this species occurs is usually associated with too early grazing when there is still very little growth of grasses and other palatable forage, while *D. bicolor* is rapidly growing and forms succulent feed.

This example illustrates a rather general rule, valid in most instances of cattle losses through plant poisoning. The rule is that poisonous or potentially poisonous plants seldom form a serious problem if sufficient palatable forage is available, and stock is not hungry. Thus, although *Triglochin maritima* is very poisonous to cattle, it is often eaten without apparent effect when it is but a small part of the diet of well fed animals.

However, severe losses have resulted in herds, driven over a considerable distance to fresh pasture where this species occurred around watering places. Thirst, and hunger combined have resulted in the death of many cattle.

Similarly, the so-called locoweeds*Oxytropis* spp., selenium accumulating species of *Astragalus*, and *Zygadenus* spp., can become problem plants under certain circumstances. Overgrazing, untimely grazing, or herding into pastures with potentially poisonous species when hungry, are usually the causes of stock poisoning and preventable by good management.

7.5.3 PASTURE DEVELOPMENT

The need for expansion of the cattle population can only be met by increased production and hence "tame pasture" and hayland. As can be seen in Table IV, introduced forage species yield considerably more, and generally are also more palatable, than most native species. This advantage is reinforced by the fact that these species are seeded in monocultures or simple grass-legume mixtures.

Without tame forage, winter pasture or hay for feeding must be obtained from native range. Ranchers utilizing B o u t e - l o u i o n g r a c i l i s prairie in good to excellent condition need a minimum of 18 ha/A.U. for a year-round feed. Using tame forage for winter feeding, this can be reduced to 12 ha/A.U. since 1.5 ha of a grass-legume mixture can provide the necessary hay of superior quality.

A further reduction in the area required can be achieved by using tame forages for spring and fall grazing. The topography or stoniness of soils seldom allow for complete conversion to tame forage, but in view of the two or three times higher yield obtained, an increase of 75% total carrying capacity is often within the realm of possibilities.

7.5.3.1 *Pasture Development in the Prairies*

In the prairies it makes little difference which vegetation type is replaced by tame forages. Yield increases are of the same order throughout this zone with adequate seedbed preparation, and the use of appropriate grass-legume mixtures. Crested wheat-grass, *Agropyron cristatum*, on light soils, and Russian wild rye-grass, *Elymus junceus* (Lawrence & Heinrichs 1966), on heavier soils, in mixture with a creeping-rooted alfalfa (Heinrichs 1963), consistently double or triple the carrying capacity of the range (Kilcher *et al.* 1956).

Crested wheatgrass pastures are long-lived under adequate management (Lodge, Smoliak & Johnston 1972). This grass remains dominant in a "climax" vegetation (Looman & Heinrichs 1973), and there are indications that Russian wild ryegrass pastures will be equally satisfactory.

Extension of the grazing season through winter pastures can increase beef production per hectare, and in this respect Altai wild ryegrass, *E. angustus*, has shown promising, and may become of great importance (Lawrence & Heinrichs 1970).

The success of many introduced species is not surprising. The most successful introductions are those from the Central Eurasian steppes, which show a great similarity of conditions to those prevailing in the prairies and parklands.

One of the most important findings of range research has been that none of the grasses and legumes of the native grasslands is suitable for cultivation. Most of the native grasses are either poor producers, or very difficult to establish, and often both.

7.5.3.2 *Pasture Development in the Parklands*

In the parklands the carrying capacity of the aspen groves increases from south to north, and with the age of the grove, with a decline in over-age groves. But at all ages and in all localities the carrying capacity rapidly declines with too heavy or too early grazing. Ranchers in this zone generally agree that the carrying capacity of the "bush pastures' seldom exceeds 20 A.U./ section (= 256 ha) per season of four to five months.

Where clearing and breaking of bush pastures is possible, seeding to appropriate grass-legume mixtures can increase the carrying capacity to 80 A.U./256 ha in the southern parts, and to 200 A.U./256 ha in the northern, Black soil areas. Much depends on the adequacy of the forage species used, and on pasture management (Fig. 10).

Often the tame forage seeded is *Bromus inermis* in monoculture (Fig. 11). The yield increase obtained with this species is always considerable. Even at a density of 50 to 60% the increase

348

Fig. 10. Tame forages can increase yields in the northern parklands tenfold. At left, cattle grazing native vegetation, at right brome-alfalfa. A narrow road separates these fields.

349

Fig. 11. Pasture development in the northern parklands. New seeding of brome-alfalfa on cleared Aspen-Poplar bushland. This first year's stand already produces more than the adjacent bush-pasture.

can be 10-fold over that of native grassland during the first few years, but in the fourth or fifth year usually drops to 5-fold or less. In mixture with *Medicago*, the yield levels off at 6- to 7-fold, but a mixture of *B. inermis*, *Phleum pratense*, *Poa pratensis*, and *Festuca rubra*, with alfalfa can maintain a 10-fold yield.

Monocultures and the simple grass-legume mixture have the disadvantage that adverse weather conditions lower the productivity, and can soon lead to weediness. At the same time, the natural limit to yield capability does not allow for a maximum yield in favorable years. Furthermore, the fertility of the soil may soon pose a problem, particularly in soils with low phosphorus.

The more complex mixture maintains a relatively stable yield in dry as well as in wet years, and because of its diversity, also is less adversely affected by fertility problems. However, in all tame pastures applications of fertilizer at a rate, approximately equalling the nutrient uptake by the crop, helps to maintain the yield at the maximum level.

350

7.5.3.3 *Economics of Pasture Development*

Further considerations in the development of tame forage crops are the economic feasibility, and the availability of water. Development of tame pastures and hayland is an expensive undertaking, particularly in the bush areas. However, the high costs there can be partly offset by using the newly cleared and broken land for the production of feed grain or a silage crop during the first two or three years. Nevertheless, satisfactory returns are largely dependent on current beef prices, and period of amortization (Wiens, Lodge & Johnston 1969; Wiens & Lodge 1972), besides the magnitude of improvement achieved.

The possibilities and costs of water development are other factors to be considered. Each additional animal requires at least 50 liters of water per day, and in the prairies a limited water supply may place a limit on pasture development. In the parklands water supplies do not usually limit the possibilities for pasture development.

7.5.4 WILDLIFE

In addition to the domestic livestock, the prairies and parklands support almost a million head of large game. Officers of the Wildlife Services estimate the numbers of *Antilocapra americana* at approximately 20,000 (Fig. 12), *Odocoileus virginianus* at 600,000, *O. hemionus* at 200,000, and *Cervus canadensis* at 50,000. Converting these numbers to the conventional animal units, this means approximately 300,000 A.U.

Most of the wild ungulates are not in competition with domestic livestock. *A. americana* and *C. canadensis* graze the grassland; the former also browses on shrubs and forbs, while the latter browses extensively on trees in the parklands. *O. virginianus* and *O. hemionus* are primarily browsers, subsisting mainly on shrubs and trees.

Wildlife numbers are maintained through regulated hunting, with the number of hunting permits issued dependent on wildlife census. These protective measures may insure that the wildlife populations do not fall below a level from which recovery is not possible. However, the quality of the populations may suffer because many of the animals killed are selected for their value as trophies. Thus, the dominant or potentially dominant males in the herd may be removed.

351

Fig. 12. The Pronghorn antelope formerly as plentiful as the buffalo, is now the least common of large game animals.

Occasionally, *Alces americana andersoni* may be encountered in the northern parkland but it is mainly an inhabitant of the boreal forests. In any case, it feeds mainly on trees and shrubs, and aquatic vegetation, and does not affect livestock management.

Buffalo no longer occur in the wild state (Fig. 13). A herd of some 12,000 plains buffalo lives in a semiwild state in the Wood Buffalo National Park in northern Alberta and adjacent Northwest Territories. A small herd of the larger Wood buffalo also occurs in this park, and has in fact interbred with the plains buffalo. To preserve the Wood buffalo, small numbers have been released in the Northwest Territories and Central Saskatchewan. Thus far, it is not known whether these experiments have been successful.

Hunting of the buffalo is allowed to Indians in the vicinity of the park, and periodically animals are shot to prevent overgrazing. Also, small numbers of animals are made available to other parks and zoological gardens, mostly through Elk Island National Park, which has the largest herd of buffalo in captivity.

Attempts have been made to crossbreed buffalo and domestic cattle, but sterility of the offspring has been a deterrent, and most breeders of "cattelo" have abandoned their projects.

352

Fig. 13. Formerly roaming the plains in herds of millions, the buffalo is now seen only in small, semiwild herds in parks.

7.6   Summary

The Canadian prairies and parklands form the northernmost part of the North American Great Plains, a vast expanse of mostly treeless grasslands. The Great Plains originated after the uplift of the Rocky Mountains, but have been greatly modified during periods of glaciation, particularly the Pleistocene.

Further significant influences on the vegetation of the Great Plains have been the vast herds of ungulates and man. The Indian and bison both are recent arrivals on the North American continent, and their greatest impact on the vegetation has been in postglacial times.

After virtual exterminations of the bison, in the late 19th century, domestic cattle almost immediately utilized the grasslands, and overgrazing was common. White settlers plowed up much of the grasslands for production of cereal crops, and at present only the least arable parts of the Great Plains remain in native grassland.

In the Canadian Prairie Provinces, vegetation science is mainly applied in the form of management of the uses of vegetation. Phytosociological and ecological range research are aimed at classifying the vegetation, establishing the productivity and carrying capacity of the vegetation units, and the capability for development of these units.

Research in pasture management is aimed at increasing beef production. Research programs include the testing of various introduced forage species, and selection of strains on the basis of adaption to climate, quality, grazing resistance and longevity.

Thus, application of vegetation science in the Canadian Prairie Provinces attempts to combine the capabilities of native vegetation with the best species from steppes elsewhere into a combination which can support a maximum number of domestic livestock as well as wildlife.

References

Anonymous, – 1939 – Fifty years of progress on Dominion Experimental Farms, 1886–1936. Can. Dep. Agr., Exp. Farms. Serv., Ottawa. 158 p.

Atherton, L., – 1972 – The cattle kings. Bison books, the Univ. of Nebraska Press, Lincoln, Nebraska. xii + 308 p.

Bird, R. D., – 1961 – Ecology of the Aspen Parkland of Western Canada. Can. Dep. Agr. Pub. 1066, x + 155 p.

Boivin, B., – 1967 – Enumeration des plantes du Canada, Univ. Laval, Quebec.

Braun-Blanquet, J., – 1972 – Pflanzensoziologie: Grundzüge der Vegetationskunde Springer Verlag Wien. 2nd ed. xi + 631 p.

Brown, D., – 1954 – Methods of surveying and measuring vegetation. Bull. 42, Commonwealth Bureau of Pastures and Field Crops. Hurley, Berkshire. 223 p.

Campbell J. B., – 1969 – Experimental methods for evaluating herbage. Can. Dep. Agr. Pub. 1315, Ottawa, Ontario. 223 p.

Campbell, J. B., – 1971 – The Wsift Current Research Station 1920-70. Historical Series No. 6. 79 p.

Campbell, J. B., R. W. Lodge, A. Johnston & S. Smoliak, – 1962 – Range management of grasslands and adjacent parklands in the Prairie Provinces. Can. Dep. Agr. Pub. 1133. 32 p.

Christiansen, A. E., – 1956 – Glacial geology of the Moose Mountain area, Saskatchewan. Rep. No. 21, Dep. of Mineral Resources, Regina, Saskatchewan. 33 p.

Christiansen, E. A., – 1961 – Geology and groundwater resources of the Regina area Saskatchewan. Sask. Res. Council, Geol. Div. Rep. No. 2. 72 p.

Christiansen E. A., – 1965 – Geology and groundwater resources of the Kindersley areas (72-N) Saskatchewan. Sask. Res. Council, Geol. Div. Rep. No. 7. 25 p.

Clarke, S. E., – 1930 – Pasture investigations on the short-grass plains of Saskatchewan and Alberta. Sci. Agr. 10: 732-749.

Clarke, S. E. & E. W. Tisdale, – 1936 – Range pasture studies in southern Alberta and Saskatchewan. Herbage Rev. 4(3): 51-64.

Clarke, S. E. & D. H. Heinrichs, – 1941 – Regrassing abandoned farms, submarginal cultivated lands and depleted pastures in the prairie areas of Western Canada. Can. Dep. Agr. Pub. 720. 23 p.

Clarke, S. E., J. A. Campbell & J. B. Campbell, – 1942 – An ecological and grazing capacity study of the native grass pastures in southern Alberta, Saskatchewan and Manitoba. Can. Dep. Agr. Bull. 44. 31 p.

Clarke, S. E., E. W. Tisdale & N. A. Skoglund, – 1943 – The effects of climate and grazing practices on short-grass prairie vegetation. Can. Dep. Agr. Bull. 46.

Clarke, S. E. & E. W. Tisdale, – 1945 – The chemical composition of native forage southern Alberta and Saskatchewan in relation to grazing practices. Can. Dep. Agr. Bull. 54.

Clarke, S. E., J. A. Campbell & W. Shevkenek, – 1950 – The identification of certain native and naturalized grasses by their vegetative characters. Can. Dep. Agr. Bull. 50.

Clements, F. E., – 1928 – Plant succession and indicators. H. W. Wilson Co., New York, xvii + 453 p.

Clements, F. E. & R. W. Chaney, – 1936 – Environmental and life in the great plains. Carnegie Supp. Pub. No. 24. pp. 1–53.

Coupland, R. T., – 1950 – Ecology of mixed prairie in Canada. Ecol. Monographs 20: 271–315.

Coupland, R. T., – 1961 – A reconsideration of grassland classification in the northern Great Plains of North America. J. Ecol. 49: 135–167.

Curtis, J. T., – 1959 – The vegetation of Wisconsin. Univ. of Wisconsin Press, Madison, Wisconsin. xi + 657 p.

Ehrlich, N. A., E. A. Poyser, L E Pratt & J. H. Ellis, – 1953 – Report of Reconnaisance Soil Survey of Winnipeg and Morris Map sheet area. Manitoba Soil Survey, Man. Dep. Agr. 109 p.

Flint, R. F., – 1955 – Glacial geology and the Pleistocene epoch. Wiley and Sons, Inc., New York. xviii + 589 p.

Flint, R. F., – 1971 – Glacial and quaternary geology. J. Wiley and Sons Inc., xii + 892 p.

Gard, W., – 1972 – The great buffalo hunt. Bison books, the Univ. of Nebraska Press, Lincoln, Nebraska vii + 324 p.

Good, R., – 1964 – The geography of the flowering plants. Longman, Green and Co., London 3rd. ed., xvi + 518 p.

Grinnell, G. B., – 1962 – (1892). Blackfoot lodge tales, the story of a prairie people. Bison books. Univ. of Nebraska Press, Lincoln, Nebraska xvii + 310 p.

Heinrichs, D. H., – 1963 – Creeping alfalfas. In Advances in Agronomy, 15: 317-337, Academic Press, New York.

Johnston, A., – 1970 – Blackfoot Indian utilization of the flora of the northwestern Great Plains. Econ. Bot. 24: 301–324.

Kilcher, M. R., R. W. Lodge, D. H. Heinrichs & J. B. Campbell, – 1956 – Pasture and hay crops for the southern Canadian prairies. Can. Dep. Agr. Pub. 980. pp. 1–21.

Lawrence, T. & D. H. Heinrichs, – 1966 – Russian wild ryegrass for Western Canada. Can. Dep. Agr. Pub. 991. 18 p.

Lawrence, T. & D. H. Heinrichs, – 1970 – Pasture attributes of Altai wild ryegrass. Can. J. Plant Sci. 50: 743–745.

Levy, E. B. & E. A. Madden, – 1933 – The point method of pasture analysis New Zealand J. Agr. 46: 267–279.

Lodge, R. W., J. B. Campbell, S. Smoliak & A. Johnston, – 1971 – Management of the Western range. Can. Dep. Agr. Pub. 1425. 34 p.

Lodge, R. W., S. Smoliak & A. Johnston, – 1972 – Managing crested wheatgrass pastures. Can. Dep. Agr. Pub. 1473. 20 p.

Looman, J. & D. H. Heinrichs, – 1973 – Stability of crested wheatgrass pasture under long-term pasture use. Can. J. Plant Sci. 53: 501–506.

Looman, J., – 1963 – Preliminary classification of grasslands in Saskatchewan. Ecology 44: 15–29.

355

Looman, J., – 1964 – Ecology of lichen and bryophyte communities in Saskatchewan. *Ecology* 45: 481–491.

Looman, J., – 1965 – Theoretical considerations of the plant association. *Neth. J. Agr. Sci.* 13: 120–128.

Looman, J., – 1969 – The fescue grasslands of Western Canada. *Vegetatio* 19: 128–145.

Malin, J. C., – 1956 – The grassland of North America: its occupance and the challenge of continuous reappraisals. In Man's Role in Changing the Face of the Early, Part I. Retrospect. pp. 350–366.

Malin, J. C., – 1957 – The North American grassland in historical perspective. *Ecology* 38: 362–363.

Malin, J. C., – 1967 – The grassland of North America: Prolegomena to its history with addenda and postscript. Gloucester, Mass. Peter Smith viii + 490 p.

Mitchell, J., H. C. Moss, J. S. Clayton & F. H. Edmunds, – 1944 – Soil Survey Report No. 12. Univ. of Saskatchewan, Saskatoon, Saskatchewan.

Moss, E. H., – 1932 – The vegetation of Alta. IV. The poplar association and related vegetation of central Alberta. *J. Ecol.* 20: 280–415.

Moss, E. H., – 1944 – The prairie and associated vegetation of southwestern Alberta. *Can. J. Res.* 22: 11–31.

Moss, E. H., – 1952 – Grassland of the Peace River region, Western Canada. *Can. J. Bot.* 30: 98–124.

Moss, E. H., & J. A. Campbell, – 1947 – The fescue grassland of Alberta. *Can. J. Res.* 25: 209–227.

Odynsky, W., J. D. Lindsay, S. W. Reeder & A. Wynnyk, – 1961 – Reconnaisance Soil Survey of the Beaverlodge and Blueberry Mountain Sheets. Alta. Res. Council, Rep. No. 81. 123 p.

Quimby, G. I., – 1968 – North American Indians. In World Book encyclopedia, Vol. 10.

Ritchie, J. C., – 1964 – Contributions to the holocene paleoecology of west central Canada. I. The Riding Mountain Area. *Can. J. Bot.* 42: 181–196.

Ritchie, J. C. & B. de Vries, – 1964 – Contributions to the holocene paleoecology of west central Canada. A late-glacial deposit from the Missouri coteau. *Can. J. Bot.* 42: 677–692.

Roe, F. G., – 1951 – The North American buffalo: a critical study of the species in its wild state. Univ. of Toronto Press, viii + 957 p.

Stuart, G. E., – 1972 – Mounds: Riddles from the Indian past. *Nat. Geog.* 142: 783–801.

Thro, E., – 1973 – Archaeology in Illinois. In Encyclopedia Science Supplement 1973: 302–305. Grolier Enterprises Ltd.

Vries, B. de & C. D. Bird, – 1965 – Bryophyte subfossils of a late glacial deposit from the Missouri Coteau, Saskatchewan. *Can. J. Bot.* 43: 947–953.

Walter, H., – 1963 – Climatic diagrams as a means to comprehend the various climatic types of ecological and agricultural purposes. In: The water relations of plants. Blackwell Sci. Pub. 3–9.

Weaver, J. E. & F. E. Clements, – 1938 – Plant ecology. McGraw-Hill Book Co., New York, 2nd ed. xxii + 601 p.

Wendorf, F., – 1961 – Paleoecology of the Llano Estacado. Fort Burgwin Res. Center, The Museum of New Mexico Press, Santa Fé, N.M. Pub. No. 1, 144 p.

Wiens, J. K. & R. W. Lodge, – 1972 – Developing bush pastures. A management and economic guide. The Saskatchewan Advisory Council on Forage Crops 72/4. 26 p.

Wiens, J. K., R. W. Lodge & A. Johnston, – 1969 – Seeding prairie rangelands – A management and economic guide. Can. Dep. Agr., Economic Branch-Research Branch 69/13. 25 p.

8 DER BEITRAG DER VEGETATIONSKUNDE ZUR REGULIERUNG DES WASSERHAUSHALTES IM GRÜNLAND

EMILIE BALÁTOVÁ-TULÁČKOVÁ & WERNER KRAUSE

Inhalt

8 DER BEITRAG DER VEGETATIONSKUNDE ZUR REGULIERUNG DES WASSERHAUSHALTES IM GRÜNLAND

8.1 Gegenstand, Konzeption

Die Grünlandvegetation wird in ihrer Struktur und räumlichen Ordnung besonders stark vom Wasserfaktor beeinflußt. Ihre Produktivität kann daher durch Eingriffe in den Wasserhaushalt wirkungsvoll in erwünschter wie unerwünschter Richtung verändert werden. Schädliche Folgen lassen sich umso sicherer vermeiden, je sorgfältiger die Planung der Hydromelioration erfolgt. Inwieweit hierbei die Vegetationskunde mitwirken kann, soll im folgenden behandelt werden. Die meisten Darlegungen beziehen sich auf das Wirtschaftsgrünland der Wald-Höhenstufe unter subozeanisch getöntem Klima in der gemäßigten Zone der Nordhemisphäre. Sie betreffen damit im wesentlichen die M o - l i n i o - A r r h e n a t h e r e t e a . Ein kurzer Blick wird auf die Überschwemmungswiesen des pannonischen Florengebietes geworfen.

Der Einsatz vegetationskundlicher Methoden im Dienst angewandt-ökologischer Planung beruht letztlich darauf, daß die Vegetation mit den Umweltfaktoren in einem dynamischen Gleichgewicht steht (Voisin 1961, Niemann 1963, 19C8 vgl. auch die Konzeption des Ökosystems bei Tansley 1935, Odum 1959, Ellenberg 1973 und der Biogeozönose bei Sukatschew 1945, 1954). Da der Standort die Zusammensetzung des Pflanzenbestandes prägt, lassen sich von letzterem Rückschlüsse auf den Standort ziehen. Ein Vorzug dieses Verfahrens liegt darin, daß ein flächenhaft erkennbares Kriterium Auskunft über Kräfte gibt, die unmittelbar oft nicht wahrnehmbar sind.

Daß vegetationskundliche Methoden am Wasserhaushalt des Grünlandes viele Einsatzmöglichkeiten finden, hat abgesehen von der formativen Kraft des Wasserfaktors eine Ursache im Artenreichtum des naturnahen Grünlandes in Mitteleuropa, der eine sehr weitgehende Gesellschaftsgliederung hervorgebracht hat. Ihr ebenso differenzierter Indikatorwert erlaubt enge Anpassung an konkrete Fragen.

Überdies gibt die vegetationskundliche Standortsansprache im Gegensatz zu bodenkundlichen, meteorologischen oder auf andere abiotische Elemente bezogenen Methoden nicht so sehr

Auskunft über einzelne isolierte Faktoren, sondern beleuchtet deren Verflechtung zum Gesamtkomplex "Standort - Vegetation", auf den es dem Wasserbauer, Meliorationstechniker und Landwirt oft ankommt (Schroeder 1958). Eine Literaturübersicht zum Thema "Vegetation und Wasserwirtschaft" gibt Seibert (1963b).

Speziell zur Verwendung pflanzensoziologischer Untersuchungen zur Beurteilung von Schäden des Grünlandes äußert sich Tüxen (1942). Allgemeines zur angewandten Vegetationskunde bringen Klapp (1949), Ellenberg (1954, 1956), Ramenskij et coll. (1956), Kayl (1965).

8.2 Aufgabengebiete angewandter Vegetationskunde im Bereich des Grünlandwasserhaushaltes

8.2.1 FLUSSREGULIERUNGEN

Bis vor kurzem erfolgten Fluß- und Bachregulierungen in erster Linie zur Abwehr des Hochwassers (Beispiele in Bauer & Weinitschke 1964). Heute und in nächster Zukunft liegt, wenigstens in Ländern mit intensiver Wirtschaft, das Schwergewicht auf der differenzierten Beeinflussung der Grundwasserstände und der Vorflut. Zugleich tritt die Speicherwirtschaft in den Vordergrund, die durch neugeschaffene Beregnungsmöglichkeiten sowie durch Änderung der Abflußdynamik vielfältigen Einfluß auf die Grünlandwirtschaft gewinnen kann.

8.2.2 BEEINFLUSSUNG DER GRUNDWASSERDYNAMIK

Als Grundwasser wird dasjenige Wasser verstanden, das im Boden oder in Lockergestein alle Hohlräume füllt, nur der Schwerkraft unterliegt und sich über einer tiefliegenden Sohlschicht in langsamer Fließbewegung befindet. Im natürlichen Verband besteht seine Oberfläche aus einem labilen, im Jahresablauf standortspezifisch schwankenden Grenzsaum zwischen verschiedenen hydrostatischen Druckbereichen (Koehne 1948, Buchholz 1961, Mückenhausen 1962). Sobald die Schwankungen den Wurzelbereich der Pflanzen erreichen, entstehen Koinzidenzbeziehungen zur Zusammensetzung der Vegetation. Für letztere ist die Höhe und der Zeitablauf der Grundwasserschwankungen wichtiger als der Mittelwert des Grundwasserstandes. Dieser komplizierte Sachverhalt macht es wünschenswert, bei der Meliorationsplanung die vereinfachende Vorstellung von "Senkungen" und "Hebungen" zu meiden und an ihrer Stelle eine gezielte Regulation der Grundwasserdynamik anzustreben.

Beispiele von Koinzidenzbeziehungen zwischen Grundwasserdynamik und Vegetation bringen Klapp (1954), Tüxen (1954b), Eskuche (1955), Zarzycki (1956), v. Müller (1956), Niemann (1973), Jovanović (1965), Balátová-Tuláčková (1966, 1968, 1972), Klötzli (1969), Ružičková (1971). Eine Bibliographie gibt Tüxen 1961.

8.2.3   STAUWASSER

Wenn der Boden nahe unter der Oberfläche verdichtet ist, staut sich in geringer Tiefe Niederschlagswasser, das bei trockener Witterung verschwindet. Staunasser Boden (Pseudogley) ist wechselnaß, luftarm, oft stark sauer; er verhärtet beim Austrocknen und nimmt bei nasser Witterung breiartige Beschaffenheit an. Damit unterscheidet er sich vom Grundwassergley, dessen ausgedehnter und tiefer, meist durch seitlichen Zufluß gespeister Wasserkörper ungeachtet aller Schwankungen relativ ausgeglichene Bedingungen gewährleistet.

8.2.4   UNTERSCHEIDUNG ZWISCHEN BEWEGTEM UND STEHENDEM
         BODENWASSER

Seitlich bewegtes Bodenwasser fördert die Produktivität stärker als stehende Vernässung. In der Ebene, wo das Fliessen unmittelbar nicht wahrnehmbar ist, vermag die Grünlandvegetation auch eine äußerst langsame Bewegung des Bodenwassers noch nachzuweisen.

8.2.5   ENTWÄSSERUNG DURCH FLÄCHENMELIORATION

Meliorative Flächenentwässerung wird erforderlich, sobald die Landwirtschaft dazu übergeht, große Flächen einheitlich zu bewirtschaften. Neben der klassischen Rohrdrainage setzen sich auf schweren Böden die Verfahren der Gefügemelioration, insbesondere die maschinelle Untergrundlockerung im Pseudogley (Gora 1964) immer mehr durch.

8.2.6   BEWÄSSERUNG

Während die früher verbreitete Grünlandberieselung (Krause 1959) künftig in Mitteleuropa keine nennenswerte Rolle spielen wird, sind Beregnung und Untergrundbewässerung im Zunehmen begriffen. Dosierung, zeitliche Abstufung der Gaben, Kombination mit Entwässerung und Grundwasserregulierung müssen an die Standorte angepaßt werden. (Lit. in Klapp 1956, 1965, s. auch Kramer 1973).

8.2.7   UMSTELLUNG DES NUTZUNGSPLANES

Flurbereinigungen im Zuge der laufenden Umstrukturierung der Landwirtschaft machen häufig die Überführung von Grünlandflächen in Acker notwendig. Das Gelingen der Umstellung ist weitgehend vom Bodenwasserhaushalt abhängig. Hier bietet sich der angewandten Vegetationskunde ein weites Betätigungsfeld.

8.2.8   EINIGE ANDERE EINGRIFFSRICHTUNGEN

Zu den umrissenen Haupteingriffen kommen speziellere Vorhaben: Beeinflussung des Wasserhaushaltes über das Lokalklima: Windschutzanlagen zur Verminderung unproduktiver Verdunstung (näheres bei Bauer & Weinitschke 1964, S. 62).

Baumpflanzung in Grünland zum Strahlungsschutz in Trockengebieten mit schweren Böden (Hundt 1958, S. 196). Verminderung unerwünschter Bodennässe durch Anbau stark transpirierender Pflanzen (biologische Entwässerung) oder durch "Parkwiesen", in denen lichtgestellte Bäume den Wasserverbrauch erhöhen (Hundt 1958, Knapp 1963, Niemann 1968).

8.3   Die gemeinsame ökologische Problematik der Eingriffe in den Wasserhaushalt

Jeder Eingriff muß der vielseitigen labilen Abhängigkeit Rechnung tragen, der das Wasser im Boden unterliegt. Die Spiegelhöhe von Flüssen bildet die Basis, auf die sich der Grundwasserstand des umliegenden Geländes stützt. In breiten Tälern reicht diese Wirkung sehr weit. Der Fluß kann auch zur Grund-

362

wasserspeisung beitragen. Überschwemmungen führen nicht nur vorübergehend Wasser zu, sondern üben mit ihren Sedimenten zeitlich unbegrenzte Nachwirkungen aus. Die seitliche Grundwasserbewegung, das Stauwasser in schwer durchlässigen Böden, die Bodenporosität und der vom Relief gelenkte oberirdische Abfluß kommen hinzu. Im Kontinentalklima rufen gelöste Salze unerwünschte Folgen hervor, sobald sich das Verhältnis zwischen Zufluß, Abfluß und Wasserverbrauch der Pflanzendecke ändert.

Die Entwirrung des komplizierten Wirkungssystems bildet eine erste Voraussetzung für das Gelingen von Meliorationen, wobei die Vegetationskunde, wie im einzelnen zu zeigen sein wird, entscheidende Hilfe bietet.

8.4 Vegetationskundliche Methoden im Dienste der Grünlandwirtschaft

8.4.1 DIE ELEMENTARE LEISTUNG DER VEGETATIONSKUNDE

Da jede Pflanzengesellschaft auf bestimmte ökologische Bedingungen angewiesen ist, zeigt sie durch ihr Vorkommen, daß auf ihrer Siedlungsfläche gerade diese und keine anderen Bedingungen herrschen. Die Unterscheidung der Pflanzengesellschaften erlaubt demnach, dem schwer faßbaren Begriff des Standortes näherzukommen, indem sie seine vielfältigen Erscheinungsformen einer reproduzierbaren Gliederung unterwirft. Diese elementare Leistung bietet zwei technische Vorteile. Erstens kann das Kriterium "Pflanzendecke" während der ganzen Vegetationsperiode flächenhaft angesprochen werden. Zweitens ermöglicht sie eine großflächige Synopsis, deren Ausgangsbasis die Vegetationskarte bildet.

Das systematische Aufsuchen und Abgrenzen der Pflanzengesellschaften regt den Bearbeiter unabweisbar dazu an, über das Beschreiben hinaus auch Zusammenhänge zu sehen und zu suchen. Diese betreffen, zumal wenn kartiert wird, zunächst die räumlichen Beziehungen. Aufschlußreich ist bereits der Verlauf der Grenzen zwischen den einzelnen Gesellschaften. Folgen sie einer Wirtschaftsfigur, z.B. einer rechtwinkelig-geradlinig begrenzten Parzelle oder einer als Kunstprodukt erkennbaren Gemeindegrenze, darf angenommen werden, daß die Bestandsunterschiede überwiegend auf Wirtschaftseinflüsse zurückgehen. Gleichen die Grenzen, was besonders im ebenen Alluvium aufschlußreich ist, den Höhenlinien, liegen meist Befeuchtungseinflüsse vor. Diese gehen entweder auf aktuelle Abstufungen des Grundwasserstandes zurück, oder sie werden von Unterschieden der Bodenqualität hervorgebracht (Niemann 1968), die ihrerseits häufig durch frühere Tätigkeit des Oberflächenwassers geschaffen wurden.

Insgesamt schärft der Blick auf die Pflanzengesellschaften und ihre Grenzen die Aufmerksamkeit für unauffällige Erscheinungen, die mit der Produktivität der Pflanzendecke in engem Zusammenhang stehen. Zu ihnen gehören kleinste Reliefunebenheiten, die auf Abstufung der Grundwasserstände oder der Bodenbeschaffenheit aufmerksam machen. Hierzu gehört auch stagnierendes Restwasser in Ackerfurchen oder Wegrinnen, das auf eine Pseudogley-Stauschicht aufmerksam macht. An Abhängen geben eingeebnete und verwachsene Terrassen Aufschluß über ehemalige Ackernutzung, die sich noch immer auf die Zusammensetzung des nachfolgenden Grünlandes auswirkt.

Feststellungen dieser Art gewinnen an Ausdruckskraft, wenn zugleich an einer Vegetationskarte ihre räumliche Ordnung zum Ausdruck kommt. Jede Karte zeigt, daß die Pflanzendecke weder ein verschwommenes Kontinuum noch ein gesetzloses Durcheinander ungleicher Teile bildet. Deutlich fallen z.B. die ökologisch aufschlußreichen Bindungen der Gesellschaften an die Geländeform ins Auge. Daneben zeigt sich, daß in geographisch definierbaren Teilgebieten manche Kontaktgesellschaften regelmäßig nebeneinander auftreten, andere ebenso regelmäßig fehlen. In Ländern mit dichter menschlicher Besiedelung macht das Zusammentreffen der Ersatzgesellschaften, d. h. der Äcker, Wiesen, Weiden, Gebüsche, Heiden, Brachflächen, die allesamt durch unterschiedliche Bewirtschaftung aus wenigen ursprünglichen Waldgesellschaften entstanden sind, auf eine im Hintergrund stehende ökologische Einheit aufmerksam.

8.5   Zum Indikatorwert der Pflanzengesellschaften

8.5.1   DIE GRUNDLAGEN

Vegetationskundliche Geländearbeit vermittelt bereits durch unmittelbare Anschauung Einblick in die Abhängigkeit der Pflanzengesellschaften von ihren Standorten. Vielfach treten allerdings zuerst die Symptome und nicht die eigentlich wirkenden Ursachen ins Bewußtsein. Das gilt z.B. für die Beziehungen zwischen Pflanzendecke und Exposition zur Himmelsrichtung, die zunächst nichts weiter bieten können als räumliche Koinzidenzen.

Genaues Hinsehen, zu dem die ersten Erfahrungen immer anregen, führt dann weiter zu den eigentlichen Ursachen im Sinne der grundlegenden Erörterung, die Pallmann (1943) zu diesem Thema mitgeteilt hat. Die Wahrnehmung einer bestimmten Exposition öffnet dann einen weiten Blick über die verschiedensten Faktoren wie z.B. Sonneneinstrahlung, Windeinfluß, Dauer der Schneebedeckung, Vorrat an Winterfeuchtigkeit, Wasserabfluß, Bodenentwicklung, Bodenmikrobiologie, deren Synthese in der Pflanzendecke zum Ausdruck kommt.

In den fünf Jahrzehnten, in denen sich die pflanzensoziologische Vegetationskunde bisher entwickelt hat, trat die Bindung zwischen Pflanzengesellschaften und Standorten immer entschiedener ins Bewußtsein. Gesellschaftsabgrenzung und -beschreibung führten an ökologische Fragen heran, deren Lösung wiederum dazu verhalf, die Gesellschaftsunterscheidung zu verfeinern. Übersichten der aus Mitteleuropa beschriebenen Grünlandgesellschaften mit bekanntem Indikatorwert geben Klapp (1965, S. 139ff.) und Niemann (1973, S. 80ff.).

Je deutlicher der ökologische Indikatorwert der Pflanzengesellschaften in das Bewußtsein der Vegetationskunde trat, desto intensiver war diese bemüht, seine Aussagen deutlich herauszuarbeiten, oder wie ein allgemein angenommener Ausdruck lautet, ihn zu "eichen".

8.5.2 DIE EICHUNG DES INDIKATORWERTES

Vorausgeschickt sei, daß als eine Hauptvoraussetzung für die Anwendung des Indikatorwertes die eindeutige Abgrenzung der Pflanzengesellschaften zu fordern ist. Hierüber berichtet Band 5, S. 6-17 ff. dieses Handbuchs. Als Pflanzengesellschaft wird im wesentlichen die Gesamtheit der in einem homogenen Bestand vorkommenden Arten verstanden. In Sonderfällen können auch enger gefaßte Artengruppen einer deutlichen ökologischen Bindung unterliegen. Die Indikatoreigenschaft von Einzelarten ist im allgemeinen zu weit, als daß sie verbindliche Aussagen ermöglicht. Ausnahmen bestehen allerdings gerade im Grünland, wo einige Cyperaceen, z.B. *Carex gracilis* und *Cladium mariscus* dichte Einartbestände bilden, die gleichwohl als Pflanzengesellschaften aufgefaßt wurden. Die Methoden zur Eichung des Indikatorwertes haben sich in drei Stufen entwickelt.

8.5.2.1 Qualitative Eichverfahren

Viele Artenkombinationen machen auf klar erkennbare, wiewohl zunächst nur qualitativ zu fassende Standortseigenschaften unmittelbar anschaulich aufmerksam. Sie zeigen z.B. die großen Abstufungen des Befeuchtungsgrades, den Gegensatz zwischen kalkhaltigem und kalkfreiem Boden, die Intensität der Bewirtschaftung. Die Ergebnisse bieten relativ hohe Sicherheit, weil Wiederholung und Kontrolle durch Generationen von Vegetationskundlern die Möglichkeiten und Grenzen des Verfahrens erprobt haben. Überdies genügt die qualitative Beurteilung in Fällen, in denen die praktische Anwendung eine Alternativentscheidung erwartet, wie es z.B. für die Beurteilung der Acker- oder Weidefähigkeit einer Wiese zutrifft.

8.5.2.2 Ordinale (Rangstufen-) Eichverfahren

Ellenberg (1950) ordnet die Feuchtigkeitsansprüche der Grünlandpflanzen in eine fünfstufige Rangfolge mit den Grenzwerten "auf sehr trockenen Standorten" und "auf nassen Standorten". Sie umfaßt überdies eine Stufe O, "Gegen den Bodenwasserhaushalt weitgehend indifferent". Die zahlenmäßige Formulierung der Wasseransprüche erlaubt arithmetische Behandlung (Ellenberg 1963, S. 750), die komprimiertere Aussagen zuläßt als die qualitative Beschreibung (Vgl. auch Ellenberg 1952b1974, Wagner 1955, De Vries et coll. 1957, Bracker 1960, Hundt 1964a, Klapp 1965, Regal 1967). Freilich lassen sich nicht alle Eigenschaften in Zahlen fassen. Ellenberg (1952b, S. 64) behält die qualitativen Bezeichnungen "Wechselfeuchtigkeitszeiger" und "Überschwemmungszeiger" bei.

8.5.2.3 Quantitative Verfahren

Durch Messung der Standortsfaktoren und den Vergleich der Ergebnisse mit der Vegetationsgliederung lassen sich Artengruppen mit monofaktorieller oder polyfaktorieller Aussagekraft ableiten (Näheres Westhoff & Van der Maarel 1973, vgl. auch Hundt 1966, sowie die ökologischen Reihen von Ramenskij 1938). Die auf den interessierenden Faktor geeichten Gesellschaften, die vielfach den hochdifferenzierten soziologischen Rangstufen der Varianten und Subvarianten angehören, können als Kartierungseinheiten dienen (Tüxen 1954b, 1958, Niemann 1963, 1973).

Die Vervollkommnung der Eichverfahren führt zwangsläufig in Richtung auf chemisch-physikalische Meßmethoden und mathematische Datenverarbeitung, die zuletzt ihre Hauptgegenstände bilden. Der Beitrag der Vegetationskunde bleibt immer der gleiche. Sie bietet nach wie vor die Kenntnis der definierten Vegetationseinheit, die den primär erfaßbaren Repräsentanten des Ökosystems darstellt (Niemann 1973, S. 10). Eine umfassende Übersicht der Koinzidenzen zwischen Grünlandgesellschaften und Grundwasser gibt Niemann (1973), der ein großes Datenmaterial mathematisch-statistisch aufbereitet hat.

Dieses hochentwickelte Eichverfahren ermöglicht volle mathematische Fassung des Indikatorwertes der Vegetation. Den Ergebnissen kommt hohe Aussagekraft und maximale Ausschöpfung der Übertragbarkeit zu. Doch verengert sich der räumliche Geltungsbereich im gleichen Maße wie die Quantifikation ins Detail geht.

Hierzu äußern sich Tüxen (1954b, 1958) und Niemann (1963, 1968, 1970, 1973). Bei der Wahl der Methode ist daher zu prüfen, ob nicht eine einfache qualitative Beurteilung, die auf breiter Fläche lückenlos anwendbar ist, die gestellte Aufgabe sachdienlicher zu lösen vermag als genaueste Messung, die sich wegen ihrer Aufwendigkeit nur in begrenztem Maße durchführen läßt.

Die Aussagekraft der Eichung läßt sich noch erweitern, wenn auch die physiologische Reaktion von Einzelarten in Bezug auf Hydratur und Evapotranspiration verfolgt wird (Arland 1968, Niemann 1972, 1974, Rychnovská et coll. 1972).

8.6   Zur kritischen Anwendung des Indikatorwertes

8.6.1   DIREKT UND INDIREKT WIRKENDE FAKTOREN AM BEISPIEL DER SOZIALBRACHE

In das hochkomplizierte System der Pflanzengesellschaften, d. h. in ein "Wirkungsgefüge im Wettbewerb um Raum, Nährstoffe, Wasser und Energie, das sich in einem soziologisch dynamischen Gleichgewicht befindet, in dem jedes auf alles wirkt" (Tüxen 1957, S. 151), greift der Mensch meistens mit einer einzigen, technisch begrenzten Maßnahme ein. Entweder verändert er den mittleren Stand des Grundwasserspiegels oder er bricht einen Stauhorizont im Boden oder er verstärkt die Niederschläge durch Beregnung.

Mit Sicherheit muß erwartet werden, daß derart einschneidende Eingriffe keinen kurzen und unverzweigten Kausalitätsablauf verursachen. Vielmehr muß mit weitreichenden und mehrseitigen Wirkungen gerechnet werden. Daher ist von jeder Auswertung der Indikatoreigenschaften von Pflanzengesellschaften zu fordern, daß sie sich des Gegensatzes zwischen direkt und indirekt wirkenden Faktoren im Sinne H. Walters (1951, S. 9ff.) bewußt bleibt. Ein Beispiel indirekter Faktorenwirkung, das erst seit kurzem in den Vordergrund getreten ist, bietet die sog. Sozialbrache.

Seitdem in Mitteleuropa die Nutzung beträchtlicher Grünlandflächen unrentabel geworden ist, werden diese nicht mehr gemäht. Das Ausbleiben der Mahd führt zu überraschenden Verschiebungen im Pflanzenbestand; aus einem feuchten A r r - h e n a t h e r e t u m kann eine C a l t h i o n - Wiese entstehen. Ein Foto, das den Kontrast zwischen genutzten und ungenutzten Wiesen längs einer Parzellengrenze eindrucksvoll zeigt, gibt Arnd (in Braun-Blanquet 1964, S. 485).

Der Pflanzenbestand auf Sozialbracheflächen kann sich zwar in unterschiedlicher Richtung verändern (vgl. Krause 1974), doch führt eine vorherrschende Tendenz zum Rückgang der A r r - h e n a t h e r e t a l i a - und zum Überhandnehmen der M o - l i n i e t a l i a - Arten, vor allem der Hochstauden *Filipendula ulmaria, Cirsium oleraceum, Angelica silvestris* sowie der Großseggen *Carex acutiformis* und *C. gracilis*. Zur Erklärung führen Stählin & Bühring (1971) die Erfahrung an, daß auf brachliegenden Wiesen die Entwässerungsanlagen nicht mehr in Ordnung gehalten werden. Doch fallen die Grenzen der kontrastierenden Bestände meist so scharf mit der Nutzungsgrenze zusammen, daß eine unmittelbare Wirkung des Brachfallens wahrscheinlich wird. Als plausible Ursache der Bestandsveränderung bietet sich eine Verschiebung im Konkurrenzverhältnis der beteiligten Artengruppen an, die aus konstitutionell gelenkten Reaktionen der Pflanzen auf unmittelbare Einwirkungen der Sozialbrache erklärt werden kann.

Dem A r r h e n a t h e r e t u m ist ein Wuchsrhythmus eigen, der mit der Zweischnittnutzung nicht kollidiert, weil er durch frühen Austrieb und kräftigen Nachtrieb den Stoffverlusten des Heuschnittes ausweicht. Zugleich wird die Gesellschaft durch Düngung entscheidend gefördert. Auf der zweischnittigen gedüngten Wiese gewinnt sie ihre höchste Konkurrenzkraft und kann sich auf Standorte ausbreiten, deren hoher Feuchtegrad nicht ihrem Optimum entspricht. Die M o l i n i e t a l i a -Hochstauden werden vom normalen Heuschnitt in einem Entwicklungsstadium getroffen, in dem sie noch nicht zur vollen Größe ausgewachsen sind. Die Mahd unterbricht ihre Stoffproduktion und hemmt damit ihre Konkurrenzkraft. Ein Beispiel bietet der Vergleich zwischen den niedrigen, schmächtigen Individuen, die *Angelica silvestris* auf regelmäßig gemähten Feuchtwiesen bildet und den massigen hohen Exemplaren dieser Pflanze an ungenutzten und ungedüngten Wald- oder Grabenrändern, die oft unmittelbar nebeneinander stehen.

Unterbleibt auf einer der häufig vorkommenden Wiesen, auf denen *Arrhenatherum* dominiert und *Molinietalia*-Stauden in geringer Menge eingestreut sind (A r r h e n a t h e r e t u m c i r - s i o s u m o l e r a c e i) die Nutzung, so verliert *Arrhenatherum* die Förderung durch die Nährstoffgaben, während den *Molinietalia*-Arten das Handicap des vorzeitigen Substanzverlustes genommen wird. Somit wirken zwei Anstösse in Richtung auf eine Bestandsumwandlung, die sich unabhängig von der Befeuchtung aus dem konstitutionell verankerten Wachstumsrhythmus und den Nährstoffansprüchen der beteiligten Pflanzen ergibt. Unmittelbar auf die Pflanzen wirkender Faktor ist demnach, obwohl der Augenschein dafür spricht, nicht die Befeuchtung, sondern das von der Nutzung gelenkte Zusammenspiel von Stoffproduktion und Stoffentzug, das über die Konkurrenzfähigkeit entscheidet. In allgemeiner Formulierung hat bereits Ellenberg (1963, S. 739) Entsprechendes gesagt.

Diese Erörterungen sollen keinesfalls den hohen und vielfach geprüften Indikatorwert der Pflanzengesellschaften für die Standortsfeuchte in Frage stellen. Sie mögen aber daran erinnern, daß bloße Koinzidenzen nicht in allen Fällen Gewähr für Sicherheit geben, solange die tatsächlich und unmittelbar wirkenden Kräfte innerhalb einer langen Kausalkette nicht bekannt sind.

Daß auf Sozialbrachflächen die Standortsbedingungen auch anderen Einflüssen unterliegen, soll nicht unerwähnt bleiben. So ist als sicher anzunehmen, daß die schwer am Boden liegende Decke der toten Pflanzenreste die Verjüngung der einzelnen Arten selektierend beeinflußt.

Erfolgreiches Eingreifen in die Pflanzendecke setzt Informationen über die abiotischen Standorte voraus, die von anderen Disziplinen als der Vegetationskunde geliefert werden müssen. Doch leistet letztere ebenfalls einen Beitrag, indem sie die ökologischen Indikatoreigenschaften der Pflanzengesellschaften auf dem Wege der unmittelbaren Anschauung und des Vergleichens zu ergründen sucht. Da ihr Gegenstand ein komplexes, in langjähriger Entwicklung entstandenes Gebilde ist, gelangt sie zu integrierenden Schlußfolgerungen über verwickelte Zusammenhänge. Sie erreicht damit in kurzer Zeit und in einem einzigen Arbeitsgang ein erstrebenswertes Fernziel. Ihre Arbeitsweise ermöglicht ihr, zugleich die ganze Pflanzendecke synoptisch in ihre Überlegungen einzubeziehen und mit hoher Sicherheit die Existenz ökologischer Grenzen zu erkennen, auch wenn sie das Zustandekommen dieser Grenzen nicht immer erklären kann. Doch muß ihr bewußt bleiben, daß sie das Endergebnis komplexer Vorgänge, jedoch nicht immer die verschlungenen Wege erkennt, auf denen es zustandekam.

Die ökologische Messung muß dagegen, wenn sie Normalzustände erfassen will, über längere Zeiträume arbeiten und apparativen Aufwand treiben. Viele Ergebnisse sind nur für punktförmig verteilte Beobachtungsstellen verbindlich, die der Interpolation bedürfen. Um die für das Grünland der Flußauen bedeutungsvollen Hochwasser- oder Eisgangsituationen zu erfassen, ist jahrelanges Abwarten erforderlich. Andererseits lassen sich die Ergebnisse zahlenmäßig fassen und sie behandeln begrenzte, nicht komplex undurchsichtige Zustände.

Das beste Ergebnis wird erreicht, wenn jede der beteiligten Disziplinen ihre spezifische Leistung erbringt und den anderen die Aufgaben überläßt, die ihnen zukommen. Zur elementaren Leistung der Vegetationskunde in Gestalt der Umgrenzung einheitlicher Standorte sei noch bemerkt, daß sie die Differenzierung sehr weit zu treiben vermag, weil die floristisch-soziologische Vielfalt alle Erwartungen übertrifft. Ein Beispiel gibt Abb. 1.

Im Sinne dieser Erörterungen versuchen Brünner & Krause (1968), sowie Krause & Müller (1973) das Ertragspotential mehrerer Grünlandgesellschaften durch kombinierte Berücksichtigung von Pflanzenbestand, Bodenqualität, Klima und Nutzung zu erklären.

8.7   Qualitative Auswertung von Vegetationskarten

8.7.1   Beispiel einer weiträumigen Flussaue in Mitteleuropa

Abb. 1 zeigt an einer mit Wiesen bestandenen Flußaue, wie die Vegetationskarte das Meer der Grashalme überschaubar zu gliedern vermag und wie schon während der Kartierung aus dem Indikatorwert der Pflanzengesellschaften, ihrer räumlichen Ordnung und aus Angaben der Besitzer zur Bewirtschaftung erste Kausalzusammenhänge sichtbar werden. Die Untersuchung (Krause 1958) wurde ausgeführt, weil eine geplante Ableitung des Flußwassers auf den Wiesen Trockenschäden erwarten ließ. Gefordert wurde eine Prognose der zu erwartenden Schäden und Anregungen für Ausgleichsmaßnahmen, z.B. die Umwandlung trockengefallenen Grünlandes in Acker oder den Ersatz der wuchsfördernden Wirkung des Wassers durch Düngemittel. Da das Projekt aufgegeben wurde, gelangte die Arbeit nicht über die Bestandsaufnahme und die ökologisch-vegetationskundliche Auswertung hinaus.

Die Karte gliedert das Grünland in fünf Gesellschaftsgruppen, deren Einzelbestände jeweils durch gemeinsame Charakterzüge zusammengehalten werden. Im Oberlauf der Flußstrecke liegen Rieselwiesen (Sign. 3–5), denen durch Gräben im Spätwinter und Sommer Flußwasser zugeführt wird. Soweit das letztere seine ursprünglich schnelle Bewegung beibehält, setzen die Wiesen sehr früh mit dem Wachstum ein und können viermal geschnitten werden. Typisch ist das A r r h e n a t h e r e t u m a l o p e c u r e t o s u m in einer Ausbildung mit *Lolium italicum*, einer in Südwestdeutschland und der Schweiz verbreiteten Gesellschaft der Rieselwiesen. Wo das Wasser stagniert, gehen die ertragreichen Wiesenpflanzen zurück, während sich lichtbedürftige Kleinkräuter (*Plantago lanceolata*) oder gegen Überflutung unempfindliche Kriechpflanzen (*Ranunculus repens*) breit machen. Der Kartierung wurde eine dreifache Abstufung zugrundegelegt. – Im Gelände bilden die Gesellschaften ein engräumiges Mosaik, das auf der zusammenfassenden Abb. 1 nicht vollständig dargestellt werden konnte.

Im ganzen ist dieser Talabschnitt einer wirksamen Durchströmung ausgesetzt, die aus der Geomorphologie verständlich wird. Am oberen Kartenrand verläßt der Fluß ein Engtal, aus dem er einen Geröllschwemmfächer getragen hat. Die Rieselwiesen stehen auf der Wurzel dieses Fächers, der mit seiner schwachen Konvexwölbung den Abfluß begünstigt.

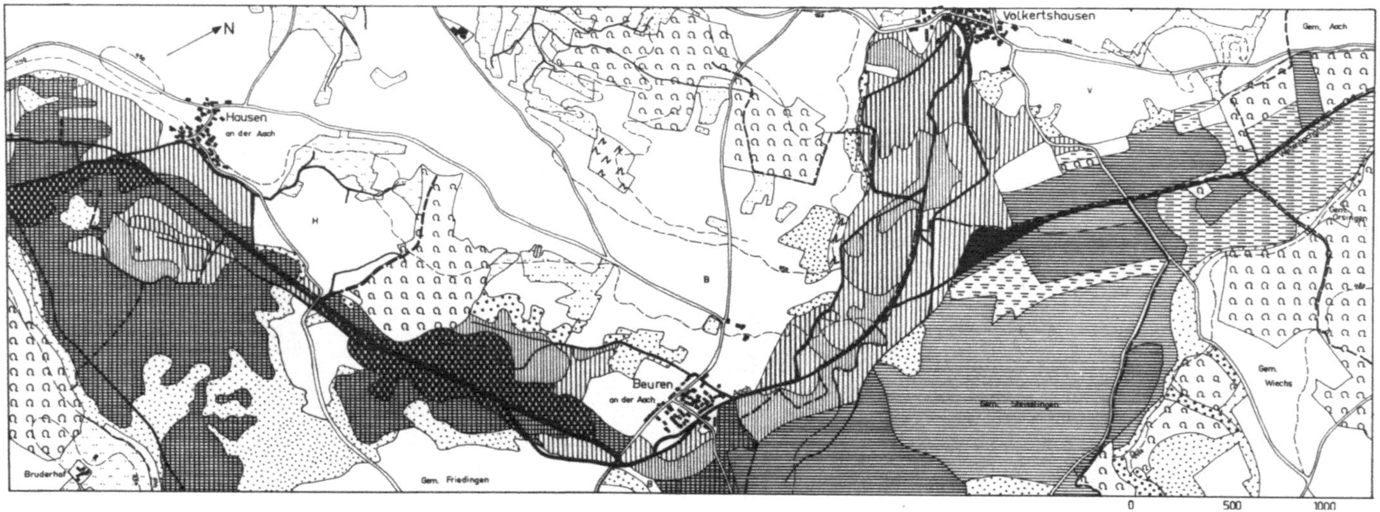

Abb. 1. Grünlandgesellschaften einer weiträumigen Flussaue in Mitteleuropa (Radolfzeller Aach am Bodensee). Fließrichtung von Volkertshausen am oberen Rand der Karte nach Hausen am linken Rand.

Gesellschaftskomplex unvernässter Wiesen

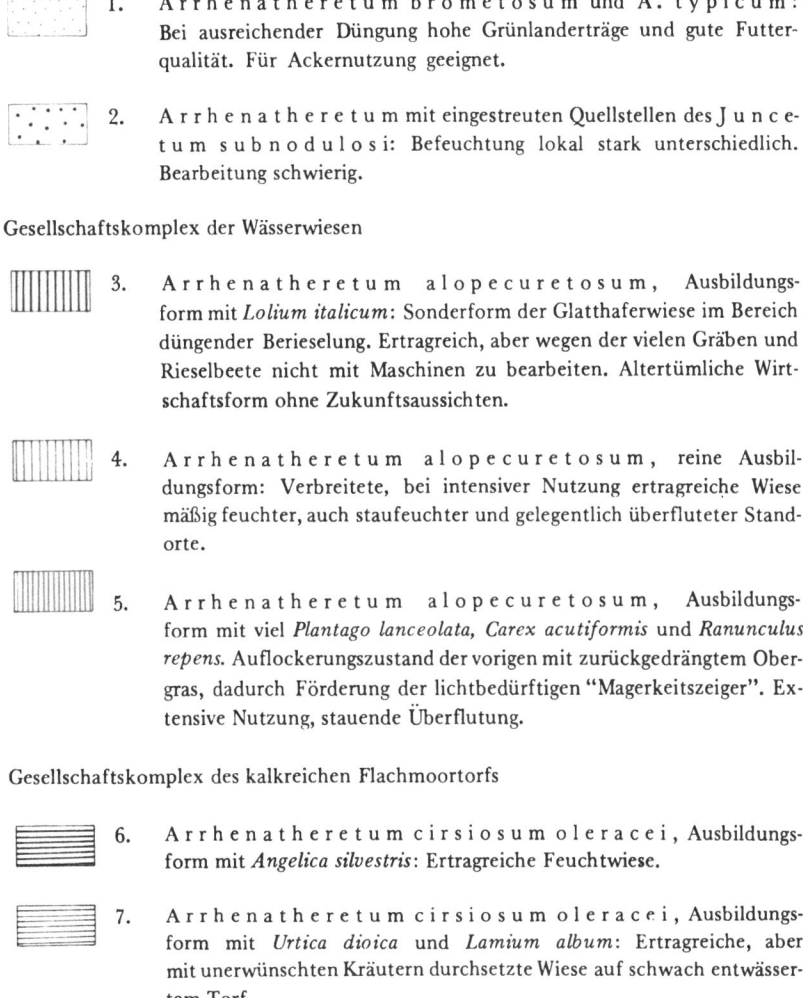

1. Arrhenatheretum brometosum und A. typicum: Bei ausreichender Düngung hohe Grünlanderträge und gute Futterqualität. Für Ackernutzung geeignet.

2. Arrhenatheretum mit eingestreuten Quellstellen des Juncetum subnodulosi: Befeuchtung lokal stark unterschiedlich. Bearbeitung schwierig.

Gesellschaftskomplex der Wässerwiesen

3. Arrhenatheretum alopecuretosum, Ausbildungsform mit *Lolium italicum*: Sonderform der Glatthaferwiese im Bereich düngender Berieselung. Ertragreich, aber wegen der vielen Gräben und Rieselbeete nicht mit Maschinen zu bearbeiten. Altertümliche Wirtschaftsform ohne Zukunftsaussichten.

4. Arrhenatheretum alopecuretosum, reine Ausbildungsform: Verbreitete, bei intensiver Nutzung ertragreiche Wiese mäßig feuchter, auch staufeuchter und gelegentlich überfluteter Standorte.

5. Arrhenatheretum alopecuretosum, Ausbildungsform mit viel *Plantago lanceolata, Carex acutiformis* und *Ranunculus repens*. Auflockerungszustand der vorigen mit zurückgedrängtem Obergras, dadurch Förderung der lichtbedürftigen "Magerkeitszeiger". Extensive Nutzung, stauende Überflutung.

Gesellschaftskomplex des kalkreichen Flachmoortorfs

6. Arrhenatheretum cirsiosum oleracei, Ausbildungsform mit *Angelica silvestris*: Ertragreiche Feuchtwiese.

7. Arrhenatheretum cirsiosum oleracei, Ausbildungsform mit *Urtica dioica* und *Lamium album*: Ertragreiche, aber mit unerwünschten Kräutern durchsetzte Wiese auf schwach entwässertem Torf.

8. Arrhenatheretum cirsiosum oleracei, Ausbildungsform mit reichlich *Achillea millefolium, Plantago lanceolata, Festuca rubra ssp. fallax*: Verarmungszustand von 6 nach übermässiger Entwässerung.

9. Torfstich, nicht landwirtschaftlich genutzt.

Gesellschaftskomplex der Naßwiesen auf Mineralboden

10. Cirsio-Polygonetum mit *Carex acutiformis*: Ertragreich, oft mit hohen Anteilen unerwünschter Pflanzen.

373

11. Cirsio-Polygonetum mit *Phalaris arundinacea* und *Rumex crispus*: Höchste Mengenleistung, geringe Futterqualität. Während der Vegetationszeit regelmässig überschwemmt und dann nicht nutzbar.

Gesellschaftskomplex der Extensivwiesen

 12. Molinietum coeruleae: Ungedüngte, erst im Spätsommer gemähte Wiese in einem abgelegenen Winkel der Gemeindeflur. Streunutzung.

Sonstiges

13. Wald

14. Wiesen, nicht kartiert.

374

Weiter ab vom Fluß deutet die großflächig-gleichmäßige Verteilung der Signaturen 6–8 auf ausgeglichene Standortsbedingungen. Der Boden besteht hier aus Flachmoortorf, also einer Stillwasserablagerung mit ebener Oberfläche. Vorherrschende Gesellschaft ist das für kalkreiches Anmoor typische A r r h e n a - t h e r e t u m   c i r s i o s u m   o l e r a c e i, das in Abhängigkeit von der Befeuchtung drei Untergesellschaften ausgebildet hat. Längs eines seitlich zufließenden tiefeingeschnittenen Grabens ist das Moor überoptimal entwässert (Sign. 8). Im lückigen Bestand dominieren magere Untergräser und niedrige Kräuter. Näher zum Fluß nehmen die Trockenschäden ab. Hier haben sich die stickstoffbedürftigen Pflanzen *Urtica dioica* und *Lamium album* eingestellt, die für schwach entwässerte Flachmoorwiesen charakteristisch sind (Walther 1950). Noch näher am Fluß treten normale Bestände des A r r h e n a t h e r e t u m   c i r s i o - s u m   o l e r a c e i auf.

Eine Sonderstellung nimmt die kleine Fläche der Sign. 12 im Norden der ausgedehnten Gemarkung Steißlingen ein. Sie liegt schwer zugänglich 4 km von der Ortschaft entfernt hinter einem breiten Graben und wird weder gedüngt noch regelmäßig genutzt. Ihr geringer Wirtschaftswert ist weniger auf naturgegebene Benachteiligung als auf die Extensivnutzung im entferntesten Winkel der Gemeindeflur zurückzuführen. In der Nähe der Ortschaft Beuren fallen noch andere Vegetationsgrenzen mit Besitzgrenzen zusammen. Teilweise liegen berieselte und unberieselte Flächen nebeneinander; an anderer Stelle grenzen intensiv und extensiv genutzte Wiesen längs einer Gemarkungsgrenze zusammen, auf deren einer Seite die ortsnahen Flächen der Gemeinde Beuren liegen, während sich daneben ein ortsferner Teil der Gemarkung Steißlingen anschließt. Hier wird ebenfalls die Abhängigkeit der Grünlandbestände von der Bewirtschaftung sichtbar.

Unterhalb der Ortschaft Beuren kommen *Phalaris arundinacea* und *Rumex crispus* in hohen Anteilen vor, die regelmäßige Überflutung mit nährstoffreichen Wasser anzeigen (Sign. 11). Als Ursache der Überflutung gilt ein Wehr, das im Knick des geradlinig laufenden Flußbettes einen Bewässerungsgraben abzweigen läßt. Vor ihm stauen sich die abschwimmenden Blattmassen von *Ranunculus fluitans*, der im Oberlauf in großer Menge wächst und gemäht wird. Außerhalb der Überflutungswiesen schließt ein Rieselwiesenkomplex an, auf dem die mageren Bestände der Signaturen 4 und 5 vorherrschen, weil die zugeführte Wassermenge nicht ausreicht, die Ertragsstufe der Sign. 3 zustandezubringen.

Hinter einer Engstelle zwischen Acker und Wald greift das Grünland nach Süden weit in die Breite aus. Hier finden sich große Flächen des trockenen A r r h e n a t h e r e t u m t y p i c u m und b r o m e t o s u m (Sign. 1) auf Gelände, das über die Flußniederung ansteigt. Seine Umrißlinien, die in Abb. 1 durch Sign. 1 angegeben sind, deuten an, daß es nicht als Ablagerung des rezenten Flußlaufs entstanden sein kann.

Signatur 1 findet sich außerdem in kleinen Flächen am Saum des kartierten Geländes, wo sie die Grenze zwischen der Wiesenniederung und dem erhöhten Ackerland bezeichnet. Stellenweise sind Quellen mit Beständen der Kalkbinse *Juncus subnodulosus* eingestreut, die aus den angrenzenden diluvialen Ablagerungen gespeist werden (Sign. 2).

Nach Süden schließt eine große Senke an, in der die Gesellschaft C i r s i o - P o l y g o n e t u m die normale Naßwiese der Tieflagen Südwestdeutschlands repräsentiert. Die Lage in größerer Entfernung vom Fluß deutet darauf, daß sich das Bodenwasser nicht schnell bewegt. Andererseits verhindert das Allgemeingefälle der Talsohle, das rd. 2 °/oo beträgt, ein ertragshemmendes Stagnieren. Inmitten dieser Feuchtwiesen liegt, durch eine Geländestufe abgesondert, eine Fläche mit Berieselungsanlage und der zugeordneten Vegetationszonierung. Die ertragreichen Wiesen der Sign. 3 stehen am Zuleiter, die schwachwüchsigen (Sign. 4 und 5) weiter abseits.

Die von der Vegetationskartierung offengelegte Landschaftsgliederung mit ihren fünf Naturräumen, die unter der herrschenden Nutzungsform durch die Rieselwiesen mit *Lolium italicum*, die Moorwiesen mit *Urtica dioica*, die Überflutungswiesen mit *Phalaris arundinacea*, die C i r s i o - P o l y g o n e t u m - Sumpfwiesen ohne besondere Zeigerpflanzen und das trockene A r r h e n a t h e r e t u m b r o m e t o s u m charakterisiert werden, vermittelt einen ersten Eindruck der ökologischen Zusammenhänge.

Offen bleibt jedoch, ob sich aus dem aktuellen Zustand der Vegetation die Folgen eines Eingriffes in den Wasserhaushalt ausreichend sicher vorhersehen lassen. Daß nach der geplanten Wasserableitung die Befeuchtung im ganzen abnehmen wird, erscheint zwar sicher. Angesichts der großen Gegensätze innerhalb des Standortsmosaiks bleibt aber zu prüfen, ob der Wasserverlust überall die gleichen Folgen nach sich ziehen wird.

Durchgehend gleichartige Reaktionen sind deswegen nicht zu erwarten, weil die Aue bisher dem hochwirksamen, alle sonstigen Standortsunterschiede in den Hintergrund drängenden Faktor der starken Befeuchtung unterworfen war. Solange diese wirken konnte, waren die Wuchsbedingungen weitgehend ausgeglichen, wenn auch nicht, wie die Vegetationskarte erweist, alle Gegensätze beseitigt waren. Nach Wasserentzug wird Kies unfruchtbarer sein als Lehm. Ausgetrockneter Ton wird wiederum andere Wuchsbedingungen bieten als entwässerter Torf. Daher konnte nicht auf Bodenuntersuchung verzichtet werden (vgl. auch Valek 1960).

Die Bodenkarte (Abb. 2), die das am stärksten betroffene flußnahe Gebiet umfaßt, gibt folgende Gliederung:

1. Bereich brauner Böden auf dem Schwemmfächer am Oberlauf des Flusses. Kleinräumiger Wechsel zwischen skelettreichem Braunem Auenboden (Braune Vega) auf alten Uferwällen des Flusses (Sign. 1), Pseudogley (Sign. 2) und grauem tonigen Grundwassergley in Senken (Sign. 3).

Ein zweiter Komplex braunen Bodens mit eingestreutem Pseudogley als nahezu kreisförmige Insel südöstlich des Dorfes Hausen, deutlich über die Umgebung erhoben.

Die dritte und größte Fläche braunen Bodens außerhalb der Aue (Mit Sign. 1 in Abb. 1 zusammenfallend. Nur in Abb. 1 eingezeichnet). Nicht als Braune Vega, sondern als ökologisch ähnlicher Brauner Waldboden aufgefaßt.

2. Bereich des Flachmoortorfs in der Randsenke des Tals, differenziert in tiefgründigen Torf (Sign. 5) und Torf über Moormergel (Sign. 6). Inseln mit anmoorigem Grundwassergley (Sign. 4) auch weiter flußabwärts.

3. Grauer bis grünlich-grauer Grundwassergley im Bereich der gebremsten Strömung am Unterlauf (Sign. 3), Oberboden z. T. > 1 m mächtig, darunter Kies. Für die landwirtschaftliche Planung ergeben sich aus der Vegetations- und der Bodenkarte folgende primäre Informationen:

zu 1. Im flußnahen Komplex am Oberlauf bestehen schon vor der geplanten Trockenlegung große Unterschiede in Boden und Bewuchs. Immerhin werden die reliefgebundenen Standortsgegensätze durch die Berieselung teilweise ausgeglichen. Selbst auf den Kiesrücken gewährleistet letztere ansehnlichen Ertrag. Wasserverlust würde die Kontraste verschärfen.

zu 2. Der Torfboden wirft die wohlbekannten Probleme aller Moorentwässerungen auf. Überdies wären tiefgründiger Torf- und Torf über Moormergel gesondert auf ihre Reaktionen zu prüfen.

zu 3. Der gleichmäßig ausgebildete Grundwassergley am Unterlauf des Flusses mit seiner mächtigen Tonschicht über Kies

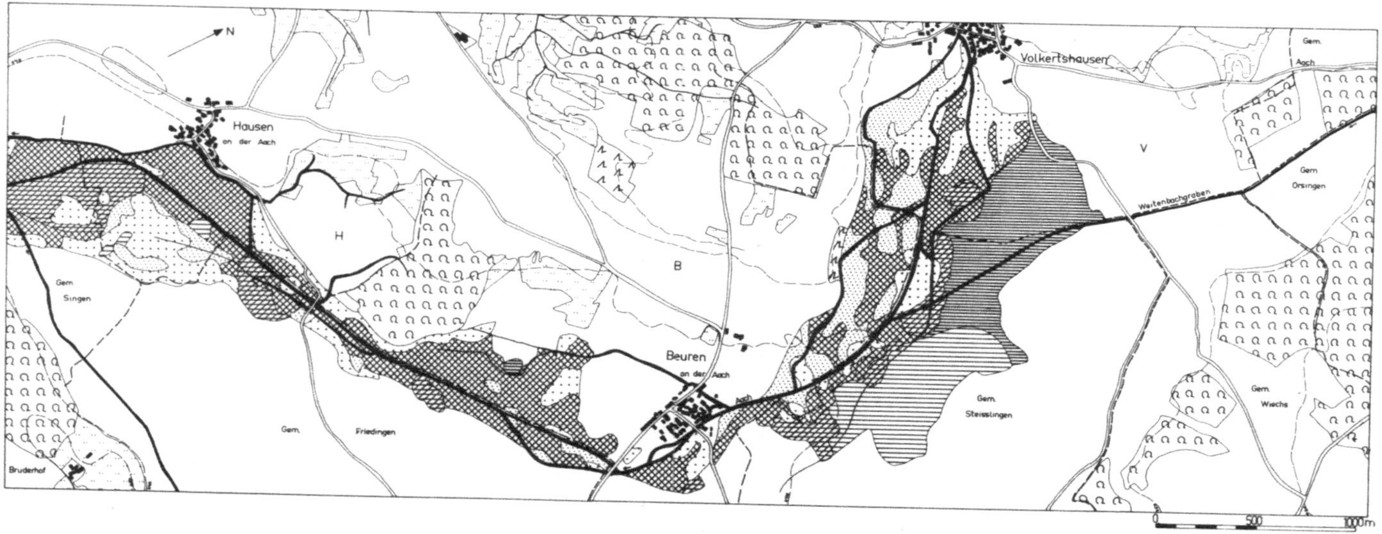

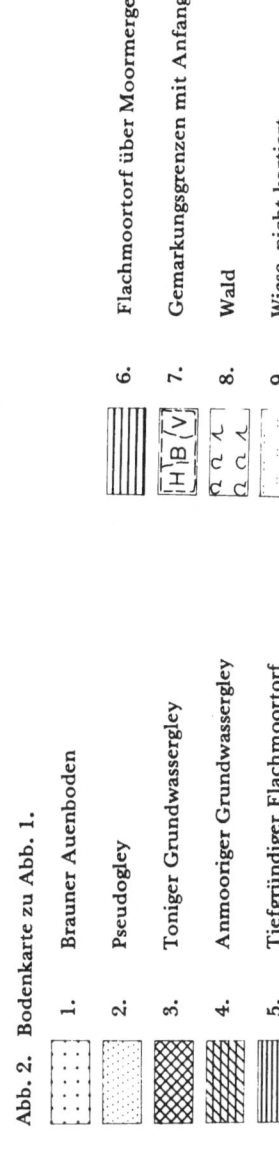

Abb. 2. Bodenkarte zu Abb. 1.

1. Brauner Auenboden
2. Pseudogley
3. Toniger Grundwassergley
4. Anmooriger Grundwassergley
5. Tiefgründiger Flachmoortorf
6. Flachmoortorf über Moormergel
7. Gemarkungsgrenzen mit Anfangsbuchstaben der Ortsnamen
8. Wald
9. Wiese, nicht kartiert

bietet ausgeglichene Wuchsbedingungen.

Aus alldem können folgende Planungsvorschläge abgeleitet werden: Auf dem Schwemmfächer mit seinen gegensatzreichen Standorten wäre zu prüfen, ob die Bewirtschaftung auch in Zukunft auf die engräumige Landschaftsgliederung Rücksicht nehmen soll, wie sie es bisher getan hat. Da die Kleinparzellierung wahrscheinlich nicht erhalten bleiben wird, muß einheitliche Bewirtschaftung über die Standortsgrenzen hinweg eingeleitet werden. Hierzu müßte die ausgleichende Wirkung verstärkter Düngergaben nutzbar gemacht werden. Die Jahresniederschläge von 750 bis 800 mm ermöglichen selbst auf kieshaltigem Braunen Auenboden noch annehmbare Grünlanderträge. Die Standortsunterschiede maschinell durch Bodenverschiebung oder Brechung von Stauhorizonten auszugleichen, sollte erst zuletzt ins Auge gefaßt werden.

Das Torfgebiet bedarf, wenn es nicht der Sozialbrache überlassen werden soll, einer speziellen moorkundlichen Untersuchung, der die Kenntnisse der beiden Meliorations-Folgegesellschaften (Abb. 1, Sign. 7 u. 8) von Nutzen sein werden.

Der tonige Boden der Sign. 10 u. 11, Abb. 1 sowie Sign. 3, Abb. 2 wird auch nach Abnahme des Wasserzuflusses durch sein erhebliches Speichervermögen dem Grünland ausreichende Befeuchtung bieten, zumal die vorherrschend flachkonkave Geländeform die Ansammlung von Oberflächenfeuchtigkeit begünstigt. Das Ausbleiben der Sommerüberflutungen im Bereich der Sign. 11 wird sogar die Nutzbarkeit der hochproduktiven Bestände verbessern. Gleichzeitig macht der hohe Tonanteil den Übergang zu Ackernutzung nicht ratsam. Wasserabzug dürfte die Feuchtestufe des A r r h e n a t h e r i o n -Verbandes zustandebringen. Zugleich ist auf dem dichtgelagerten Boden mit besonderer Förderung von *Alopecurus pratensis* und der Ansiedlung des G a l i o - A l o p e c u r e t u m oder des A r r h e n a t h e r e t u m a l o p e c u r e t o s u m zu rechnen. Beide Gesellschaften sind dankbar für Düngung. Demnach kann auch nach Ableitung des Flusses die Möglichkeit intensiver Grünlandnutzung erwartet werden.

Die Mitteilungen sollten die unmittelbare spezifische Leistung der Vegetationskunde darstellen, die in der übersichtlichen, auf das Kriterium der Pflanzengesellschaften gestützten Gliederung einer zunächst unübersehbaren Mannigfaltigkeit besteht. An diesem elementaren Ergebnis schärft sich der Blick für die Ursachenzusammenhänge zwischen pflanzlicher Produktion und Standort. Die gewonnenen Einsichten werfen weiterführende

Fragen auf, die oft nur durch messende und rechnende Arbeitsmethoden gelöst werden können, deren Lösung aber durch die vegetationskundlichen Ergebnisse gefördert wird.

Einen kurzen Überblick über die Vegetationskartierung auf Grünland in den Niederlanden gibt De Boer (1954).

8.7.2   ZUM ERTRAGSPOTENTIAL AUF DEN AUENWIESEN DER SUBKONTINENTALEN TROCKENGEBIETE

In den Flußauen der kontinental getönten Gebiete Europas, z.B. den slowakischen Anteilen der Donau- und Theissniederung, wo die Niederschläge 550 bis 600 mm im Jahr betragen und die Verdunstung hoch ist, bildet das Wasser den ökologischen Minimumfaktor. Die Wiesen werden dort mehrmals im Jahr überschwemmt, leiden aber in Trockenzeiten an Wassermangel. In den überwiegend tonigen Böden bilden sich dann polygonale Risse mit Salzüberzügen. Das Überschwemmungswasser liefert mit seinen Schlick- und Planktonablagerungen außer der Befeuchtung auch große Nährstoffmengen.

In Abhängigkeit vom Relief der Flußaue, das über Tiefe, Strömungsgeschwindigkeit und Dauer des Hochwassers entscheidet, bilden die Artengruppen der Naß-, Feucht- und Frischwiesen eine charakteristische Abstufung (Abb. 3). In den Senken dominieren Phragmitetea-Gesellschaften, vor allem das P h a l a r i - d e t u m   a r u n d i n a c e a e , an besonders tiefen Stellen auch das G l y c e r i e t u m   m a x i m a e . Auf den Rücken bilden M o l i n i o n -   und M e s o b r o m i o n - Arten mehrere wechseltrockene Gesellschaften, z.B. das S e r r a t u - l o - F e s t u c e t u m   c o m m u t a t a e in der Fazies von *Festuca sulcata.* Das mittlere Niveau wird vom C n i d i o n - und A g r o p y r o - R u m i c i o n - Verband, z.B. von der G r a - t i o l a   o f f i c i n a l i s - C a r e x   p r a e c o x - s u z a e - Assoziation besetzt. Zum C n i d i o n - Verband vgl. Balátová- Tulačková (1969b). Hier gelangen häufig die landwirtschaftlich hochwertigen Gräser *Alopecurus pratensis* und *Festuca pratensis* zur Dominanz. Die *Festuca sulcata-* Facies des S e r r a t u l o - F e - s t u c e t u m   c o m m u t a t a e wird auf den beweideten, schwach salzhaltigen Standorten der Ost-Slovakei durch das A l o p e c u r o - F e s t u c e t u m   p s e u d o v i n a e ersetzt.

Die verschiedenen Artengruppen stehen miteinander in einem dynamischen Gleichgewicht, das in den Jahren mit mehrfacher und langdauernder Überschwemmung die Bewohner der

380

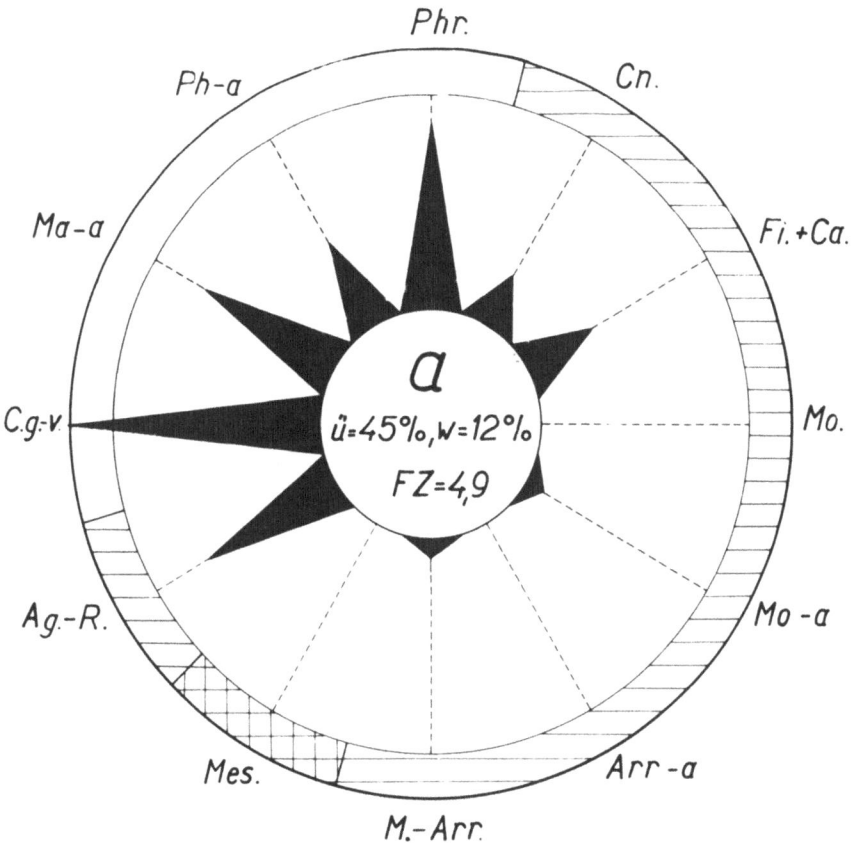

Abb. 3a

Abb. 3. Anteil der soziologisch bedeutenden Artengruppen in vier Gesellschaften (a-d) der Überschwemmungswiesen von Lanžhot/Südmähren. Die erste (a) bildet das feuchteste, die vierte (d) das trockenste Glied derselben ökologischen Feuchtigkeitsreihe; a) Phalaridetum arundinaceae Libbert 1931; b) Gratiola officinalis — Carex praecox-suzae-Ass. rorippetosum sil—vestris Bal.—Tul. 1963c;) Gratiolaofficinalis — Carex praecox-suzae-Ass. galietosum borealis Bal.-Tul. 1963d;) Serratulo-Festucetum commutatae Bal.-Tul. 1963. - Pflanzensoziologische Artengruppen, Cn. = Cnidion; Fi. + Ca. = Filipendulion + Calthion; Mo. = Molinion; Mo — a = Molinietalia; Arr — a = Arrhenatherion u. Arrhenatheretalia; M — Arr. = Molinio-Arrhenatheretea; Mes-(Vi.) = Mesobromion (Violion caninae); Ag.-R. = Agropyro-Rumicion crispi; C.g.-V = Caricion gracilis-vulpinae; Ma — a = Magnocaricetalia; Ph — a = Phragmitetalia; Phr = Phragmitetea. Zum Indikatorwert, FZ = Feuchtezahl nach Ellenberg; ü = Anteil der Überschwemmungszeiger (%); w = Anteil der Wechselfeuchtigkeitszeiger (%).

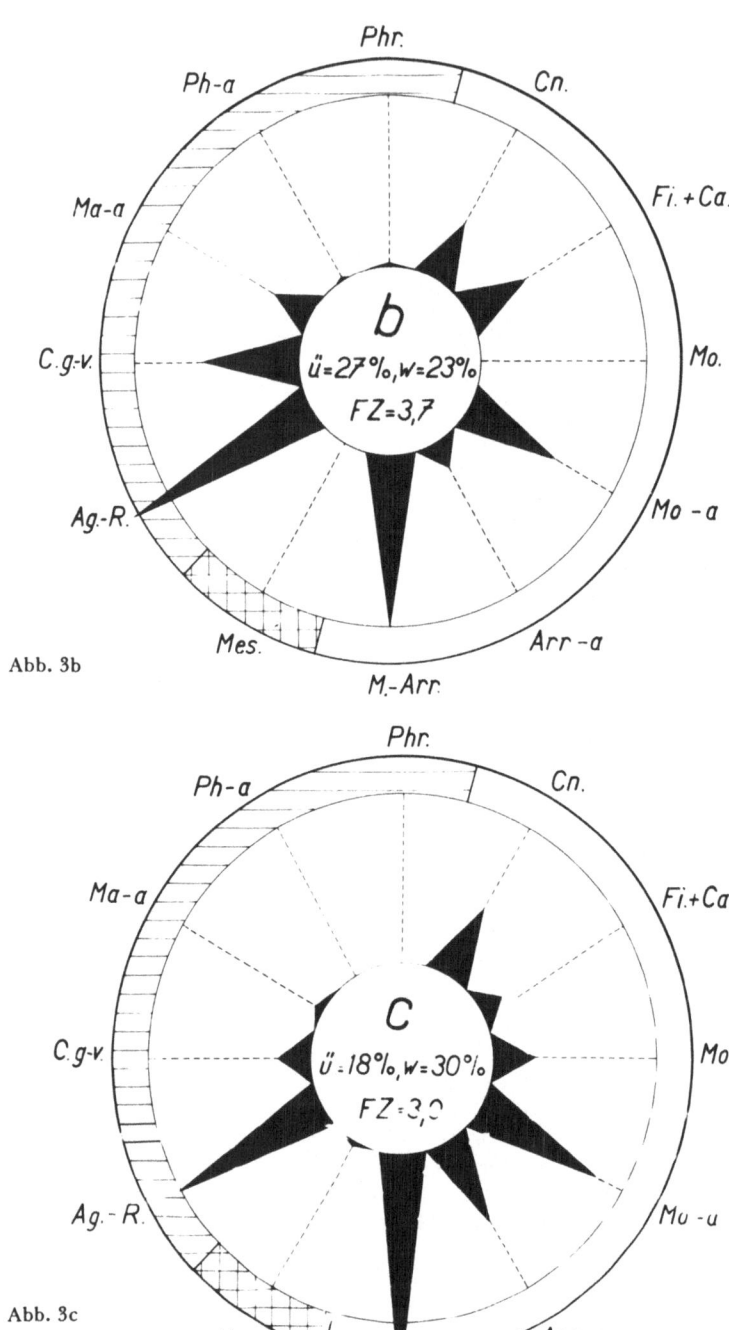

Abb. 3b

Abb. 3c

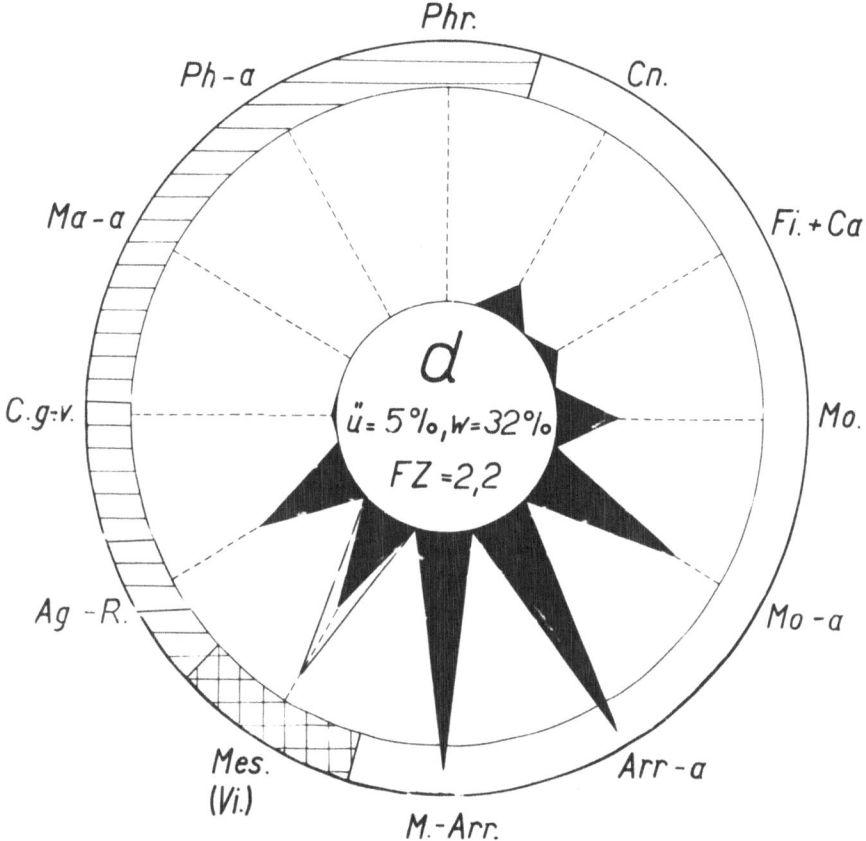

Phr.

Ph-a

Cn.

Ma-a

Fi.+Ca.

C.g-v.

d

ü= 5%, w=32%

FZ =2,2

Mo.

Ag-R.

Mo-a

Mes.
(Vi.)

Arr-a

M.-Arr.

Abb. 3d

Naßstandorte, in den Jahren mit niedrigem Wasserstand die trockenheitsfesten Pflanzen fördert (Rabotnov 1974 S. 262, Balátová-Tuláčková 1972a). Auf den Lanžhot-Wiesen in Südmähren erreichte im Jahre 1963 *Alopecurus pratensis* in der G r a t i o - l a o f f i c i n a l i s - C a r e x p r a e c o x - s u z a e Ass. im ersten Schnitt 170 cm Höhe und absolute Dominanz mit einem Ertrag von 60 q/ha (Abb. 4). In diesem Jahr hatte das nährstoffreiche Überschwemmungswasser im Frühjahr etwa 40 Tage auf den Wiesen gestanden und eine größte Tiefe von ca. 40 cm erreicht, während die mittlere maximale Tiefe ca. 15 cm beträgt. In Jahren mit langandauerndem Katastrophenhochwasser wird *Alopecurus* auf schwerem Boden von *Phalaris arundinacea* überwachsen (Abb. 5). Dagegen bleibt er ebenso wie die weniger feuchtigkeitsbedürftige *Festuca pratensis* in Jahren ohne Überschwemmung klein, blüht nicht und erreicht keinen vollen

Bestandschluß (Abb. 6). Der Rückgang verstärkt sich, wenn die Überschwemmungen durch Flußkorrektion ausgeschaltet werden. Danach nimmt zunächst der Wasser- und Nährstoffvorrat im Boden ab und die Tätigkeit der Mikroorganismen geht zurück. Darüber hinaus werden in den von Tonmineralien der Illit- und Kaolinitgruppe gebildeten Böden die physikalischen Eigenschaften durch die niedrige Bodenfeuchtigkeit zu Beginn des Austreibens und das bald anschließende Austrocknen bis zum Welkungsgrad ungünstig beeinflußt. Gleichzeitig gehen die ertragreichen Pflanzen der M o l i n i e t a l i a zurück, während kurz- und schmalblättrige *Festuca*-Arten, die keinen vollen Narbenschluß zustandebringen, immer mehr zunehmen. Beweidung verschlimmert die Situation noch weiter.

Abb. 4. *Alopecurus pratensis* — Fazies der G r a t i o l a o f f i c i n a l i s - C a r e x p r a e c o x - s u z a e Assoziation. Höchsterträge nach langdauernder Frühjahrsüberschwemmung. Lanžhot-Wiesen, Südmähren. Phot. E. Bálatová-Tuláčková.

Abb. 5. Phalaridetum arundinaceae Veränderter Bestand nach lang-
dauerndem Katastrophenhochwasser. Lanžhot-Wiesen, Südmähren. Phot. E. Bálatová-
Tuláčková.

Abb. 6. Kleinwüchsiger Bestand der G r a t i o l a o f f i c i n a l i s - C a r e x p r a e ‚
c o x - s u z a e Assoziation. Nach mehrjärigem Ausbleiben der Überschwemmung, Ost-
slowakei. Phot. E. Bálatová-Tuláčková.

Weitere Ausführungen zum Verhältnis zwischen Befeuchtung und Produktivität der kontinentalen Überschwemmungswiesen macht Balátová-Tuláčková (1966, 1968, 1969, 1970, 1972a,
1973). Über Standort und Bewuchs vergleichbarer Wiesen in Un-
garn berichtet Kovács (1968a, b). Material aus Kroatien bringt
Ilijanić (1962).

Im humiden Mitteleuropa, wo keine klimabedingte Salzan-
reicherung stattfindet, kann Abwasser aus Kalibergwerken zu
Schädigungen des Grünlandes führen. Wie sich diese durch vegetationskundliche Wege nachweisen lassen, zeigen Speidel (1963)
sowie Krisch (1967, 1968) am Beispiel der Werra-Aue in Mittel-
deutschland. Die Ergebnisse erleichtern sowohl die Schadensver-
hütung als auch die landwirtschaftliche Planung.

Im Gebirge ist das Mosaik der Standorte und Gesellschaften oft so kleinflächig aufgeteilt (vgl. auch Niemann 1962; Hundt 1964c), daß eine untragbare Zersplitterung der Wirtschaftsmaßnahmen zustande käme, wenn jede Gesellschaft einer besonderen Meliorationsmaßnahme unterworfen würde. Die Planungsempfehlung muß diejenigen Flächen zusammenfassen, die trotz floristischer Unterschiede bezüglich der gestellten Frage gleichwertig sind.

Ein Beispiel bieten die Wiesen eines Schwarzwaldtales, deren natürliche, vom Relief bestimmte Gliederung durch künstliche Wasserzufuhr extrem gesteigert ist (Abb. 7). Eine vegetationskundliche Bestandsaufnahme sollte die ökologischen Voraussetzungen für eine durchgreifende, die Kleinparzellierung überwindende Wirtschaftsumstellung klären.

Die Kartierung ließ eine Großgliederung in A r r h e n a t h e r e t u m   m o n t a n u m , A n g e l i c a - P o l y g o n e t u m und C r e p i d o - J u n c e t u m   a c u t i f l o r i erkennen. Der Vergleich benachbarter Parzellen in einheitlicher Relieflage, also auf ähnlichen natürlichen Standorten, ergab dann, daß jede dieser Gesellschaften durch lokale Wirkungen des Rieselwassers in starkwüchsige, mäßig produktionsstarke und äußerst ertragsarme Fazies aufgeteilt war. Vegetationskarte, Reliefgliederung und Bodenuntersuchung (Beckmann 1965) ließen zuletzt in der Menge der Einzelbestände die ökologische Ordnung erkennen, auf der die Wirtschaftsplanung aufzubauen hatte.

Die zum A r r h e n a t h e r e t u m gehörenden Wiesen konzentrieren sich auf die drainierten Talhänge und die flachkonvexen Ablagerungen eines Geröllschwemmfächers. Diese Standorte tragen Braunerde. Die A n g e l i c a - P o l y g o n u m   b i s t o r t a -Gesellschaft, die in einer *Holcus lanatus*-Fazies entwickelt ist, bewohnt Unterhänge mit Wasseraustritt sowie den feuchten Saum des Schwemmfächers. Der Boden ist Hangwassergley.

Das C r e p i d o - J u n c e t u m steht in Mulden, in denen das Allgemeingefälle noch langsame Wasserbewegung hervorruft. An den untersten Talhängen gewinnt *Holcus lanatus* hohe Anteile neben *Juncus acutiflorus*. In den tiefsten Senken treten Kleinseggen des C a r i c i o n   c a n e s c e n t i s - f u s c a e in den Vordergrund. Diese Gesellschaften besiedeln sauren Anmoorgley.

Die Befunde erlauben es, eine Auswertungskarte zu entwer-

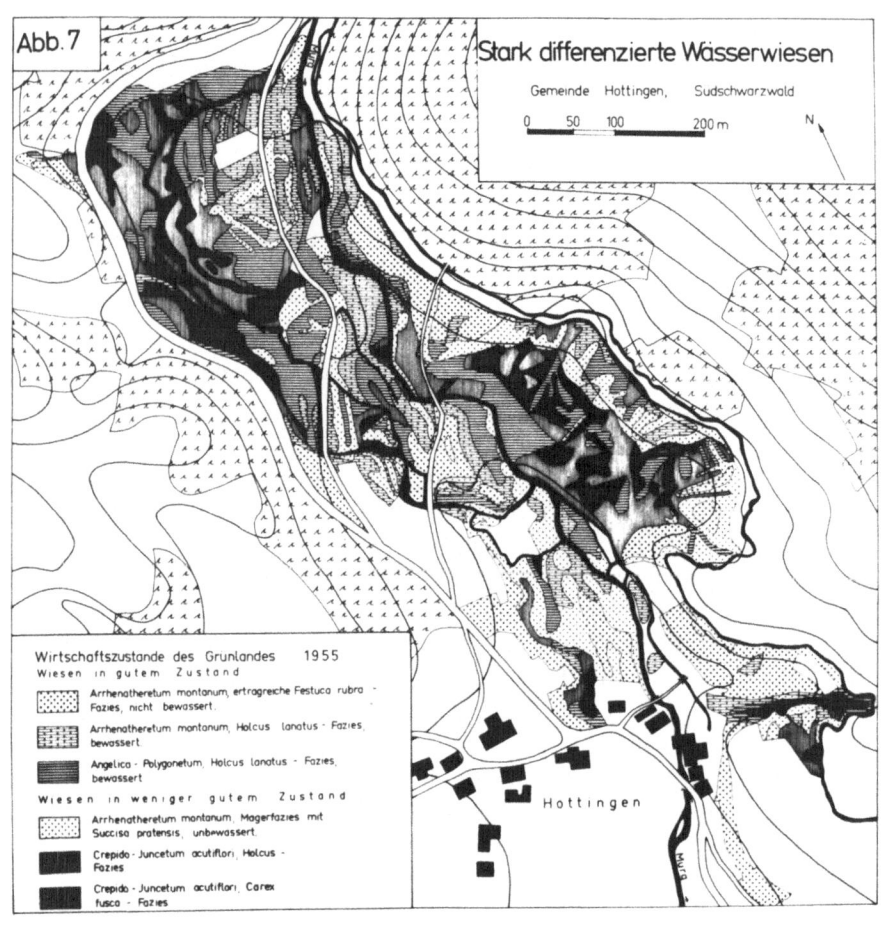

Abb. 7. Vegetationskarte des Grünlandes in einem engen Schwarzwaldtal. Tatsächlicher Zustand 1955.

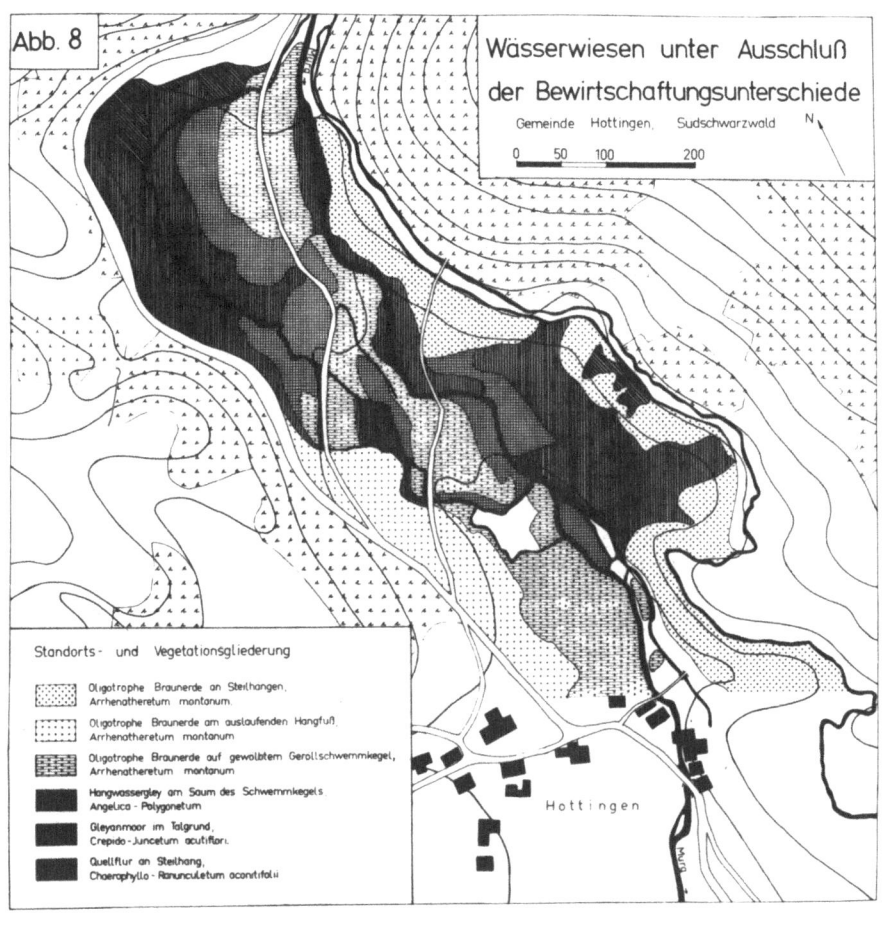

Abb. 8. Vegetationskarte des Grünlandes in einem engen Schwarzwaldtal. Natürliche Standortsgliederung ohne Wirtschaftseinflüsse.

fen, die von der Vielfalt des Bewuchses abstrahiert und großräumige ökologische Einheiten abgrenzt, die zugleich der Wirtschaftsplanung nutzbar gemacht werden können (Abb. 8). D a s A r r h e n a t h e r e t u m bietet ertragreiche, für Düngung dankbare Standorte (Brünner o.J.), deren fester Mineralboden auch schwere Maschinen trägt. Deren Einsatz kann durch die Hangneigung verhindert werden, weshalb diese in Abb. 8 berücksichtigt wurde. Das A n g e l i c a - P o l y g o n e t u m besiedelt produktive Feuchtwiesenstandorte, deren weicher Boden allerdings moderne Bewirtschaftung unmöglich macht. Das C r e p i d o - J u n c e t u m , das am wenigsten zur Intensivnutzung verlockt, wird voraussichtlich der Sozialbrache anheimfallen. Sollte weiterhin Interesse an einer Nutzung bestehen, wäre zu prüfen, ob nicht nach Ausbleiben des Rieselwassers die Vernässung zurückgehen wird.

Ein anderes Beispiel der Zusammenfassung engräumig verzahnter Feucht- und Naßwiesen beschreiben Halva & Lesak (1974) aus dem Bergland der Böhmisch-Mährischen Höhe. Durch Düngung und Produktionssteigerung wurde biologische Entwässerung mit Angleichung der Feuchtigkeitsgrade erreicht. Mit Hilfe zusammenfassender, die Vegetation berücksichtigender Kartierungseinheiten konnte Kielhauser (1958) die Grundlagen für Frostfrei-Verlegung von Bewässerungszubringern im höheren Gebirge, die Bemessung der Dränrohrdimensionen und ihren Abstand, sowie die Führung von Entwässerungsgräben festlegen.

Schließlich sei die Bewertung der Weidefähigkeit bisheriger Mähwiesen angeführt. Sie wird weitgehend von der Trittfestigkeit des Bodens bestimmt, die ihrerseits von der Durchfeuchtung abhängt. Mit ihrem Indikatorwert für die Bodenfeuchte erlaubt die Vegetation diejenigen Vernässungsgrade abzugrenzen, die eine Beweidung ausschließen oder mit einem noch tragbaren Risiko verbinden oder uneingeschränkt ermöglichen. Unwesentlich ist, welche Pflanzenbestände innerhalb einer Feuchtestufe auftreten. Die staudenreiche T r o l l i u s   e u r o p a e u s - P o l y g o n u m   b i s t o r t a — Gesellschaft ist ebensowenig weidefest wie der hohe Binsenbestand des C r e p i d o - J u n c e t u m a c u t i f l o r i oder der nahezu ertraglose Kleinseggenrasen des C a r i c i   c a n e s c e n t i s - A g r o s t i d e t u m   c a n i n a e . Andererseits können ungedüngtes N a r d e t u m und gedüngtes T r i s e t e t u m gleichermaßen ohne Trittschäden beweidet werden. Die Auswertungskarte, die der Planung dienen soll, unterscheidet nur drei Feuchtestufen, in denen die Vielzahl der vorhandenen Gesellschaften aufgegangen ist.

390

Die Notwendigkeit, eine unkomplizierte Auskunft zu geben, verlockt dazu, von Anfang an mit einer vereinfachten Skala zu arbeiten. Damit wäre das Ergebnis vorweggenommen, bevor es gefunden ist. Recht verstandene angewandte Vegetationskunde bringt zuerst die tatsächlich vorhandene Vielfalt zum Bewußtsein und leitet erst nachträglich an ihr die Auswertung ab, die nach größtmöglicher Vereinfachung streben darf und muß.

8.7.4 NACHWEIS VON STANDORTSVERÄNDERUNGEN UND DEREN PROGNOSE

Große technische Eingriffe in den Wasserhaushalt der Landschaft, z.B. Kraftwerksbau oder umfangreiche Brauchwasserentnahme fordern die exakte Feststellung ihrer Folgen für die Landwirtschaft. Am nächsten liegt es, den Zustand vor Beginn der Veränderung zu dokumentieren und zum Maßstab der Umstellung zu machen. Eine elementare Grundlage der Beweissicherung bietet die Vegetationskartierung. Werden nach einer angemessenen, zur Stabilisierung des neuen Zustandes erforderlichen Wartezeit keine Veränderungen an den Pflanzenbeständen erkennbar, dürfen Schäden im Wurzelbereich der Nutzpflanzen ausgeschlossen werden. Nach Grundwassersenkung war die Umstellung von einer nassen Kohldistelwiese zu einer typischen Glatthaferwiese auf Niedermoorboden nach 7 Jahren abgeschlossen (Ellenberg 1952a). Wiedenroth (1961) fand, daß die Veränderung in gleicher Richtung bis zu 30 Jahre dauern kann, während sie bei zunehmender Vernässung meist schon nach 2–3 Jahren sichtbar wird (vgl. auch Ilijanić 1971).

Wenn die Beständigkeit der Vegetation beurteilt werden soll, bedürfen auch kleine Verschiebungen des Artenbestandes der Beachtung. Wasserentzug führt keineswegs immer zur Entstehung einer anderen Assoziation. Nicht selten veranlaßt er nur den Übergang in eine andere Variante der gleichen Assoziation, wobei sich trotz geringer floristischer Unterschiede erhebliche Ertragsänderungen ergeben können. Ein Beispiel bieten die Signaturen 4 und 5 in Abb. 1. Beide repräsentieren das A r r h e n a - t h e r e t u m a l o p e c u r e t o s u m unter verschiedenen Feuchtebedingungen. Zu Sign. 4 gehört ein dichter, ertragreicher, zu Sign. 5 ein aufgelichteter schwachwüchsiger Bestand.

Eine Kartierung, die mit gleicher Intensität und Methodik vor und nach dem verändernden Eingriff erfolgte, beschreibt Ellenberg (1952a). Zwei Karten, die am gleichen Gelände mit 19 Jahren Abstand vor und nach einem Wasseraufstau durchgeführt wurden, vergleicht Hundt (1961). Allgemein zum Thema "Entwässerung und Fruchtbarkeit der Landschaft" äussert sich Balátová-Tuláčková (1969a).

Soll aus einer einmaligen Untersuchung eine Prognose abgeleitet werden, ist "große Vorsicht geboten" (Hundt 1969b, S. 154). Die Pflanzen reagieren nicht geradlinig, wenn Beeinflussungen wie z.B. Wasseraufstau oder -abzug ± weit entfernt vom Wurzelbereich angreifen und erst über ein System vermittelnder Faktoren auf Umwegen einwirken. Zwischenmedien sind horizontale Strömungsgeschwindigkeit, Hangdruckwassereinfluss, Nährstoff- und Sauerstoffgehalt des Grundwassers, Salzgehalt; Abgrenzung von Grundwasserströmen verschiedener Ausstattung mit derartigen Eigenschaften, Rückstaueinflüsse (Kielhauser 1958; Seibert 1958, 1963a; Niemann 1963, 1964; Vollrath 1965). Seibert (1963a) baute auf dieser Grundlage "das Verfahren der Grundwasserstufenkarte mit Darstellung verschiedener Wassereigenschaften" auf, das auch Auskunft darüber gibt, wodurch in den verschiedenen Kartierungseinheiten Wassermangel oder -überschuss zustande kommen (vgl. auch Ellenberg 1956).

Noch einen Schritt weiter gehen die nach Anforderungen der Praxis zusammengefassten, stets ökologisch-geobotanisch charakterisierbaren Wasserhaushaltsstufen und Wasserhaushaltszonen (Buchwald 1954, 1968, Hügin 1963) die nicht selten mit den Einzugsgebieten großer oder kleiner Fliessgewässer zusammenfallen. Ziel dieser Arbeitsrichtung ist es, die wirkenden Ursachen kennenzulernen, die von der Gesamtlandschaft ausgehen. Da eine Landschaft stets durchschaubar gesetzmäßig aufgebaut ist, bietet die Einsicht in ihren Bauplan zugleich eine geordnete Fülle von Kenntnissen, die auch für die Grünlandwirtschaft nutzbar gemacht werden können (Chardabellas 1957, Wiedenroth 1961). Eines der sichersten Bindeglieder, das hinter den einzelnen Landschaftselementen das Ganze erkennbar macht, ist die Vegetation. Eine umfassende, abiotischen Standort, Mensch und Vegetation berücksichtigende Landschaftskunde, deren Anfänge auf Tüxen (1939, 1968) zurückgehen (vgl. noch Krause 1950, 1952, 1957, 1963), ist demnach, womit an den Titel dieser Abhandlung erinnert sei, der eigentliche, wiewohl erst unvollkommen erreichte, aber doch erreichbare "Beitrag der Vegetationskunde zur Regelung des Wasserhaushaltes im Grünland" (vgl. auch betreffende Arbeiten in Tüxen (ed.) 1968 u. Bibliographie von Bosse, Buchwald & Robbel 1971).

Ansatzpunkte bieten auch einige regionale Grünlandwuchs-bezirksgliederungen (z.B. Schreiber 1960, Hundt 1964c, Niemann 1968). Für den konkreten Einzelfall muß dann allerdings auf großmaßstäbliche Vegetationskarten zurückgegriffen werden. Je genauer die qualitativen und quantitativen Beziehungen der Vegetationseinheiten zu hydrologischen Parametern bekannt sind, um so aussagefähiger ist eine auf Grundlage von Vegetationskartierungen vorgenommene "hydrologisch-geobotanische Gebietsdifferenzierung für landeskulturelle Zwecke" (Niemann 1968, 1973). Koinzidenzanalysen vermögen hierzu wesentliches Material zu erschließen.

Weitere Beispiele: Aus Vegetationskarten leiteten Tüxen (1952), Meisel (1954), Meisel & Wattendorf (1962), Walther (1960) durch Eichung Auswertungskarten verschiedenen Inhaltes ab. Einige geben Auskunft über Einzelfaktoren, z.B. über Grundwasserstufen Tüxen (1954a, 1969), vgl. auch Walther (1960), Seibert (1963a), Niemann (1963, 1964), Hundt (1968). Andere machen komplexe Aussagen z.B. über das langjährige mittlere Ertragspotential: Meisel (1960), De Boer (1963), vgl. auch Stählin (1957), Hundt (1956, 1961, 1964b, 1969a+b), Almadi (1970), so wie die Bibliographie von Klausing, Lohmeyer & Walther 1963, Hundt (1972) erarbeitete eine siebenstufige, an Pflanzengesellschaften orientierte Skala der Eignung von Grünlandstandorten für Abwasserverregnung mit Angaben zur Bodentextur, Befeuchtung, Nährstoffversorgung, Ent- und Bewässerungsbedürftigkeit, Nutzungseignung. Kuntze (1967) behandelt die besondere Problematik des Verhältnisses zwischen Grünlandvegetation und Grundwasser im humiden Klima.

Auf der Grundlage seiner Feuchtezahlen für die einzelnen Grünlandgesellschaften erstellte Ellenberg (1952b, S. 67) eine sechsteilige Skala der Feuchtegrenzen, die über die Nutzungsmöglichkeiten eines Geländes als Wiese, Umtriebsweide oder Acker entscheiden. Die Grenzen der Meliorationsbedürftigkeit sowie die Eignung der einzelnen Dränverfahren hat Tüxen (1958) für Niedermoorstandorte nach der Koinzidenz zwischen bodenphysikalischen Parametern und den Feuchtigkeitsansprüchen der Pflanzengesellschaften festgestellt. Für die Projektierung, die große Rücksicht auf die Torfprofile nehmen muß (Wojan & Illner 1962), wurde ein Bestimmungsschlüssel aus den Gesellschaften des S e n e c i o n r a c e m o s i erstellt.

Weitere Beispiele der Eichung von Grünlandgesellschaften mittels quantitativer geobotanisch ökologischer Komplexverfahren geben Andersson (1970), Klötzli (1969), Niemann (1963,

1970, 1973), Bracker (1972). Vgl. im übrigen auch die Biblio-
graphien von Tüxen (1961, 1969), Klausing, Lohmeyer, Walther
(1963), Seibert (1963b), Bosse, Buchwald & Robbel (1971).

Möglichkeiten der Prüfung und Differenzierung von Scha-
denersatzansprüchen der Landwirtschaft oder von Leistungen der
Wasserwirtschaft gegenüber der Landwirtschaft behandeln Baeu-
mer (1962) und Meisel (1963, 1968). Weiter ausgebaut sind die
Wasserhaushaltszonen von Buchwald (1954, 1968), in denen die
Wasserstufen mit anderen, den Wasserhaushalt beeinflussenden
Größen wie Bodendurchlässigkeit, Windeinfluß, Geländemorpho-
logie kombiniert sind.

Auf ähnliche Weise können kulturtechnisch und landwirtschaftlich wichtige Grenzen an Pelosolen herausgearbeitet werden, wenn deren Wasserzügigkeit (Eskuche 1955, 1958), Infiltra-
tionsgeschwindigkeit (Niemann 1967) oder der Wechselfeuchtig-
keitsgrad an der Vegetation fassbar sind (Bochert 1958, Niemann
1968). Niemann (1968, 1973) hat aus der Koinzidenz zwischen
Vegetationsstruktur, Bodenfeuchte und Grundwasserdynamik
(Größe der Grundwasserschwankungsamplitude, Ausgeglichen-
heitsgrad des Bodenfeuchteganges, Eintrittszeiten der Maxima
und Minima des Grundwassers) vegetationskundlich indizierbare
Formen der Wechselfeuchte herausgearbeitet und ist damit in
einen schwierigen und bei weitem noch nicht ausgeschöpften An-
wendungsbereich vorgedrungen.

Von besonderem Interesse bei den von Niemann untersuch-
ten Varianten des D e s c h a m p s i o - S i l a e t u m ist die
Tatsache, daß die an ihren Standorten herrschende Wechselfeuch-
tigkeit der Tonböden erst nach mehreren Jahren der Einwirkung
ökologisches Gewicht gewinnt, weil mit ihr noch andere Standortseigenschaften gekoppelt sind. Nach Almadi (1970) wird
durch den Wasserfaktor vor allem die zellulolytische Aktivität
gesteuert. Wichtige Ergebnisse bezüglich der Bodenmikroflora
und der Bodenfeuchtigkeitsdynamik auf den Überschwemmungswiesen der Trockengebiete (Zusammenhang mit der Geschwindig-
keit des Nährstoffkreislaufes, die auf diese Weise durch die Vegetation indirekt indiziert werden kann) bieten auch die Arbeiten
von Kovács (1968a + b), Tesařová & Úlehlová (1968) und Úlehlová (1973).

Allerdings bestehen gerade bei Moor- und Tonböden auch
Grenzen der Anwendbarkeit vegetationskundlicher Methoden.
Hat ein Moor tiefgreifende Änderungen des Wasserhaushaltes er-
fahren, darf das Verfahren erst angewendet werden, wenn sich
die Pflanzendecke auf die neuen Verhältnisse eingestellt hat, was

5–10 Jahre dauern kann. An Pelosolen ist oft nicht die bloße Tatsache der Wechselnässe, sondern der Zeitablauf des Übergangs von der nassen zur trockenen Phase von Bedeutung. Er muß zunächst durch unmittelbare Beobachtung geklärt sein, bevor nach Koinzidenzen zur Vegetation gesucht werden kann.

Eingeschränkt ist die Aussagekraft stark gedüngter Bestände aus zwei oder drei hochproduktiven und konkurrenzkräftigen Arten oder Zuchtsorten, die in der modernen Landwirtschaft zusehends an Bedeutung gewinnen. Sie sind nahezu unkrautfrei, dulden also keine charakteristische Artenkombination mit definiertem Indikatorwert. Eine Möglichkeit, trotzdem zu vegetationskundlichen Aussagen zu gelangen, bietet die synthetisierende Betrachtungsweise, die den Aufbau der Gesamtlandschaft ins Auge faßt und die einzelne Parzelle als abhängiges Glied dieser Landschaft versteht.

Als in Norddeutschland Reinbestände aus *Dactylis glomerata* zur Saatgutgewinnung angelegt wurden, ergaben sich teils gute, teils unbefriedigende Wirtschaftsergebnisse, deren Zustandekommen zunächst unerklärbar waren. Mit pflanzensoziologischen Methoden Einsicht in die Ursachen zu gewinnen, erschien zunächst aussichtslos, weil der dichte Obergrasbestand keine anderen Pflanzen duldete. Eine landschaftsökologische Analyse (Kolpak 1941) ergab dann, daß die befriedigenden Bestände auf basenreichem Geschiebemergel, die mißlungenen auf reinem Quarzsand stehen. Zugleich zeigte ein Blick auf die Kontaktgesellschaften der Raine und Wegränder neben den *Dactylis*-Beständen, daß letztere auf Geschiebemergel stets von einer A r r h e n a - t h e r i o n -Gesellschaft gesäumt sind, während auf Sand ebenso regelmäßig C o r y n e p h o r i o n oder A g r o s t i d i o n neben ihnen wächst. Hiernach war es ein leichtes, an der *Anthriscus*-Fazies des A r r h e n a t h e r i o n im Mai eine zutreffende Standortskartierung vom fahrenden Auto aus zu machen.

8.8   Zusammenfassung

Zunächst werden einleitende Bemerkungen zu praktischen Problemen der Grünlandwirtschaft gegeben, zu deren Lösung die Vegetationskunde beitragen kann. Hauptthema ist anschliessend der ökologische Indikatorwert der Grünlandgesellschaften ((S. 364 ff.). Seine elementare Leistung besteht in der Aufdeckung verborgener Standortsgrenzen mit Hilfe der leicht sichtbaren Vegetation. Im Grundsatz wird zuerst die Existenz dieser

Grenzen, noch nicht ihre ökologische Bedeutung angezeigt. Doch ergeben sich in der Praxis vielseitig differenzierte qualitative Aussagen über die Standorte, die es erlauben, die räumlichen Beziehungen zwischen den Pflanzengesellschaften sowie die Beziehungen zwischen Pflanzendecke und Umwelt zu erkennen. Vegetationskarten erweisen sich hierzu besonders nützlich.

Die wertvollen Dienste, die der Indikatorwert der Vegetation zur praktischen Standortsbeurteilung leistet, geben Anlaß, ihn genau zu präzisieren, wozu qualitative wie quantitative Methoden erarbeitet wurden. Dem allgemeinen Trend nach Quantifizierung folgend wird den letzteren besondere Aufmerksamkeit gewidmet (S. 365 ff.).

Der Kausalzusammenhang zwischen den lenkenden Umweltfaktoren und den Pflanzengesellschaften, der die Grundlage des Indikatorwertes bildet, ist in den meisten Fällen lang und kompliziert. Daher sind Fehlschlüsse möglich, sobald allein das Anfangs- und das Endglied der Kausalkette beachtet wird. Um Irrtümer zu vermeiden, ist die Beachtung des Gegensatzes zwischen den direkt und den indirekt wirkenden Standortsfaktoren von Nutzen. Ein Beispiel wird an der neuerdings stetig zunehmenden Sozialbrache abgeleitet (S. 367 ff.).

Anschließend folgen mehrere ausführliche, zum Teil mit Vegetationskarten belegte Beispiele qualitativer Anwendung des Indikatorwertes der Pflanzengesellschaften in der Grünlandwirtschaft. Als erstes (S. 371 ff.) wird eine großräumig aufgeteilte Flußaue in Mitteleuropa behandelt, anschliessend (S. 380 ff.) die ebenfalls weit ausgedehnte Auenvegetation im kontinentalen SO-Europa, zuletzt (S. 387 ff.) das engräumig differenzierte Grünland eines mitteleuropäischen Gebirgstales. In allen Beispielen, wie überhaupt in diesem Beitrag, wird der Wert der integrierenden Betrachtungsweise hervorgehoben, die über die einzelnen Wirtschaftsparzellen hinaus den ganzen Landschaftsbau ins Auge faßt. Nicht wenige Eigenschaften des Einzelstandortes lassen sich erst deutlich erkennen, wenn bekannt ist, in welchem Abhängigkeitsgefüge das Teil zum Ganzen steht.

Den Abschluß bildet eine Aufzählung kurzgefaßter Beispiele und Literaturhinweise zur Anwendung der Vegetationskunde in der Grünlandwirtschaft.

Literatur

Almadi, L., – 1970 – Anwendung ökologisch-geobotanischer Untersuchungen zur Beurteilung der Wiesen und ihrer Standorte im Rückhaltebecken bei Straussfurt. Diss. (nat.) Univ. Halle-Wittenberg.

Andersson, F., – 1970 – Ecological studies in a Scanian woodland and meadow area, Southern Sweden. I. Vegetational and environmental structure. *Opera Botanica*, Lund, 27: 1–190.

Arland, A., (ed.), – 1968 – Die Anwelkmethode im Dienste des Landbaues. Die Transpirationsintensität der Pflanzen im Blickfelde der Ermittlung optimaler Kulturmassnahmen. Ber. wiss. Tagung Inst. Acker- Pflanzenbau Leipzig, Berlin, 154 pp.

Baeumer, K., – 1962 – Die Wasserstufen-Karte der Wümme-Niederung. *Abh. naturwiss. Ver. Bremen* 36: 118–168.

Balátová-Tuláčková, E., – 1966 – Synökologische Charakteristik der südmährischen Überschwemmungswiesen. *Rozpravy ČSAV, Ř. mat.-přír.*, Praha, 76: 1–42.

Balátová-Tuláčková, E., – 1968 – Grundwasserganglinien und Wiesengesellschaften. (Vergleichende Studie der Wiesen aus Südmähren und der SW.-Slowakei.) *Acta Sci. Natur. Acad. Sci. Bohemoslov. Brno*, Praha, 2 (2): 1–38.

Balátová-Tuláčková, E., – 1969a – Entwässerung und Fruchtbarkeit der Landschaft. *In:* "Experimentelle Pflanzensoziologie", ed. R. Tüxen. Ber. int. Symp. Int. Ver. Vegetationskunde Rinteln 1965, pp. 110-116. Den Haag.

Balátová-Tuláčková, E., – 1969b – Beitrag zur Kenntnis der tschechoslowakischen Cnidion venosi-Wiesen. *Vegetatio*, 17: 200–207.

Balátová-Tuláčková, E., – 1970 – K problematice odvodňování mokrých a vlhkých luk. (On the Problems of Draining Waterlogged and Wet Meadows.) *Rostl. Výroba*, Praha, 16: 1261–1273.

Balátová-Tuláčková, E., – 1972 – Flachmoorwiesen im mittleren und unteren Opava-Tal (Schlesien). *Vegetace ČSSR* A 4. Academia, Praha. 201 pp.

Balátová-Tuláčková, E., – 1972a – Dynamics of the Plant Cover in Inundated Meadows of Southern Moravia. In: Rychnovská, M. (ed.): Czechosl. IBP/PT-PP Report No. 2, Brno.

Balátová-Tuláčková, E., – 1973 – Zur Problematik des Erhaltens der hochproduktiven Überschwemmungswiesen in Trockengebieten. In "Anwendung der Landschaftsökologie in der Praxis", ed. M. Ružička, *Questiones geobiol.*, Bratislava, 11: 41–54.

Bauer, L. & H. Weinitschke, – 1964 – Landschaftspflege und Naturschutz. Eine Einführung in ihre Grundlagen und Aufgaben. Jena. 194 pp.

Beckmann, W., – 1965 – Untersuchungen zum Landschaftshaushalt in Auen der Hauensteiner Murg. Hamburger Geographische Studien 19.

Bochert, H., – 1958 – Standörtliche Gliederung der Nutheniederungen am südwestlichen Flämingrand auf Grund vegetationskundlicher, hydrologischer und bodenkundlicher Untersuchungen. Diss. (forstwiss.), Eberswalde.

Boer, Th. A. de, – 1954 – Grünlandvegetationskartierung in den Niederlanden. *Angew. Pfl. Soziol.*, Wien, Festschr. E. Aichinger, 2: 1232–1234.

Boer, Th. A. de, – 1963 – Die Grünlandvegetationskartierung als Grundlage für die landwirtschaftliche Verbesserung ganzer Gebiete in den Niederlanden. In "Bericht über das Internationale Symposium für Vegetationskartierung vom 23.–26.3.1959 in Stolzenau/Weser", ed. R. Tüxen, pp. 60–63. Weinheim.

*Bosse, M., K. Buchwald & H. Robbel, – 1971 – Pflanzensoziologie als Grundlage für Landschaftspflege und Naturschutz. *Exc. bot.*, *Sectio B.*, Stuttgart, 11: 1–85.

Bracker, H. H., – 1960 – Die grünlandsoziologische Erfassung von Feuchtestufen zur Kennzeichnung des natürlichen Standorts. *Kulturtechn.*, Berlin, 48: 34–59.

Bracker, H. H., — 1972 — Grünlandnutzung künftig ökologisch differenzieren. *Z. Kulturtechn. Flurbereinig.*, Berlin, 13: 137—141.

Braun-Blanquet, J., — 1964 — Pflanzensoziologie. 3 Aufl. Wien. 865 pp.

Brünner, F., — (o.J.) — Die Wirkung von Bewässerung und Düngung auf Schnittwiesen des Südschwarzwaldes. Forschung und Beratung Reihe B H. 10.

Brünner, F. & W. Krause, — 1968 — Über den Einfluss der natürlichen Nährstoffvorräte und der Befeuchtung auf den Ertrag von Dauerwiesen, dargestellt an Wechseldüngungsversuchen in Südwestdeutschland. *Z. Acker- u. Pflanzenbau*, Berlin, 128: 151—173.

Buchholz, F., — 1961 — Vorgänge in grundwasserbeeinflussten Sandböden und ihre Bedeutung für die Standortsbeurteilung. *Arch. Forstwesen*, Berlin, 10: 430—438.

Buchwald, K., — 1954 — Wasserhaushaltsstufen und -zonen des Wassermangelgebietes Unteres Illertal. *Angew. Pfl. Soziol.*, Stolzenau/Weser, 8: 37—55.

Buchwald, K., — 1968 — Die Austrocknung von Flusstälern nach wasserbaulichen Massnahmen, Gesundungsplanung und deren Ausführung. Handb. Landschaftspf. Naturschutz, München, 2.

Chardabellas, P., — 1957 — Aufschlüsselung der Grundwasserleistung eines Flussgebietes auf Teilgebiete unter Zugrundelegung der geologischen und pflanzensoziologischen Gliederung des Einzugsgebietes. Mskr., n.p., Berlin.

Ellenberg, H., — 1950 — Landwirtschaftliche Pflanzensoziologie I. Unkrautgemeinschaften als Zeiger für Klima und Boden. Stuttgart. 141 pp.

Ellenberg, H., — 1952a — Auswirkungen der Grundwassersenkung auf die Wiesengesellschaften am Seitenkanal westlich Braunschweig. *Angew. Pflanzensoziol.*, Stolzenau/Weser, 6: 1—46.

Ellenberg, H., — 1952b — Landwirtschaftliche Pflanzensoziologie II. Wiesen und Weiden und ihre standörtliche Bewertung. Stuttgart. 143 pp.

Ellenberg, H., — 1954 — Landwirtschaftliche Pflanzensoziologie III. Naturgemässe Anbauplanung, Melioration und Landespflege. Stuttgart. 109 pp.

Ellenberg, H., — 1956 — Grundlagen der Vegetationsgliederung I. Aufgaben und Methoden der Vegetationskunde. In "Einführung in die Phytologie IV", ed. H. Walter. Stuttgart. 136 pp.

Ellenberg, H., — 1963 — Vegetation Mitteleuropas mit den Alpen. In "Einführung in die Phytologie IV—2", ed. H. Walter. Stuttgart. 945 pp.

Ellenberg, H., (ed.) — 1973 — Ökosystemforschung. Berlin. 280 pp.

Ellenberg, H., — 1974 — Zeigerwerte der Gefäßpflanzen Mitteleuropas. *Scripta Geobotanica*, 9, Göttingen.

Eskuche, U., — 1955 — Vergleichende Standortuntersuchungen an Wiesen im Donauried bei Herbertingen. *Veröff. Württ. Landesst. Naturschutz Landschaftspfl.*, Ludwigsburg, 23: 33—135.

Eskuche, U., — 1958 — Über Messungen der Wasserbeweglichkeit an verschiedenen Böden und Pflanzengesellschaften. *Angew. Pfl. Soziol.*, Stolzenau/Weser, 15: 145—160.

Gora, A., — 1964 — Feld- und Laboruntersuchungen an Pseudogleyen verschiedenen geologischen Ausgangsmaterials als Grundlage für deren standortsgerechte Meliorationen. Albrecht Thaer-Archiv, 8, 6/7.

Halva, E. & J. Lesák, — 1974 — Některé výsledky vědeckovýzkumných prací v oblasti výživy lučních porostů/Einige Resultate über die wissenschaftliche Untersuchungen in der Ernährung von Wiesenbeständen/ (Tschechisch). *Stundijní informace* 5: 1—33. Landw. Hochschule Brno.

Hügin, G., — 1963 — Wesen und Wandlung der Landschaft am Oberrhein. *Beitr. Landespflege*, Stuttgart, Festschr. H. Wiepking, 1: 185—250.

Hundt, R., – 1956 – Grünlandvegetationskartierung im Unstruttal bei Straussfurt. *Wiss. Ztschr. Martin-Luther-Univ. Halle-Wittenberg, Math.-nat. R.*, 1955/56, 5: 1291–1306.

Hundt, R., – 1958 – Beiträge zur Wiesenvegetation Mitteleuropas I. Die Auenwiesen der Elbe, Saale und Mulde. *Nova Acta Leopoldina NF*, Leipzig, 20 (135): 1–206.

Hundt, R., – 1961 – Die Auswirkung der Saaletalsperren auf die Grünlandvegetation des mittleren Saaletales. *Mitt. Inst. Wasserwirtsch.*, Berlin, 14: 20–60.

Hundt, R., – 1964a – Vegetationskundliche Verfahren zur Bestimmung der Wasserstufen im Grünland. *Z. Landeskultur*, Berlin, 5: 161–186.

Hundt, R., – 1964b – Vegetation, Feuchtigkeitsverhältnisse und Ertragsverhältnisse der Wiesenflächen im Luhne-Rückhaltebecken bei Lengefeld (Thüringen). *Wiss. Z. Univ. Halle*, Sonderheft: 149–170.

Hundt, R., – 1964c – Die Bergwiesen des Harzes, Thüringer Waldes und Erzgebirges. Pflanzensoziologie 14. Jena. 284 pp.

Hundt, R., – 1966 – Ökologisch-geobotanische Untersuchungen an Pflanzen der mitteleuropäischen Wiesenvegetation. *Bot. Studien* 16. Jena. 176 pp.

Hundt, R., – 1968 – Vegetation und Wasserstufen der Wiesenflächen eines Dränversuches bei Sittendorf am Kyffhäuser. *Wiss. Z. Univ. Halle* 17: 93–123.

Hundt, R., – 1969a – Wiesenvegetation, Wasserverhältnisse und Ertragsverhältnisse im Rückhaltebecken bei Kelbra an der Helme. *Mitt. Inst. Wasserwirtsch.*, Berlin, 30: 3–99.

Hundt, R., – 1969b – Vegetation, Wasserstufen und Bodendurchfeuchtung der Wiesenflächen eines Grabeneinstauversuches bei Edersleben. *Mitt. Inst. Wasserwirtsch.*, Berlin, 30: 103–162.

Hundt, R., – 1972 – Ökologisch-geobotanische Untersuchungen in Biozönosen, dargestellt am Wasserfaktor. *Biol. Rundschau* 10: 170–187.

Ilijanić, Lj., – 1962 – Prilog poznavanju ekologije nekih tipova nizinskih livada Hrvatske (Beitrag zut Kenntnis der Ökologie einiger Niederungswiesentypen Kroatiens). *Acta bot. Croatica*, Zagreb, 20/21: 95–167.

Ilijanić, Lj., – 1971 – Istraživanje utjecaja sniženja podzemne vode u asocijaciji Deschampsietum caespitodae H–ić u okolici Sesveta (Untersuchunge über die Auswirkung der Grundwassersenkung auf das Deschampsietum caespitosae H–ić in der Umgebung von Zagreb). *Spomenica uz 70 god. prof. Gračanina*, Zagreb, 1971: 257–267.

Jovanović, R., – 1965 – Zavisnost močvarnih i livadskih fitocenoza od visine podzemne vode u dolini Velike Morave (Abhängigkeit der Sumpf- und Wiesengesellschaften von der Höhe des Grundwassers im Tale der Grossen Morava). *Conserv. Nature*, Beograd, 29/30: 25–49.

Kayl, R., – 1965 – Verbreitung, Entwicklungsgeschichte und standörtliche Bewertung von Kulturrasen- und Ödlandpflanzengesellschaften. Diss. Bonn. 318 pp.

Kielhauser, G. E., – 1958 – Die ingenieurbiologische Standortskarte im Wasserbau. *Österr. Wasserwirt.*, Wien, 10: 251–254.

Klapp, E., – 1949 – Landwirtschaftliche Anwendung der Pflanzensoziologie. Stuttgart. 56 pp.

Klapp, E., – 1954 – Erträge von Pflanzengesellschaften in Beziehung zu Grundwasser und Nährstoffversorgung. In "Pflanzensoziologie als Brücke zwischen Land- und Wasserwirtschaft", ed. R. Tüxen. *Angew. Pfl. Soziol.*, Stolzenau/Weser, 8: 137–148.

Klapp, E., – 1956 – Wiesen und Weiden. Behandlung, Verbesserung und Nutzung von Grünlandflächen. 2. Auf. Berlin. 519 pp.

Klapp, E., – 1965 – Grünlandvegetation und Standort. Berlin. 384 pp.

*Klausing, O., W. Lohmeyer & K. Walther, – 1963 – Bibliographie zum Thema Produktionspotentiale von Pflanzengesellschaften. *Exc. bot., Sectio B*, Stuttgart, 5: 81–102.

Klötzli, F., – 1969 – Die Grundwasserbeziehungen der Streu- und Moorwiesen im nördlichen Schweizer Mittelland. *Beitr. geobot. Landesaufn.*, Bern, 52: 1–296.

Knapp, R., – 1963 – Die Vegetation des Odenwaldes, *Schriftenr. Inst. Naturschutz*, Darmstadt 6: 11–150.

Koehne, W., – 1948 – Grundwasserkunde. 2. Aufl. Stuttgart. 314 pp.

Kolpak, W., – 1941 – Untersuchungen über die Standortsfrage im Knaulgras – Samenbau. *Der Futtersaatbau 1*, H 1, Leipzig.

Kovács, M., – 1968a – Die Vegetation im Überschwemmungsgebiet des Ipoly (Eipel) Flusses II. Die ökologischen Verhältnisse der Pflanzengesellschaften. *Acta bot. Acad. Sci. Hung.*, Budapest, 14: 77–112.

Kovács, M., – 1968b – Nitrification Capacity of the Soils of Marshy and Hay Meadows. *Acta Agronom. Sci. Hung.*, Budapest, 17: 25–36.

Kramer, D., – 1973 – Spezielle wasserwirtschaftliche Aspekte zur landwirtschaftlichen Bewässerung. *Wasserwirtschaft-Wassertechnik*, Berlin, 23: 118–120.

Krause, W., – 1950 – Über Vegetationskarten als Hilfsmittel kausalanalytischer Untersuchung der Pflanzendecke. *Planta*, Berlin, 38: 296–323.

Krause, W., – 1952 – Das Mosaik der Pflanzengesellschaften und seine Bedeutung für die Vegetationskunde. *Planta*, Berlin, 41: 240–289.

Krause, W., – 1957 – Die Untersuchung der Pflanzengesellschaften des Grünlandes im Dienste der Wirtschaftsplanung. "Kalium Symposium", ed. Intern. Kali-Inst., pp. 29–40. Bern.

Krause, W., – 1958 – Methoden und Ergebnisse der Vegetationskartierung. *Umschau in Wissenschaft und Technik 58*, H 19: 595–598.

Krause, W., – 1959 – Über die natürlichen Bedingungen der Grünlandberieselung in verschiedenen Landschaften Südbadens mit Ausblick auf den Wirtschaftserfolg. *Z. Acker u. Pflanzenbau.* Berlin 107 H 3: 245–274.

Krause, W., – 1963 – Eine Grünland- Vegetationskarte der Südbadischen Rheinebene und ihre landschaftsökologische Aussage. *Arb. rhein. Landeskunde*, Bonn, 20.

Krause, W. & A. Müller, – 1973 – Möglichkeiten und Grenzen einer Nutzungsintensivierung auf Dauerwiesen in Abhängigkeit vom Standort. *Wirtsch.-eig. Futter*, Frankfurt a.M., 19: 174–194.

Krause, W., – 1974 – Bestandsveränderungen auf brachliegenden Wiesen. *Wirtsch.-eig. Futter*, Frankfurt a.M., 20: 51–65.

Kreeb, K., – 1966 – Die Registrierung des Wasserzustandes über die elektrische Leitfähigkeit der Blätter. *Ber. Dt. Bot. Ges.*, Stuttgart, 79: 150–162.

Krisch, H., – 1967 – Die Grünland- und Salzpflanzengesellschaften der Werraaue bei Bad Salzungen I. Die Grünlandgesellschaften. *Hercynia N.F.* Leipzig, 4: 375–413.

Krisch, H., – 1968 – Die Grünland- und Salzpflanzengesellschaften der Werraaue bei Bad Salzungen II. Die salzbeeinflussten Pflanzengesellschaften. *Hercynia N.F.*, Leipzig, 5: 49–95.

Kuntze, H., – 1967 – Wasserhaushalts- und Grünlandforschung bei Grundwasserböden im humiden Klima. *Z. Kulturtechnik Flurbereinig.*, Berlin, 8: 142–162.

Meisel, K., – 1954 – Anwendung der Pflanzensoziologie zur Beurteilung von Wasserschäden in der Landwirtschaft. *Angew. Pfl. Soziol.*, Stolzenau/Weser, 8: 127–129.

Meisel, K., – 1960 – Die Auswirkung der Grundwassersenkung auf die Pflanzengesellschaften im Gebiet um Moers (Niederrhein). *Arb. Bundesanst. Veg. Kartierung*, Stolzenau/Weser, 1960: 1–105.

Meisel, K., – 1963 – Die Vegetationskarte als Grundlage für die Beurteilung von

Wasserschäden. In "Ber. Intern. Symp. Veg. Kartierung, 23.–26.3.1959 Stolzenau/Weser", ed. R. Tüxen, pp. 423–430. Weinheim.

Meisel, K., – 1968 – Vegetationsuntersuchungen als wesentlicher Bestandteil der Beweissicherung bei Eingriffen in den Wasserhaushalt der Landschaft. *Natur u. Landschaft*, Bad Godesberg, 43: 167–170.

Meisel, K. & J. Wattendorf, – 1962 – Über eine von der Wirtschaftsart unabhängige Wasserstufenkarte. *Mitt. flor.-soz. Arbeitsgem.*, Stolzenau/Weser, N.F. 9: 230–238.

Mückenhausen, E., – 1962 – Entstehung, Eigenschaften und Systematik der Böden der Bundesrepublik Deutschland. Frankfurt/Main. 148 pp.

Müller, A. von, – 1956 – Über die Bodenwasser-Bewegung unter einigen Grünlandgesellschaften des mittleren Wesertales und seiner Randgebiete. *Angew. Pfl. Soziol.*, Stolzenau/Weser, 12.

Niemann, E., – 1962 – Vergleichende Untersuchungen zur Vegetationsdifferenzierung in Mittelgebirgstälern – dargestellt am Beispiel eines Querschnitts durch den mittleren Thüringer Wald. Diss. TU Dresden.

Niemann, E., – 1963 – Beziehungen zwischen Vegetation und Grundwasser. Ein Beitrag zur Präzisierung des ökologischen Zeigerwertes von Pflanze und Pflanzengesellschaft. *Arch. Nat. Schutz Landsch. Forsch.*, Berlin, 3: 3–36.

Niemann, E., – 1964 – Beiträge zur Vegetations- und Standortgeographie in einem Gebirgsquerschnitt über den mittleren Thüringer Wald. *Arch. Nat. Schutz Landsch. Forsch.*, Berlin, 4: 3–50.

Niemann, E., – 1967 – Infiltrationsmessungen an verbreiteten Pflanzenstandorten des Thüringer Waldes. *Limnologica*, Berlin, 5: 251–272.

Niemann, E., – 1968 – Zur Rolle des Grundwasserfaktors im Vegetationsgefüge. Ökologisch-geobotanische Untersuchungen an grundwasserbeeinflussten Vegetationseinheiten. Habil.-Schr. TU Dresden. 231 pp.

Niemann, E., – 1970 – Vegetationsdifferenzierung und Wasserfaktor. *Arch. Naturschutz Landschaftsforsch.*, Berlin, 10: 111–130.

Niemann, E., – 1972 – Physiologisch-ökologische und experimentell-ökologische Methoden zur Erfassung des Wasserfaktors am Pflanzenstandort. *Petermanns Geogr. Mitt.* 116: 186–197.

Niemann, E., – 1973 – Grundwasser und Vegetationsgefüge. *Nova Acta Leopold.*, Leipzig, Suppl., 38 (6).

Niemann, E., – 1974 – Die Hydraturmessung als physiologisch-ökologische Methode im Rahmen geobotanischer Wasserhaushalts-Untersuchungen. *Limnologica*, Berlin, 9 (i.Dr.).

Odum, E. P., – 1959 – Fundamentals of ecology. 2. Aufl. Saunders, Philadelphia. 546 pp.

Pallmann, H., – 1943 – Über Waldböden. *Beih. Z. schweiz. Forstverein* 21: 113–140.

Rabotnov, T. A., – 1974 – Lugovedenie. Moskva. 384 pp. (russ.)

Ramenskij, L. G., – 1938 – Vvedenije v kompleksnoje počvennogeo-botaničeskoje issledovanije zemel (Einführung in die komplexe bodengeobotanische Untersuchung von Böden). (Russisch). Moskva. 620 pp.

Ramenskij, L. G., J. A. Cacenkin, O. N. Čižkov & N. A. Antipin, – 1956 – Ekologičeskaja ocenka kormovych ugodij po rastitelnom pokrovu (Die ökologische Beurteilung von Futterflächen nach der Pflanzendecke. (Russisch). Moskva. 472 pp.

Ružičková, H., – 1971 – Rostlinné spoločenstva lúk a slatín v povodí Čiernej vody. (Východoslovenská nížina) (Pflanzengesellschaften der Wiesen und Niedermoore in Einzugsgebiet des Flusses Čierna voda). *Biol. Práce*, Bratislava, 17 (7): 1–136.

Regal, V., – 1967 – Ekologické indikační hodnoty nejrozšířenějších lučních rostlin, ČSSR (Ökologische Indikationswerte der verbreitetsten Wiesenpflanzen der ČSSR). *Rostl. výroba*, Praha, N.F. 13: 77–88.

Rychnovská, M., J. Květ, J. Gloser & J. Jakrlová, – 1972 – Plant Water Relations in three Zones of Grassland. *Acta Sci. Nat. Acad. Sci. Bohemoslov. Brno*, Praha, 6 (5): 1–38.

Schreiber, K. F., – 1960 – Über die standortsbedingte und geographische Variabilität der Glatthaferwiesen in Südwestdeutschland, Diss. Landw. Hochschule Hohenheim.

Schröder, G., – 1958 – Landwirtschaftlicher Wasserbau, 3 Aufl. Berlin. 551 pp.

Seibert, P., – 1958 – Die Pflanzengesellschaften im Naturschutzgebiet "Pupplinger Au". *Landschaftspflege Veg.kde*, München, 1: 1–79.

Seibert, P., – 1963a – Über eine Grundwasserstufenkarte mit Darstellung verschiedener Wassereigenschaften. *Mitt. Flor.–soz. Arbeitsgem.*, Stolzenau/Weser, N.F. 10: 223–231.

*Seibert, P., – 1963b – Bibliographie der Arbeiten über das Zusammenwirken zwischen Pflanzensoziologie, Wasserwirtschaft und Wasserbau. *Exc. Bot., Sectio B*, Stuttgart, 5: 81–102.

Speidel, B., – 1963 – Vegetationskartierung als Grundlage zur Melioration salzgeschädigter Wiesen an der Werra. In "Ber. Int. Symp. Vegetationskartierung 23.–26.3.1959 Stolzenau/Weser," ed. R. Tüxen, pp. 457–468. Weinheim.

Stählin, A., – 1957 – Über die Schädigung von Grünland bei Überschwemmungen und in Rückhaltebecken. *Grünland*, Hannover, 6: 93–95.

Stählin, A. & H. Büring – 1971 – Sozialbrache auf Äckern und Wiesen in pflanzensoziologischer und ökologischer Sicht. *Z. Acker u. Pflanzenbau*, Berlin 133: 200–214.

Sukatschew, W. N. (Sukačev, V. N.), – 1945 – Biogeocenologija i fitocenologija. *Dokl. AN SSSR*, Moskva, 47 (6): 447–449.

Sukatschew, W. N. (Sukačev, V. N.), – 1954 – Die Grundlagen der Waldtypen. *Angew. Pfl. Soziol.*, Wien, Festschrift E. Aichinger, 2: 956–964.

Tansley, A. G., – 1935 – The use and abuse of vegetational concepts and terms. *Ecology*, Durham, 16: 284–307.

Tesařová, M. & B. Úlehlová, – 1968 – Abbau der Zellulose unter einigen Wiesengesellschaften. In "Mineralisation der Zellulose", *Tagungsber. DALW*, Berlin, 98: 277–287.

Tüxen, R., – 1939 – Die Pflanzendecke Nordwestdeutschlands in ihren Beziehungen zu Klima, Gesteinen, Böden und Mensch. *Dt. Geogr. Blätter* 42. Bremen.

Tüxen, R., – 1942 – Über die Verwendung pflanzensoziologischer Untersuchungen zur Beurteilung von Schäden des Grünlandes. *Dt. Wasserwirtsch.*, München, 37: 455–459, 501–505.

Tüxen, R., – 1952 – Ein einfacher Weg zur nachträglichen Feststellung von Entwässerungsschäden. *Mitt. Florist.–soziol. Arbeitsgem.*, Stolzenau/Weser, N.F. 3: 128-129.

Tüxen, R., – 1954a – Die Wasserstufenkarte und ihre Bedeutung für die nachträgliche Feststellung von Änderungen im Wasserhaushalt einer Landschaft. *Angew. Pflanzensoziol.*, Stolzenau/Weser, 8: 31–36.

Tüxen, R., – 1954b – Pflanzengesellschaften und Grundwasser-Ganglinien. *Angew. Pflanzensoziol.*, Stolzenau/Weser, 8: 64–98.

Tüxen, R., – 1957 – Entwurf einer Definition der Pflanzengesellschaft (Lebensgemeinschaft). *Mitt. Flor.-Soz. Arbgem.* N.F. 6/7, 151.

Tüxen, R., – 1958 – Die Eichung von Pflanzengesellschaften auf Torfprofiltypen. Ein Beitrag zur Koinzidenzmethode in der Pflanzensoziologie. *Angew. Pfl. Soziol.*, Stolzenau/Weser, 15: 131–141.

*Tüxen, R., – 1961 – Bibliographie der Arbeiten über Grundwasserganglinien unter Pflanzengesellschaften. *Exc. Bot., Sectio B*, Stuttgart, 3: 237–240.

Tüxen, R., (ed.), — 1968 — Pflanzensoziologie und Landschaftsökologie. Ber. Int. Symp. int. Ver. Vegetationskunde Stolzenau/Weser 1963. Den Haag. 426 pp.

*Tüxen, R., — 1969 — Bibliographie der Arbeiten über pflanzensoziologisch bestimmte Wasserstufen und Wasserstufen-Karten. *Exc. Bot., Sectio B*, Stuttgart, 10: 93—96.

Úlehlová, B., — 1973 — Alluvial Grassland Ecosystems-Microorganisms and decay processes. *Acta Sci. Nat. Acad. Sci. Bohemoslovacae Brno*, Praha, NS 7 (5): 1—44.

Válek, B., — 1960 — Pedologické a hydropedologické vlastnosti lučních půd ve vztahu k jejich porostům II. (Pedologische und hydropedologische Eigenschaften der Wiesenböden in Beziehung zu ihren Beständen). *Vědec. Pr. Výzk. ústavu Melioraci ČSAZV Praha*, 1960: 117—167.

Voisin, A., — 1961 — Lebendige Grasnarbe (übers. aus franzöz. R. Wecke). München. 245 pp.

Vollrath, H., — 1965 — Das Vegetationsgefüge der Itzaue als Ausdruck hydrologischen und sedimentologischen Geschehens. *Landschaftspflege u. Vegetationskunde*, München, 4: 1-128.

Vries, D. M. de, A. A. Kruijne & H. Mooi, — 1957 — Veelvuldigheid van graslandplanten en hun aanwijzing van milieu-eigenschappen (Frequency of occurrence of herbage plants and their indication of environmental conditions). *Jaarboek van het I.B.S.*, Wageningen, 1957: 183–191.

Wagner, H., — 1955 — Die Bewertung der Wasserstufen in der Bodenschätzung des Grünlandes. *Bodenkultur*, Wien, 8: 133—150.

Walter, H., — 1951 — Standortslehre. In "Einführung in die Phytologie III/I", ed. H. Walter. Stuttgart. 525 pp.

Walther, K., — 1950 — Unkraut-Herden als Zeiger grundwassergeschädigter Grünlandgesellschaften auf Niedermoorboden. *Mitt. Flor.-soz. Arbgem. N.F.* 2, 43—51. Stolzenau/Weser.

Walther, K., — 1960 — Pflanzensoziologie und Kulturtechnik. *Z. Kulturtechn.*, Berlin, 1: 65—76.

Westhoff, V. & E. van der Maarel, — 1973 — The Braun-Blanquet approach. In "Ordination and Classification of Vegetation". Handbook of Vegetation Science V, ed. R.W. Whittaker, pp. 617—726. The Hague.

Wiedenroth, E. M., — 1961 — Der Aussagewert der pflanzensoziologischen Gebietsanalyse und -kartierung für die Aufgaben der wasserwirtschaftlichen Praxis. *Mitt. Inst. Wasserwirtschaft*, Berlin, 14: 1—19.

Wojahn, E. & K. Illner, — 1962 — Die Standorttypen der Niedermoore als Grundlage der Meliorationsplanung und -projektierung. *Wasserwirtschaft-Wassertechnik*, Berlin, 12: 139—147.

Zarzycki, K., — 1956 — Meadow Associations and the Ground-Water Level. *Bull. Acad. polon. Sci. Cl. II*, 4/5: 183—187.

9 VEGETATIONSKUNDE ALS GRUNDLAGE DER VERBESSERUNG DES GRASLANDES IN DEN ALPEN.
Zur landwirtschaftlichen Bedeutung des Graslandes in den Alpen

W. DIETL

Inhalt

VEGETATIONSKUNDE ALS GRUNDLAGE DER VERBESSERUNG DES GRASLANDES IN DEN ALPEN

9.1 Probleme der Landwirtschaft in den Alpen

Geländeform, hohe Lage und rauhes Klima erschweren die landwirtschaftliche Nutzung des Graslandes im Gebirge auf vielfältige Weise. Steilhänge und kupierte Flächen eignen sich nicht für intensive Bodennutzung. Viele Alpweiden und Bergwiesen sind mit Transportmitteln überhaupt nicht oder nur über schlechte Verbindungswege zu erreichen. Mit zunehmender Höhe über Meer verkürzt sich die Vegetationszeit und die Leistungsfähigkeit des Graslandes* geht zurück. Je Großvieheinheit (GVE) werden immer größere Flächen benötigt und alle produktionsfördernden Maßnahmen verteuern sich.

Die sozio-ökonomische Entwicklung der neueren Zeit bereitet überdies den Bergbauern zunehmende Schwierigkeiten, wenn sie die bisher übliche Nutzungsform beibehalten. Für neue arbeitsintensive Wirtschaftsmethoden fehlen die Voraussetzungen. Es müssen daher standortsgemäße Nutzungsformen angestrebt werden, die zugleich weitgehend wirtschaftlich sind.

9.2 Die physiographischen Grundlagen

9.2.1 KLIMA

Die Temperatur beträgt im Jahresmittel auf 1000 m ü.M. etwa 6,0°C; mit 100 m Höhenzunahme sinkt sie durchschnittlich um 0,5°C. In 2000 m Meereshöhe liegt sie bei rd. 1°C. Die mittlere Julitemperatur beträgt auf 1500 m ü.M. rd. 11°C. Weiter oberhalb kann Frost jederzeit auftreten. Die Niederschläge, die im Sommer ihr Maximum erreichen, belaufen sich in den feuchten Gebieten der Alpennordseite in Höhen über 1500 m auf mehr als 1800 mm im Jahresmittel. In den inneralpinen Trockentälern (z.B. Wallis, Engadin, Vintschgau) werden diese Mengen erst oberhalb 2000 m ü.M. erreicht. Im Mittelland und in den tieferen

---

* Synonyme: Wiesen = Wiesland, Grasland, Grünland, Rasen; Matte = gemähte Wiese, Mähwiese; Weide = beweidete Wiese.

Lagen der Täler fallen rund 10% der Niederschläge als Schnee. Auf 1000 m Meereshöhe beträgt dessen Anteil etwa 30%, auf 2000 m etwa 60% und oberhalb 3500 m nahezu 100%. Klimadiagramme nach Walter (1955) für zwei Stationen der Nordalpen gibt Fig. 1.

9.2.2   GESTEINE

Die Alpen sind in der zentralen Zone aus kristallinen Silikatgesteinen (Granit, Gneis, kristalline Schiefer) aufgebaut, und werden im Norden und Süden von Kalk-Sedimentgesteinen flankiert (Kalksteine, Mergel, Dolomit sowie Sandsteine mit kalkhaltigem Bindemittel). Kalkarm sind die Verrucano-Sandsteine und -Konglomerate. Auch quartäre Moränen und Schotter treten als Muttergesteine häufig auf. Die jüngeren Sedimente der Molasse haben am Bau der eigentlichen Alpen keinen Anteil; sie wurden von den alpinen Schüben nicht mehr verfrachtet.

9.2.3   BODEN

Der Boden, soweit er sich über lange Zeiträume entwickeln konnte und nicht ständiger Erosion oder Aufschüttung, Vernässung oder Überschwemmung ausgesetzt war, hat seine Ausprägung in erster Linie vom Klima erhalten. In der hochmontanen Buchen-Tannenstufe ist Braunpodsol entstanden, in der darüber folgenden subalpinen *Picea*- oder *Larix-Pinus cembra*-Stufe Podsol, im Bereich der alpinen *Carex curvula*-Rasen Humussilikatboden (alpiner Ranker), in der obersten alpinen Stufe Rohboden, der Schuttgesellschaften trägt. Diese Böden entwickeln sich in der entsprechenden Höhenstufe nur auf leicht verwitternden und mineralisch nicht einseitig zusammengesetzten Gesteinen. Sie werden als Klimax- oder Normalböden bezeichnet. Auf harten Kalksteinen konnte sich die Klimax nicht einstellen (vgl. Bach in Hess & Landolt 1967). Hier herrschen humusreiche und steinige Humuskarbonatböden, die in der Regel wenig Wasser speichern. Auf ihnen stehen *Pinus montana*-Wälder oder *Sesleria-Carex sempervirens*-Rasen und *Carex firma*-Gesellschaften.

Aus Mergeln haben sich tonreiche, humusarme Rendzinen gebildet, auf denen in der subalpinen und alpinen Region häufig das C a r i c e t u m   f e r r u g i n e a e siedelt. In den höheren, humiden Lagen werden schwer durchlässige bindige Böden, die

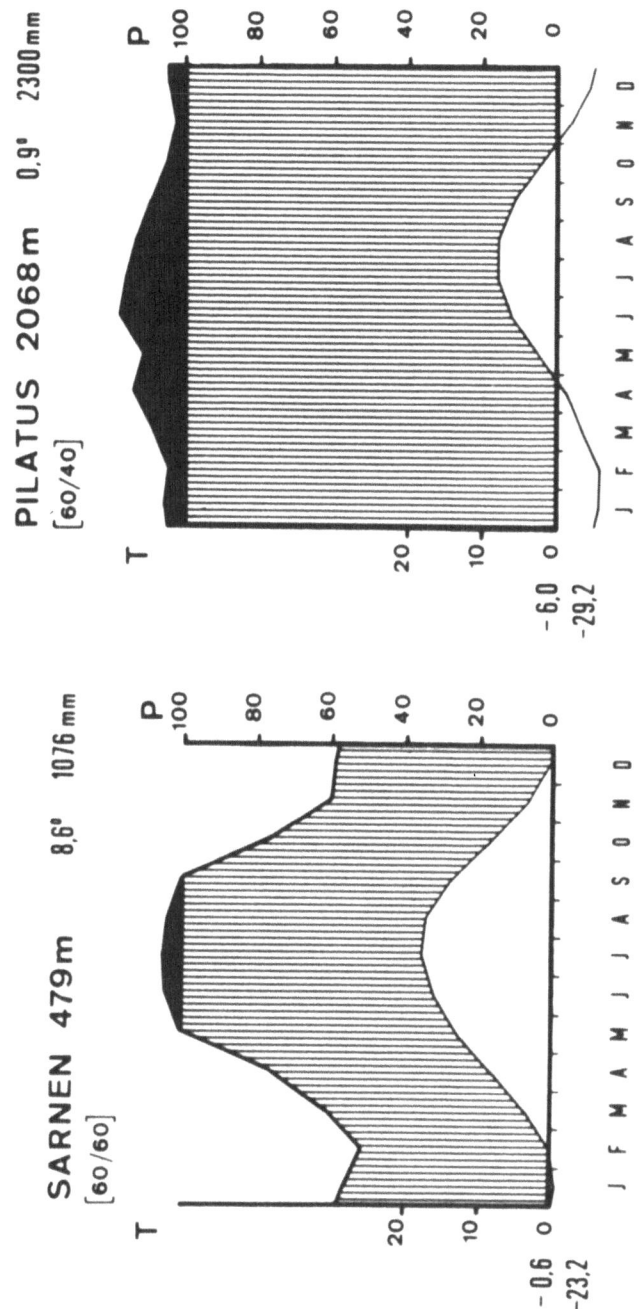

Fig. 1. Klimadiagramme aus unterschiedlicher Meereshöhe der Nordalpen.

hauptsächlich aus Flyschgesteinen und tonigem Bündnerschiefer entstanden sind, häufig durch Hang- oder Quellwasser vernässt. Auf ihnen bilden sich Anmoorgleye mit basiphiler Flachmoorvegetation (z.B. C a r i c e t u m d a v a l l i a n a e s.l.). Erfolgt die Befeuchtung überwiegend durch basenarmes Niederschlagswasser, so entstehen oligotrophe Flachmoore oder Hochmoore.

Die Böden der Alpen waren seit langem Gegenstand grundlegender Untersuchungen, von denen als herausragende Beispiele Braun-Blanquet & Jenny (1926), Pallmann (1943), Braun-Blanquet, Pallmann & Bach (1954) genannt seien.

9.3 Die Wechselbeziehungen zwischen Boden, Vegetation und Weidenutzung

9.3.1 DER EINFLUSS DER BÄUERLICHEN BEWIRTSCHAFTUNG AUF BODEN UND VEGETATION

Schon vor mehr als 2000 Jahren wurde in den Alpen Weidewirtschaft betrieben. Wie Kral (1971) durch Pollenanalyse für das Dachsteingebiet nachweisen konnte, drückte damals der Alpbetrieb die Waldgrenze dort um fast 200 m nach unten. Die weitere Verringerung des Waldareals durch Mensch und Tier erfolgte schubweise während längerdauernder günstiger Klimaperioden, so besonders im späteren Mittelalter. Kral vermutet, daß im Dachsteingebiet das Absinken der Waldgrenze auf den heutigen Stand nur etwa zu 40% auf klimatische Ursachen, zu 60% auf die Alpwirtschaft zurückzuführen ist.

Von den Siedlungen aus wurde der Wald durch Mensch und Tier von unten her ausgehöhlt und zurückgedrängt. Es entstanden ausgedehnte Allmendweiden, die heute mehr und mehr wieder verbuschen und verwalden, weil die Weidenutzung zurückgeht oder sich auf die Mähwiesen verlagert.

Wo der Wald oder das *Rhododendron-Vaccinium*-Gebüsch in Weideland oder Heumatten übergeführt worden ist, verbraunen die Podsole, weil keine Streu mehr anfällt und die Rohhumusauflage verschwindet. Es entstand das sekundäre Braunpodsol, das gewöhnlich sauer ist, aber trotzdem recht fruchtbar sein kann. Je nach Bewirtschaftungsintensität trägt es magere N a r d e t a l i a - oder wüchsige C y n o s u r i o n -weiden. Durch gute Nährstoffversorgung und landwirtschaftliche Nutzung lassen sich auch die basiphilen Magerrasen auf Kalkstein- und Mergelrendzinen in Fettrasen überführen (vgl. Lüdi, 1936). Diese Massnahmen för-

410

dern gleichzeitig die Verbraunung der karbonatreichen Böden. Infolge jahrhundertelanger Alpwirtschaft sind so auch subalpine Braunerden entstanden, die einen mullreichen obersten Bodenhorizont besitzen und meistens eine saure Reaktion aufweisen (vgl. Jäggli in Dietl, 1972).

Langdauernde alpwirtschaftliche Nutzung und sorgfältige Pflege des Weidelandes, insbesondere die gleichmäßige Verteilung des alpeigenen Düngers, führte zur Ausbildung von Fettrasen, in denen wertvolle Weidepflanzen stark vertreten sind. Sie dürften spontan auf natürlichen Anreicherungsstandorten in Hangmulden vorgekommen sein, die von Natur aus nährstoffreich und frischfeucht sind (vgl. Marschall, 1958). Von dort aus konnten sie in die anthropogenen Fettweiden einwandern.

Auf nährstoffarmen und meistens auch sehr sauren Standorten gedeihen Magerrasen. Sie nehmen im Gebiet der Alpweiden die weitaus grösste Fläche ein. In der Regel herrscht in diesen Pflanzenbeständen das Borstgras (*Nardus stricta*) vor. Neben Nährstoffarmut und saurer Bodenreaktion hat die "selektive Unterbeweidung" (Klapp, 1951) infolge des üblichen freien Weideganges die Ausbreitung der ertragsschwachen und qualitativ ungenügenden Nardusrasen gefördert.

Eine ebenso wirksame Beeinträchtigung folgte aus dem weitverbreiteten Vieheintrieb in den Wald, der zu unbefriedigenden Beständen führte. Weidewälder weisen in der Regel sehr geringen Holzwuchs auf. Dazu sind sie meist überaltert, weil sie sich kaum noch natürlich verjüngen. Auch die Aesung entspricht nach Menge und Qualität nicht den Ansprüchen unserer anspruchsvollen Nutztiere. In der Kraut- und Strauchschicht der subalpinen Fichten- und Arvenwälder sind hauptsächlich *Deschampsia flexuosa, Calamagrostis villosa*, Vaccinien und Farngewächse zu finden. Sehr oft ist der Boden nur von einer Moosschicht bedeckt. Die Milchleistung der Kühe und die Gewichtszunahme der Jungtiere sind deshalb ungenügend.

9.3.2   DAS BEMÜHEN UM NEUE NUTZUNGSFORMEN

Unter Alp oder Alm verstehen wir einen während der Weidezeit (Sömmerung) weitgehend selbständigen Wirtschaftsbetrieb (Alpwirtschaftsbetrieb), der in der Regel mit den Heimgütern in einem untrennbaren wirtschaftlichen Zusammenhang steht (vgl. Bäbler & Strebel, 1968).

Je nach der Art der Bestossung können Kuh- und Jungviehalpen unterschieden werden. In bestimmten Gegenden gibt es auch gemischte Alpen, auf denen Kühe und Galtvieh gemeinsam gesömmert werden. Auch Ross-, Ochsen- und Stieralpen haben in manchen Gebieten eine grosse Bedeutung. Im Hochgebirge, oft im Bereich der alpinen Rasengrenze, gibt es noch häufig Schafalpen. Kuh- und Jungviehalpen sind je nach der Art der Besiedlung und Tradition ein- oder mehrstafelig, das heisst, der Alpbetrieb besteht aus einer Einheit oder nach der Höhenlage geordneten mehreren Alpeinheiten; meistens zwei bis drei. So wie der Frühling in die Berge einzieht, steigt der Hirt mit seiner Herde vom Vorsäss (Voralp) über das Mittelsäss auf das Obersäss (Hochalp). Dies entspricht einem ganz natürlichen Weidewechsel.

Welche Bedeutung der Weidegang seiner Rinder, Schafe und Ziegen für den Bergbauern bereits im Mittelalter erlangte, wird durch zahlreiche Alp- und Weideverordnungen bezeugt, die von einer weitgehenden Ordnung der Weiderechte im Alpgebiet berichten. Die Alpweiden spielen für den Bergbauernbetrieb auch heute noch eine wesentliche Rolle; sie dienen der Futterproduktion für die Sommermonate, bringen dem Bergbauern Arbeitsentlastung während dieser Zeit und sichern die Gesundheit der Tiere.

Alte Rechte, Verordnungen und Bräuche in bezug auf Be-

Fig. 2. Durch Beweidung aufgelichteter subalpiner Fichtenwald (H o m o g y n o - P i c e e t u m) im Obwaldener Flyschgebiet zwischen 1600 und 2000 m ü.M. (Zentralschweiz).

stossung, Nutzung und Pflege der Alpweiden werden den heutigen Verhältnissen nicht mehr gerecht. Es ist daher notwendig, dass wir der Alpwirtschaft zu neuen Formen verhelfen, indem wir sie strukturell und organisatorisch den neuen land- und volkswirtschaftlichen Erfordernissen anpassen (Spatz, 1970). Voraussetzung hierfür ist eine durchgreifende Strukturverbesserung, deren Grundlage die Trennung von Wald und Weide bildet. Um diese Neugestaltung bemühen sich Wald- und Futterbauer zwar schon seit langem (Aichinger, 1951; Gayl, 1951; Koblet, 1954; Rieben, 1957). Doch werden bis heute noch viele Wälder beweidet, deren Standorte erst dann Befriedigendes leisten können, wenn sie im Hinblick auf ihre spezifische Eignung entweder ganz als Wald oder ganz als Weide genutzt werden. Einen Weidewald in unbefriedigendem Zustand zeigt Fig. 2.

Anders zu beurteilen sind die mit gutgewachsenen Einzelbäumen oder Baumgruppen bestockten Weideflächen, die "Wytweiden" des Jura und die Lärchenwiesen der inneralpinen Trockentäler. Diese schönen, landwirtschaftlich nutzbaren Naturparkgebiete bilden ökologisch wertvolle Bestandteile der betreffenden Landschaften.

Voraussetzung für eine dauerhafte, die landwirtschaftlichen Erträge sichernde Raumordnung der Alpgebiete ist weiterhin die standortsgemässe Nutzung des Bodens. Aufgrund der Kenntnis von Vegetation, Boden und Geländegestalt müssen die nutzungswürdigen Flächen vom ungeeigneten Land (Wald, Zwergstrauchheiden, Riedwiesen, Schutzgebiete, Oedland) getrennt werden. Ebenso sind allzu kleine oder aus anderen betriebswirtschaftlichen Gründen nicht rationell nutzbare Flächen auszuscheiden. An die Konzentration der Weidewirtschaft auf die bestgeeigneten Flächen muss sich die planmässige und den örtlichen Verhältnissen angepasste Intensivierung des Weidebetriebes anschliessen. Beispielsweise erfordert der Übergang zur Koppelweide die Anlage neuer Tränkestellen, weil die Tiere ihre bisherige Tränke nicht mehr erreichen können. Zugleich sind überkommene, auf überholte Wirtschaftsformen zugeschnittene Weiderechte den neuen Bedingungen anzupassen.

Erst die Gesamtheit dieser aufeinander abgestimmten Maßnahmen gewährleistet durchgreifende Verbesserung. Technische Teilmeliorationen wie Drainagen, Wege- und Hüttenbau für sich allein genügen nicht mehr. Zugleich sichert eine derart umfassende Planung den Bestand der gesunden und naturnahen Bewaldung, die an Steilhängen und nahe der Waldgrenze Schutz gegen Lawinen bietet und das Landschaftsbild verschönt.

Um das Ziel einer lohnenden und zugleich naturnahen Nutzung erreichen zu können, muss bekannt sein, welche Pflanzengesellschaften mit wirtschaftlich tragbarem Aufwand zu hoher Leistung gebracht werden können und welche anderen den erforderlichen Aufwand nicht lohnen. Auf zahlreichen Versuchsflächen konnte nachgewiesen werden, dass manchen Gesellschaften, deren Weideleistung unter der bisherigen Nutzungsform minimal ist, ein grosses Leistungspotential innewohnt, das durch zweckentsprechende Bewirtschaftung rationell nutzbar gemacht werden kann. Magere Ausbildungen von Fettweiden und viele Borstgrasrasen konnten mit Erfolg in ertragreiche Fettweiden übergeführt werden. So berichtet die Kali AG (1966) von hohen Erträgen an Trockenmasse, Eiweiss und Stärkeeinheiten, die durch Phosphat-Kali- und Volldüngung in mageren Ausbildungen des C r e p i d o - C y n o s u r i o n erzielt wurden (vgl. auch Bourqui & Schmid, 1968/69; Dietl, 1972).

Tabelle 1.
Bodenprofil eines milden *Nardetum*

Ort	:	Giswil, Alp Fontanen. Höhe ü.M. 1690 m Exp.: ESE, Neigung 30%.
Bodentyp	:	Initiales Braunpodsol.
Bodenform	:	Skelettarmer, schwach toniger Lehm, frisch, subalpin.
Muttergestein	:	Mergelschiefer.
Pflanzengesellschaft	:	P o l y g o n o - N a r d e t u m .

Horizont	Tiefe in cm	pH (H$_2$O)	Körnung % Ton	Bodenart	Org. Substanz %	Bodengefüge
Amo	0– 10	5,5	30	schwach toniger Lehm	23,0	Schwammgefüge
Aeh/E	10– 20	5,4	38	toniger Lehm	10,0	kugeliges Krümelgefüge
Bfe	20– 40	5,4	40	toniger Lehm	2,9	kugeliges Krümelgefüge
Bch	40– 80	5,9	21	schwach sandiger Lehm	1,2	kleinpolyedrisches Klumpengefüge
BC	80–100	6,7	15	sandiger Lehm		Primitivgefüge
R	ab 100					

Die pflanzliche Produktion ist ebensosehr vom naturgegebenen Standort wie von der Bewirtschaftung abhängig. Während für die Pflanze kein prinzipieller Unterschied zwischen Natur- und Wirtschaftsfaktoren besteht, muss die Planung beide getrennt ins Auge fassen. Dies ist nötig, weil viele Naturfaktoren unveränderlich oder nur mit hohem Aufwand beeinflussbar sind, während die Bewirtschaftung den Bedürfnissen angepasst werden kann. Wenn daher bekannt ist, ob eine Fläche aufgrund des ungünstigen naturgegebenen Standortes (steile Hanglage, flachgründige oder podsolige Böden, hoher Grundwasserstand, rauhes Lokalklima u.ä.) wenig leistet, oder ob Bewirtschaftungsfehler (Nährstoffmangel, falsche Weideführung) überwiegen (vgl. Krause, 1964), sind wertvolle Planungsunterlagen gewonnen.

Um zum Ziel zu gelangen, bedarf es vielseitiger Untersuchungen der Standortsqualität, der Wirtschaftseinflüsse und der aus beiden resultierenden Pflanzendecke. Sie verursacht zu An-

Tabelle 2.
Bodenprofil eines strengen *Nardetum*

Ort : Sarnen, Alp Hohnegg, Höhe ü.M. 1550 m Exp.: SE, Neigung 30%.
Bodentyp : Gleyartiges Eisenpodsol.
Bodenform : Skelettarmer, schwach sandiger Lehm, hangfeucht, subalpin.
Muttergestein : Flyschsandstein.
Pflanzengesellschaft : C a r i c e t o p i l u l i f e r o - N a r d e t u m.

Horizont	Tiefe in cm	pH (H_2O)	Körnung % Ton	Bodenart	org. Substanz %	Bodengefüge
Ol				modrige Streueauflage		
Amo	0— 5	4,2	19,5	schwach sandiger Lehm	21,3	Schwammgefüge
Ehg	5— 15	4,2	22,5	schwach sandiger Lehm	10,2	Krümelgefüge
Ife	15— 20	4,3	23,3	schwach sandiger Lehm	3,3	Krümelgefüge
Bfe	20— 40	4,4	18,9	schwach sandiger Lehm	3,2	Krümelgefüge
Bg	40—100	4,3	39,6	toniger Lehm		Krümelgefüge

Der wesentliche Unterschied zum Braunpodsol (Tab. 1) besteht im Auftreten eines Eluvialhorizontes (Ehg) und eines Horizontes mit Eisenhydroxyd-Umhüllung der Sandkörner (Ife). Zu den Abkürzungen vgl. FREI (1975)

fang hohen Arbeitsaufwand, besonders wenn schwer fassbare Faktoren aufgeklärt werden sollen. Da aber die Pflanzendecke das Resultat des Zusammenwirkens aller Einzelfaktoren darstellt, kann von ihr aus, sobald die Zusammenhänge an repräsentativen Probeflächen geklärt sind, auf die Standortseigenschaften geschlossen werden, die das Leistungspotential des Graslandes bestimmen. Dabei ergibt sich der Vorteil, dass das flächenhaft sichtbare Kriterium "Vegetation", über ± verborgene Kräfte des Bodens, des Klimas, der Nährstoffversorgung Auskunft gibt. In der Aufklärung des Leistungspotentials extensiv genutzter Flächen liegt einer der wichtigsten Förderungsbeiträge, den die Vegetationskunde der alpinen Landwirtschaft leisten kann.

Innerhalb der alpwirtschaftlich wichtigen Rasengesellschaften bestehen Magerweiden, Nassweiden und Fettweiden. Magerweiden zeichnen sich durch hohe Anteile schwachwüchsiger, oft verholzter und kleinblättriger Pflanzen aus, von denen manche vom Vieh zwar gern gefressen werden, aber unbefriedigenden Ertrag liefern. Sie gedeihen am besten in schwach genutzten Flächen. In den Nassweiden gewinnen trittfeste, meist hochwüchsige, aber vom Vieh verschmähte Binsen die Oberhand. In den Fettweiden dominieren wuchskräftige, weich- und grossblättrige Pflanzen mit hohem Futterwert, die zugleich grosse Ansprüche an die Nährstoffversorgung stellen und zum optimalen Gedeihen der regelmäßigen Nutzung bedürfen (Tab. 3, S. 426). Die Pflanzenbestände der Alpweiden sind bereits ausführlich beschrieben worden (z.B. Braun-Blanquet 1948 – 1950, 1969; Oberdorfer 1950; Marschall 1958, 1963; G. & R. Knapp 1953; Knapp 1962; & Spatz 1970; Dietl 1972; Dietl & Jäggli 1972; Marschall & Dietl 1974). In einer Abhandlung über landwirtschaftliche Vegetationskunde muss es genügen, ihre Artenzusammenstellung kurz zu rekapitulieren.

Die grösste Bedeutung für die Alpwirtschaft hat die Verbesserung der Magerrasen. Der Erfolg hängt von folgenden standörtlichen und wirtschaftlichen Bedingungen ab:

— gute Futterpflanzen müssen in den Magerweiden bereits in geringer Menge vorhanden sein (vgl. Klapp 1951, Marschall 1964);
— die Weide muss bezüglich Hangneigung und Geländeform intensivierungsfähig sein;
— Düngung und futterbaulich richtige Nutzungsweise (Umtriebsweide) müssen sich ergänzen.

Steilere Hänge dürfen nicht zu stark gedüngt werden, sonst könnte der Rasen zu mürbe werden. *Nardus* und *Festuca rubra* sorgen für eine geschlossene, trittfeste Weidenarbe. Besonders hohe Stickstoff- oder Hofdüngergaben vermögen diese Arten zurückzudrängen und es können dann leichter Viehpfade und Trittlöcher entstehen, die die Erosionsgefahr erhöhen. Auf frischen, feuchten Standorten in flacher oder nordexponierter Lage entwickeln sich in der alpinen Region unter dem Einfluß der Beweidung häufig Höckerlandschaften. Solche Weiden sind in der Regel nicht verbesserungswürdig.

9.3.3.1 *Die Verbesserung der Weiden auf saurem Boden*

Die für die Alpwirtschaft bedeutendsten Magerrasen auf sauren Böden — der pH-Wert in der Krume liegt meistens zwischen 4 und 5 — sind die Borstgrasweiden. Marschall & Dietl (1974) gliederten die vielfältigen Borstgrasrasen der Schweizer Alpen im wesentlichen in 4 Gesellschaften. Die Bestände auf vorwiegend kieselhaltigem Muttergestein (Granit, Gneis, z.T. auch Flysch) sind in ihren reinen Ausbildungen sehr artenarm und futterbaulich schwer zu verbessern; es sind sogenannte strenge Nardeten. Sie werden in ihrem typischen Verbreitungsgebiet, in der oberen subalpinen Region, als T r i f o l i o a l p i n o - N a r d e t u m , im montan-subalpinen Bereich als C a r i c e t o p i l u l i f e r o - N a r d e t u m bezeichnet (Fig. 4).
Auch über kalkreichem Muttergestein ist in den Alpen häufig saurer Boden entwickelt, sei es als Folge der starken klimatischen Auswaschung, sei es durch Überdeckung mit Fremdmaterial. Als analoge Gesellschaften auf derartigen Böden kommen im Hochalpengebiet das L u z u l o s p i c a t a e - N a r d e t u m , in der unteren subalpinen Region das P o l y g o n o b i s t o r t a e - N a r d e t u m vor. Diese Bestände sind in der Regel durch Bewirtschaftungsmaßnahmen verbesserbar, weil sie fast immer wertvolle Weidepflanzen tragen, wie *Trifolium pratense, T. badium, T. repens, Poa alpina, Phleum alpinum* und *Ligusticum mutellina*. Es sind sogenannte milde Nardeten (Fig. 3).
Borstgrasweiden stocken am häufigsten auf Braunpodsolen, sehr strenge Ausbildungen mitunter auf Eisenpodsolen und Podsolgleyen. Zur kennzeichnenden Artenkombination zählen *Nardus stricta* als vorherrschende Art; weiters *Arnica montana, Leontodon helveticus, Campanula barbata, Gentiana kochiana, Geum montanum, Trifolium alpinum*, sowie die Zwergsträucher *Calluna vulgaris, Vaccinium myrtillus* und *Vaccinium uliginosum*. Seit den Anfängen der Vegetationskunde haben die weitverbrei-

teten, eintönigen Nardusrasen besondere Beachtung gefunden. Seit Kerner v. Marilaun sowie Stebler & Schröter im letzten Jahrhundert haben sich zahlreiche Pflanzensoziologen damit befaßt. Aus den Alpen wurden zahlreiche Arbeiten von Braun-Blanquet und Oberdorfer bekannt; in den Dinariden haben sich I. Horvat und in den Karpaten Szafer, Pawlowski und Puscaru dem Borstgrasproblem gewidmet. Marschall & Dietl (1974) geben in einer neueren Arbeit einen Überblick über die verschiedenen Borstgrasgesellschaften der Schweizer Alpen.

Die ausgedehntesten Nardusrasen stehen auf dem Boden ehemaliger, durch die Beweidung verdrängter Wälder, sie gehören also nicht den alpinen "Urwiesen" des C a r i c i o n c u r v u l a e an. Sie besiedeln sauren, sehr nährstoffarmen Boden und haben ihre floristische Zusammensetzung unter dem Einfluss selektiver Unterbeweidung gewonnen.

Den ersten erfolgreichen Versuch zur Melioration von Borstgrasweiden beschreiben Stebler & Schröter (1887). Anhand vollständiger Artenlisten mit genauen Gewichtsbestimmungen zeigen sie, "wie durch Mist-Düngung eine Borstgraswiese", deren Nardusanteil 69 Gewichtsprozent ausmachte, "in eine gute Matte umgewandelt werden kann, indem *Nardus* völlig verschwindet und besseren Gräsern Platz macht". Weitere Mitteilungen machen Stebler & Schröter 1888, Aichinger 1933, Schneiter 1934, Lüdi 1936, Klapp 1951, Koblet, Frei & Marschall 1953, Marschall 1964, Krause 1954, 1962, Speidel 1963. Nachhaltig konnten Nardeten jedoch nur verbessert werden, wenn sich Düngung und verbesserte Weideführung ergänzen. Klapp (1951) konnte nachweisen, dass sich Borstgrasbestände erst durch das Zusammenwirken von Düngung und Umtriebsweide auf kleinen Koppeln rasch umbilden. In Versuchen über Borstgrasbekämpfung fand Lüdi (1936) im Gebiet der "Schynige Platte" bei Interlaken im Berner Oberland (2000 m ü.M.), dass eine Volldüngung ohne Kalkgaben "die vollständige Umwandlung des Rasens nach Abundanz, Dominanz und Vitalität" bewirkte. Die Erträge erhöhten sich um mehr als das fünffache. Gefördert wurden vor allem die wertvollen Futterpflanzen, während *Nardus* deutlich zurückging. Andere Magerkeitszeiger verschwanden nach einigen Jahren.

Koblet, Frei & Marschall gaben 1953 die Ergebnisse langjähriger Düngungsversuche im L u z u l o - s p i c a t a e - N a r d e t u m auf Braunpodsol ob Arosa (2050 m ü.M.) bekannt. Der Trockensubstanzertrag der gedüngten Flächen hatte sich mehr als verdoppelt. Die Magerrasenarten wurden zugunsten der erwünschten Futterpflanzen stark zurückgedrängt. Die Versuche zeigen, dass die durch Nährstoffgaben ausgelösten Veränderungen im Pflanzenbestand für die Futterproduktion der Alpweiden von grösster Bedeutung sind.

Die Beobachtung, daß ein ertragsschwacher, hartrasiger Borstgrasrasen durch Düngung und intensivierte Nutzung in relativ kurzer Zeit und mit tragbarem Aufwand in eine gutgräsige hochwertige Weide verwandelt werden kann, läßt erkennen, daß seinem Standort nicht ausschließlich ungünstige wuchshemmende Eigenschaften innewohnen, wie nach dem Erscheinungsbild zu vermuten wäre. Eine Bestätigung gibt das Bodenprofil, das in der hochmontanen Stufe des Schwarzwaldes (Krause 1962, S. 73) unter *Nardus*-Beständen regelmäßig als tiefgründige locker gekrümelte steinarme Braunerde ausgebildet ist.

Im Alpgebiet konnten wir durch vergleichende vegetationskundlich-ökologische Untersuchungen feststellen, daß sich durch sachgerechte Bewirtschaftung auf Braunerde, Braunerde-Gley, Braunpodsol (Tab. 1) und Mergelrendzina, ertragreiche Fettweiden ausbilden können (Fig. 3). Andererseits fanden wir auf denselben Böden verschiedene futterbaulich ungenügende Ausbildungen der Fettweiden und sogar Borstgrasrasen. (Vgl. Dietl 1972; Dietl & Jäggli 1972). Dies zeigt deutlich, daß durch angepasste

Fig. 3. Durch Kali-Phosphat-Düngung verbessertes P o l y g o n o - N a r d e t u m (sog. mildes N a r d e t u m) auf Braunpodsol (vgl. Tabelle 7). Starke Zunahme des Klees, Reste des Magerbestandes (*Arnica montana*).

Bewirtschaftung die natürliche Fruchtbarkeit der Böden ausgenutzt werden kann.

Das Beispiel eines Borstgrasrasens, der sich wahrscheinlich erst nach vielen Jahren geregelter Düngung und Nutzung in eine Fettweide überführen läßt, gibt Geering (1968). In einem strengen, reinen T r i f o l i o a l p i n i - N a r d e t u m im Urgesteinsgebiet der Zentralalpen auf Humuspodsol (2060 m ü.M.) hatte Düngung, selbst mit sehr hohen Nährstoffgaben, nur geringen Erfolg. Im Ausgangsbestand fehlten die Pflanzen der Fettweiden; zudem bildete die Alp ein ausgedehntes Borstgrasweiden-Gebiet. Ein Habitusbild gibt Fig. 4, zum Boden vgl. Tab. 2.

Schon früher wurde mehrmals versucht, durch Umbruch und Neuansaat oder durch Einsatz von Herbiziden, strenge Borstgrasrasen zu verbessern. Diesen Bemühungen war, wie Lüdi (1936), Schechtner & Wagner (1963) und Guyer (mdl. Mitt.) übereinstimmend berichten, kein Erfolg beschieden.

Hingegen ist es möglich, ungepflegte Weiden, die einen nutzbaren Rasenbestand aufweisen, aber von Farnpflanzen (*Pteridium aquilinum, Thelypteris limbosperma = Dryopteris montana*) oder Sträuchern (*Genista tinctoria, Sarothamnus scoparius*) überwachsen sind, mit Herbiziden zu säubern (Krause mdl. Mitt.). Farn-

Fig. 4. C a r i c e t o p i l u l i f e r a e - N a r d e t u m (sog. strenges N a r d e t u m) auf saurem Eisenpodsol (vgl. Tab. 2). Durch Düngung nicht oder nur schwer zu verbessern.

420

kräuter können auch durch gute Düngung und häufigen Schnitt zurückgedrängt werden.

Anders zu beurteilen sind reine Zwergstrauchbestände, in den Alpen z.B. R h o d o d e n d r o - V a c c i n i e t u m Braun-Blanquet 1927 (an Nordhängen), J u n i p e r o - A r c - t o s t a p h y l e t u m Braun-Blanquet 1926, Haffter 1939 (an Südhängen), E m p e t r o - V a c c i n i e t u m Braun-Blanquet 1926 (im Schwarzwald als Reliktgesellschaft) und L o i s e - l e u r i o - C e t r a r i e t u m Braun-Blanquet 1926. Eine neuere umfassende soziologisch-ökologische Darstellung gibt Schwein- gruber 1972. Im Schwarzwald ist das C a l l u n o - V a c c i - n i e t u m Büker 1942, das zu den atlantischen Zwergstrauch- heiden gehört, die bedeutendste Zwergstrauchgesellschaft. Alle diese Heiden sind in der Regel kaum futterbaulich verbesserungs- fähig und noch seltener verbesserungswürdig.

In den Südalpen stockt an armen, humusarmen, felsigen Steilhängen in der subalpinen und alpinen Stufe das *Festucetum variae* Brockmann-Jerosch 1907. Es ist eine azidophile Trocken- rasengesellschaft. Gelegentlich dient sie als schüttere Kleinvieh- weide (vgl. Braun-Blanquet 1969). Futterbaulich sind diese Hal- den nicht verbesserbar.

9.3.3.2 *Die Verbesserung der Weiden auf kalkreichem Boden*

An steilen, früh ausapernden Südhängen sind häufig auf Rendzinaböden lückige, treppenartig aufgebaute, basiphile Ma- gerweiden des S e s l e r i o n zu finden. Solange in ihnen *Sesle- ria coerulea* und *Carex sempervirens* vorherrschen, liefern sie eine wenig geschätzte Aesung; gelegentlich werden sie gemäht. Werden diese Rasen regelmässig beweidet, so können sich zu den ur- sprünglichen Arten auch Fettwiesenpflanzen mit hohen Nähr- stoffansprüchen gesellen, z.B. *Trifolium pratense, Poa alpina, Phleum alpinum, Cynosurus cristatus* und *Crepis aurea.* G. & R. Knapp (1953) bezeichnen diesen Weidetyp als S e s l e r i o - S e m p e r v i r e t u m t r i f o l i e t o s u m . Diese Gesellschaft erreicht in den Alpen (vgl. auch Knapp 1962, Dietl 1972) und auch im Jura grosse Verbreitung. Der Anteil der Legumino- sen im Pflanzenbestand ist in der Regel sehr hoch. In den regen- reichen Gebieten der Alpen, in denen eine langdauernde Aus- trocknung der flachgründigen Humuskarbonatböden kaum zu be- fürchten ist, können die S e s l e r i e t a l i a - Magerrasen durch angepasste Phosphat-Kali-Düngung in wertvolle, frühzeitig zu nutzende Weiden umgewandelt werden (vgl. Lüdi 1936).

Eine S e s l e r i o n - Gesellschaft ist auch das C a r i -
c e t u m   f i r m a e (Kerner) Braun-Blanquet 1926, dessen kurzer, hartblättriger Rasen von den Weidetieren kaum beachtet
wird. Es gedeiht in der subalpinen und alpinen Stufe auf humusarmen Kalkgesteinsböden an steilen Halden oder windgefegten
Rücken. Es ist nicht verbesserungsfähig. Ähnliches gilt auch für
das E l y n e t u m Braun-Blanquet 1913.

Über die Ökologie subalpiner und alpiner Magerrasen und die
Literatur geben Albrecht (1969), Rehder (1970) und Gigon
(1971) umfassende Auskunft.

9.3.3.3   Die Verbesserung der Weiden auf nassem Boden

Tonreiche, bindige Böden lassen das Wasser schlecht durch-
sickern und vernässen deshalb in niederschlagsreichen Gebieten.
So sind besonders im Flyschgebiet der nördlichen Voralpen auf
Gleyen und Anmoorgleyen Naßweiden stark verbreitet. Regelmäßig dominiert *Juncus effusus* in diesen Flächen (vgl. Oberdorfer 1957; Berset 1969; Dietl & Guyer 1974). Es sind meistens C a l t h i o n - Gesellschaften. In manchen Gegenden werden sogar Flachmoore, z.B. *Carex davalliana*-Rieder beweidet.

Die eigentlichen Moore und Riedflächen sollen mit Rücksicht auf ihre große landschaftsökologische Bedeutung in ihrer
ursprünglichen Form erhalten bleiben. Daneben gibt es aber in
der Flyschzone der nördlichen Voralpen große Riedlandschaften
mit ausgedehnten Naßweiden. In den meist wassergesättigten,
schlecht durchlässigen Gleyböden mit einer täglichen seitlichen
Wasserbewegung von einem halben bis einigen wenigen Zenti-
metern erwiesen sich Drainageanlagen häufig als wirkungslos.
Fig. 5 zeigt ein ehemaliges *Caricetum davallianae* nach einem Ent-
wässerungsversuch und anschließender Düngung. An die Stelle
der nahezu ertragslosen *Carex* ist ein Bestand des hochwüchsigen,
aber als Futter minderwertigen *Scirpus silvaticus* getreten.

Indessen konnten in den *Juncus effusus*-Weiden der
Flyschgebiete durch angemessene Nährstoffversorgung und ge-
regeltes schonendes Beweiden beachtliche futterbauliche Erfolge
erzielt werden (vgl. Dietl & Guyer 1974). Die Erträge stiegen
teilweise um mehr als das Doppelte (auf über 50 q TS/ha in
900 m ü.M.) und der Anteil an wertvollen Futterpflanzen, besonders der Leguminosen und Gramineen, hat sich auf Kosten der
Juncus- und *Carex*-Arten stark erhöht. Da der Weidegang die tritt-

empfindlichen Naßböden stark verdichten kann, ist eine sorgfältige Bestossung anzustreben.

Meistens sind die wertvollen Futterpflanzen sehr konkurrenzkräftig, wenn sie gut genährt und richtig genutzt werden. Besteht kein Mangel an den direkt wirkenden Wachstumsfaktoren Wasser und Nährstoff, dann sind die Ansprüche vieler Fettwiesenpflanzen an die Durchlüftung dank ihres Anpassungsvermögens nicht sehr groß.

Petrascheck (1973) konnte in dem von Dietl & Guyer (1974) beschriebenen Versuch nachweisen, daß durch die höheren Futtererträge dem Boden nicht mehr Wasser entzogen wurde als vor Einsetzen der Düngung. Es darf deshalb angenommen werden, daß die "biologische Entwässerung" durch Evapotranspiration auf mageren und fetten Naßwiesen zumindest annähernd gleich sein muß. Sie ist auf undurchlässigen "Flyschböden" der entscheidende Faktor für den Bodenwasserhaushalt (vgl. Puffe 1973).

Bevor nasses Wiesland entwässert wird, müssen die Vernässungsursachen bekannt sein. Die Vernässung kann durch hohen Grundwasserstand (Fremdvernässung) oder starke Wasserbindung (Sorption) im Boden bedingt sein. Letzteres führt zur Eigenvernässung (durch Niederschlagswasser), deren Ursache beim

Fig. 5. Naßstandort auf tonreich-bindigem Boden im Flyschgebiet. Ursprünglich C a - r i c e t u m   d a v a l l i a n a e , nach versuchter Entwässerung und Düngung in einen wirtschaftlich wertlosen Bestand von *Scirpus silvaticus* verwandelt.

Korngerüst und damit bei den Bodeneigenschaften liegt. Tonreiche Böden mit hohem Anteil an Feinporen und großer Sorptionskraft lassen das Wasser schlecht durch und sind deshalb kaum entwässerbar (vgl. Richard 1963, Petrascheck 1973, 1974).

9.3.3.4 *Verbesserung der Lägerfluren*

Durch tierische Ausscheidungen stark eutrophierte Standorte tragen hochwüchsige Staudenfluren, in denen fallweise *Rumex alpinus, Senecio alpinus, Ranunculus aconitifolius, Chenopodium bonus henricus* und *Urtica dioica*, in höheren Lagen auch *Rumex arifolius, Cirsium spinosissimum, Alchemilla vulgaris* und *Deschampsia caespitosa* vorherrschen. Es sind größtenteils lästige Unkräuter, die durch ihr Massenauftreten das beste Weideland entwerten. Es gibt verunkrautete Weideflächen, die unter dem Blätterdach der Hochstauden eine geschlossene Rasendecke aufweisen und solche, deren Boden offen und zerstampft ist (vgl. Dietl 1972).

Immer wurde versucht, durch Einsatz von Herbiziden vor allem den *Rumex alpinus* zu vernichten. Der Erfolg war sehr unterschiedlich.

Auf verunkrauteten Weiden mit einem intakten Rasen haben sich selektive, klee- und grasschonende Wuchsstoffherbizide gut bewährt (Ammon, mdl. Mitt.). Grasfreie Lägerbestände können auch mit Totalherbiziden behandelt werden. Es wird empfohlen, in den kahlen Flächen nach dem Verstreichen der Nachbaukarenzfrist, im nächsten Jahr eine weitgehend standortsgemäße Weidemischung umbruchlos einzusäen (Neururer 1969).

Neben der direkten Bekämpfung sind indirekte, präventive Maßnahmen unerläßlich. Geordnete Düngerwirtschaft (Güllegruben; gleichmäßige Ausbringung des Düngers) und geregelter Weidegang (Umtriebsweide) verhindern von allem Anfang an die Entstehung größerer Lägerplätze. Ohne diese Vorkehren ist der direkten chemischen oder mechanischen Bekämpfung kein bleibender Erfolg beschieden.

9.3.3.5 *Die Fettweiden als Ergebnis standortsgemäßer Weidewirtschaft*

Langdauernde alpwirtschaftliche Nutzung und sorgfältige Pflege des Weidelandes, insbesondere die gleichmäßige Verteilung

Fig. 6. C r e p i d o - C y n o s u r e t u m in 1600 m ü.M. Frühsommeraspekt von *Poa alpina vivipara*. Trockensubstanzertrag 40–60 q/ha.

Fig. 7. P o o - P r u n e l l e t u m , die kraut- und kleereiche Fettweide der Hochalpen über 1600 m ü.M. Trockensubstanzertrag 20 bis 40 q/ha.

des alpeigenen Düngers führte zur Ausbildung von Fettrasen, in denen wertvolle Weidepflanzen hohe Anteile erreichen. Fettweiden können aber auch auf natürlich nährstoffreichen, frischen Standorten, beispielsweise in Mulden entstehen, die als lokale Ausbreitungszentren anzusehen sind, von denen aus ihre charakteristischen Arten in die anthropogenen Fettweiden auswandern konnten (vgl. Marschall 1958). Die wichtigsten Gesellschaften für die alpwirtschaftliche Futterproduktion sind das A l c h e m i l - l o - C y n o s u r e t u m Müller 1967 (Bergkammgrasweide) der

Tabelle 3.
Kennzeichnende Artenkombination und Trennartengruppen der alpwirtschaftlich wichtigsten Fettweidegesellschaften.

	Alchemillo-Cynosuretum	Crepido-Cynosuretum	Poo-Prunelletum
Trifolium repens	x	x	x
Trifolium pratense	x	x	x
Festuca rubra	x	x	x
Alchemilla vulgaris aggr.	x	x	x
Leontodon autumnalis	x	x	x
Leontodon hispidus	x	x	x
Veronica serpyllifolia	x	x	x
Cynosurus cristatus	x	x	
Festuca pratensis	x	x	
Poa trivialis	x	x	
Plantago lanceolata	x	x	
Carum carvi	x	x	
Lolium perenne	x		
Phleum pratense	x		
Poa pratensis	x		
Crepis aurea		x	x
Poa alpina		x	x
Phleum alpinum		x	x
Trifolium badium			x
Trifolium thalii			x
Ligusticum mutellina			x

Montanstufe, das C r e p i d o - C y n o s u r e t u m Dietl 1972 (Goldpippau-Kammgrasweide) in der subalpinen Region und das P o o - P r u n e l l e t u m Oberdorfer 1950 (Milchkrautweide, Stebler & Schröter 1892; Marschall 1958) in Höhenlagen bis 2400 m ü.M. In Tabelle 3 sind die kennzeichnende Artenkombination und die Trennartengruppen dargestellt. Die Tabelle zeigt von links nach rechts, wie mit zunehmender Meereshöhe wärmebedürftige Tieflandarten verschwinden, echte Alpenpflanzen hinzutreten. Fig. 6 und 7 geben Habitusbilder der beiden wichtigsten alpinen Weidegesellschaften.

Neuere Arbeiten, die sich mit der soziologischen Gliederung und alpwirtschaftlichen Bedeutung der Fettweiden befassen, geben u.a. Oberdorfer (1950), Marschall (1958, 1963), G. & R. Knapp (1953), Knapp (1962), Spatz (1970), Dietl (1972), Dietl & Jäggli (1972).

In intensiv genutzten Weiderasen tritt die Wirkung der ursprünglichen Standortsfaktoren, Boden und Klima, gegenüber dem Einfluss der Bewirtschaftung zurück. Die wirtschaftsbedingten Standortsfaktoren gewinnen aber leichter die Oberhand, wenn die Naturfaktoren günstig sind. Gute Fettweiden sind deshalb leichter und billiger durch Wirtschaftsmaßnahmen zu erzielen und zu erhalten, wenn die Natur nachhilft (vgl. Brünner & Krause 1968).

Die Fettweidegemeinschaften können einen breiten ökologischen Bereich einnehmen. Dies trifft vor allem für die montanen und subalpinen Kammgrasweiden zu. Am häufigsten besiedeln sie Braunerdeböden, Braunpodsole, Braunerdegleye; seltener Gleye und Mergelrendzinen. Nässe- oder Trockenheitszeiger, Basen-, oder Säure- und Magerkeitszeiger, sogar Ruderalpflanzen gesellen sich deshalb fallweise zur Gruppe der Fettwiesenpflanzen (vgl. Dietl & Jäggli 1972, S. 484). Die futterwüchsigsten Fettweiden sind auf frischen und feuchten Standorten zu finden.

Das P o o - P r u n e l l e t u m , das in die unwirtliche Urrasenstufe hineinreicht, ist an enger begrenzte, geschützte Standorte gebunden. In Mulden und an leicht geneigten Hängen mit langer Schneebedeckung kann es grössere Flächen einnehmen. Da die Auswaschung der Böden durch den lange dauernden Frost vermindert wird, entstehen aus basenreichem Muttergestein alpine Braunerden, die wertvolle Milchkrautweiden tragen. Bei guter Bewirtschaftung kann sich das P o o - P r u n e l l e - t u m auch auf Braunpodsolen und Mergelrendzinen ausbreiten.

9.3.3.6 Das Leistungspotential der Weidegesellschaften in quantitativer Hinsicht

In Abhängigkeit von der ungleichen ortsüblichen Nutzung erbringen verschiedene Einzelbestände der gleichen Graslandgesellschaften unterschiedliche Erträge. Sobald sie optimal genutzt werden, steigern sie sich zu einem Höchstwert, der mit erträglichem Aufwand nicht mehr überboten werden kann. Er ist jedoch nicht unbedingt an eine bestimmte Pflanzengesellschaft, sondern weitgehend an die naturgegebene Produktivität des Standorts gebunden. Wenn diese, wie vom N a r d e t u m bekannt ist, nicht ausgeschöpft wird, kann der Ertrag minimal sein. Wenn sie voll in Anspruch genommen wird, steigert er sich um ein vielfaches. Doch bringt den hohen Ertrag nicht mehr ein "produktiv gemachter *Nardus*-Rasen", sondern die an seine Stelle getretene P o i o n - bzw. C y n o s u r i o n - Weide. Die Erfahrung, dass in manchen für den Landwirt höchst unansehnlichen Magerrasen hohes Leistungspotential verborgen ist, das er nutzbar machen kann, ist eine spezifische Leistung der Vegetationskunde im Bereich der landwirtschaftlichen Verbesserung extensiv genutzter Landschaften, also auch der alpinen Weiden.

In den Alpen wird, sofern die ökologischen Bedingungen nicht extrem sind, d. h. auf tiefgründigen, unvernäßten Böden in Hanglagen, die 40% Neigung nicht überschreiten, das Leistungspotential der Weiden weitgehend von der Höhenlage bestimmt. Bis hinauf zur potentiellen Waldgrenze ist mit einem Ertragsabfall der Größenordnung von 4−6% je 100 m Höhenzuwachs zu rechnen (vgl. Caputa und Schechtner 1970, Spatz 1970). Die tägliche Futterproduktion während der Vegetationszeit bleibt in allen Höhenlagen bis zu dieser Grenze mindestens gleich oder nimmt sogar zu und kann 100 kg TS/ha erreichen. Der Rückgang der Leistung wird durch die Verkürzung der Wachstumszeit verursacht (vgl. auch Caputa 1966). Im Bereich der natürlichen Waldgrenze und darüber läßt die Produktionskraft zunehmend stärker nach. In einer neungliedrigen Versuchsserie hat die Eidg. Forschungsanstalt Zürich-Reckenholz im Gebiet des Hinterrheins folgende Trockensubstanzerträge ermittelt:

1000 m ü.M.	rd. 90 q TS/ha	im A l c h e m i l l o - A r r h e n a t h e r e t u m
1500 m ü.M.	rd. 70 q TS/ha	im A n t h r i s c o - T r i s e t e t u m
1900 m ü.M.	rd. 55 q TS/ha	im P h l e o - T r i s e t e t u m
2100 m ü.M.	rd. 28 q TS/ha	in einer *Ligusticum mutellina-Festuca violacea-* Naturfettmatte

428

In diesem Beispiel nimmt die TS-Produktion zwischen 1000 und 1900 m ü.M. je 100 m Anstieg um knapp 4,5% ab. Zwischen 1900 und 2100 m ü.M. beträgt der Abfall rd. 15% je 100 m Höhenzunahme.

Da die Fettweidengesellschaften über mehrere hundert Meter Höhendifferenz vorkommen, läßt sich der Ertragsabfall von 4,5% je 100 m Höhendifferenz, den eine intensive Landwirtschaft nicht vernachlässigen kann, am Vorkommen einer bestimmten Pflanzengesellschaft nicht ablesen. Immerhin läßt sich aufgrund zahlreicher Ertragserhebungen im gesamten Alpenraum folgende Beziehung zwischen Gesellschaftszugehörigkeit und Ertrag angemessen gedüngter und geregelt genutzter Weiden erkennen (vgl. auch Bourqui & Schmid 1968/69, Caputa & Schechtner 1970):

Alchemillo-Cynosuretum	60–80 q TS/ha
(Höhen über 1000 m)	
Crepido-Cynosuretum	40–60 q TS/ha
Poo-Prunelletum	20–40 q TS/ha

Die Futterqualität wird – gemessen an der Klappschen Bestandeswertzahl – durch die Höhenlage nicht beeinflusst (Spatz 1970, Spatz & Voigtländer 1971, vgl. auch Stebler & Schröter 1889, Caputa 1966, Caputa & Schechtner 1970, Caputa 1973 b). Gut bewirtschaftete Kammgrasweiden und kleereiche Milchkrautweiden erreichen häufig Bestandeswertzahlen um 6. Dagegen ist die Futterqualität der Mager- und Naßweiden meist sehr gering. Die Bestandeswertzahlen liegen meistens unter 3, häufig sogar unter 2. Wenn die Tiere können, meiden sie diese Flächen. Da der Anteil der brauchbaren Futterpflanzen normalerweise sehr klein ist, werden sie selektiv beweidet und der wirkliche Futterwert muß deshalb äußerst tief eingeschätzt werden. Auch Spatz (1970) kam bei der Untersuchung von Allgäuer Alpweiden zum Schluss "dass die quantitativ geringen Bestände im Durchschnitt auch qualitativ am schlechtesten sind". Dazu haben zahlreiche Düngungsversuche in mageren Weiden bewiesen, dass zwischen Rohprotein und TS-Erträgen eine signifikante positive Korrelation besteht.

Der Futterertrag der Alpweiden lässt sich erst über die tierischen Veredlungsprodukte wirtschaftlich ermessen. Guyer (mdl. Mitt.) konnte mit fünfjährigen umfassenden Leistungskontrollen auf einer zwischen 2000 und 2500 m ü.M. gelegenen, gedüngten und unterteilten Alpweide in Graubünden an einer Milchkuhherde von rd. 130 Tieren im Tagesdurchschnitt von 90 Alptagen über 7 kg Milch und bis zu 500 g Lebendgewichtszunahmen je Kuh ermitteln (vgl. auch Guyer 1970). In Stärkeeinheiten gemessen stellt die Gewichtszunahme eine höhere Leistung dar als die produzierte Milch.

9.3.3.7   *Organisation der Weidenutzung*

Neben dem Ertrag und dem Wachstumsverlauf des Weiderasens bestimmen die topographischen Verhältnisse entscheidend die Organisation der Nutzung der Alpweide.

Mit steigender Höhe über Meer nimmt die Wuchsschnelligkeit der Pflanzenbestände im Frühling zu. Die Vegetation setzt im Gebirge ohne Anlaufzeit mit voller Kraft ein, weil der Winter mit seiner Schneedecke meistens ohne Uebergang von futterwüchsiger sommerlicher Witterung abgelöst wird. Weidereifes Futter, d.h. einen Ertrag von 15 q TS/ha konnte Caputa (1966) ab Vegetationsbeginn auf 430 m Meereshöhe nach 45 Tagen, auf 1200 m nach 32 Tagen und auf einer Alpweide 1900 m ü.M. nach 13 Tagen feststellen (Fig. 8). Andererseits geht der produktive Futterzuwachs in Höhenlagen über 1500 m bereits im August merkbar zurück.

Der hohe Futteranfall in kurzer Zeit im Laufe des Frühlings verlangt frühzeitige Alpfahrt und schnellen Weidewechsel. Auf

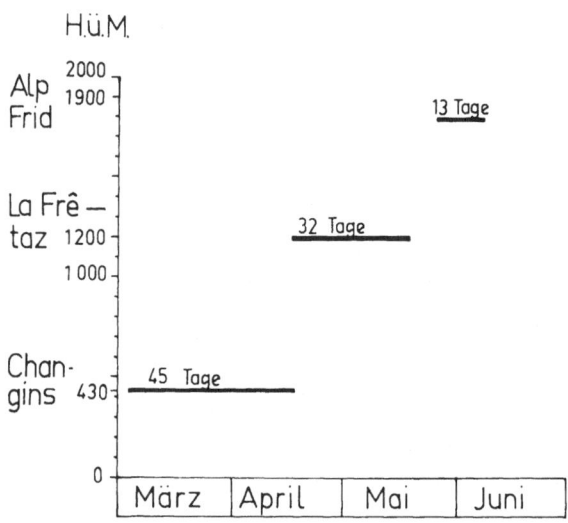

Fig. 8 Das Verhältnis zwischen Meereshöhe und Wachstumsgeschwindigkeit auf Weiden. Zahl der Tage, in denen nach Vegetationsbeginn 15 q TS/ha aufwachsen.

430

ertragreichen Weiden können trotzdem im Vorsommer bedeutende Futterüberschüsse entstehen. Die geeignetsten Koppelteile sollen zu dieser Zeit gemäht und das Gras getrocknet werden. Läßt der Futterwuchs im Spätsommer dann nach oder schneit es einmal, so verfügt der Alpbetrieb über nützliche Heureserven.

Die Verlängerung der Sömmerungszeit auf den Bergweiden läßt sich in erster Linie durch den frühen Weidebeginn und durch einen den Verhältnissen angepassten Weidewechsel erreichen. Durch geregelten Weidegang können höhere Erträge erzielt werden, weil das Gras während der Ruhepause ungestört wachsen kann. Nach dem 8.–12. Tag nach der Weideruhe haben Rasen den größten Ertragszuwachs. Deshalb ist auf den Alpen eine Weideruhe von 3–5 Wochen anzustreben.

Caputa (1973 a) prüfte im Jura auf 1100 m ü.M. den Einfluß der Zahl der Schläge auf die Ertragsfähigkeit einer Weide. Die Lebendgewichtzunahme bei Mastochsen betrug während einer Weidezeit von knapp 120 Tagen im Durchschnitt von 4 Jahren auf der Standweide 318,5 kg/ha, auf Weiden mit 4–14 Schlägen 438,5 bis 492,5 kg/ha. Durch Weidewechsel erhöht sich auch die tägliche Lebendgewichtzunahme je Tier beachtlich. Auf der Standweide nahm jedes Tier pro Tag durchschnittlich 592 g, auf den Koppelweiden 765 bis 828 g zu. Aufgrund der Erfahrungen von Caputa dürften auf Bergweiden 6 Schläge angemessen sein.

Der Weidebesatz kann nach Caputa (1966) in allen Höhenlagen dieselbe Stärke erreichen, wenn Pflanzenbestände, sowie Düngung und Nutzung gleichwertig sind, weil die tägliche Futterproduktion im Durchschnitt ungefähr gleich bleibt; sie beträgt etwa 55 kg TS/ha.

Diese Menge würde bei einem Futterverzehr von 13 kg TS pro GVE und Tag während der Vegetationszeit einen theoretischen Weidebesatz von nahezu 4 GVE/ha möglich machen. Da aber mit einem Beweidungsverlust von rund 20% gerechnet werden muss, erachtet Caputa für die Praxis 3 GVE als optimale Besatzstärke. Magerweiden vermögen häufig nur eine GVE/ha oder noch weniger zu ernähren.

Die Weidedauer hängt von der Vegetationszeit ab und wird mit steigender Meereshöhe kürzer. Je nach Klimagebiet und Bewirtschaftung können auf Alpen, deren mittlere Höhenlage 1200 m ist, die Tiere 120–140 Tage weiden; auf durchschnittlich 1500 m sind etwa 110 Weidetage möglich, auf Hochalpen zwischen 1800–2300 m beträgt die Sömmerungszeit 50–90 Tage.

Der geregelte Weidegang wirkt sich nicht nur auf den Ertrag, sondern auch auf die botanische Zusammensetzung des Rasens günstig aus. Da die selektive Beweidung verhindert wird, können sich überbeanspruchte Trittrasen erholen und in unterbeweideten Nardeten läßt sich das Borstgras wirksam bekämpfen. Auf Koppelweiden ist auch eine sorgfältige Weidepflege (Mähen von Unkraut) und ein gezielter Düngereinsatz (Gülle, mineralischer Stickstoff) möglich.

Auf Alpen, die mit Milchkühen und Jungrindern bestossen werden, kann es vorteilhaft sein, die Herden zu trennen. Die Ansprüche des Aufzuchtrindes bezüglich Energie- und Eiweissgehalt des Weidefutters sind geringer als diejenigen der laktierenden Kuh (Schürch 1967). Deshalb sollen den Kühen die nahen, ertragreichen Weiden zugeteilt werden; die ärmeren, weitläufigen Alpteile kann das Jungvieh nutzen (vgl. Fig. 16, S. 452).

In Hanglagen ist die Form der Schläge für die Weidetechnik wichtig. Guyer (in Frei & Guyer 1968) schlägt vor, die Unterteilung an einem Hang so vorzunehmen, "daß ein Schlag die Form eines auf der Schmalseite stehenden Rhomboides erhält". So lassen sich die Bildung von Viehpfaden in der Schichtlinie und das Herunterspringen am Hang in der Fallinie einschränken.

Um die Nutzung einer Alpweide optimal gestalten zu können, ist es notwendig, daß sie von außen her und auch innerhalb der Gemarkung gut mit Fahr- und Triebwegen erschlossen ist.

9.4 Standorte und Pflanzengesellschaften der Alpinen Mähwiesen

In vielen Gegenden der Alpen besteht eine Dreiteilung des landwirtschaftlichen Wirtschaftsraumes in Heimgüter, Bergwiesen und Alpweiden. Hinzu kommen die Wildheuwiesen ausserhalb des bewirtschafteten Geländes. Auf den Matten der Heimgüter und auf den Bergwiesen wird das Winterfutter bereitet; auf den Weiden werden die Tiere gesömmert. Auf schwer zugänglichen Wildheuplanggen sammelten die Aelpler bis vor kurzem noch häufig das Notfutter für die Alpzeit. Die Verteilung dieser Pflanzengesellschaften und Wirtschaftseinheiten ist in Fig. 9 dargestellt.

9.4.1 BERGWIESEN

Sie liegen — häufig von Wald umsäumt — oberhalb der Dörfer und Höfe, aber auch über der aktuellen Waldgrenze auf sanften Terrassen und steileren Hängen. Die Siedlungen, die in diesen

Gebieten entstanden sind, werden im alemannischen Sprachraum als Maiensässe, im bajuwarischen als Schwaigen oder Asten, im rätoromanischen als Misés oder Cuolms und im italienischen als Monti bezeichnet.

Das Heu, das auf den Bergwiesen geerntet wird, vergrössert die Winterfutterbasis. Vor allem in Gebieten, die in tieferen Lagen im Sommer unter Trockenheit leiden, können die Bergwiesen eine wirksame Sicherung bieten.

In der Regel wird das Heu im Herbst und im Frühling vor der Alpfahrt auf dem Maiensäss selbst verfüttert. Der anfallende Dünger wird grösstenteils in der Nähe der Maiensäss-Siedlungen auf den Matten verteilt, weil das oft ausgedehnte Bergwiesengebiet meistens schlecht durch Flurwege erschlossen ist. Auf den gedüngten Flächen stellte sich das P h l e o - T r i s e t e - t u m (Marschall 1947, Dietl 1972) als subalpine Fettmatte ein. Neben allgemein verbreiteten A r r h e n a t h e r e t a l i a - und T r i s e t i o n - Arten umfasst die kennzeichnende Artengruppe *Phleum alpinum, Poa alpina, Trifolium badium, T. nivale, Rumex arifolius, Crepis blattarioides, C. aurea, Soldanella alpina* als Gebirgspflanzen. In den Ausbildungsformen der Gesellschaft, die keine extrem ungünstigen Standortsbedingungen bieten und optimal bewirtschaftet werden, herrscht in der Regel *Trisetum flavescens* vor. *Agrostis tenuis*, das im A n t h r i s c o - T r i - s e t e t u m in Höhen um 1200 m zur Vorherrschaft kommt, ist

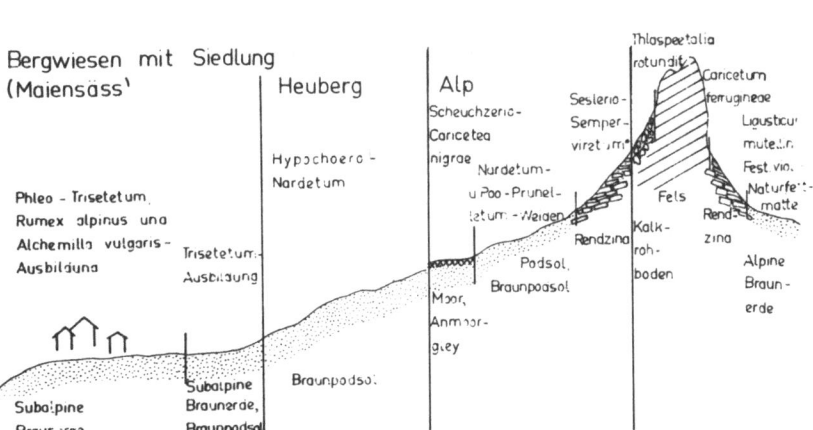

Fig. 9. Die Vegetationsverhältnisse im Bereich der Bergwiesen (Heualpen) und Alpweiden in Abhängigkeit von naturgegebenen oder wirtschaftsbedingten Standortsfaktoren (schematisch).

im Phleo-Trisetetum bei rd. 1800 m Höhe nicht mehr konkurrenzstark (Vgl. auch Stebler & Schröter 1887, S. 132).

Die Flächen werden jährlich einmal geschnitten und ein- bis zweimal abgeweidet. Die Erträge sind auf guten Matten beachtlich und können zwischen 1500 und 1900 m ü.M. 40–60 q TS/ha erreichen. Bewirtschaftungsfehler — starke Hofdüngergaben und reine Mähnutzung können *Alchemilla vulgaris, Geranium silvaticum* und *Rumex arifolius,* manchmal auch *Rumex alpinus* derart überhand nehmen lassen, daß diese unerwünschten Stauden die wertvollen Gräser fast vollständig verdrängen (vgl. Stebler & Schröter 1892, Dietl & Lauer 1973). Durch gleichmäßige Verteilung des anfallenden Düngers und durch Ueberweiden im Frühling läßt sich das "Entarten" der Pflanzenbestände vermeiden.

9.4.2   HEUBERGE

Die Bergwiesen, die von den Maiensässen weiter entfernt sind, werden vielfach als Heuberge bezeichnet. Es sind vermarchte*, in der Regel alle 2–3 Jahre gemähte (halbschürige) Magermatten. Bisher wurden sie kaum gedüngt. Die alternierende Nutzung ermöglichte einen weitgehend ausgeglichenen Nährstoffkreislauf. Der regelmäßige Schnitt verhinderte die Verheidung und Wiederbewaldung und damit das Ansammeln von Rohhumus.

Die bedeutendste Pflanzengesellschaft der Schweizer Heuberge ist das Hypochoero-Nardetum Marschall & Dietl 1974. Dieses sogenannte Mähnardetum unterscheidet sich von den beweideten Borstgrasrasen ökologisch durch die Schnittnutzung, floristisch und damit auch physiognomisch durch eine Reihe typischer Magermattenpflanzen wie *Hypochoeris uniflora, Crepis conyzifolia, Pedicularis tuberosa, Pulsatilla sulphurea, P. vernalis* (vgl. Marschall & Dietl 1974). Von der kennzeichnenden Artenkombination der subalpinen Nardeten treten besonders *Nardus stricta, Helictotrichon* (= *Avena*) *versicolor, Deschampsia flexuosa, Leontodon helveticus, Arnica montana, Geum montanum, Campanula barbata* und *Trifolium alpinum* sehr regelmäßig auf. Die typische Ausbildung der Gesellschaft gedeiht in mäßig steiler, sonniger Lage über tiefgründigem, saurem (pH 4,5–5) Braunpodsol. Wo Hangneigung über 60% die Humusanhäufung verhindert, kann *Carex sempervirens* (oft gemeinsam mit *Trifo-*

* Im Gelände abgegrenztes Grundstück

434

lium alpinum) in Beständen dominieren, die sich im übrigen wenig von den typischen unterscheiden (vgl. Stebler & Schröter 1892, Braun-Blanquet 1969). Auf kalkreichem Muttergestein wird die Gesellschaft vom S e s l e r i o - S e m p e r v i r e - t u m abgelöst, das ebenfalls zu den verbreiteten Magermatten der Heuberge gehört.

Auf weniger sauren, humusreichen Böden in sonniger Lage (mittlerer pH-Wert in 43 Proben 5,8, vgl. Dietl & Lauer 1973) so vor allem in Kalkschiefergebieten treten häufig *Festuca violacea* (oft dominant), *Helianthemum grandiflorum* und *Trifolium montanum* als Differentialarten in die Bestände ein. Diese Ausbildung vermittelt bereits zu den basiphilen Magerrasen und auch zu den Naturfettmatten, weil sich der Violette Schwingel nur auf basen- und nährstoffreichen Standorten stark ausbreiten kann.

9.4.3 NATURFETTMATTEN

Bereits Stebler & Schröter (1892) beschreiben eine "Mutternwiese", die auf "frischeren, tiefgründigeren Stellen der Heuberge", besonders im Gebiet der Mergelschiefer, "eine häufige Form alpiner Fettmatten" bildet. Solche ursprüngliche, vom natürlichen Nährstoffreichtum des Bodens zehrende halbschürige "Urfettmatten" sind relativ häufig in den Alpen zu finden. Ihr bevorzugter Standort sind geschützte natürliche Anreicherungs-standorte, also Mulden und Hangfüsse mit langer Schneebedeckung. Als weitere bestandesbildende Arten dieser L i g u - s t i c u m m u t e l l i n a - F e s t u c a v i o l a c e a - Wiese kommen *Phleum alpinum, Trifolium pratense, T. nivale, T. badium, Geranium silvaticum, Rumex arifolius, Plantago atrata, P. alpina* und *Leontodon hispidus* vor. Der pH-Wert der Böden lag in 8 Proben zwischen 5,2 bis 6,4; die natürlichen Vorräte an Phosphat und Kali können nach dem für gedüngte Wiesen geltenden Maßstab als mäßig bezeichnet werden (vgl. Dietl & Lauer 1973). Für eine Naturmatte ist das relativ hoch.

9.4.4 WILDHEUPLANGGEN

"Die unvermarchten, hochgelegenen schwer zugänglichen Halden, Grasbänder und Rasenflecken der Alpen, deren Nutzung oft frei und keine regelmäßige ist", werden als Wildheuplanggen bezeichnet (Stebler & Schröter 1892). In der Regel handelt es sich um Urrasenflächen, die aufgrund der extremen Hangneigung oder Höhenlage von Natur aus baumfrei sind. Auf hartem Kalkgestein in sonniger Lage dominiert S e s l e r i o - S e m p e r - v i r e t u m, auf frischen Mergelrendzinaböden C a r i c e - t u m f e r r u g i n e a e. Beherbergen diese Bestände grössere Anteile wertvoller Futterpflanzen, wie etwa *Ligusticum mutellina, Plantago atrata, Trifolium badium, Hedysarum obscurum* und *Astragalus*-Arten, so liefern sie ein gutes Heu; seggenreiche Halden geben hartes, wenig geschätztes Futter.

Solange eine große Zahl Hirten das Vieh auf den Alpen versorgte, wurden diese Rasenflächen, die häufig an schwer erreichbaren, gefährlichen, felsigen Orten lagen, fleißig gemäht. Heute bleiben sie weitgehend ungenutzt. Indessen hat sich gezeigt, daß die dauernde Brachlegung der früheren Wildheuwiesen nicht ohne Einfluß auf die Umwelt bleibt. *Carex ferruginea* und *Carex sempervirens* schaffen mit ihren langen, schmalen, einseitig überhängenden Blättern eine Gleitunterlage für den Schnee an Stellen, an denen früher die Lawinengefahr geringer war. Gefrieren aufrechtstehende Pflanzenteile und Schnee zusammen, so kann dieser beim Abrutschen auch ganze Rasenstücke in die Tiefe reißen. In diesen Wunden setzt die Bodenerosion ein. Mit dem Rückzug des Bauern von diesen abgelegenen Futterflächen haben die Lawinen- und Erosionsschäden in vielen Gebieten der Alpen zugenommen, wo nicht durch teure Lawinenverbauung die neuen Gefahren gebannt werden.

9.4.5   RIEDWIESEN

Im ganzen Alpengebiet, vor allem aber in der Flyschzone, bestehen Hangsümpfe und flache Torfmoore (*Caricetea nigrae*-Gesellschaften), auf denen im Spätsommer Streue gemäht wird. Da in reinen Graslandwirtschaften das Stroh fehlt, ist die Riedstreue an vielen Orten sehr geschätzt. Sie wird im Alpbetrieb wie auch im Heimgut verwendet. Die meisten Streuewiesen der höheren Berglagen sind von *Carex davalliana*-Gesellschaften besiedelt. Auch *Scirpus silvaticus*-Fluren, *Carex nigra*-Sümpfe und *Trichophorum caespitosum*-Moore sind in manchen Gebieten verbreitet (Koch 1928, Höhn 1936, Wagner 1965, Berset 1969, Dietl 1972, Dietl & Jäggli 1972). Durch die regelmäßige Mahd wird die sozio-ökologische Sukzession verhindert. So bleiben diese floristisch interessanten und ökologisch wertvollen und reichhaltigen Biotope erhalten. Nach Ausbleiben der landwirt-

436

schaftlichen Nutzung werden nasse, nährstoffarme Moorböden vom S p h a g n o - P i c e e t u m oder vom S p h a g n o - P i n e t u m m o n t a n a e besiedelt; auf feuchten Gleyböden stellt sich natürlicherweise das E q u i s e t o - A b i e t e - t u m ein (vgl. Ellenberg & Klötzli 1972).

9.5   **Nutzungsplanung des Futterbaues im Berggebiet (unter Mitarbeit von W. Krause)**

9.5.1   DIE BEDEUTUNG DER VEGETATIONSKARTE FÜR DIE STANDORTSBEURTEILUNG

Der Mangel an bäuerlichen Arbeitskräften und die allgemeine gesellschaftlich-wirtschaftliche Lage der Bergbevölkerung hat in den letzten Jahrzehnten Anlaß gegeben, die wenigen Äcker des Alpengebietes und viele Matten in Weiden überzuführen oder große Teile des Kulturlandes der Dauerbrache anheimfallen zu lassen (vgl. Surber, Amiet & Kobert 1973). Um den Gesamtertrag des verkleinerten Kulturlandes auf der alten Höhe halten zu können, bedarf es der verbesserten Nutzung, die ihrerseits auf sorgfältige Planung angewiesen ist. Letztere hat gleichermaßen den technischen wie den ökologischen Erfordernissen gerecht zu werden. Vor allem sollten die getroffenen Maßnahmen im Einklang mit der naturgegebenen Nutzungseignung stehen, so daß Natur und Technik einander nicht entgegenarbeiten, sondern beide dem Wirtschaftsziel zustreben.

Der Bearbeiter landwirtschaftlicher Nutzungsplanung hat nach Kriterien zu suchen, die eine schnelle und großflächige, aber auch zuverlässige Standortsbeurteilung ermöglichen. Als Methode hat sich die Vegetationskartierung bewährt (für Gebirgsgrünland vgl. Krause 1954, 1962, 1964; Wagner 1965; Dietl & Jäggli 1972). Ihre Anwendung ist wenig zeitaufwendig, wiewohl sie einen großräumigen Überblick vermittelt. Zugleich findet sie engen Kontakt zu anderen, teils messenden, teils experimentierenden Disziplinen. Damit vermag sie ihre eigenen Aussagen ebenso zu sichern, wie sie die Verallgemeinerung von Messungen und Experimenten anderer Arbeitsrichtungen zu klären hilft.

Weil sie dem Bearbeiter in allen Winkeln seines Gebietes Aufmerksamkeit abfordert, lenkt die Vegetationskartierung seinen Blick auf andere unmittelbar sichtbare und bedeutungsvolle Kriterien, zu denen vor allem die Geländemorphologie gehört. Deren herausragenden Einfluß auf Bodenentwicklung, Befeuchtungsgrad, Mikroklima sowie andere wuchsbeeinflussende Faktoren hat zuerst Pallmann (1943) systematisch untersucht und im Begriffspaar der Standortsanreicherung und -aushagerung zusammengefaßt. Für die Planungsarbeit ist vor allem wichtig, daß schon ein physiognomisch kaum wahrnehmbarer Wechsel zwischen konvexen und konkaven Oberflächenformen, der häufig tiefgehende Standortsgegensätze bewirkt, durch die Vegetation deutlich angezeigt wird.

Einer der wertvollsten Aufschlüsse, den die synoptische Betrachtung von Vegetation und Geländeformen im Gebirge zu bieten hat, liegt in der Klärung des Einflusses der Schneebedeckung auf die Produktivität des Graslandes (Krause 1962). Auch ein ökologisch bedeutungsvoller Wechsel des Gesteinsuntergrundes wird von der Vegetation angezeigt. Kalkführende und saure Gesteine unterscheiden sich durch das Vorkommen basiphiler bzw. azidophiler Pflanzen voneinander. Wasserstauende, zwischenhinein austrocknende Tone und Mergel sind durch Arten ausgezeichnet, die wechselfeuchte Standorte besiedeln. Die gesuchten Indikatorpflanzen treten aber nicht allein in der landwirtschaftlichen Nutzfläche auf. Die aufschlußreichsten unter ihnen gehören oft den angrenzenden Kontaktgesellschaften an, auch wenn diese sehr kleine Flächen bedecken. Hierher gehören flachgründige Plätze mit Felsbewohnern wie *Silene rupestris* im Grasland auf dysgeogenen Hartgesteinen, Naßstellen mit quellbegleitenden Arten wie *Juncus inflexus* auf wasserstauenden Schichten.

Aufschlußreich ist endlich der Grenzverlauf zwischen den Gesellschaften. Fällt ihr Vorkommen mit einer Wirtschaftsfigur, z.B. einer rechtwinkelig-geradlinig begrenzten Parzelle zusammen, deutet dies auf die Wirkung eines durchgreifenden Wirtschaftseinflusses. Sind die Grenzen ähnlich den Höhenlinien weich gerundet, weisen sie auf Natureinflüsse, häufig auf Abstufung der Bodenfeuchte hin. Einen wichtigen anthropogenen Faktor bildet die Verkehrserschließung. Hochwertige Bestände, die stets intensiv bewirtschaftet werden, liegen entweder in der Nähe der Wirtschaftsgebäude oder sie sind durch gut befahrbare Wege zugänglich gemacht. Für vernachlässigte Magerrasen gilt das umgekehrte Verhältnis. Der Aufwand an Zeit und Mühe, den die Zufahrt erfordert, entscheidet über den Wirtschaftsaufwand, der einer Fläche zuteil wird.

Insgesamt fördert eine Vegetationskarte die Synopsis und Synthese der Einzelfaktoren. Schon ihre Aufnahme im Gelände bringt Einsichten in den Zusammenhang zwischen Pflanzengesellschaften einerseits, natürlichen und anthropogenen Standortseigenschaften andererseits. Die Kenntnis der Landschaftsstruktur, die z.B. das geordnete Zusammentreffen bestimmter Gesellschaften zu Gesellschaftskomplexen klarstellt (vgl. nochmals Fig. 9), bietet anschließend neue Informationen und Fragestellungen. So macht die gegenseitige Zuordnung von Fettweiden und Magerrasen, wenn sie mit der Verkehrserschließung, dem Relief und der vom Relief bestimmten Bodenqualität in Beziehung gesetzt wird, auf die Möglichkeiten zur Weideverbesserung aufmerksam. Vor allem stellt sich die Frage nach dem Leistungspotential der Magerweiden. Es zeigt sich bald, daß diese fundamental wichtige Größe nicht unmittelbar aus den Pflanzenbeständen, wohl aber aus dem Bodenprofil und dem Wasserhaushalt zu klären ist, wobei jedoch den Pflanzenbeständen eine Indikatoreigenschaft für die eigentlich wirkenden Kräfte zukommt.

Diese zunächst durch Anschauung qualitativ erfaßten Zusammenhänge geben Anlaß zu gezielten und erfolgversprechenden Spezialuntersuchungen. So läßt sich die aufwendige Anlage tiefer Bodenprofilgruben zweckentsprechend und kostensparend steuern, wenn die Profile in die Verbreitungszentren wohldefinierter Gesellschaften, nicht in untypische Grenzbereiche gelegt werden oder wenn sie sich an den Intensitätsstufen der Bewirtschaftung orientieren. An Gesellschaften gekoppelte quantitative Untersuchungen vermögen zuletzt den Indikatorwert dieser Gesellschaften ebenfalls quantitativ zu fassen oder zu "eichen", so daß die Feststellung der Gesellschaft ausreicht, um voll zutreffende Aussagen über den Standort zu machen.

Der Gewinn, den die Vegetationskartierung vermittelt, weil sie den Bearbeiter an die funktionale Struktur der Landschaft heranführt, reicht oft weit über die Aufklärung des Mosaiks der Pflanzendecke hinaus. Wenn z.B. in alluvialen Ablagerungen basiphile und azidophile Gesellschaften nebeneinander gefunden werden, muß unterschiedliche Herkunft des Bodenmaterials angenommen werden. Wenn die Herkunft durch geologische Erkundung geklärt ist, können auf Grund der Gesteinsbeschaffenheit scharf lokalisierte Aussagen über Boden, Wasserhaushalt und Produktivität der Standorte abgeleitet werden. Diese Eigenschaften sind unmittelbar oft schwer zu erkennen, sie treten aber an der Indikatoreigenschaft der Pflanzengesellschaften indirekt auf das deutlichste zutage. Als Voraussetzung für zutreffende Aussagen müssen die lokalen Gültigkeitsgrenzen der benutzten Koinzidenzen geprüft werden, was innerhalb begrenzter Gebiete meist ohne erheblichen Aufwand möglich ist.

Eine wichtige Aufgabe der Standortsbeurteilung besteht darin, an der ökologisch interpretierten Vegetationskarte eine Nutzungskarte abzuleiten, die als Planungsgrundlage dienen kann. Ihre Erstellung setzt die klare Definition der Aufgabe voraus, die von der Vegetationskunde beantwortet werden soll. Im alpinen Bereich vorrangig ist die Frage, welche derzeit unbefriedigenden Bestände mit tragbarem Aufwand merklich verbessert werden können und welche anderen keinerlei Investitionen rentabel machen würden. Mit dieser Unterscheidung ist eine Grundlage für die unerläßliche Trennung von Wald und Weide geschaffen. Genaue Planung wird weiterhin nach Menge und Qualität der Erträge fragen, die durch Verbesserung eines Magerbestandes erzielt werden kann. Bei allem wird es nicht nur auf die in Zahlen ausgedrückte Größe der Flächen, sondern ebenso auf deren Umrißformen und Lage ankommen. Zipfel und Streifen oder abgelegene Teilstücke sind geringer zu bewerten als große geschlossene Komplexe.

Auch wird die Ertragssicherheit zu prüfen sein. Für Zeiten großer Sommertrockenheit, in denen die stark beanspruchten Hochleistungsflächen versagen können, sollten extensiv bewirtschaftete Reserveflächen zur Verfügung stehen, deren Wert erst in Notfällen sichtbar wird. Ebenso kann die Gleichmäßigkeit der Erträge durch zweckmäßige Berücksichtigung von Sonnen- und Schattenhängen gesichert werden.

Der Übergang von der Mäh- zur Weidenutzung, den der Mangel an Arbeitskräften erzwingt, erfordert die Feststellung der erzielbaren Weidedauer und der Trittfestigkeit von Feuchtstandorten. Auch bei Umstellung auf Weidebetrieb müssen ausreichende Flächen für Mähnutzung freigehalten werden, damit auf der Alp ein Heuvorrat angelegt werden kann, der den Tieren über Zeiten unvorhergesehenen Schneefalls hinweghilft.

Die Zahl der Fragen, die von der Planung gestellt werden, ist stets geringer als die Zahl der Pflanzengesellschaften des kartierten Gebietes. Infolgedessen weisen Eignungskarten weniger Signaturen auf als Vegetationskarten. Ein Verhältnis 50:5 ist bei großen Vorhaben normal. Doch würde der Versuch, von Anfang an nur die gefragten Einheiten, z.B. die verbesserungswürdigen Bestände aufzunehmen, leicht zu Fehlschlüssen führen, da die Vereinfachung erst dann verantwortet werden kann, wenn der Überblick über die vorhandene Vielfalt gewonnen ist.

Ein Beispiel aus dem Hochschwarzwald, das in die Höhenstufe des subalpinen N a r d e t u m hineinreicht, geben Fig. 10 und 11. Unter den Pflanzengesellschaften (Fig. 10) hat auf wenig steilen, meist ortsnahen Flächen das F e s t u c o - C y n o s u - r e t u m t r i f o l i e t o s u m große Ausdehnung gewonnen (Sign. 1). Es repräsentiert das lokale Optimum der Weideverbesserung. In geringerer Ausdehnung kommt die Gesellschaft ohne *Trifolium repens* vor (Sign. 2). Über die ganze Weide verteilt sind kleine *Nardus*-Bestände im F e s t u c o - C y n o s u r e t u m, während der Bergrücken am NNO-Rande der Gemarkung in ganzer Ausdehnung vom L e o n t o d o n h e l v e t i c u s - N a r - d e t u m beherrscht wird. Diese Gesellschaft tritt in zwei Ausbildungsformen auf, von denen die eine durch *Vaccinium myrtillus*, die andere durch *Calluna vulgaris* ausgezeichnet ist (Sign. 3 u. 4). Die erste besiedelt tiefgründig-frischen Boden in ± geschützter Lage, die zweite steinige, zur Austrocknung neigende exponierte Kuppen.

Verbreitet sind Quellfluren des C h a e r o p h y l l o - R a - n u n c u l e t u m a c o n i t i f o l i i, in denen *Juncus effusus* und *Cirsium palustre* zur Vorherrschaft gelangen, wenn sie beweidet werden. Zwischengeschaltet sind Siedlungen des N a r - d o - J u n c e t u m s q u a r r o s i auf wechselfeuchtem Boden und des F e s t u c o - C y n o s u r e t u m sowie des trockenen N a r d e t u m auf den lokalen Erhebungen der unebenen Quellfluren. Ein zweites enges Nebeneinander mehrerer Gesellschaften hat sich zwischen den Rieselgräben gebildet, die eine alte, im Schwarzwald verbreitete Form der Grünlandbewirtschaftung repräsentieren.

Die Planung fragt nach dem Leistungspotential der einzelnen Bestände, von dem der wirtschaftliche Erfolg einer Weideverbesserung abhängt. Die Auswertungskarte (Fig. 11) kennzeichnet zunächst die Sumpfwiesen und Quellfluren unter den zusammenfassenden Signaturen 3 u. 4 als nicht verbesserungswürdig. Ebenso wenig wird die steinige *Calluna-Nardus*-Fläche im NNO der Karte (Sign. 2) Verbesserungsaufwand lohnen. Als Weidereserve in futterarmen Jahren wird sie trotzdem nützlich sein. Überdies werden einige kleine, meist steinige *Nardus*-Bestände von der Verbesserung auszuschließen sein. Alles andere (Sign. 1) befindet sich entweder in gutem Zustand oder wird sich mit tragbarem Aufwand verbessern lassen. – Für die bewässerten Flächen stellt sich die Frage, ob die Ertragssteigerung, die das Rieselwasser zustandebringt (Krause 1959), nicht aufgezehrt wird durch den Flächenverlust, den die Gräben verursachen. Hier wäre zu prüfen, ob durch Einebnen der Gräben und Düngeranwendung ein besserer Wirtschaftserfolg zu erzielen ist.

Das zersplitterte Bild der aktuellen Vegetation in Fig. 10 läßt sich somit auf wenige, pflanzensoziologisch und ökologisch charakterisierte Einheiten zurückführen, die gleichzeitig wirtschaftlichen Inhalt haben.

Ausschlaggebend für den Grad der Vereinfachung ist die wirtschaftliche Fragestellung. Sie soll allerdings Konsequenzen, die von der Vegetation ausgehen, nicht in den Hintergrund drängen. Wenn z.B. generell nach verbesserungswürdigen Flächen gefragt wird, wenn diese aber in unterschiedlicher, wirtschaftlich belangreicher Ausprägung auftreten, müssen sie getrennt behan-

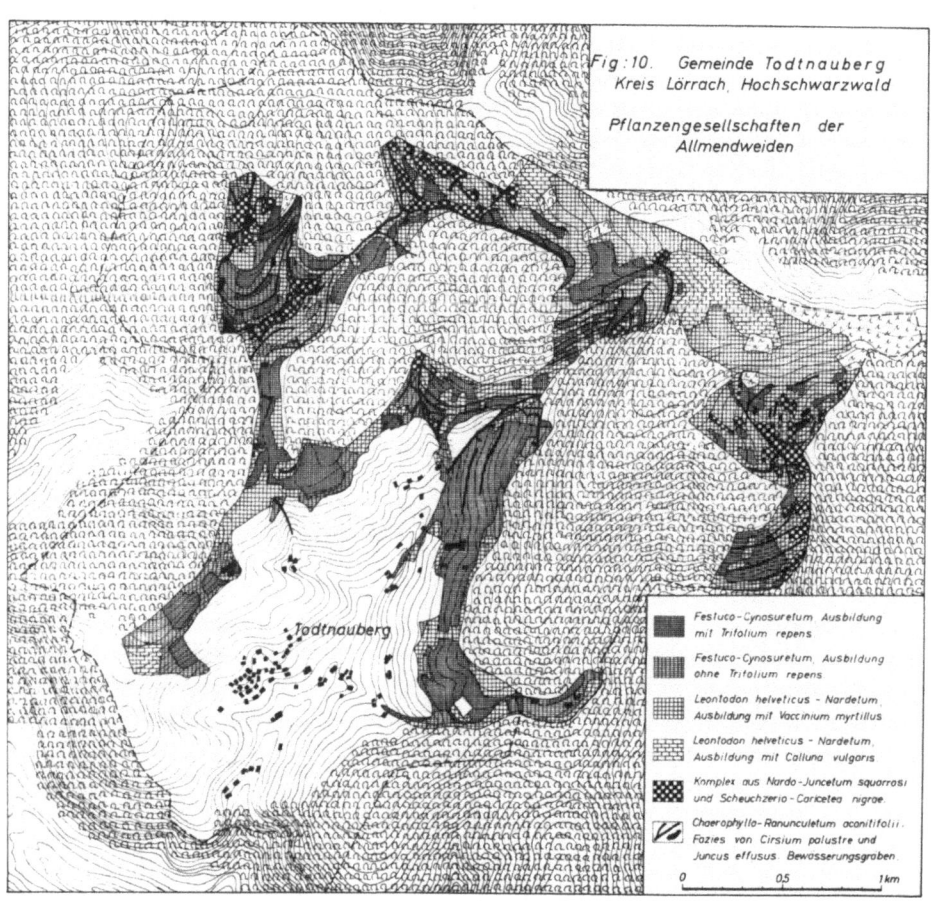

Fig. 10. Gemeinde Todtnauberg, Kreis Lörrach, Hochschwarzwald Pflanzengesellschaften der Allmendweiden. Im Text werden die Signaturen in ihrer Reihenfolge von oben nach unten numeriert genannt.

delt werden. Dies trifft zu, wenn eine große Weide Süd- und Nordhänge oder beträchtliche Höhenunterschiede umfaßt, so daß manche Teilflächen früh, andere spät weidereif werden. Doch empfiehlt sich, die Zahl der Einheiten so klein zu halten, wie eben noch vertretbar ist. Die Planung wird regelmäßig durch juristische, technische und finanzielle Probleme so belastet, daß jede Erleichterung der Prozedur dienlich wird.

Schließlich sollte die Eignungskarte auch Angaben machen, die aus der Vegetation allein nicht ableitbar sind. Dazu gehört die Hangneigung, die im Zeitalter zunehmenden Maschineneinsatzes oft darüber entscheidet, ob eine Fläche auch als Mähweide genutzt werden kann (vgl. Dietl, Guyer & Stadler 1974).

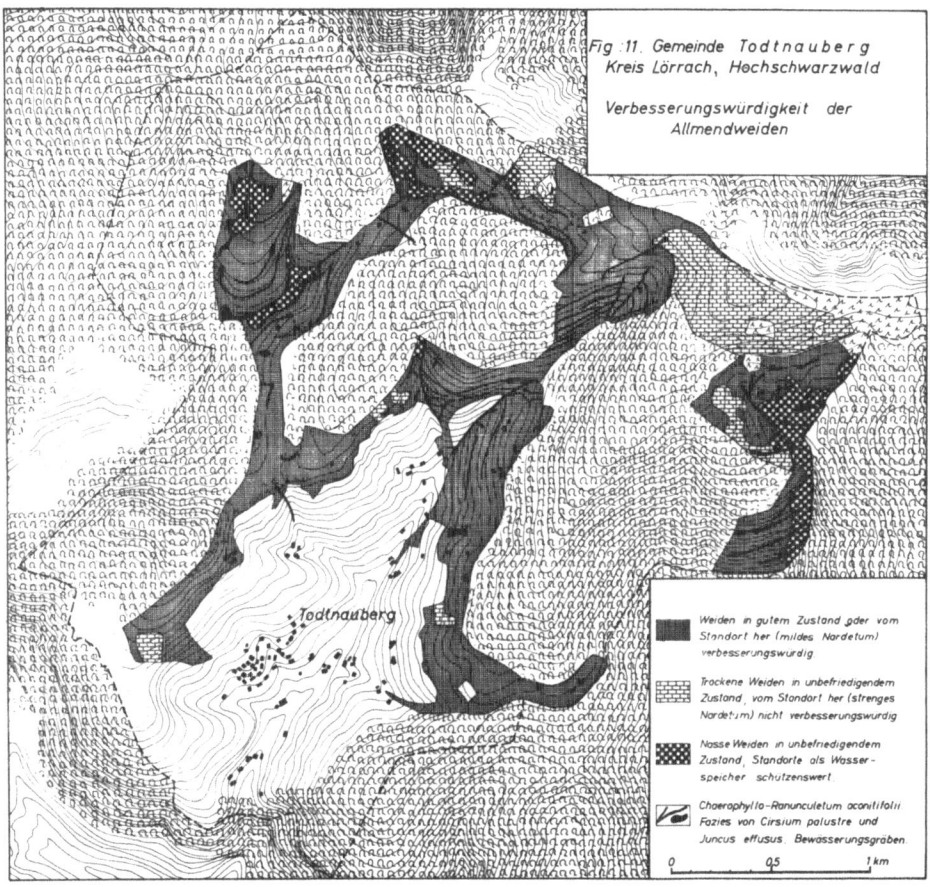

Fig. 11. Gemeinde Todtnauberg, Kreis Lörrach, Hochschwarzwald Verbesserungswürdigkeit der Allmendweiden. Im Text werden die Signaturen in ihrer Reihenfolge von oben nach unten numeriert genannt.

443

Fig. 12. Übersichtsbild der Alp Glaubenbüelen. In Bildmitte der Rotspitz (1787 m).

9.5.3 NUTZUNGSPLANUNG AUF GRUND DER VEGETATIONSKARTE (ALP GLAUBENBÜELEN, OBWALDEN, SCHWEIZ)

Die Alp Glaubenbüelen (Fig. 12, 13) bietet ein Beispiel der Hochweiden in den Schweizer Nordalpen. Sie liegt in der Kontaktzone zwischen Kalk- und Flyschgebiet ungefähr 30 km südwestlich Luzern in einer Höhe zwischen 1490 m und 1787 m ü.M. Abgesehen von ungleichmäßig verteilten Fichtengruppen ist sie unbewaldet. Außen grenzen nach mehreren Seiten Weidewälder an. Die Bewirtschaftung erfolgt aus zwei festen Gebäuden "Lochhütte" und "Obere Hütte" (Fig. 16), die durch Fahrwege an das Straßennetz angeschlossen sind.

Zu Beginn der Planungsarbeit herrschten große Unterschiede der Pflanzenbestände und des Nutzwertes (Fig. 14). Die Legende zu dieser Karte umfaßt 15 Signaturen für Grünland, von denen manche nur kleine bis kleinste Flächen umfassen So starke Unterteilung, die bei der Planung nicht voll berücksichtigt werden kann, verlangt nach Zusammenfassung unter wirtschaftlichen Gesichtspunkten. Fig. 15, die diese Zusammenfassung bringt, enthält nur noch 8 Signaturen für Grünland. Unverändert übernommen wurden die großflächig auftretenden und wirtschaftlich bedeutungsvollen Gesellschaften, zusammengefaßt oder in die angrenzende

444

Fig. 13. Alp Glaubenbüelen, Umgrenzung und Relief.

445

Signatur einbezogen die für attraktive Verbesserung nicht geeigneten Flächen.

Den höchsten Weideertrag bietet die typische, frischfette Subassoziation der Kammgrasweide, C r e p i d o a u r e a e - C y n o s u r e t u m t y p i c u m (Fig. 14 u. 15, Sign. 1, Tab. 3, Spalte 2) mit einer Leistung von 40-60 q TS/ha (vgl. auch S. 428 ff.). Den Figuren 13, 14 und 15 ist zu entnehmen, daß diese Ausbildungsform vorwiegend auf den hochgelegenen Teilen der Alp vorkommt, die ausgedehnte, gleichmäßig geneigte Abhänge ohne ausgeprägte Kleinmodellierung bilden. Diese, der Bewirtschaftung günstige Topographie mag seit altersher Anlaß für relativ intensive Weidepflege gegeben haben, die ihrerseits das

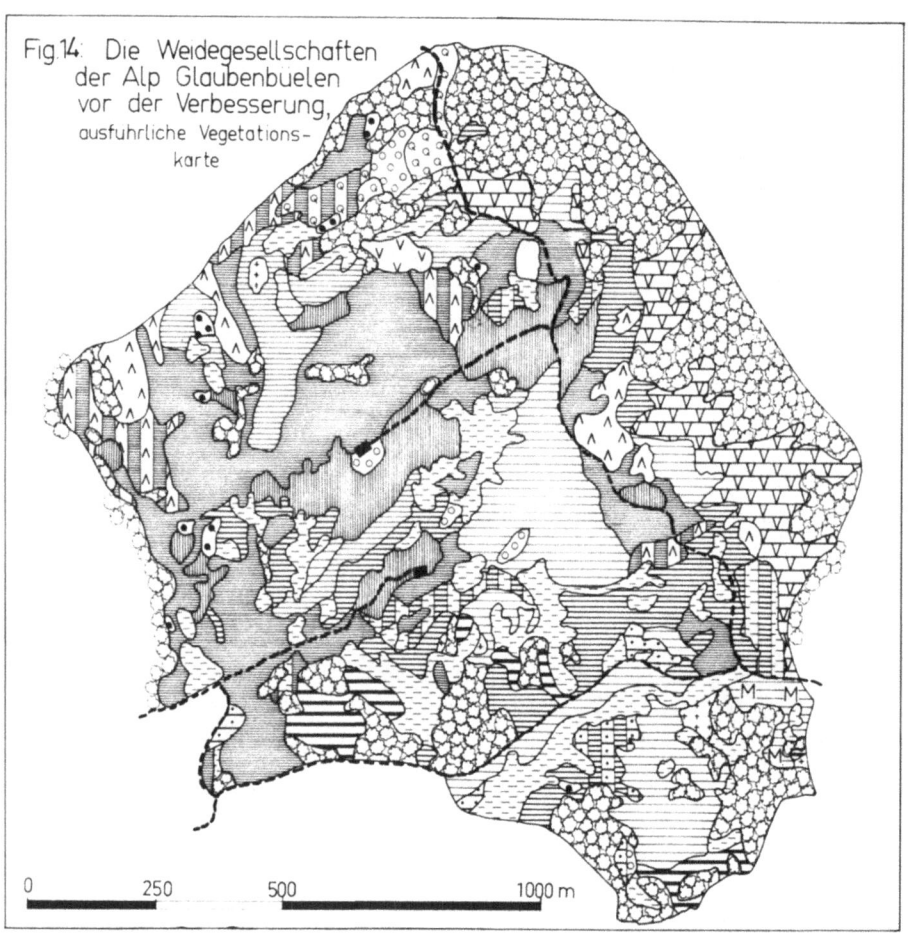

Fig. 14. Alp Glaubenbüelen, ausführliche Vegetationskarte.

446

Fig. 14   Die Weidegesellschaften der Alp Glaubenbüelen vor der Verbesserung

ausführliche Vegetationskarte

Fettweiden
Kammgrasweiden

C r e p i d o   a u r e a e - C y n o s u r e t u m

1. Crepido aureae-Cynosuretum typicum, subalpine Fettweide

2. Crepido aureae-Cynosuretum typicum, Var. v. *Rumex obtusifolius.* Überdüngte Trittvariante von 1.

3. Crepido aureae-Cynosuretum, Subass. v. *Succisa pratensis.* Wechselfeuchte Ausbildung von 1.

4. Crepido aureae-Cynosuretum, Subass. v. *Nardus stricta.* Magere Ausbildung von 1.

5. Crepido aureae-Cynosuretum, Subass. v. *Nardus stricta,* Var. v. *Succisa pratensis.* Mager-wechselfeuchte Ausbildung von 1.

Milchkrautweiden

P o o - P r u n e l l e t u m

6. Ausbildung mit *Gentiana purpurea* (mager)

Magerweiden

7. S e s l e r i o - S e m p e r v i r e t u m (basiphil, trocken). Ausbildung mit *Trifolium pratense* (leicht eutrophiert)

8. C a r i c e t u m   f e r r u g i n e a e (basiphil, frisch)

9. N a r d e t u m (Azidophil, frisch)

Heumatten, Läger- und Hochstaudenfluren

10. T r i s e t e t u m   f l a v e s c e n t i s

11. S e n e c i e t u m   a l p i n i

12. A l n e t u m   v i r i d i s (inklusive gemähte Hochstaudenfluren)

447

Feuchtwiesen, Sümpfe und Moore

C a r i c e t u m   d a v a l l i a n a e

13. Reine Ausbildung (basenreich, oligotroph, quellnass), − nicht verbesserbar −
Ausbildung mit *Trichophorum caespitosum* (basenarm, oligotroph, staunass), − nicht verbesserbar −
Ausbildung mit *Cirsium salisburgense* und *Caltha* (basenarm, eutroph, staunass). − nicht verbesserbar −

14. Ausbildung mit *Cirsium salisburgense* und *Carex pulicaris* (wechselnass) − verbesserbar −

15. Vegetationskomplex von *Cirsium salisburgense*
Ausbildung mit C a r i c e t u m   f e r r u g i n e a e − nicht verbesserbar −

16. Hochmoorkomplex (S p h a g n o   c o m p a c t i - T r i c h o p h o r e - t u m)

17. Wälder wurden nicht kartiert. Nach Ellenberg & Klötzli (1972) stocken im Gebiet hauptsächlich subalpine Fichtenwälder (S p h a g n o - P i - c e e t u m   t y p i c u m   und   c a l a m a g r o s t i e t o s u m   v i l - l o s a e).

Entstehen der wirtschaftlich optimalen Subassoziation der Kammgrasweide förderte. Eingeschlossen von den Beständen der typischen Weide sind zwei ansehnliche Flächen der mageren Subassoziation mit *Nardus stricta* (Fig. 14 u. 15, Sign. 4), deren größte einen Steilhang einnimmt, wie aus Fig. 13 zu ersehen ist. Hier darf naturgegebene Beeinträchtigung der Produktivität durch verstärkte Bodenabtragung angenommen werden.

Die wesentlich kleineren Flächen, die das C r e p i d o - C y n o s u r e t u m   t y p i c u m an der Ostgrenze der Alp einnimmt, liegen nach der topographischen Karte auf lokalen Verebnungen in der Nähe der Zufahrtsstraße, d. h. in einer dem Entstehen der Gesellschaft günstigen Situation. Halbkreisförmig umschlossen vom typischen C r e p i d o - C y n o s u r e - t u m herrschen im Zentrum der Alp andere Gesellschaften (Sign. 3 u. 4), die durch *Succisa pratensis* als wechselfeucht, durch *Nardus stricta* als mager, durch *Rumex obtusifolius* (Sign. 2) als überweidet und überdüngt charakterisiert werden. Ausgehend von den beiden Hütten erstreckt sich peripher zum Zentrum ein breiter Streifen der Variante von *Rumex obtusifolius* in einer Lage, die auf übermäßige Verweildauer des Viehes deutet, womit ein Nutzungsfehler angezeigt wird. Das Relief unterscheidet sich nicht prinzipiell von den darüberliegenden Flächen der typischen Weide, weshalb eine durchgehend einheitliche ökologische Grundstruktur angenommen werden darf.

Am Ostrand der zentralen Fläche und in größerer Entfernung von den Hütten, wo das Vieh seltener hingelangt, nehmen Ausbildungen des C r e p i d o - C y n o s u r e t u m mit *Succisa pratensis* überhand, die gleichermaßen auf Nährstoffmangel und auf leicht erhöhte Befeuchtung durch vorübergehenden Stau von Niederschlagswasser über einer undurchlässigen Bodenschicht schließen lassen. Die Hangneigung ist gering, doch verursachen flache Einkerbungen eine unruhige Oberfläche.

Unterhalb der Lochhütte besteht ein Mosaik aller beschriebenen Gesellschaften mit einer Vielzahl kleiner bis kleinster Parzellen, die der Heuwerbung oder der Streuegewinnung dienen. Die Ursachen der Aufsplitterung liegen in der unterschiedlichen Verwitterung des Muttergesteins, das aus Gips und dolomitischem Mergel besteht. Es entstanden Dolinen und flache Mulden. Immerhin zeigt Fig. 14, daß keine anderen Gesellschaften auftreten als in den übrigen Teilen der Alp. Demnach ist, abgesehen von der störenden Kleinparzellierung nicht mit zusätzlicher Beeinträchtigung der Nutzungsmöglichkeit zu rechnen.

Koppeleinteilung bestand nicht. Das Vieh bewegte sich frei.

Kühe und Jungvieh weideten gemeinsam auf guten Flächen sowie auf Riedwiesen und im Wald. Während der Alpzeit ging die Milchleistung stark zurück, weil der Futterwuchs im Spätsommer abnahm und die täglich zurückzulegenden Wege für die Tiere immer länger wurden.

In diesem Gelände wurde im Jahre 1969 der Nutzungsplan für eine umfassende Weideverbesserung mit Koppeleinteilung, verstärkter Düngung, Trennung von Wald und Weide, Tränkestellenbau und Einrichtung von Unterständen aufgestellt und verwirklicht. Die Koppeleinteilung, die das Fundament der Verbesserung bildet, stützt sich auf die Wirtschaftsgebäude der Oberen Hütte und der Lochhütte. Abgegrenzt wurden jeweils fünf Kop-

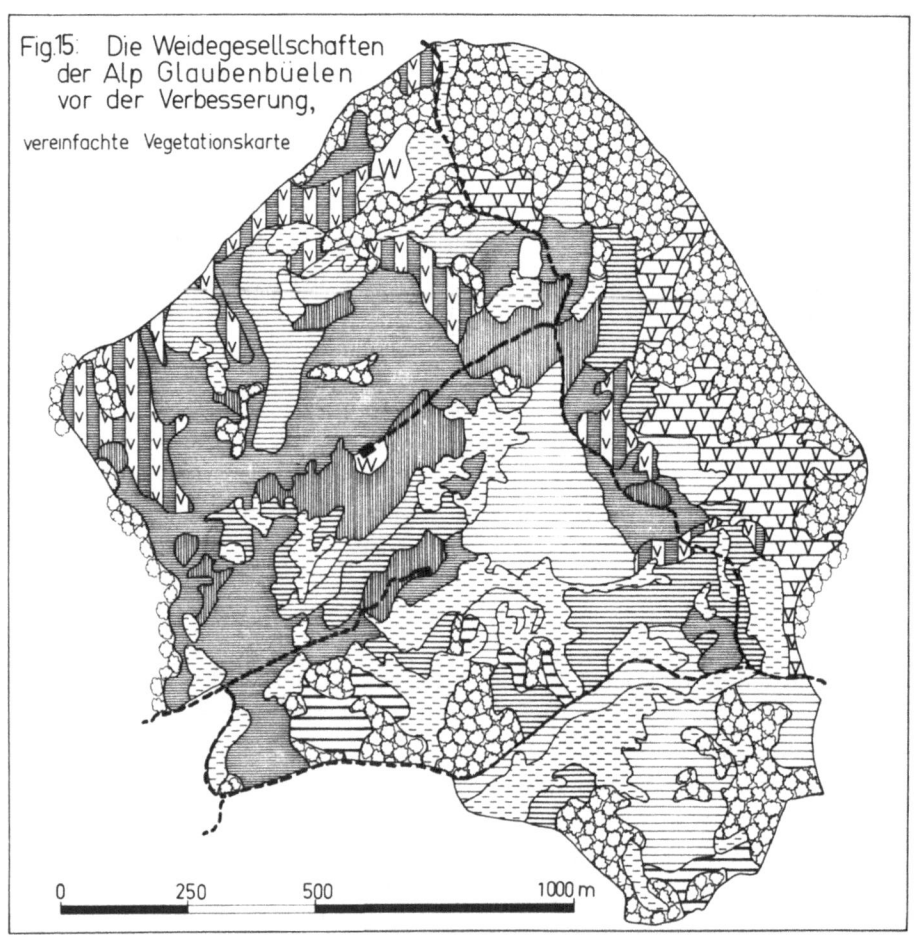

Fig. 15. Alp Glaubenbüelen, vereinfachte Vegetationskarte.

Fig. 15   Die Weidegesellschaften der Alp Glaubenbüelen vor der Verbesserung

            vereinfachte Vegetationskarte

            Fettweiden

1.   Crepido aureae-Cynosuretum typicum, subal-
            pine Fettweide

2.   Crepido aureae-Cynosuretum typicum, Var. v.
            Rumex obtusifolius. Überdüngte Trittvariante von 1.

3.   Crepido aureae-Cynosuretum, Subass. v. *Succisa pra-*
            tensis Wechselfeuchte Ausbildung von 1.

4.   Crepido aureae-Cynosuretum, Subass. v. *Nardus stric-*
            ta. Magere Ausbildung von 1.

5.   Crepido aureae-Cynosuretum, Subass. v. *Nardus stric-*
            ta, Var. v. *Succisa pratensis.* Mager-wechselfeuchte Ausbildung von 1.

            Magerweiden

6.   Crepido Cynosuretum im Komplex mit nicht verbesse-
            rungsfähigen Magerweiden (Seslerio-Semperviretum,
            Caricetum ferrugineae, Alnetum viridis).

7.   Flachmoor, vgl. Fig. 14, Sign. 13

8.   Hochmoorkomplex (Sphagno compacti-Trichophore-
            tum)

9.   Vegetationskomplex von *Cirsium salisburgense*
            Ausbildung mit Caricetum ferrugineae — nicht verbes-
            serbar —

10.   Wald

11.   Wiese (*Trisetetum flavescentis*)

451

peln für Milchkühe in der Nähe der Wirtschaftsgebäude und 1—2 Koppeln für Jungvieh (Rinder) in den höchsten und entferntesten Teilen der Gesamtfläche (Fig. 16).

Wie der Vergleich zwischen den Karten ergibt, überwiegen auf den Koppeln der Oberen Hütte Weiderasen in gutem

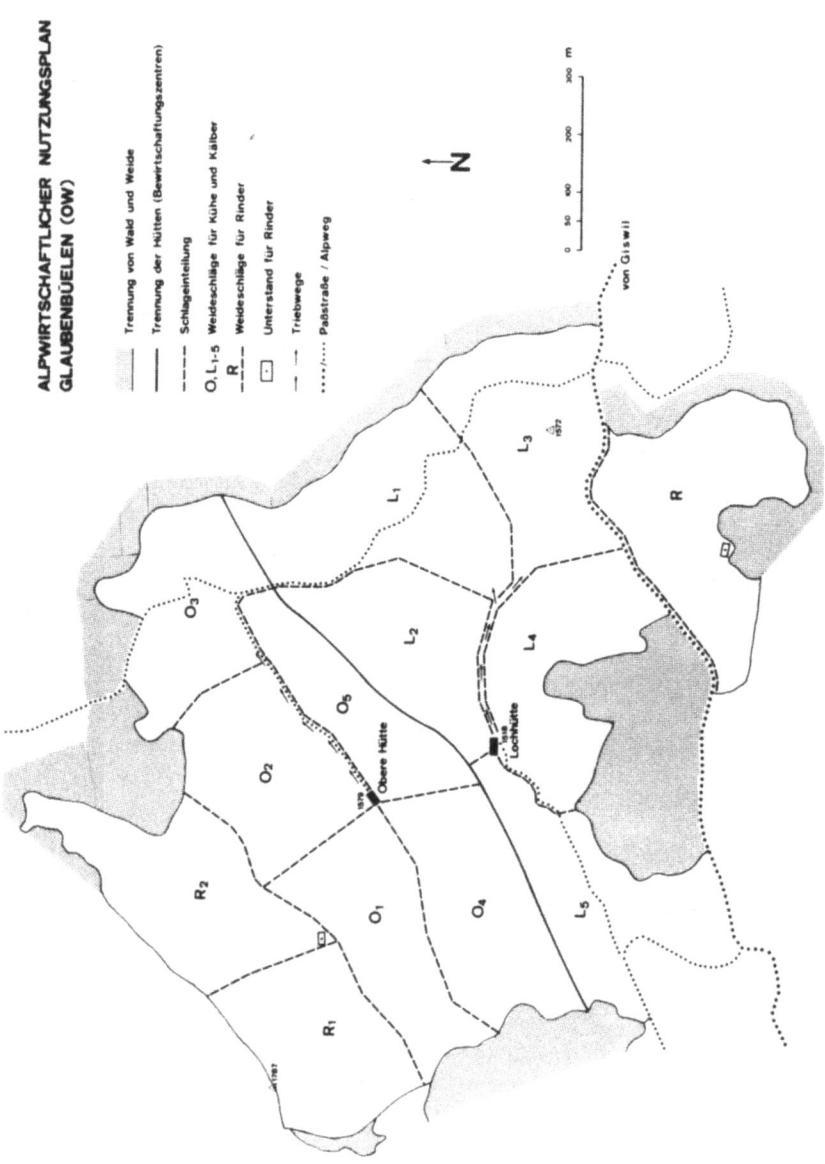

Fig. 16. Alp Glaubenbüelen, Nutzungsplan.

Wirtschafts- oder Ernährungszustand. Das typische C r e p i -
d o - C y n o s u r e t u m repräsentiert den optimalen Ernäh-
rungs- und Pflegezustand, die *Rumex obtusifolius*-Variante ist
ertragreich, aber verunkrautet. Aufgabe der Bewirtschaftung wird
es sein, die wertvollen Bestände zu erhalten, wenn möglich noch
zu verbessern, die verunkrauteten durch Herbizide und zweckmäßige Bewirtschaftung von *Rumex obtusifolius* zu säubern.

Die Lochhütte ist mit einer weniger günstigen Ausstattung
in die Verbesserung gegangen. Den Hauptteil ihrer Flächen nehmen die mageren Subassoziationen von *Nardus stricta* und *Succisa pratensis* ein. Hinzu kommt die Aufsplitterung der Parzelle L 4 und L 5 in unregelmäßig begrenzte Kleinstandorte ungleicher Beschaffenheit. Demnach muß hier die Verbesserung des
Rasens durch Düngung und Weidetechnik im Vordergrund
stehen. Die Erfahrungen über das Leistungspotential unbefriedi-
gender Pflanzenbestände machen dieses Vorhaben aussichtsreich,
denn die Subassoziationen mit *Nardus* und *Succisa* des C r e -
p i d o - C y n o s u r e t u m ähneln ökologisch einem milden
Nardetum, über dessen Verbesserungsfähigkeit auf S. 419 berichtet wurde.

Die Vegetationskarte ermutigt demnach zur Durchführung
der Weideverbesserung und läßt im Bereich beider Hütten ein
gutes Ergebnis erwarten. Doch macht sie auf einen Unterschied in
der Ausgangsposition aufmerksam, indem sie die Pflanzengesellschaften der Oberen Hütte von Anfang an als wirtschaftlich wert-
voll, die der unteren als wenig befriedigend, aber verbesserungsfähig ausweist. Im Bereich der Lochhütte werden also die größe-
ren Anstrengungen nötig sein, um das Wirtschaftsziel zu erreichen.

Im übrigen vermitteln Fig. 16 eine Vorstellung des Unter-
nehmens "Trennung von Wald und Weide". Dunkel schraffiert
sind die Flächen außerhalb der Alp, die in Zukunft als Wald
bewirtschaftet werden sollen. Eingeschlossen sind größere Bestände des nicht verbesserungsfähigen C a r i c e t u m f e r -
r u g i n e a e - Komplexes (Sign. 15, Fig. 14). Der überalterte
Weidewald wird durch Pflegemaßnahmen in einen naturnahen
Bergwald (Subalpiner Fichtenwald) übergeführt. Entfernen mor-
scher Stämme bringt Licht in den Bestand, so daß lokale Ver-
jüngungszentren entstehen. Innerhalb der Weide ist keine bewal-
dete Fläche ausgewiesen. Baumgruppen bleiben bestehen.

Klima und Topographie erschweren die Bewirtschaftung des Graslandes in den Alpen. Große Flächen bringen geringe Erträge. Da wirtschaftliche Erfordernisse neuerdings nach Produktionssteigerung verlangen, muß die überkommene Nutzungsweise durch verbesserte Methoden, insbesondere Düngeranwendung und Einrichtung von Koppelweiden ersetzt werden.

Seit langem gilt als Erfahrungstatsache, daß manche bisher unproduktive Pflanzenbestände auf Düngung mit hoher Ertragssteigerung reagieren, während andere ihr niederes Niveau beibehalten. Das Leistungspotential eines ertragsschwachen Weiderasens bereits am unverbesserten Bestand zu erkennen, gehört zu den wesentlichen Aufgaben der landwirtschaftlichen Planung in den Alpen. Ein sicheres und leicht festzustellendes Merkmal der Verbesserungsfähigkeit bietet die Gesellschaftszugehörigkeit, die dank umfangreicher floristisch-soziologischer Vorarbeit bereits routinemässig gefunden werden kann.

Die verbreitetste Weidegesellschaft mit derzeit niederem Ertrag, aber hohem Leistungspotential ist das P o l y g o n o b i s t o r t a e - N a r d e t u m. Es besiedelt günstig strukturierte Braunerde oder schwach podsolierte, mäßig saure Braunerde von hoher Produktionskraft. Die Standorte sind wenig steil oder liegen in Hangmulden, so daß sich Feinerde ansammeln konnte. Die geringen Erträge des P o l y g o n o - N a r d e t u m sind weit mehr auf unzweckmäßige Bewirtschaftung als auf Ungunst der Standorte zurückzuführen. Die Feststellung großflächig entwickelter Bestände dieser Gesellschaft bietet ein sicheres Anzeichen für die Möglichkeit beträchtlicher Leistungssteigerung bei geringem Wirtschaftsaufwand. Der *Nardus*-Rasen ändert nach Düngung und veränderter Nutzung seine floristische Zusammensetzung und entwickelt sich zum C r e p i d o - C y n o s u r e t u m, der ertragreichsten Weidegesellschaft der subalpinen Stufe.

Andere Rasen, z.B. das C a r i c e t o p i l u l i f e r a e - N a r d e t u m, das auf stark saurem Eisenpodsol siedelt, sind nicht oder nicht leicht verbesserbar. Hierher gehört noch das C a r i c e t u m f i r m a e und mehrere Zwergstrauchgesellschaften, z.B. das R h o d o r e t o - V a c c i n i e t u m. Durch Vegetationskartierung läßt sich der Flächenanteil verbesserungsfähiger Magerweiden in einem Meliorationsgebiet feststellen und als Planungsgrundlage nutzen. Die Karte zeigt zugleich die räumliche Verteilung der Gesellschaften und die Größe der Einzelbestände, die ebenfalls für das Gelingen der Verbesserung wichtig sind. An Beispielen wird die Anwendung der Vegetationskartierung für die Wirtschaftsplanung gezeigt.

Auch zu den alpinen Mähwiesen erfolgen Angaben über die Bestandszusammensetzung sowie über den Einfluß der Siedlungsnähe und der Zugänglichkeit auf die Erträge.

Literatur

Aichinger, E. — 1933 — Vegetationskunde der Karawanken. Pflanzensoziologie 2. Jena, 329 S.

Aichinger, E. — 1951 — Vegetationskundliche Vorarbeiten zur Ordnung von Wald und Weide. *Angew. Pflanzensoziologie* 2: 53-127, Klagenfurt.

Albrecht, J. — 1969 — Soziologische und ökologische Untersuchungen alpiner Rasengesellschaften insbesondere an Standorten auf Kalk-Silikat-Gesteinen. Diss. Botanicae 5, Lehre, 91 S.

Bäbler, R. & E. Strebel — 1968 — Alp- und Weidewirtschaft. Huber, Frauenfeld, 138 S.

Berset, J. — 1969 — Pâturages, prairies et marais montagnards et subalpins des Préalpes fribourgeoises. (Dt. Zus.) *Bull Soc. Frib. Nat.* 58: 1-55.

Bourqui, P. & R. Schmid — 1968/1969 — Die neuzeitliche Düngung der Alpen. *Alpwirtsch. Monatsbl* 102: 490-502, 103: 8-22.

Braun-Blanquet, J. — 1913 — Die Vegetationsverhältnisse der Schneestufe in den Rätisch-Lepontischen Alpen. *N. Denkschr. d. Schweiz. Nat. Ges.* 48.

Braun-Blanquet, J. — 1948-1950 — Uebersicht der Pflanzengesellschaften Rätiens. *Vegetatio* 1: 29-41, 129-146, 285-316; 2: 20-37, 214-237, 341-360.

Braun-Blanquet, J. — 1969 — Die Pflanzengesellschaften der rätischen Alpen im Rahmen ihrer Gesamtverbreitung. 1. Teil, Bischofberger, Chur 100 S.

Braun-Blanquet, J. & H. Jenny — 1926 — Vegetationsentwicklung und Bodenbildung in der alpinen Stufe der Zentralalpen. *Denkschr. schweiz. naturf. Ges.* 63: 183-349.

Braun-Blanquet, J., H. Pallmann & R. Bach — 1954 — Pflanzensoziologische und bodenkundliche Untersuchungen im schweizerischen Nationalpark und seinen Nachbargebieten. Vegetation und Böden der Wald und Zwergstrauchgesellschaften (Vaccinio-Piceetalia). Erg. wiss. Unters. schweiz. Nationalpark 4, NF 28, 200 S.

Brockmann-Jerosch, H. — 1907 — Die Flora des Puschlav und ihre Pflanzengesellschaften. Leipzig.

Brünner, F. & W. Krause — 1968 — Über den Einfluss der natürlichen Nährstoffvorräte und der Befeuchtung auf den Ertrag von Dauerwiesen, dargestellt an Wechseldüngungsversuchen in Süddeutschland. (Engl. summ.). *Z. Acker- und Pflanzenbau* 128: 151-173.

Caputa, J. — 1966 — Contribution à l'étude de la croissance du gazon des pâturages à différentes altitudes. (Dt. Zus., Engl. summ.). *Rech. agron. Suisse* 5: 393-426.

Caputa, J. — 1973 a — Influence de nombre des parcs sur la productivité d'un pâturage d'altitude; résultats d'essais 1968 à 1971. (Dt. Zus., Ital. rias.). *Arb. Gebiete Futterb.* 16: 16-39.

Caputa, J. — 1973 b — Potentiel de production du sol à différentes altitudes. (Dt. Zus., Ital. rias.). *Arb. Gebiete Futterb.* 17: 11-18.

Caputa, J. & G. Schechtner — 1970 — Wachstumsrhythmus und Stickstoffwirkung auf natürlichen Beständen der Bergweiden. (Engl. summ., Franz. rés.). *Das wirtschaftseigene Futter* 16/3: 165-182.

Dietl, W. – 1972 – Die Vegetationskartierung als Grundlage für die Planung einer umfassenden Alpverbesserung im Raume von Glaubenbüelen (Obwalden). (Franz. rés., Engl. abstr.). Diss ETH Zürich, Sarnen 152 S.

Dietl W. & F. Jäggli – 1972 – Die Kartierung von Vegetation und Boden als Planungsgrundlage für eine umfassende Alpverbesserung. (Franz. rés., Engl. summ.). *Schweiz, landw. Forsch.* 11/4: 475-520.

Dietl, W. & C.F. Lauer – 1973 – Natürliche Grundlagen der futterbaulichen Nutzungseignung im Berggebiet. *Schweiz. landw. Zeitschr.* 101/21: 747-759.

Dietl, W. & H. Guyer – 1974 – Pflanzenbestände, Bewirtschaftung und Produktivität von einigen Standorten im Flyschgebiet der Schweiz. (Franz. rés., Engl. summ.). *Schweiz. landw. Forsch.* (Festschrift Koblet) 13: 101-113.

Dietl, W., Guyer, H. & F. Stadler – 1974 – Pflanzenstandort- und Eignungskarten für futterbauliche Nutzungseignung. *Mitt. Schweiz. Landw.* 22: 109-132.

Ellenberg, H. – 1952 – Wiesen und Weiden und ihre standortliche Bewertung. Landw. Pflanzensoz. II, Ulmer, Stuttgart 143 S.

Ellenberg, H. – 1958 – Ueber die Beziehung zwischen Pflanzengesellschaft, Standort, Bodenprofil und Bodentyp. *Angew. Pflanzensoz.* 15: 14-18, Stolzenau.

Ellenberg, H. – 1963 – Vegetation Mitteleuropas mit den Alpen. In H. Walter, Einführung in die Phytologie IV/2, Ulmer Stuttgart 943 S.

Ellenberg, H. & F. Klötzli – 1972 – Waldgesellschaften und Waldstandorte der Schweiz. (Franz. Explications préliminaires). *Mitt. Schweiz. Anst. forstl. Versuchsw.* 4814: 587-930.

Frei, E. – 1975 – Die Horizontbezeichnung am Bodenprofil. *Mitt. Eidg. Anstalt forstl. Versuchswesen* 51, H 1: 215-224.

Frei, E. & H. Guyer – 1968 – Die landbauliche Beurteilung der Sauren Braunerde im Voralpengebiet unter besonderer Berücksichtigung der Nutzung als Intensivweide. (Franz. rés., Engl. summ.). *Schweiz. landw. Forsch.* 7, H 3/4: 352-370.

Gayl, A. – 1951 – Ordnung von Wald und Weide im Bereich der Almen. *Angew. Pflanzensoziologie* 2: 5-40, Klagenfurt.

Geering, J. – 1968 – Ueber die Ausnützung und die Wirtschaftlichkeit von Handelsdüngern im Naturfutterbau (Franz. rés., Engl. summ.). *Schweiz. landw. Forsch.* 7, H 314: 266-293.

Gigon, A. – 1971 – Vergleich alpiner Rasen auf Silikat und auf Karbonatboden. (Franz. rés., Engl. summ.). *Veröff. Geobot. Inst. ETH Zürich, Stiftung Rübel* 48: 1-163.

Guyer, H. – 1967 – Ueber die Futterproduktion auf Sömmerungsweiden. *Mitt. Arbeitsgem. Förd. Futterb.* 70: 1-6.

Guyer, H. – 1970 – Die Düngung von Alpen mit dem Flugzeug. (Franz. rés., Ital. rias.). *Arb. Gebiete Futterb.* 13: 30-38.

Hess, E. & E. Landolt – 1967 – Flora der Schweiz. Bd 1: Pteridophyta bis Caryophyllaceae. Birkhäuser, Basel und Stuttgart 858 S.

Höhn, W. – 1936 – Vegetationsstudien in Ober-Iberg (Schwyz). Die hydrophilen Pflanzengesellschaften. *Ber. Schweiz. Bot. Ges. .Festbd. Rübel* 46: 365-411.

Klapp, E. – 1951 – Borstgrasheiden der Mittelgebirge. (Engl. summ.). *Z. Acker- und Pflanzenbau* 93: 400-444.

Knapp, G. & R. – 1953 – Ueber Pflanzengesellschaften und Almwirtschaft im Ober-Allgäu und angrenzenden Vorarlberg. Landw. Jahrb. Bayern 30 H. 9/10: 548-588.

Knapp, R. – 1962 – Die Vegetation des Kleinen Walsertales, Vorarlberg. *Geobot. Mitt.* 12: 1-53.

Koblet, R. – 1954 – Ueber die Probleme der Waldweide. *Schweiz Z. Forstw.* 105, H. 8; 468-472.

Koblet, R. – 1965 – Der landwirtschaftliche Pflanzenbau. Birkhäuser, Basel, 829 S.

Koblet, R., Frei, E. & F. Marschall – 1953 – Untersuchungen über die Wirkung der Düngung auf Boden und Pflanzenbestand von Alpweiden. (Franz. rés.). *Landw. Jahrb. Schweiz* 67: 597-658.

Koch, W. – 1928 – Die höhere Vegetation der subalpinen Seen und Moorgebiete des Val Piora (St. Gotthard-Massiv). *Zeitschr. Hydrobiologie* 4: 131-175.

Kral, F. – 1971 – Zur Vegetationsgeschichte der Höhenstufen im Dachsteingebiet. (Engl. summ.). *Ber. Deutsch. Bot. Ges.* 85, H. 1-4: 137-151.

Krause, W. – 1954 – Zur ökologischen und landwirtschaftlichen Auswertung von Vegetationskarten der Allmendweiden im Hoch-Schwarzwald. *Angew. Pflanzensoz.* (Festschrift Aichinger) 2: 1076-1100, Klagenfurt.

Krause, W. – 1959 – Über die natürlichen Bedingungen der Grünlandberieselung in verschiedenen Landschaften Südbadens mit Ausblick auf den Wirtschaftserfolg. *Z. Acker-u. Pflanzenbau* 107: 245-274.

Krause, W. – 1962 – Über das Leistungspotential der Allmendweiden des Hochschwarzwaldes. In D. Lieth, Die Stoffproduktion der Pflanzendecke, 67-116, Gustav Fischer, Stuttgart.

Krause, W. – 1964 – Grossräumige Auswertung einer Vegetationskarte der Allmendweiden des Hochschwarzwaldes. (Engl. summ., Franz. rés.). *D. wirtschaftseigene Futter* 10: 101–112.

Krause, W. – 1974 – Bestandesveränderungen auf brachliegenden Wiesen. (Engl. summ., Franz. rés.). *D. wirtschaftseigene Futter* 20: 51-65.

Krause, W. & J. Frei – 1965 – Die Verbesserung der Allmendweiden im Südschwarzwald, dargestellt an der Gemeinde Schönenberg (Kreis Lörrach). (Engl. summ., Franz. rés.). *D. wirtschaftseigene Futter* 11: 191-200.

Landwirtschaftlicher Beratungsdienst der Kali AG – 1966 – Naturfutterbau/Versuchsergebnisse und Beobachtungen. Bern, 115 S.

Lüdi, W. – 1936 – Experimentelle Untersuchungen an alpiner Vegetation. *Ber. Schweiz. Bot. Ges.* 46 (Festband Rübel): 623-681.

Marschall, F. – 1947 – Die Goldhaferwiese (Trisetetum flavescentis) der Schweiz. Beitrag geobot. Landesaufn. Schweiz 26, Huber Bern 168 S.

Marschall, F. – 1958 – Die Milchkrautweide, ein Beitrag zur botanischen Klassifikation der Alpweiden. (Franz. rés.). *Landw. Jb. Schweiz* 72: 81-87.

Marschall, F. – 1963 – Die Grundlagenforschung im Naturfutterbau mit besonderer Berücksichtigung der schweizerischen Verhältnisse. (Franz. rés., Engl. summ.). Ber. Europ. Konf. Naturfutterb. Berglagen Chur, 1962: 15-37.

Marschall, F. – 1964 – Weitere Untersuchungsergebnisse über die Alpweide-Düngungsversuche auf Churer Alpen bei Arosa. *Schweiz. landw. Forsch.* 3: 93-98.

Marschall, F. & W. Dietl – 1974 – Beiträge zur Kenntnis der Borstgrasrasen der Schweiz. (Franz. rés., Engl. summ.). *Schweiz. landw. Forsch.* (Festschrift Koblet) 13: 115-127.

Neururer, H. – 1969 – Möglichkeiten zur chemischen Bekämpfung des Almampfers (*Rumex alpinus*). (Engl. summ.). *Pflanzenschutzberichte* 40: 49-59.

Oberdorfer, E. – 1950 – Beitrag zur Vegetationskunde des Allgäu. *Beitr. naturkundl. Forsch. SW-Deutschl.* 9: 29-98.

Oberdorfer, E. – 1957 – Süddeutsche Pflanzengesellschaften. *Pflanzensoziologie* 10. Jena, 564 S.

Oberdorfer, E. – 1959 – Borstgras- und Krummseggenrasen in den Alpen. *Beitr. naturkundl. Forsch. SW-Deutschl.* 18, H. 1: 117-143.

Pallmann, H. – 1943 – Über Waldböden. *Z. schweiz. Forstverein* 21: 113-140.

Petrascheck, A. — 1973 — Über die Wirkung systematischer Entwässerungen in Hanglagen. (Engl. summ.). Diss ETH Zürich, 171 S.

Petrascheck, A. — 1974 — Zur Frage der Bemessungskriterien bei Grünlandentwässerungen. (Engl. summ.). Z. *Kulturtech. Flurbereinigungen* 15, H. 2: 102-117.

Puffe, D. — 1973 — Untersuchungen zur Bodenfeuchte einer Vielschnittwiese mit differenzierter Stickstoff-Düngung. (Engl. summ.). Z. *Kulturtech. Flurbereinigung* 14, H. 4: 247-255.

Rehder, H. — 1970 — Zur Ökologie insbesondere Stickstoffversorgung subalpiner und alpiner Pflanzengesellschaften im Naturschutzgebiet Schachen (Wettersteingebirge). Diss. Botanicae 6, Lehre, 90 S.

Richard, F. — 1963 — Wasserhaushalt und Entwässerung von Weideböden. (Franz. rés., Ital. rias., Engl. summ.). *Mitt. Schweiz. Anst. forstl. Versuchsw.* 39, H. 5: 247-269.

Rieben, E. — 1957 — La forêt et l'économie pastorale (Dt. Zus., Ital. rias., Engl. summ.). Thèse EPF-Zurich, Vallorbe, 250 S.

Schechtner, G. & H. Wagner — 1963 — Pflanzenbestandesveränderungen in den Borstgrasbekämpfungsversuchen Kaiserau. (Engl. summ., Franz. rés.). Ber. Europ. Konf. Naturfutterb. Berglagen Chur, 1962: 95-109.

Schmid, A. — 1973 — Alpung und Leistungszucht / L'alpage et la productivité du bétail bovin. *Alpwirtsch. Monatsbl.* 107: 317-321 / 321-325.

Schneiter, F. — 1934 — Ein Rindergülle- und Kotdüngungsversuch auf einer oststeirischen Bürstlingshochweide und das vorläufige Ergebnis. Verhandlungsber. 3. Grünlandkongress der nord- und mitteleurop. Länder, 296-307.

Schürch, A. — 1967 — Sömmerungsweide und Fütterung. *Mitt. Arbeitsgem. Förd. Futterb.* 70: 6-14.

Schweingruber, F.H. — 1972 — Die subalpinen Zwergstrauchgesellschaften im Einzugsgebiet der Aare (Schweizerische nordwestliche Randalpen). (Franz. rés., Ital. rias., Engl. summ.). *Mitt. Schweiz, Inst. forstl. Versuchsw.* 48/2: 195-504.

Spatz, G. — 1970 — Pflanzengesellschaften, Leistungen und Leistungspotential von Allgäuer Alpweiden in Abhängigkeit von Standort und Bewirtschaftung. Diss TH München, Freising-Weihenstephan, 159 S.

Spatz, G. & G. Voigtländer — 1971 — Leistungen und Leistungsreserven von Allgäuer Alpweiden. (Engl. summ.). Z. *Acker- und Pflanzenbau* 133: 233-259.

Speidel, B. — 1963 — Das Grünland, die Grundlage der bäuerlichen Betriebe auf dem Vogelsberg. *Bodenverband Vogelsberg* 3: 1-68.

Stebler, F.G. & C. Schröter — 1887 bis 1892 — Beiträge zur Kenntnis der Matten und Weiden der Schweiz. 2. Teil — 1887 — Untersuchungen über den Einfluss der Düngung auf die Zusammensetzung der Grasnarbe. *Landw. Jb. Schweiz* 1: 93-148. 7. Teil — 1888 — Das Borstgras (*Nardus stricta* L.) ein schlimmer Feind unserer Alpwirtschaft. *Landw. Jb. Schweiz* 2: 139-150. 10. Teil — 1892 — Versuch einer Übersicht über die Wiesentypen der Schweiz. *Landw. Jb. Schweiz* 6: 95-212.

Stebler, F.G. & C. Schröter — 1889 — Die besten Futterpflanzen. 3. Teil: Die Alpen-Futterpflanzen. Wyss, Bern, 193 S.

Surber, E., Amiet, R. & H. Kobert — 1973 — Das Brachlandproblem in der Schweiz. (Franz. rés., Engl. summ., Ital. rias.). Ber. Eidg. Anst. forstl. Versuchsw., Birmensdorf 112, 138 S.

Wacker, F. — 1954 — Beispiele für die Leistung und Grenzen angewandter Pflanzensoziologie. *Angew. Pflanzensoz.* 8: 99-101, Stolzenau.

Wagner, H. — 1965 — Die Pflanzendecke der Kamperdellalm im Tirol. (Franz. rés.). Documents Carte Végétation Alpes II. Grenoble, 59 S.

Walter, H. — 1955 — Die Klimagramme als Mittel zur Beurteilung der Klimaverhältnisse für ökologische, vegetationskundliche und landwirtschaftliche Zwecke. *Ber. D. Bot. Ges.* 68: 331—

10 THE INFLUENCE OF FERTILIZERS ON THE PLANT COMMUNITIES OF MESOPHYTIC GRASSLANDS

T. A. RABOTNOV

Contents

10.1 On the History of Study of Influence of Fertilizers on Grassland Ecosystems

Organic fertilizers (manure, gulle) have long been used for improvement of grasslands in a number of West-European countries, in the Lake Baikal regions (Prebaikalye and Transbaikalye), in some mountainous regions of the Caucasus and Middle Asia (in arid areas in conjunction with irrigation). On the basis of a comparative study in field conditions of adjacent meadows, fertilized and unfertilized, Stebler & Schröter (1887) established the main regularities of the change of meadow vegetation under the influence of fertilizers.

The first scientific experiment aimed at assessing the effect of mineral and organic fertilizers was started by Lawes & Gilbert at the Rothamsted Experimental Station (England) in 1856 and is still in progress. The results of the experimental investigations of the first twenty years were published as three monographs (Lawes & Gilbert 1880, 1890, Lawes et al. 1882). The results of later observations, which are of great value, were published in a number of works (Brenchley 1924, 1925, 1926, 1930, 1935, 1937, Brenchley & Heintze 1923, Brenchley & Warington 1958, Cashen 1947, Hall 1905, Thurston 1969a, b, Warren & Johnston 1964[1]). Beginning with the twenties and thirties of this century, thousands of experiments have been carried out in European countries, including the USSR with the purpose of determining the efficiency of fertilizers. In these experiments main attention was given to the effect of fertilizers on the yield and the ratio of main agricultural groups of plants (grasses, legumes, herbs). In fewer experiments account was taken of the influence of fertilizers on single species, the structure of swards, the root systems, etc. The results of these experiments are of great importance for some studies on the phytocoenology and ecology of plants. The experiments with fertilizers on meadows have been carried

---

1. In Great Britain extensive studies on the influence of fertilizers on meadow phytocoenoses were carried out also at the Welsh Experimental Station (Davies & Jones 1932, Jones 1934, Milton 1938, 1940, 1947).

out in all continents, on different types of grasslands. Owing to the limited volume of this review, we shall consider here only the data on the grasslands of the temperate zone (mesophytic grasslands-meadows), mainly those of Europe including the USSR, primarily the results of the studies on the influence of combinations of fertilizers most widely used in meadows: PK and NPK.

10.2 General Directions of Influence of Fertilizers on Grassland Ecosystems

Meadows represent ecosystems (biogeocoenoses), their components are the organisms forming the biocoenosis and the environment (biotope, community habitat or site). The biocoenoses consist of autotrophic plants (in meadows of the temperate zone, mainly perennial mesophilous herbaceous species) and heterotrophs (animals, fungi, bacteria, actinomycetes). The biotope consists of the aboveground part (aerotope) and the soil part (edaphotope). Fertilizers influence all the components of grassland ecosystems: autotrophs, heterotrophs, including animals, edaphotope and aerotope (phytoclimate). The immediate effect of fertilizers on herbaceous plants is of very little importance. It is associated with the herbicide properties of some kinds of fertilizers. On meadows with moss cover fertilizers can act as muscicides. The main effects of fertilizers on grassland herbaceous plants involve changes in the conditions of the plant growth as the result of 1. introduction into soil of one or more nutrients contained in the fertilizer, mainly NPK; 2 introduction into soil of substances contained in fertilizers, other than main nutrients. Some of them, e.g. microelements, have a positive effect on plants. An example of the secondary, unfavourable effect of some fertilizers on the yield is a progressing acidification of soil in the case of yearly application of $(NH_4)_2SO_4$; 3. the changes in the phytoclimate due to increase in the density and height of swards; 4. the changes in the numbers and activity of other biocomponents of grassland biocoenoses, including ammonifiers, nitrifiers, nitrogen-fixers, cellulose-lytic organisms, mycorrhiza-forming fungi, earthworms, etc.; 5. the changes in the utilization of meadows by man, generally increased defoliation due to higher productivity.

The influence of fertilizers on grassland vegetation depends on the rates, forms and dates of application of fertilizers as well as on the kind of utilization of meadows (for hay or for grazing).

462

Of great importance are also: 1. provision of plants with appropriate nutrients prior to the addition of fertilizers; 2. participation in phytocoenoses of species capable of utilizing to the best advantage their improved provision with N, P, K, etc.; 3. presence or absence of other conditions limiting the increase of productivity of the plants potentially capable of responding to the fertilizers, primarily soil moisture.

Particular types of meadows differ significantly with respect to the conditions of growth of plants (climate, hydrology, soils) and the composition of phytocoenoses. Therefore, it is natural that application of the same fertilizers at the same rates and with the same form of utilization of a meadow has different effect on different types of meadows, which manifests itself mainly in unequal increase of yields. This fact has been established by analysis of the results of numerous experiments in different countries (Antipin 1951; Davies & Jones 1932; Krause & Müller 1973; Milton 1947; Onoshko 1934, 1936; Romashev 1941, 1949, 1963; Siebold 1958; Vorhauer 1958; Zürn 1961, etc.).

Application of fertilizers enhances the yields of swards, which is accompanied by a change of their structure (height, vertical distribution of the mass of aboveground organs, leaf area index, number of tillers, etc). In the case of utilization for hay, not only does the height of meadows swards increase under the influence of fertilizers, but the vertical distribution of the mass of aboveground organs in them also changes: it becomes more uniform, which is particularly true for swards formed by species with the concentration of their foliage near to soil surface (Table 1).

Table 1.

Vertical distribution of mass of aboveground organs (in % of total mass) in swards of subalpine meadow (North Caucasus) depending on fertilizers. (Rabotnov 1950).

Horizons cm	Treatments				
	O	N	P	NP	NPK
0—10	56,5	47,6	35,7	35,5	32,4
10—20	36,2	27,2	18,4	19,5	22,4
20—30	6,2	18,1	21,7	18,8	18,3
30—40	0,9	5,6	15,7	14,6	14,9
40—50	0,2	1,2	7,1	6,4	4,8
50—60		0,3	0,9	2,5	2,9
60—70			0,4	1,2	1,3
70—80			0,1	0,7	1,9
80—90				0,6	0,6
90—100				0,2	0,5

10.3 The Effect of Fertilizers on the Structure of Grassland Plant Communities

10.3.1 THE EFFECT ON THE SWARDS (YIELDS, HEIGHT, DENSITY, QUANTITY OF TILLERS, LEAF AREA INDEX)

As has been established by experiments with single species of grasses, their improved provision with nutrients, especially nitrogen, leads to increase of the number of tillers per unit area. Thus, in one experiment the number of tillers of *Phleum pratense* per plot was as follows: without fertilizers, 605 with 19% generative tillers and with application of NPK, 790, 33% of which were generative tillers (Smelov 1947). However, in meadow phytocoenoses formed by many species, responding differently to application of fertilizers, in the case of their utilization for hay, as a rule, the number of tillers per unit area decreases, which is usually the result, especially when nitrogen fertilizers are applied, of greater participation in swards by tall grasses and a lesser one by short grasses. Thus, on a meadow in the flood valley of the Oka river, inundated for only a short time and not every year, as the result of ten years' application of $N_{60}P_{60}K_{60}$, the total number of tillers decreased from 6730 (control) to 3765 per 1 m^2 (NPK). In this case the number of tillers of short grasses (*Poa pratensis, Festuca rubra*) decreased from 5130 to 2005 per 1 m^2 and that of tall grasses (*Elytrigia repens, Agrostis stolonifera, Alopecurus pratensis, Bromus inermis, Dactylis glomerata, Festuca pratensis, Phleum pratense*) rose from 550 to 1190 per 1 m^2 (Rabotnov & Demin 1971, 1974). However, the strength (productivity) of the tillers increases under the influence of fertilizers (Stebler & Schröter 1887 *et al.*). On pastures, where as a result of frequent defoliation, swards do not reach great height and are formed mainly by short grasses (*Lolium perenne, Poa pratensis, Festuca rubra*), application of fertilizers, especially those with nitrogen, can lead to considerable increase of the number of tillers per unit area. Only if very high rates of nitrogen fertilizers are applied does the number of tillers decrease. In spite of a decreased number of tiller, under the influence of fertilizers, the leaf-area index increases 1.3–2.0 times (Alekseenko 1967, Geyger 1971, Rabotnov 1970). The density of swards becomes greater and this results in a change of the phytoclimate (lesser light penetration to the soil surface, decrease of the air temperature, increase of its humidity).

464

In the absence of the competition between plants (pot cultures, etc.), application of fertilizers, especially with nitrogen, increases the number of grass roots. This is due to the appearance of numerous tillers from which adventive roots arise and often also to larger numbers of roots per tiller. In this case the strength of roots usually rises, as evidenced by a relatively larger increase of their volum. For instance under the influence of fertilizers, in the case of *Phleum pratense* and *Festuca pratensis* in the second year of their life the number of roots became twice as large and the volume rose accordingly 3.5 and almost 4 times (Lebedev 1968). In the case of rhizomatous species, high nitrogen rates increase the number of rhizomes and their total length (Lebedev 1968). It has been found in many experiments that application of fertilizers enhances the mass of underground organs much less than that of the aboveground ones (Remy & Fasters 1931, Bronzova 1940). In many-component meadow phytocoenoses, in which application of fertilizers alters the quantitative relationship between species the number of adventitious roots decreases, this being due to smaller number of tillers per unit area. In the experiment in the flood valley of the Oka river mentioned above, in which the number of tillers decreased as a result of long-term yearly application of $N_{60}P_{60}K_{60}$, the number of adventitious roots dropped from 95800 (control) to 64555 per 1 m^2 (NPK) and among them that of grasses and *Carex praecox* from 87555 to 60410. At the same time, the total length of rhizomes and aboveground stolons fell from 13390 to 7350 cm/m^2. The reduction of the number of adventitious roots occurred mainly at the expense of *Festuca rubra* and *Poa pratensis* (from 74265 to 20125 per m^2), while number of roots of tall grasses (*Elytrigia repens, Agrostis stolonifera, Alopecurus pratensis, Bromus inermis, Dactylis, Phleum*) rose from 7035 to 27540 per m^2. Owing to the reduction in the number of tillers, that of the adventitious roots from the bases of living tillers dropped sharply, whereas the number of roots of the dead tillers remained almost unchanged. As a result of the decrease of the number of roots and total length of rhizomes, penetration by them of the upper (0–10 cm) layer of the soil diminished (Rabotnov & Demin 1971). Similar data were obtained by Speidel & Weiss (1973), who found that due to application of fertilizers, the number of root tips per unit soil volume showed a decrease: without fertilizers 186, PK 146,

NPK 96, which seems to be associated with the reduction of the number of adventitious roots.

In analyzing numerous, often contradictory data on the influence of fertilizers on the total mass of underground organs of meadow plants, it is necessary to bear in mind that on natural, and also on old seeded meadows the mass of underground organs, which is usually significantly (several times) larger then that of aboveground ones, increases when the conditions of growth of plants deteriorate, in particular, when provision of plants with nutrients diminishes. For P_2O_5 this fact was established by Kmoch (1952).

The increase in the mass of underground organs on meadows in the case of their insufficient provision with nutrients is explained by the fact that under such conditions, the mineralization of dead roots is a slow process, but with the use of the existing methods of determination of the mass of underground organs, account is taken both of living and dead roots and rhizomes.

Moreover the duration of alive roots may increase when the conditions of growth deteriorate. The results of the observations on two types of meadows in the flood valley of the Oka river are evidence of the effect of fertilizers on the mass of living and dead underground organs (Demin 1970). On a drier meadow, inundated for a short period not every year, the long-term application of $N_{60}P_{60}K_{60}$ led to increase in the mass of underground organs from 1177 (control) to 1509 g/m^2 (NPK), almost exclusively at the expense of the mass of dead roots since the mass of alive roots changed insignificantly (from 839 to 878 g/m^2). It was otherwise on a more moist meadow, inundated almost every year, where under the influence of long-term application of $N_{60}P_{60}K_{60}$, the total mass of underground organs rose from 1426 to 2051 g/m^2, the mass of living roots and rhizomes from 905 to 1286 g/m^2 (in the horizon 0–10 cm from 615 to 783 g/m^2).

The sharp increase of the content of dead roots (from 28.7 to 41.7% of the total mass of underground organs) on the first type of meadows was associated with the change in the quantitative relationships of species in the phytocoenoses, with lesser participation of legumes and herbs (whose dead underground organs become mineralized quickly) and with a considerable increase of participation of grasses, whose dead roots decompose more slowly. On a more moist type of meadow, where grasses were dominant, the difference in the mass of dead roots on the variants "control" and "NPK" was not so great (521 and 770 g/m^2) as on a drier meadow (338 and 631 g/m^2).

Thus, the influence of fertilizers on the mass of dead and alive underground organs is associated with the change in the quantitative relationships of species in the phytocoenoses. When under the influence of fertilizers, the participation of the species whose dead underground organs become quickly mineralized increases, for instance when the content of legumes under the influence of PK becomes greater, we can assume that the participation of dead roots in the total mass of underground organs decreases and this can lead to decrease of the latter. This is illustrated by the results of the experiment in which the mass of underground organs (g/m^2) varied for different treatments as follows: control 2000, PK 1739, NPK 2173 (Schulze & Mues 1961).

The decrease of the mass of underground organs for the variant PK can be explained by sharp increase of the participation of legumes, whose roots become quickly mineralized. As a rule, under the influence of NPK, the total mass of underground organs increases, but usually not too much. An increase in the mass of underground organs of 1.5–2.0 times is observed more rarely (Lopatin & Zaikova 1966). A significant decrease in the mass of underground organs upon application of NPK, observed in the experiment carried out by Speidel & Weiss (1971, 1973), in which the following data (averages over 3 years) were obtained (g/m^2): control 946, $P_{90}K_{120}$ 873, $P_{90}K_{120}$ N_{200} 738 is undoubtedly associated with the fact that among the treatments the herbage was cut twice on the control and PK and three times on the NPK during the vegetation period.

Some part of the root system of every meadow plant is located in the most superficial soil horizon. However, some species (for example, grasses) have a large quantity of absorbing roots in this horizon unlike some other species (taprooted ones) with few absorbing roots in the superficial horizon. Some species are able to change their rootsystems under the influence of fertilizers and thus have more absorbing roots in the superficial soil horizon. Sometimes a change of the type of root system occurs. For example, under longterm fertilization by NPK the taprooted *Heracleum sibiricum* forms a root system of the racemoseous type (Demin 1969). As a result of the application of NPK many species increase both the horizontal distribution of their roots and the concentration of absorbing roots in the superficial soil horizon as well as the quantity of hairlike rootlets on the roots

467

(Demin 1969). The changeability of root systems is an essential adaptive quality for successful competition with other species under the influence of fertilizers.

Significant changes in the number and composition of soil organisms and in the conditions of growth of herbaceous plants occur on meadows with marked moss cover, where under the influence of fertilizers mosses disappear, which results in strong improvement in the gas- and heat exchanges between soil and atmosphere as well as in a change in the water regime of the soil.

10.3.3   THE EFFECT ON THE SOIL MICROORGANISMS

Increase in the mass of underground organs upon application of fertilizers is accompanied by increased addition of dead roots to the soil. If they do not accumulate (e.g. when the soil is acidified), this leads to increase in the number of microorganisms in the soil. This is shown by the results of the experiment in which fertilizers were applied to a meadow for a period of twelve years (Table 2). In this case the number of bacteria in the soil was in direct proportion to the mass of roots. The application of NPKCa in the experiments in Karelia led to considerable increase of the total number of microorganisms in the soils of the 6 types of meadows studied. On all types of meadows, except one, the amount of aerobic cellulose lytic bacteria rose, but that of cellulose lytic fungi dropped (Ershov 1963, Ershov & Kuz'mina 1965). The same fact was observed in England, where application of Ca, and especially CaNP, to acid soil enhanced the amount of bacteria, but diminished that of fungi (Kirkwood 1964).

It has been established that legumes stimulate the activity of cellulose lytic microorganisms, as indicated by increase of their

Table 2.

Influence of 12 year application of fertilizers on mass of underground organs and on number of bacteria (Könekamp & Weise 1961).

Treatments	Mass of underground organs		Total number of bacteria (average per year) in millions/g of soil		Number of bacteria in milliards for 1 g of dry roots
	g/m²	%		%	
Control (without fertilizers	451	100	4.8	100	1.07
NPK	533	127	6.1	127	1.06
NPK + manure	607	134	6.7	139	1.10

468

number in the soils of meadows upon application of PK, which favours increase of the amount of legumes (Lauzne 1960, Lauzne & Unitas 1958). Application of fertilizers, if not accompanied by soil acidification, stimulates the activity of soil microorganisms, accelerating the mineralization of dead underground organs. The application of very high rates of nitrogen fertilizers, can cause unfavourable changes in the ratio of the physiological groups of soil microorganisms (Gurfel 1974).

Thus, fertilizers, acting on plants and other organisms forming part of meadow biocoenoses bring about various changes both in the phytoclimate and in the soil. Application of fertilizers, especially nitrogen, leads to changes in the relationships between meadow herbaceous plants and their symbiotrophic and parasitic consorts, which in its turn alters the competitive relationships of species in meadow phytocoenoses. Application of nitrogen fertilizers decreases the fixation of atmospheric nitrogen by nodule bacteria of legumes, which weakens the competitive ability of legumes. The majority of meadow herbaceous plants are mycotrophs possessing endomycorrhiza (Rabotnov 1974) and application of NPK diminishes development of their mycosymbiont (Kirillova 1968, Demin 1971).

Nitrogen fertilizers enhance the susceptibility of meadow plants to the action of fungus parasites.

Under the influence of long-term application of fertilizers, the structure of the underground organs of plants, their ability to reproduce by seeds and vegetatively undergo considerable changes. The observations in the flood valley of the Oka river (Demin 1969) show that under the influence of long-term application of $N_{60}P_{60}K_{60}$ for many species of meadow plants the root system changes significantly, becoming more adapted for better utilization of the nutrients contained in fertilizers applied to the surface of meadows. This manifests itself in decrease of the depth of the root system (for most species), in greater width of spreading of the roots (for all species except grasses), in greater number of hair rootlets covering the roots.

10.3.4   THE EFFECT ON THE REPRODUCTION OF GRASSLAND SPECIES
BY SEEDS AND VEGETATIVELY

The ability of plants to reproduce by seeds is also significantly affected by fertilizers. Depending on their response to fertilizers, the seed productivity of species either rises or falls. A manifold increase of the yield of the seeds of valuable forage plants; grasses (upon application of NPK) and legumes (PK) served as the basis for recommendations on the creation of natural sources for seeds of forage plants by application of fertilizers to some types of meadows (Rabotnov 1947, Smelov & Terekhova 1957, 1959). However, fertilized meadows are cut at earlier dates and therefore the possibility of seeding of plants decreases sharply, though some plants manage to go to seed, usually in the second cutting, but on a relatively small scale. As a rule fertilized pastures are used intensively, and forage uneaten by cattle is cut after grazing, which also excludes or strongly limits natural seeding. The conditions of the seedlings taking root on fertilized hay meadows are also less favourable than on unfertilized meadows owing to greater shading. Therefore, when fertilizers are applied the number of juvenile plants and seedings is reduced (Stebler & Schröter 1887, Rabotnov 1948, Zaikova 1966, Kotelina 1967). Only in one case, when manure was applied did the number of seedlings rise (Stebler & Schröter 1887). The participation in meadow swards of some short-lived species reproducing exclusively by seeds drops rather soon or they disappear entirely. Other species grow successfully with two cutting utilization of meadows. Thus, at the Rothamsted Experimental Station, after 90 years application of NPK, on the plots where nitrogen was applied as $NaNO_3$, the plants reproducing exclusively by seeds (*Arrhenatherum elatius, Dactylis glomerata* etc) gave more than 50% of the yield, whereas on the plots, where $(NH_3)_2SO_4$ was applied, they gave over 90% (mainly *Holcus lanatus*) (Brenchley & Warrington 1958). In the experiment in the flood valley of the Oka river after 10 years application of NPK, the plants reproducing exclusively by seeds accounted for 29% of the yields against 39% in the control (Rabotnov 1965). In the experiment in the Leningrad region with yearly application of NPK, when *Arrhenatherum elatius* was over-sown additionally it became one of predominant species in the sward (Makarevich 1968). Thus, we see that reproduction of meadow plants by seeds is quite possible also when fertilizers are applied to meadows, at least under certain conditions.

The species reproducing themselves vegetatively respond especially favourably to NPK. Therefore, in the meadows with swards containing long- and short-rhizomatous grasses (*Bromus inermis, Elytrigia repens, Alopecurus pratensis* et al.) the productivity of these species is greatly increased by nitrogen fertilizers.

10.4 The Effects of Fertilizers on the Response of Grassland Species to Water, Aeration and Temperature

10.4.1 MESOPHYTIZATION OF DRY AND WET GRASSLANDS

Improved provision with nutrients alters the response of meadow plants to other ecological factors. For instance, in Ellenberg's experiment (1954) the maximum yield of *Arrhenatherum elatius* was observed at a low level of provision with nitrogen when ground water was at the depth of 50 cm, and at a high level of provision with nitrogen at the depth of 100 cm (Table 3). It was found in the experiment with *Dactylis glomerata* and *Agrostis tenuis* that at a low level of provision with NPK the soil pH was of importance for both species. Their yield rose considerably when liming was used, whereas at high rates of NPK the pH of the soil (and hence application of lime) did not affect their yield (van den Bergh 1968). There are numerous data on the lower winter hardiness of meadow plants when high rates of nitrogen are applied. The degree of the decrease of winter hardiness differs for different species and cultivars within the same species (Baker & David 1963, Howell & Jung 1965, Smith 1964, Huokuna 1974, Jung & Kocher 1974).

Of great interest are the phenomena occurring in meadow phytocoenoses under the influence of fertilizers, especially nitrogen. They show that increased provision of plants with nutrients changes the response of some species of meadow plants to provision with water, to heat, to soil aeration. First of all, here belongs "mesophytization" of dry meadows, increased participation of mesophytes in their swards (Stebler & Schröter 1892, Ellenberg 1952, Klapp 1965, Sannikova 1963).

Table 3.

Yields of grasses (g/m^2) depending on depth of ground water and on providing with nitrogen (Ellenberg 1954).

Species	Providing with nitrogen	Depth of ground water, cm				
		5	25	50	75	100
Bromus erectus	poor	29	71	100	101	87
Bromus erectus	rich	46	124	138	111	114
Arrhenatherum elatius	poor	72	197	348	253	167
Arrhenatherum elatius	rich	79	222	542	772	961

Increase in the yield when fertilizers are applied means that the available water reserves on meadows can ensure higher yields as compared with those on unfertilized meadows. This can be explained both by more complete and more economical utilization of available water. Of particular importance is a more economical water utilization since on fertilized meadows the water consumption for evapotranspiration does not increase at all, or but slightly, in spite of a significant increase of the yield (Klapp 1962c, 1965, Wind 1954). (Table 4).

The absence of an appreciable increase of water consumption when fertilizers are applied can be due both to more economical water consumption by meadow plants for transpiration and to decrease of the water loss by evaporation from the surface (and on slopes also by runoff). Many investigators found the coefficient of transpiration to decrease when fertilizers are applied. Thus, in the pot culture with *Dactylis glomerata* and *Phalaris arundinacea* application of NPK led to the decrease of this coefficient by 27–28% (Koenekamp 1930). In the experiment with *Lolium multiflorum* the water consumption per unit yield decreased 1.5–2 times when nitrogen fertilizers were applied (Amberger 1955). Moreover, on dry and moderately wet meadows application of NPK enhanced the participation of grasses in swards, including their participation in the formation of leaf-area. Thus, on a dry meadow in the flood valley of the Oka river the participation of grasses in the total leaf-area upon application of NPK rose from 30 to 61.5% (Rabotnov 1970), while the coefficients of transpiration of grasses seem to be lower than those of legumes and herbs.

Thus, the economy in water consumption by meadow swards can result from: 1) greater participation in swards of the species with lower coefficients of transpiration; 2) decrease of

Table 4.

Change of evapotranspiration of meadow depending on rates of nitrogen fertilizer

Applied nitrogen (kg/ha)	Yields g/m²	Evapotranspiration mm	Coefficient of evapotranspiration
0	717	451.6	630
70	806	444.4	550
120	980	474.7	480
270	1236	458.7	370
520	1623	465.8	290

the coefficient of transpiration of the same species on fertilized meadows as compared with the unfertilized ones. Nevertheless, as the result of a significant increase of the yield, the total water consumption for transpiration on fertilized meadows rises. However, due to decreased evaporation from the soil surface, the total water consumption (evapotranspiration) remains at the former level, or rises insignificantly. On fertilized meadows in the case of water deficiency, plants utilize more thoroughly the water of deeper-lying horizons (Lawes & Gilbert 1880, Gliemeroth 1951).

When fertilizers are applied the amount of water supplied to the meadow does not increase. Therefore, the mesophytization of dry meadows under the influence of nitrogen (organic or mineral) fertilizers is to be explained, if not completely then to a considerable extent, by the higher competitive ability of more mesophilous species when they are better provided with nitrogen. This is confirmed by the results of Ellenberg's experiment (1954), which showed that at the ground water level 100 cm, with improved provision with nitrogen the yield of a xerophyte *Bromus erectus* rose only insignificantly (from 87 to 114 g/cm^2), whereas the yield of a mesophyte *Arrhenatherum elatius* rose almost 6 times from 167 to 961 g/m^2 (Table 3). Therefore, there are reasons to conclude that on dry meadows mesophytes are suppressed not only because they lack water but also, and even more so, because with poor provision with nitrogen they cannot, among other things, compete for water with more xerophylous species which at the same time are less nitrogen-demanding. This is the reason why in Western Europe the transformation of the swards with *Bromus erectus* into communities with *Arrhenatherum elatius* (Ellenberg 1963, Klapp 1965) is so common. Mesophytization can occur naturally only on those types of dry meadows whose swards contain, even though in a suppressed state, mesophilous plants capable of positive response to increased provision with nitrogen.

Under the influence of nitrogen fertilizers, on wet meadows including those with sedge and rush swards, phytocoenoses can be formed with dominance of *Arrhenatherum elatius, Alopecurus pratensis* and other highly productive mesophilous grasses. This phenomenon became known as "biological drying" (König 1955, Klapp 1965, Kreuz 1965). An example of this is the transformation, due to application NPK, of a meadow with *Juncus acutiflorus, Festuca rubra* and *Carex (nigra, leporina, panicea* et al.) dominant into A l o p e c u r e t u m   p r a t e n s i s with a simultaneous three-fold increase of the yield (Klapp 1965). The degree of moistening determining by means of Klapp's scale decreased from 7.0 to 5.9, and judging from the sward composition, the changeability of moisture during vegetation season also decreased.

"Biological drying" occurs not only on wet meadows but even on sphagnum bogs. A case of the transformation of a sphagnum bog with *Eriophorum vaginatum* into a meadow with dominant *Holcus lanatus*, resulting from the four year application of NPKCa, was reported by Bradshaw (1969). In an experiment carried out in the Moscow region on a spagnum bog with complete cover of *Sphagnum fallax*, after application of NPK and especially NPKCa, Sphagnum died off and a closed high sward with dominant *Calamagrostis canescens* was formed. However, in Tamm's experiment in Sweden (1954) application of fertilizers on a sphagnum bog was not accompanied by mesophytization (*Eriophorum vaginatum* becoming dominant), which possibly can be accounted for by the fact that there were no mesophytes in this phytocoenosis, capable of responding more favourably to fertilizers than *Eriophorum*.

Application of nitrogen fertilizers to moist and wet meadows is not always accompanied by "biological drying". According to Bommer (1964), the application of 200 kg/ha N led to spreading of *Deschampsia cuespitosa* and *Rumex acetosa*, and sometimes of *Polygonum bistorta*. But no unfavourable changes were observed in the swards containing a large amount of *Alopecurus pratensis*. In the experiment of Kresnil (1974) when high nitrogen rates were applied *Alopecurus pratensis, Poa pratensis, Festuca pratensis* and *Festuca rubra* dominated only the first 3—4 years, whereupon they were displaced by *Deschampsia caespitosa* and other species. Application of NPK to a moist meadow in the Moscow region led to dominance of *Filipendula ulmaria*. In the experiment in the USA the change from sedges and rushes to grasses was observed only if very high nitrogen rates were applied. In the fourth year of application of fertilizers the content of *Carex* and *Juncus* in the yield was: control 62%, 225 k/ha N 54%, 450 k/ha N 12%, 675 k/ha N 4%, and the content of grasses, 20, 44, 88, 96%, respectively (Rumberg & Cooper 1961).

In some cases when NPK was applied, "biological drying" occurred only after elimination of the existing vegetation by mechanical treatment and subsequent sowing of valuable grasses. Thus, on a meadow with dominant *Molinia coerulea* and *Schoenus ferrugineus*, after harrowing and sowing of grasses in the 8 th year of application of NPK, a sward where grasses were dominant

(*Poa pratensis, Poa trivialis, Festuca pratensis*) was formed and the moisture level on Ellenberg's (1952) scale dropped from 4.5 to 3.3–3.4 (Finckh 1960). Thus, the "biological drying" seems to be possible only if swards contain mesophilous grasses capable of responding positively to improved provision with nitrogen and if more competitive hydrophylous species (*Deschampsia caespitosa*, et al.) are absent.

There are reasons to believe that in the case of "biological drying" the moisture conditions of wet meadows change under the influence of fertilizers. This can be associated with increase of:1) water percolating capacity of the upper horizon of soil, which diminishes the possibility of water stagnation and leads to less changeability of moisture during the vegetation season; 2) water consumption for evapotranspiration.

There are some data available on a tenfold increase of the water percolating capacity of the upper soil horizon of a marsh meadow after 5 years application of NPK (Kuntze, cit. from Kreuz 1965), which possibly was the result of increased intensity of mineralization of dead organs of plants, accumulating usually in considerable amounts in the soils of wet meadows.

In the example given above of "biological drying" of a swampy meadow, according to Klapp (1965), due to application of NPK, the mass of underground organs in the soil layer 0–10 cm decreased from 6332 to 3732 g/m^2, and in the layer 10–15 cm from 810 to 276 g/m^2, which was evidently the result of mineralization of dead roots and rhizomes as a result of a lower C : N ratio in them. It is quite possible that the increase of the water percolating capacity of the upper soil layer could be caused by enhancement of earthworms. A higher water consumption for transpiration is evidently associated with a greater mass of transpiring organs (in Klapp's experiment the yield rose 3 times) and also with the change from species with a relatively low transpiration coefficients to those with higher ones. Thus, it can be assumed that in Klapp's experiment *Alopecurus pratensis* showed greater transpiration than *Juncus acutiflorus, Festuca rubra, Carex spp.* which have narrower leaves. Among the 8 species of grasses (*Phleum pratense, Festuca pratensis, Festuca rubra, Dactylis glomerata, Arrhenatherum elatius, Bromus inermis, Lolium perenne, Lolium multiflorum*) studied by Lampeter (1959–1960), *Festuca rubra* had the lowest transpiration coefficient. In the experiments with another group of 8 grasses (*Lolium perenne, Festuca pratensis, Arrhenatherum elatius, Phleum pratense, Dactylis glomerata, Poa pratensis, Phalaris arundinacea,*

Alopecurus pratensis), Alopecurus showed the highest transpiration coefficient on an average of 3—4 years observations (Schwarz 1932). Evidently the increase of water consumption for transpiration is much greater than the decrease of evaporation from the soil surface because of the greater density of the sward. Of great importance for increase of water consumption is also the transition from one to two or three cuttings in the utilization of the meadow.

10.4.2   THE INCREASE OF THE CIRCUMOPTIMAL SYNECOLOGICAL
AREA OF GRASSLAND SPECIES

In accordance with this, when fertilizers are applied on dry meadows the soil moisture (judging by the composition of phytocoenoses) increases, on wet meadows it decreases and on medium moist meadows remains unchanged (Gigon 1968). It is clear from these data that it is necessary to introduce corrections into the determination of the soil moisture by means of the scales of Ellenberg, Klapp *et al.*, depending on the provision of plants with nitrogen. Under the influence of nitrogen fertilizers, the circumoptimal region of the synecological (ecological according to Ellenberg 1953) area of mesophilous grasses of the type of *Arrhenatherum elatius* and *Alopecurus pratensis*, spreads in the direction of both drier and wetter habitats. According to Ellenberg (1963), the area of meadows with *Arrhenatherum elatius* expands from dry meadows with *Bromus erectus* to M a g n o c a - r i c e t u m . This can be considered proof that aeration and water provision are of less importance (within of course, certain limits) in the determination of the competitive ability of mesophilous grasses than provision with nutrients, primarily with nitrogen. It should be also borne in mind that the unfavourable aeration conditions of the soil of wet meadows affect negatively some species of meadow plants, not so much directly as indirectly, hindering mineralization of dead organs of plants. Moreover, of importance also is the ability of mesophytes on wet meadows to form in the underground organs well-developed aerenchyme. This property is particularly pronounced in the case of some mesophytes, e.g. *Alopecurus pratensis* (Ellenberg 1963). By this means some, and possible many, species can adapt themselves to insufficient soil aeration, but they can grow successfully under these conditions only if they are sufficiently supplied with nitrogen. This accounts for the phenomenon, which at first glance

476

seems paradoxical, viz. that "nitrogen replaces oxygen" (Ellenberg 1963).

Under the influence of fertilizers, the circumoptimal synecological area of meadow plants expands owing to their increased participation in coenoses confined to habitats with more severe temperature conditions. Even Stebler & Schröter (1887) established the fact that fertilizers "help" lowland species to move high into the mountains.The results of subsequent experiments with the application of mineral fertilizers on high-mountain meadows showed that there are changes in the swards in the direction of greater participation in them of more "thermophilous" species, and that these changes depend completely on the transition of their individuals from suppressed to normal states owing to their better provision with nutrients, primarily with nitrogen. Thus, the experiments with application of fertilizers on meadows showed how important for meadow plants is their provision with accessible forms of nitrogen, and also that the content of mineral forms of nitrogen in the soil defines the response of meadow plants to moisture, aeration, and temperature.

10.5 The Effects of Fertilizers on the Floristic Composition and on Quantitative Ratio of Components of the Grassland Phytocoenoses

10.5.1 THE FACTORS DETERMINING THE COMPETITIVE ABILITY OF GRASSLAND SPECIES UNDER APPLICATION OF FERTILIZERS

Every plant species is peculiar in its response to fertilizers. As a result, fertilization of many component meadow phytocoenoses is always followed by changes in the quantitative relationships of their components and often by change of dominants. Different responses of species to fertilizers are partly determined by their ecological individuality. This was demonstrated by the results of pot experiments with single species stands at different rates of supply with nutrients (N, P, Ca) (Williams 1968, Bradshaw *et al.* 1958, 1960, 1964). For example, in the pot experiment (sand, cultures), in which the influence of different rates of nitrogen supply (from 1 mg to 243 mg N per 1 kg of sand) on the yields of seven species of grasses was studied, the highest yields were obtained at the following rates: *Nardus stricta* − 23 mg, *Festuca ovina* and *Cynosurus cristatus* − 81 mg, *Agrostis tenuis*,

A.stolonifera and *Lolium perenne* — 243 mg (Bradshaw et al. 1964).

From the comparison of the results of experiments carried out both on meadows with many component swards and in sown singlespecies swards it can be concluded that the changes in the phytocoenoses under the influence of fertilizers depend greatly on those in the competitive interrelationships between species. For example, in pot culture *Anthoxanthum odoratum* in pure stand responded to increasing rates of nitrogen as well as *Alopecurus pratensis*, a meadow grass with a very good response to nitrogen (Remy & Vasters 1931). Meanwhile, in the many component meadow phytocoenoses *Anthoxanthum odoratum* usually shows a negative response to nitrogen fertilizers. Its response to nitrogen varies, however with the content of other nutrients in the soils. As an average of the results of numerous experiments, its productivity on the variant "NPK" on meadows with rather fertile soils was 38% and on meadows with poor soils 61% of the productivity on the control variant (Klapp 1962d). Thus, the competitive ability of *Anthoxanthum odoratum* under the influence of NPK was higher on poor soils than on fertile ones.

The results of a pot experiment with a mixture *Anthoxanthum odoratum* + *Lolium perenne* confirms this conclusion. On poor soil the competitive ability of *Anthoxanthum odoratum* was higher than that of *Lolium perenne*, while on fertile soil *Anthoxanthum odoratum* was suppresed by *Lolium perenne* (Berg & Elberse 1962). Thus, *Anthoxanthum odoratum* in single species swards responds positively to NPK and increasing rates of N, but in the many component meadow phytocoenoses under the influence of NPK its participation decreases, the decrease being the greater the more fertile is the soil. On the meadows where fertilizers are applied the competitive ability of species is determined by their ability: 1) to absorb nutrients of fertilizers, which primarily depend on the extent of absorbing roots in the most superficial soil horizon into which fertilizers penetrate; 2) to utilize effectively the absorbed nutrients for building their organs; 3) to grow successfully in the swards with a changed microclimate, especially with decreased light intensity.

Of considerable importance also is the fact that the ability of species to utilize nutrients of fertilizers for building their organs differs. It is known that the aboveground organs of herbs contain approximately half again as much N as grasses; the underground organs of herbs are also richer in nitrogen than those of grasses. Thus, grasses utilize nitrogen more effectively for build-

478

ing their organs than herbs, which undoubtedly increases their competitive ability.

While the competitive ability of plants for nutrients is determined by the extent of absorbing roots in the superficial soil horizon, the extent of leaf area and its vertical distribution plays an important role in the competition for light. The competition for light becames stronger under the influence of fertilizers, as a result of the formation of taller and closer swards. Of particular importance in the competition for light is the ability of a species to develop rapidly a large leaf area and to transfer its foliage upward by growth as the height of swards increases.

The observations carried out in the flood valley of the Oka river illustrate different changes in the leaf area of two biologically similar long rhizomatous species of grasses: *Elytrigia repens* and *Bromus inermis*. Before the application of fertilizers, the participation of *Elytrigia repens* in the sward was higher than that of *Bromus inermis*. Both species responded positively to NPK but the response of *Bromus inermis* was greater. Therefore after long-term application of $N_{60}P_{60}K_{60}$, comparing the NPK plots the controls, the number of tillers of *Elytrigia* increased twice, that of *Bromus* 25 times; the leaf area of Elytrigia increased 3.5 times (15100 as compared with 4300 cm^2/m^2), that of Bromus 62 times (36000 as compared with 580 cm^2/m^2); the volume of the aboveground space where leaves of *Elytrigia* are distributed rose from 600 to 1000 dm^3/m^2, that of *Bromus* from 400 dm^3/m^2 to 1000 dm^3/m^2; the average leaf area per dm^3 of *Elytrigia* increased from 7,2 to 15,1 cm^2/dm^3, that of *Bromus* from 1,5 to 36,0 cm^2/dm^3, the content of chlorophylls of *Elytrigia* rose from 3,4 to 4,1 mg/g fresh substances, that of *Bromus* from 2,4 to 4,7 (Skripko 1971). Thus, in the meadow studied *Bromus inermis* responded to NPK more positively than *Elytrigia repens*, developing larger leaf area, which undoubtedly increased its competitive ability.

The increase of height and closeness of swards as a result of fertilizer application creates for low-growing plants unfourable conditions of light supply, and the shading decreases the ability of these plants to compete for nutrients (Donald 1963). Therefore, many low-growing species of low shade resistance respond negatively to fertilizers, especially to nitrogen in meadows utilized for hay. Many low-growing species respond positively to fertilizers only on meadows with swards where taller species that respond favourably to fertilizers are absent, or where frequent defoliation prevents their long-term shading. Many indicators of infertile soils grow successfully on fertilized meadows if the height of swards is kept low (Lüdi 1959). However, among low-growing meadow species there are shade resistant ones, for example *Glechoma hederacea*, which respond to NPK positively or indifferently. Some species adapt to changed light intensity in swards by increasing their chlorophyll content.

Long-term fertilization of meadows is followed by changes in the floristic composition of phytocoenoses. In the phytocoenoses with moss cover mosses gradually disappear (Rabotnov 1948, Kalda & Kannukene 1966, Kotelina 1967, Makarevich 1968 *et al.*) chiefly as a result of changes in the light climate in closed swards. Low-growing, light-demanding species and other species of low competitive ability, such as orchids, which respond negatively to fertilizers also disappear. Therefore, as a result of long-term application of fertilizers, and especially nitrogen, the number of species in meadow phytocoenoses usually decreases. For example, in Rothamsted experiment, after 90 years of fertilizer application the number of herbaceous species was: on control −36; on PK−30; on NPK (N as $NaNO_3$) −16; on NPK (N as $(NH_4)_2SO_4$) −6 (Brenchley & Warington 1958). Fertilization of meadows is followed by considerable changes in the quantitative relationship of components of the phytocoenoses and often by a change of dominants. The results of 10 years experiment in the flood valley of the Oka river (Table 5) give an idea of the main trends of changes in the meadow phytocoenoses under the influence of fertilizers (Rabotnov 1973). Long-term application of NPK resulted in considerable increase in the grasses (absolute yield more than 3 times: % participation 1,5 times), and decrease of legumes (yield decreased almost twice, % participation more than 3 times). The percentage participation of herbs decreased, but their yield increased 1,5 times.

10.5.2 THE EFFECT OF FERTILIZERS ON LEGUMES

Under the influence of PK, a great increase both in yield and % participation of legumes was observed; the yield of grasses rose 1½ times, while that of herbs did not change; the % participation of grasses did not change, while that of herbs decreased. Similar changes were observed in numerous experiments carried out in different countries (Brenchley & Warinhton 1958, König 1950, Klapp 1962d, 1965, Künzli 1968, Sturm 1958 *et al.*). It has been established that due to increase of the total yields under

480

the influence of fertilizers, it is impossible to determine the responses of species from their percentage participation in swards.

A considerable increase in the participation of legumes when PK (sometimes only P or K) are used is connected with the inability of legumes, in spite of good nitrogen supply by nodule bacteria, to compete successfully with grasses due to the deficiency in soil of available P and K and sometimes S, since legumes have less ramified root systems than grasses (Romashev 1936, 1939, Kydd 1965, Jackman & Mouat 1972). The difference in

Table 5.

Participation of different species and groups of plants in the total yields (for period from 1955 to 1964) of first cut in shortinundated meadow under application PK and NPK (Rabotnov 1973)

	Control yields g/m²	%	PK yields g/m²	%	difference in yields compared with control g/m²	NPK yields g/m²	%	difference in yields compared with control g/m²
Grasses including:	1328	38.5	2006	41.7	678	4390	62.0	3062
Elytrigia repens	132	3.8	228	4.7	96	731	10.3	599
Bromus inermis	151	4.4	359	7.5	208	1196	16.9	1045
Dactylis glomerata	83	2.4	96	2.0	13	206	2.9	123
Festuca pratensis	313	9.1	458	9.5	145	688	9.7	321
Phleum pratense	114	3.3	107	2.2	−07	439	6.2	325
Poa pratensis (angustifolia)	364	10.6	556	11.6	192	685	9.7	321
Legumes including:	587	17.0	1299	27.0	712	325	4.6	−262
Trifolium pratense	127	3.7	256	5.3	129	13	0.2	−114
T.hybridum	49	1.4	90	1.9	41	08	0.1	−41
Lathyrus pratensis	248	7.2	669	13.9	421	232	3.3	16
Herbs including:	1535	44.5	1505	31.3	−30	2365	33.4	830
Achillea millefolium	216	6.3	222	4.6	06	249	3.5	33
Galium verum	59	1.7	97	2.0	38	145	2.0	81
Geranium pratense	69	2.0	94	2.0	25	287	4.1	218
Heracleum sibiricum	25	0.7	23	0.5	−02	145	2.0	120
Silene vulgaris	75	2.2	35	2.0	20	185	2.6	110
Stellaria graminea	51	1.5	32	0.7	−19	190	2.7	139

the cation exchange ability of the roots of grasses and legumes is of essential importance for competition for potassium (Gray *et al.* 1953, Mouat & Walker 1959).

Not all species of legumes respond identically to PK: *Trifolium repens* is one of the species most responsive to PK on pasture. *Trifolium pratense* and *T. hybridium* respond to PK more favourably than *Lathyrus pratensis* and *Vicia cracca* (Rabotnov 1973). However *Lotus corniculatus* in any case under certain conditions shows an indifferent response to PK (Klapp 1962d).

The participation of legumes in meadow swards undergoes considerable fluctuation from year to year. The application of PK does not eliminate these fluctuations. Under the influence of PK the participation of legumes in swards often increases to 30–40% of total yield. But subsequently their yield decreases greatly (Klapp 1957). According to the observations carried out in the USSR, considerable fluctuations in the yields of *Trifolium pratense* and *T.hybridium* are caused by the peculiarities of their life cycles. A better supply with P and K when PK is applied results both in the acceleration of their development and in the transformation of polycarpic individuals into mono-dicarpic ones, and also in the mass transition of weakly developed clover individuals into luxuriantly developed plants with many generative shoots. The yields of clovers increase greatly. However, the luxuriant development of the individuals is followed by their mass dying off as a result of the completion of their life cycle (Rabotnov 1960). New populations of clovers are formed from seeds which are present in soil or enter into the soil surface as a result of self seeding or sowing and subsequently the "clover cycle" repeats itself.

One of the causes of an abrupt decrease of the participation of clovers can be the mass reproduction of parasitic nematodes and fungi. In an experiment on the variant "PK", where the participation of clovers was especially large, the number of individuals of the clover nematode *(Heterodera trifolii)* was 30–50 times greater than on the "control" (Krall 1962). Nematodes decrease the ability of legumes to absorb phosphates (Healy & Widdowson 1973, Widdowson & Jeates 1973) and facilitate penetration of the fungus infection into the roots of legumes. It was shown that by means of nematocides and fungicides, it is possible to increase greatly the yields of *Trifolium repens* (Ennik *et al.* 1964, 1965, 1970, Widdowson *et al.* 1973). In some years the participation of clovers in swards of meadows fertilized by PK is greatly decreased by voles. The fluctuations from year to year are

482

observed not only in the participation of clovers reproducing by seeds, but also in that of legumes reproducing vegetatively (*Lathyrus pratensis, Vicia cracca, V.sepium, Trifolium repens*) and the application of PK does not eliminate these fluctuations (Klapp 1965, 1971, Rabotnov *et al.* 1960, 1973). It was shown in one experiment (pot cultures) that clover nematode (*Heterodera trifolii*) decreased considerably the yields of *Trifolium repens* and *Lathyrus pratensis* (Krall & Riysnere 1966). It is possible that one of the causes of the fluctuations of vegetatively reproducing legumes is the influence of parasitic nematodes.

Different species of grasses respond differently to PK. For example, in the experiment on dry meadow in the flood valley of the Oka river among 8 species of grasses, *Bromus inermis, Poa pratensis* and *Elytrigia repens* responded the most favourably to PK, while on moist meadows the greatest response to PK was shown by *Alopecurus pratensis* and to a lesser extent by *Agrostis stolonifera* (Rabotnov 1973).

The increase of the productivity of grasses under the influence of PK can be due to their better provision not only with P and K but also with N as a result of greater participation of legumes in swards. However, even in the absence of legumes, in some cases with sufficient nitrogen supply some species of grasses respond positively to PK (or P or K alone), replacing other species requiring P or K to a lesser degree. Some cases are known of the transformation of M o l i n i e t u m into A r r h e n a - t h e r e t u m (Klapp 1965).

Under the influence of nitrogen fertilizers, the participation of legumes in swards decreases to the greater degree, the higher are the rates of nitrogen supply and the longer is the fertilization period. On dry meadows the negative influence of nitrogen fertilizers is observed to a lesser extent than on moist ones (Künzli 1967). In the single species stands legumes do not respond unfavourably to nitrogen fertilizers. Their negative response to nitrogen fertilizers in meadow phytocoenoses is caused by decrease of their ability to compete with grasses since with better supply of mineral forms of nitrogen, the ability of nodule bacteria to fix atmospheric nitrogen decreases. Moreover the competition of legumes with grasses for K and P rises as a result of increased absorption of the available forms of potassium and phosphates by grasses due to increase of their yield.

For some species of legumes, in particular, for *Trifolium repens*, on meadows fertilized by NPK the decrease of the light intensity in tall closed swards is of essential importance. Different species of legumes respond to NPK differently. Vegetatively reproducing legumes (*Lathyrus, Vicia*) are more resistant to nitrogen fertilizers than the legumes reproducing only by seeds (*Trifolium pratense, T.hybridum*). The following series with increasing resistance to $N_{60} P_{60} K_{60}$ was observed in the flood valley of the Oka river: *Trifolium pratense* et *hybridum* → *Vicia cracca* → *Lathyrus pratensis* → *Vicia sepium* (Rabotnov & Krylova 1963, Rabotnov 1973).

10.5.3 THE EFFECT OF FERTILIZERS ON GRASSES AND HERBS

Long ago (Lawes *et al.* 1882, Stebler & Schröter 1887) it was established that nitrogen fertilizers (organic, mineral) influence especially favourably many species of grasses and some herbs. On the meadows of Western Europe the following species are the most responsive to nitrogenous fertilizers — the grasses: *Arrhenatherum elatius, Alopecurus pratensis, Trisetum flavescens, Dactylis glomerata, Festuca pratensis, Lolium perenne, Holcus lanatus, Poa trivialis, P. annua, Elytrigia repens;* the herbs: *Anthriscus sylvestris, Heracleum sphondylium, Rumex crispus, R. obtusifolius, Taraxacum officinale* (Brenchley & Warrington 1958, Klapp 1962d, 1965, 1971, Minderhoud *et al.* 1974, Aperdannier 1959 *et al.*) Manure influences especially favourable some herbs: *Heracleum sphondylium, Anthriscus sylvestris, Taraxacum officinale, Rumex obtusifolius* (Klapp *et al.* 1971, Williams 1968, Künzli *et al.* 1967).

In the USSR, depending on the type of meadows, positive response to application of NPK is shown by: *Bromus inermis, Alopecurus pratensis, Poa pratensis, Festuca pratensis, F. rubra, Elytrigia repens, Agrostis stolonifera, Phleum pratense, Polygonum bistorta, Anthriscus sylvestris, Filipendula ulmaria* et al. (Magaeva 1951, 1953, Sannikova 1962, 1965, Zapolsky 1939, Rabotnov 1964, 1973, Lyiv & Khein 1974).

It is possible that the positive response of some species to NPK is not connected with better nitrogen supply. For example according to many observations in different countries in the meadow phytocoenoses NPK influences favourably *Taraxacum officinale* while in single species stands Taraxacum responds only weakly to N, but its yield increased when potassium was applied (Hofer 1970). A high responsiveness of *Taraxacum officinale* to K was also observed in the Rothamsted experiment (Brenchley & Warrington 1958, Thurston 1959 a,b). Therefore, it is possible that the positive response of *Taraxacum officinale* to NPK is due to its high responsiveness to K and also to formation of places suitable for establishment of seedings of *Taraxacum officinale*, which is connected with lesser winter hardness of grasses on meadows fertilized with high nitrogen rates.

10.6 On the Causes of the Different Responses of the same Species on Fertilizers.

The same species can respond differently to the same kind of fertilizers depending on 1) initial composition of swards; 2) quantitative participation of species in the swards; 3) environment; 4) meteorological and hydrological peculiarities of the period of fertilizers; 5) form and rates of fertilizers; 6) form of utilization of meadow. The same species can respond differently to the same fertilizer depending on the participation in the phytocoenoses of species more responsive to fertilizers. For example, *Agrostis tenuis* and *Festuca rubra* in Nardeta on poor acid soils respond favourably to NPK and especially favourably to NPKCa, replacing gradually *Nardus stricta* (Klapp 1951, 1962a, Lüdi 1936, 1959, Milton 1947 *et al.*). On more fertile soils where swards contain tall grasses more responsive to nitrogen (*Phleum pratense, Festuca pratensis, Dactylis glomerata* et al.) *Agrostis tenuis* and *Festuca rubra* respond to NPK negatively. This conclusion is confirmed by the results of numerous (198) experiments (Klapp 1962d). In the phytocoenoses where *Agrostis tenuis* is dominant and where the species more responsive to NPK are absent not only does it preserve its dominance but under the influence of NPK even increases its participation in swards. For example, in the experiment carried out in Serbia on a meadow with pseudogley acid soils (900 m.a.s.l) on the control the participation of *Agrostis tenuis* was 38% of total (413 g/m²) yield, while on application of $N_{150} P_{150} K_{150}$ its participation was 66% of total (957 g/m²) yield. Thus its absolute yield rose from 157 g/m² on the control to 632 g/m² on NPK (Mijatovič & Pavesič-Popovič 1974).

A quantitative participation in swards before fertilization is of importance for species whose competitive ability is decreased as a result of the application of fertilizers. For example, NPK acts favourably on *Bromus erectus* in meadows where this species is dominant and negatively in the phytocoenoses where its participation is insignificant (Klapp 1962d).

485

The environmental conditions including water regime, contents of available nutrients in soil, pH etc. are of considerable importance for the responsiveness of different species to fertilizers. In particular, in Western Europe *Alopecurus pratensis* on moist soils responds to NPK more favourably than *Arrhenatherum elatius*, whereas on less moist soils under the influence of NPK the latter species dominates (Klapp 1965). Therefore, when characterizing the response of species to fertilizers it is necessary to indicate their habitat (Künzli 1967). The responsiveness of species to fertilizers can vary from year to year owing to different meteorological and hydrological conditions. The precipitation, temperature, depth of ground water, inundation etc. are of importance. In the years unfavourable for more demanding tall species, NPK influences positively the low-growing and less demanding species, while in the years with favourable weather conditions such tall grasses as *Arrhenatherum elatius, Dactylis glomerata, Trisetum flavescens* respond positively to NPK (Künzli 1967). The response of some species to fertilizers can change in the years when phytocoenoses are disturbed. For example, NPK influences *Geranium pratense* in the flood valley of the Oka river positively but after an abrupt decrease of the participation of this species in phytocoenoses due to the formation of ice cover, its response to NPK became negative owing to increase of the competitive ability of grasses (Rabotnov 1973). On the contrary, *Ranunculus repens* responded to NPK negatively in closed swards of *Alopecurus pratensis* and *Phalaris arundinacea* and very positively when it became dominant after mass dying off these grasses as a result of long-term stagnation of water on the soil surface (Rabotnov 1966). In the experiments in the flood valley of the Oka river, the long-term application of $N_{60}P_{60}K_{60}$ did not eliminate the changes of yields and quantitative relationships of species from year to year, but on PK and NPK these fluctuations differed from these on the control (Rabotnov 1973). The responses of species to fertilizers can change with increase in the period of the fertilizer application, owing to the changes in the quantitative relationships of phytocoenose components and in their competitive interrelations. Therefore, the results of one- to two-year fertilizations can give erroneous ideas about the responsiveness of species to fertilizers in a particular phytocoenosis. The transformation of plant communities under the influence or regular application of fertilizers occur at different rates, sometimes, for example in Nardeta, very slowly (Marshall 1962). The transformation of phytocoenoses is accelerated by grazing use (Klapp 1951, Milton 1947 *et al.*). In the experiment at the Welsh Experimental Station (Great Britain) on a meadow with dominance (77%) of *Molinia coerulea* and *Nardus stricta*, as a result of liming and grazing after 8 years Molinia and Nardus disappeared and an invasion of *Trifolium repens* was observed, *Festuca rubra* and *F.ovina* became dominant. Only after 16 years a sward, typical for lowland pastures (dominance of *Trifolium repens, Poa annua, P.trivialis, Holcus lanatus* et al.) was formed (Jones 1959).

The responses of species to the same kind of fertilizer can differ depending on whether they are applied singly or in combination with other fertilizers. There are many examples of such differences. In one experiment addition of phosphates to NK was followed both by considerable increase of participation of *Festuca pratensis, Trisetum flavescens, Poa pratensis, Trifolium pratense* and by significant decrease of *Plantago lanceolata, Carum carvi, Ranunculus acris, R.repens* (Moser & Klatz 1962). *Holcus lanatus* was more competitive than *Anthoxanthum odoratum* when nitrogen fertilizer was applied in combination with gypsum, while when nitrogen alone was used *Anthoxanthum odoratum* had an advantage over *Holcus lanatus* (Vartha 1963). In the experiment on pseudogley soil basic slag increased the yield of *Trifolium repens* if potassium was simultaneously applied, while without potassium basic slag decreased the white clover yield (Dhein 1962–1963). *Festuca rubra* (Table 6) responded positively to an increase of nitrogen rates and the number of defoliations but its participation decreased greatly if PK was added to N and the PK rates were increased (Mott 1962).

The importance of different forms of nitrogen fertilizers was demonstrated in the Rothamsted experiment where as a result of long term application of NPK with $(NH_4)_2SO_4$ or $NaNO_3$ as

Table 6.

Changes in participation of *Festuca rubra* in yield (%) depending on number of cuts, rates N and application of PK (Mott 1962).

Number of cuts	Rates N kg/ha	Applied PK		
		without PK	$P_{50}K_{50}$	$P_{100}K_{100}$
2 cuts	40	7.9	3.6	2.5
3 cuts	60	18.3	7.7	7.6
4 cuts	80	36.1	17.7	14.1

source of nitrogen, different swards were formed: in the former variant a floristically poor (6 species) sward with dominance of *Holcus lanatus*, while on the latter variant a floristically richer one (16 species) with dominance of *Dactylis glomerata* and *Arrhenatherum elatius* (Brenchley & Warrington 1958). In this experiment significant acidification of the soil as a result of the application of the physiologically acid $(NH_4)_2SO_4$ was of major importance. However, there are some data which show that some species respond differently to NH_4 and NO_3 as a source of nitrogen (Tyler 1967), though in other experiments no significant difference in the response of some species to NO_3 or NH_4 was observed (Gigon 1971, Gigon & Rorison 1972). The importance of the forms of phophate fertilizers was demonstrated in the experiment carried out on mountain N a r d e t u m in Switzerland. Here in the 8th year of application of fertilizers of participation (% by weight) of some species and groups of species was as follows: on control — *Nardus stricta* — 36,1, other grasses 18,2, legumes 1,5, *Carex sempervirens* 15,7, *Ericales* 6,3; for the variant "application of superphosphate" — 17,4, 41,4, 19,2, 3,1, 0,6 respectively and for the variant "basic slag" 1,6, 65,2, 24,4, 1,9, 0,1 respectively. The yields (g/m² air-dry substance) were: *Nardus stricta* — "control" 33, "superphosphate" 25, "basic slag" 04; other grasses 16,7, 60, 166 respectively; legumes 1,4, 27,8, 62,2 respectively (Koblet *et al.* 1953). Thus basic slag was more favourable than superphosphate in replacing species of little value by those of high forage value.

There are many data on the changes in the responses of species to nitrogen fertilizers used in increased rates. For example, in an experiment with the rates of N up to 720 k/ha, when the rates were increased the participation of *Lolium perenne* and *Trifolium repens* decreased, while the portion of *Elytrigia repens* and *Poa pratensis* was greater (Kreil *et al.* 1964, 1966).

Great differences are observed in the responses of species to fertilizers depending on the form of utilization of meadows (for hay or for grazing). In pastures short grasses (*Lolium perenne, Poa pratensis, Festuca rubra*) respond to NPK especially favourably. The results of numerous experiments with fertilization of meadows show how important is the supply of meadow plants with nutrients especially with nitrogen for determination of the productivity, structure and composition of the meadow phytocoenoses.

10.7   Summary

Results of extensive experiments with fertilization of meadows and pastures in the USSR and Western Europe are summarized. Effects of fertilization are complex and variable, affected by interactions between species and between the community and its biotope. Fertilization increases the number of tillers in single species swards and pastures, but can decrease the number in mixed meadows. Fertilization can leed to increased microbial populations and decomposition in the soil; but N pertilization decreases N fixation by nodule bacteria and decreases competitive ability of legumes. Fertilization can increase seed production by many species, but also the increased growth and shade can lead to decrease or disappearance of short-lived species dependent on reproduction by seeds. Fertilization often produces "mesophytization" of opposite direction in dry and moist meadows. In dry meadows fertilization increases the proportion of less xerophytic species (and increases the participation of grasses in the community), because of the higher competitive ability of the more mesophytic species when they are better provided with N. In moist meadows fertilization can produce a "biological drying," with dominance of the community by less hydrophytic species because the increased water percolation and evapotranspiration accompanying increased productivity make the biotope more suitable for mesophytes. Ecological ranges of species adapted to intermediate moisture conditions may thus increase toward both extremes with fertilization. Responses of species to fertilization can, however, be quite different according to whether they are grown alone or in competition with other species. With long-term fertilization the coverage of mosses, proportion of legumes, and number of vascular plant species generally decrease. Responses of communities are, however, different according to time and form of fertilization, presence or absence of species able to take advantage of the added fertilizer, and meadow use or harvest practice, as well as different characteristics of biotopes.

Bibliography

Alekseenko, L. W., — 1967 — Productivity of meadow plants according to environment. Leningrad.

Amberger, A. — 1955 — Zur Frage des Wasserverbrauches unter dem Einfluss der Düngung. Z. Acker- u. Pflbau 99: 45—54.

489

Antipin, N. A. – 1951 – To the division into districts of application of fertilizers on meadows of the forest zone in the European part of the USSR. *Voprosy kormodobyvania.* 3: 50–58.

Aperdannier, R. – 1959 – Uber die ökologischen Grenzen der Glatthaferwiese (A r r h e n a t h e r e t u m e l a t o r i s) im Vogelsberg. *Z. Acker- u. Pflbau* 107: 371–390.

Berg, van den, J. P. & W. Th. Elberse – 1962 – Competition between *Lolium perenne* L. and *Anthoxanthum odoratum* L. at two levels of phosphate and potash. *J. Ecol.* 50: 87–95.

Bommer, D. – 1964 – Zur Frage von Stickstoffdüngung und Schnitthäufigkeit auf der Wiese. *Landw. Forschung* 17: 252–259.

Bradshaw, A. D. – 1969 – An ecologist's viewpoint. "Ecological aspects of the mineral nutrition of plants", (ed.) J. H. Rorison. 400–427 p. Blackwell, Oxford and Edinburgh.

Bradshaw, A. D., M. J. Chadwick, D. Jowett & R. W. Snaydon – 1964 – Experimental investigations into the mineral nutrition of several grass species. IV. Nitrogen level. *J. Ecol.* 52: 665–676.

Bradshaw, A. D., D. Jowett, R. W. Lodge & R. W. Snaydon – 1960 – Experimental investigations into the mineral nutrition of several grass species. III. Phosphate level. *J. Ecol.* 48: 631–637.

Bradshaw, A. D., R. W. Lodge, D. Jowett & M. J. Chadwick – 1959 – Experimental investigations into the mineral nutrition of several grass species. II. ph and calcium level. *J. Ecol.* 48: 143–150.

Brenchley, W. E. – 1924 – Manuring of grassland for hay. The Rothamsted monographs on agricultural science. London.

Brenchley, W. E. – 1925 – The effect of light and heavy dressings of lime on grassland. *J.min.agric.* 32: 504–512.

Brenchley, W. E. – 1926 – Die Rothamsteder Wiesendüngsversuche von 1856 bis 1919. Berlin

Brenchley, W. E. – 1930 – The varying effect of lime in grassland with different schemes of manuring. *J. min. agric.* 37: 663–673.

Brenchley, W. E. – 1935 – The influence of season and the application of lime on the botanical composition of grassland herbage. *Ann. appl. biol.* 22: 183–207.

Brenchley, W. E. – 1937 – Correlation of manuring and botanical composition of continuous hay crops. *Forth Report Int. Grassland Congr.* 441–446.

Brenchley, W. E. & S. G. Heintze, – 1933 – Colonization by Epilobium augustifolium. *J. Ecol.* 21: 101–102.

Brenchley, W. E. & K. Warington, – 1958 – The Park Grass Plots at Rothamsted, 1956–1949, Harpenden.

Bronzova, G. Ja. – 1940 – Influence of fertilizers on development of root system of meadow grasses and on accumulation in it of carbohydrates. *Vestnik kormodobyvania* 4: 29–38.

Cashen, R. O. – 1947 – The influence of rainfall on the yield and botanical composition of permanent grass. *J. agric. Sci.* 37: 1–10.

Davies, W. & T. E. Jones, – 1932 – The yield and response to manures of contrasting pasture types. *Welsh J. agric.* 8: 170–192.

Demin, A. P. – 1969 – Influence of fertilizers on the morphological structure of underground organs of some meadow species. *Bjull. Mosk. O.J. Prir. O. Biol.* 74 (6): 94–108.

Demin, A. P. – 1970 – Underground mass of meadow vegetation in the flood valley of the Oka river and influence of fertilizers on it. *Bjull. Mosk. O.J. Prir. O. Biol.* 75(6): 79–85.

490

Demin, A. P. – 1971 – Mycotrophy of meadow plants depending on application of fertilizers. *Bjull. Mosk. O.J. Prir. O. Biol.* 76(6): 128–134.

Dhein, A. – 1962–1963 – Einfluss der Phosphat- und Kalidüngung auf Ertrag, Bestandzusammensetzung und Manganschädigung des Grünlandes auf staunassem Boden (Pseudogley). *Z. Acker- u. Pflbau*, 116: 75–87.

Donald, C. M. – 1963 – Competition among crop and pasture plants. *Advanc.Agron* 15: 1–118.

Ellenberg, H. – 1952 – Wiesen und Weiden und ihre standörtliche Bewertung. Stuttgart.

Ellenberg, H. – 1953 – Physiologisches und ökologisches Verhalten derselben Pflanzenarten. *Ber. deutsch. bot. Ges.* 65: 351–362.

Ellenberg, H. – 1954 – Über einige Fortschritte der kausalen Vegetationkunde. *Vegetatio* 5–6: 199–211.

Ellenberg, H. – 1963 – Vegetation Mitteleuropas mit den Alpen in kausaler, dynamischer und historischer Sicht. Stuttgart.

Ennik, G. C., J. Kort & B. Lussink, – 1964 – The influence of soil disinfection with D.D. and some other compounds with nematocidal activity on the growth of white clover. *Neth. J. Pl. path.* 70: 117–135.

Ennik, G. C., C. J. Kort & C. F. van de Bund, – 1965 – The clover cyst nematode (*Heterodera trifolii* Goffart) as the probable cause of death of white clover in a sward. *J. Br. Grassl. Soc.* 20: 258–262.

Ennik, G. C., J. Kort & A. M. van Doorn, – 1970 – Effect of seed and soil disinfectants on establishment, growth and mutual relations of white clover and grass in leys. *Agric. res. reports Wageningen* 741: 1–27.

Ershov, V. V. – 1966 – Influence of fertilizers on microbiological processes in the meadow soils of the Karelia. *Microbiologia* 35 (1): 168–173.

Ershov, V. V. & T. S. Kuzmina, – 1965 – Microbiological processes in meadow soils and their changes under the influence of fertilizers. *Uchen. Zap. Petrozavodsk. Gos. Univers.* 13: 164–170.

Finckh, B. – 1960 – Umbruchlose Verbesserung ertragsarmer Streuwiesen. *Bayer. landw. Jb.* 37: 91–119.

Geyger, E. – 1971 – Green area indices of grassland communities and agricultural crops under different fertilizing conditions. In "Integrated experimental Ecology," ed. H. Ellenberg 68–71. Springer, Berlin. Heidelberg. New York.

Gigon, A. – 1967 – Stickstoff- und Wasserversorgung von Trespen-Halbtrockenrasen (M e s o b r o m i o n) im Jura bei Basel. *Ber. geobot. Inst. Rübel.* 38: 28–85.

Gigon, A. – 1971 – Vergleich alpiner Rasen auf Silikat- und Karbonatboden. *Veröff. Geobot. Inst. Rübel.* 48. 159.

Gigon, A. & I. H. Rorison, – 1972 – The response of some ecologically distinct plant species to nitrate and to ammonium nitrogen. *J. Ecol.* 60: 93–102.

Gliemeroth, G. – 1951 – Der Einfluss von Düngung auf den Wasserentzug der Pflanzen aus den Unterbodentiefen, *Z. Pflanzenern. Düng. Bodenk.* 52: 21–41.

Gray, B., M. Drake, & W. G. Colthy, – 1953 – Potassium competition in grass-legume associations as a function of root cation exchange capacity. *Soil Sci. Soc. Am. Proc.* 17: 235–239.

Gurfel, D. – 1974 – On the microbiological processes in meadow soils under different forms of utilization and intensive fertilization. *Sborn. trud. Eston. nauchno-issled. Instit. zemled. i melioras.* 33: 80–88.

Hall, A. D. – 1905 – Experiments upon grassland mown for hay every year. The, Rothamst. experiments: 150–159.

Healy, W. B., J. B. Widdowson, & G. W. Jeates, – 1973 – Effect of root nematodes on the growth of seedling and established white clover on a yellow-brown loam. *New Zeal. J. Agric. Res.* 16: 70—76.

Hofer, H. — 1970 — Über die Zusammenhänge zwischen der Düngung und der Konkurrenzfähigkeit ausgewählter Naturwiesenpflanzen. Diss. N 4500. E.T.H. Zürich.

Howell, J. H. & G. A. Jung, — 1965 — Cold resistance of Potomac Ochardgrass as related to cutting management, nitrogen fertilization and mineral levels in the plant sap. *Agron. J.* 57: 525—529.

Huokuna, E. — 1974 — Wintering of heavily fertilized grasslands. *XII. Internat. grassl. Congr. Sect. — chemicalization of grassl. farming*: 218—223.

Jackman, R. H. & M. C. H. Mouat, — 1972 — Competition between grass and clover for phosphate. II. Effect of root activity efficiency of response to phosphate and soil moisture. *New Zeal. J. agric. res.* 15: 667—673.

Jones, L. I. — 1960 — Grassland agronomy. *Rep. Welsh Plant breeding Station* for 1959: 43—51.

Jones, T. E. — 1934 — The influence of manuring on the yield and botanical composition of lowland pastures: A, under controlled grazing by sheep. B, under hay conditions. *Welsh J. agric.* 10: 223—235.

Jung, G. A. & R. E. Koscher, — 1974 — Influence of applied nitrogen and clipping treatments on winter survival of perennial cool-season grasses. *Agron. Journ.* 66: 62—65.

Kalda, A. & L. Kannukene, — 1966 — Changes in the moss cover due to the fertilization of meadows. *Izvest. Acad. Nauk Eston. SSR* 15, Ser. biol. 1: 46—60.

Kirillova, V. P. — 1968 — Mycotrophy of plants in shortgrassherb community. *Trudy Bot. Inst. Akad. Nauk USSR, Ser. III, Geob.* 18: 223—240.

Kirkwood, R. C. — 1964 — A study of some factors causing mat formation of reseeded hill pasture swards. *J. Brit. Grassl. Soc.* 19: 387—395.

Klapp, E. — 1931 — Über den allgemeinen Düngungserfolg auf Wiesenland. *Ernähr. d. Pflanze* 27: 321—329.

Klapp, E. — 1951 — Borstgrasheiden der Mittelgebirge, Entstehung, Standort, Wert und Verbesserung. *Z. Acker- u. Pflbau* 93: 400—444.

Klapp, E. — 1957 — Einfluss der Kalidüngung auf der Zustand und die botanische Zusammensetzung des Grünlandes. *Potassium Symposium*: 101—111.

Klapp, E. — 1962a — Einige Grundlagen der Verbesserung von Naturgrünland im Gebirge. *Bericht über d. Europ. Konf. f. Naturfutterbau in Berglagen.* Chur: 63—70.

Klapp, E. — 1962b — Ertragsfähigkeit und Düngungsreaktion von Wiesenpflanzen-Gesellschaften. *Z. Acker- u. Pflbau* 114: 81—98.

Klapp, E. — 1962c — Ertragsteigerung und Wasserverbrauch landwirtschaftlicher Kulturen. *Z. Kulturtechn.* 3: 1—5.

Klapp, E. — 1962d — Über das Verhalten der Wiesenpflanzen bei verschiedener Düngung unter besonderer Berücksichtigung der Stickstoffwirkungen von Düngung und Standort. *Bayer. Landwirtsch. Jb* 39: 515—527.

Klapp, E. — 1965 — Grünlandvegetation und Standort nach Beispielen aus West-, Mittel- und Süddeutschland. Berlin-Hamburg.

Klapp, E. — 1971 — Wiesen und Weiden. Eine Grünlandlehre 4 Aufl. Berlin u. Hamburg.

Klesnil, A. & F. Turek, — 1974 — The effect of increased N, P and K rates on the botanical composition, yielding capacity and hay quality in conditions of mesophytic and mesohydrophytic meadows. *XII Intern. Grassl. Congr. — Chemicalization of grassl. farming*: 269—275.

Kmoch, H. G. — 1952 — Über den Umfang einiger Gesetzmässigkeiten der Wurzelmassenbildung unter Grasnarben. *Z. Acker- u. Pflbau* 95: 363—380.

Koblet, R., Frey, E. & F. Marschall – 1953 – Untersuchungen über die Wirkung der Düngung auf Boden und Pflanzenbestand von Alpweiden. *Landw. Jahrb. Schweiz*, 67: 597–658.

Könekamp, A. – 1930 – Berichte über die Tätigkeit des Instituts für Grünlandwirtschaft. *Landw. Jahrbuch*. 72 *Erg. Bände.*: 35–6I.

Könekamp, A. H. & F. Weise, – 1961 – Bakterienleben unter Dauergrünland. "Neue Ergebnisse futterbaulicher Forschung". Frankfurt/Main, 22–31.

König, F. – 1950 – Die Rolle der Nährstoffversorgung bei der Leistungssteigerung der Wiese. *Landw. Jahrb. Bayern*, 27. Sonderheft, 3–209.

König, F. – 1955 – Die Sprache der Grünlandpflanzen.

Kotelina, N. S. – 1967 – Dynamics of meadow vegetation in valley of the Vychegda river. Leningrad.

Krall, E. L. – 1962 – On the experimental study of the oscillation of populations of some Heteroderids in the soils of Estonia. *Tezisy dokl. nauchn. konfer. Vsesoyuzn. obtsh. gelmintol.*

Krall, E. & U. Riysnere, – 1966 – Investigation of host-parasite relationships of Heterodera trifolii on legumes. *Izvest. Acad. Nauk Eston. SSR XV Ser. biol.* 1: 83–89

Krause, W. & A. Müller, – 1973 – Möglichkeiten und Grenzen einer Nutzungsintensivierung auf Dauerwiesen in Abhängigkeit vom Standort. *Wirtschaftseig. Futter* 19: 174–194.

Kreil, W., H.Kaltofen, & G. Wacker, – 1964 – Weitere Versuchungsergebnisse über die Düngung einer Weide mit verschiedenen hohen N-Gaben (1961–1963). *Z. Landeskultur* 5: 221–244.

Kreil, W., G. Wacker, H. Kaltofen & E. Hey, – 1966 – The effects of heavy applications of nitrogen fertilizer on the yield, botanical and chemical composition of pasture grass. – Nitrogen and Grassl. *Proc. 1st Gen. Meeting Europ. Grassl. Feder.*: 127–132.

Kreuz, E. – 1965 – Wasserhaushalt und Wasserregulierung auf dem Grünland in ihrer Wirkung auf Boden und Pflanze. – *Schriftenr. Deutsch. Agrarwiss. Ges., Bezirksverband Leipzig* 5: 63–77.

Künzli, W. – 1967 – Über die Wirkung von Hof- und Handelsdüngern auf Pflanzenbestand, Ertrag und Futterqualität der Fromentalwiese. *Schweiz. landw. Forsch.* 6: 34–130.

Künzli, W. – 1968 – Pflanzenbestand und Ertrag der Fromentalwiese in Abhängigkeit vom Standort und Düngung. *Arbeit a.d. Gebiete d. Futterbaues* 10: 9–27.

Kuntze, H. – 1965 – Die Marschen schwererer Böden in der landwirtschaftlichen Evolution.

Kydd, D. D. – 1965 – Fertilizer and management responses on permanent downland pasture grazed by cattle. *Journ. Agric. Sci.* 64: 335–342.

Lampeter, W. – 1959–60 – Gegenseitige Beeinflussung höherer Pflanzen in Bezug auf Spross- und Wurzelwachstum, Mineralstoffgehalt und Wasserverbrauch untersucht an einigen wirtschaftlich wichtigen Futterpflanzen. *Wiss. Z. Karl-Marx-Univers. math. -nat. Reihe* 4: 611–722.

Lawes, J. B. & J. H. Gilbert, – 1880 – Agricultural, botanical and chemical results of experiments on the mixed herbage of permanent meadow, conducted for more than twenty years in succession on the same land. Part 1. *Philos. Transact. Roy. Soc. London*, 171: 289–416.

Lawes, J. B. & J. H. Gilbert – 1890 – Agricultural, botanical and chemical results of experiments on the mixed herbage of permanent grassland. Part III. The chemical results. *Philos. Transact. Roy. Soc. London* 192: 139–210.

Lawes, J. B., J. H. Gilbert, & M. T. Masters, – 1882 – Agricultural, botanical and chemical results of experiments on the mixed herbage of permanent meadow conducted for more than twenty years in succession on the same land. Part II. The botanical results. *Philos. Transact. Roy. Soc. London* 173: 1181–1413.

Lebedev, P. V. – 1968 – Pecularities of morphogenesis of meadow grasses as a function of factors of environment. *Uchen. zap. Ural. Gos. Univers.* 73, *Ser. biolog.* 4: 155–210.

Lopatin, V. D. & V. A. Zaykova, – 1966 – Analysis of changeability of meadows and prognosis of effectiveness of fertilizers with the use of Sukachev's principle of ecological-phytocoenotical series. *Bot. Zh.* 51: 309–321.

Lüdi, W. – 1936 – Experimentelle Untersuchungen an alpiner Vegetation. *Ber. Schweiz. Bot. Ges.* 46: 632–681.

Lüdi, W. – 1959 – Versuche zur Alpweidenverbesserung auf der Schynigen Platte bei Interlaken. Interlaken: 1–8.

Lyiv, Ja. & V. Khein, – 1974 – Changes of floristical composition and productivity of natural meadows under changes of environment. *Sborn. nauch. trudov Eston. nauchn. -issled. Inst. zemled. i. melioras.* XXXIII: 33–43.

Lyiv, Ja. & Kh. Krall, – 1974 – Formation of swards and productivity of perennial cultivated meadow under fertilization of PK and NPK on sod-gley soils of Saaremaa Isle as a function of different methods of establishment and utilization. *Sborn. nauch. trudor Eston. nauch. -issled. Inst. zemled. i melioras.* XXXIII: 44–64.

Magaeva, A. D. – 1953 – Yields of grasses in the valley of the Oka river under different rates of nitrogen fertilizers. *Bot. Z.* 38: 805–816.

Magaeva, A. D. – 1954 – Influence of nitrogen fertilizers on changes of botanical composition and yields of natural meadows of flood valley. *Voprosy proisvodstva kormov* 4: 32–43.

Makarevich, V. N. – 1968 – On the influence of different treatments on the participation of some species in the sward of short-grass-herb meadow. *Problemy Bot.* X: 228–238.

Marschall, F. – 1961 – Die Grundlagenforschung im Naturfutterbau mit besonderer Berücksichtigung der schweizerischen Verhältnisse. *Ber. Europ. Konf. Naturfutterbau in Berglagen.* Chur: 15–37.

Marschall, F. – 1964 – Mineraldüngung und botanische Zusammensetzung der Grasnarbe. *Das Kalium und Qualität landw. Produkte*: 45–54.

Mijatović, M. & J. Pavesič-Popovič, – 1974 – The effect of high rates of NPK fertilizers on the yielding capacity and quality of some native meadows in the hill-mountain region of the S.R. of Serbia – *XII. Intern. Grassl. Congr. – Chemicalization of grassl. farming*: 404–413.

Milton, W. E. J. – 1934 – The effect of controlled grazing and manuring on natural hill pastures. *Welsh J. Agric.* 10: 196–211.

Milton, W. E. J. – 1934 – The relative palatability of seed mixtures and a study of the influence of fertilizers on natural hill pastures. *Emp. J. exp. agric.* 2: 51–64.

Milton, W. E. J. – 1938 – The composition of natural hill pastures under controlled and free grazing, cutting and manuring. *Welsh J. agric.* 14: 182–195.

Milton, W. E. J. – 1940 – The effect of manuring, grazing and cutting on the yield, botanical and chemical composition of natural hill pastures. 1. Yield and botanical section. *J. Ecol.* 28: 326–356.

Milton, W. E. J. – 1947 – The yield, botanical and chemical composition of natural hill herbage under manuring, controlled grazing and hay conditions. I Yield and botanical section. *J. Ecol.* 35: 65–89.

Moser, G. & L. Klotz, – 1962 – Phosphorsäure-Wirkung auf Dauergrünland. Ergebnisse langjähriger Hyperphos-Versuche auf Wiese und Weide. *Bayer. Landw. Jb.* 39: 337–346.

Mott, N. – 1962 – Der Einfluss der Schnitthäufigkeit auf Ertrag und Pflanzenbestand der Fuchsschwanzwiese bei unterschiedlicher N- und PK-Düngung. *Bayer. Landw. Jb.* 39: 311–336.

Mouat, M. C. H. & T. W. Walker, – 1955 – Competition for nutrients between grasses and white clover. II. Effect of root cation-exchange capacity and rate of emergence of associated species. *Plant a. soil.* XI: 41–52.

Onoshko, B. D. – 1934 – Necessity of meadow and peat soils in fertilizers. *Trudy Vsesoyuz. nauch.-issledov. Inst. Kormov* 2: 3–53.

Onoshko, B. D. – 1936 – Ferilization of meadows and pastures. Moscow.

Rabotnov, T. A. – 1947 – Formation of natural swards for seed production of legumes, *Selektsia a. semenovodstvo* 1: 65–66.

Rabotnov, T. A. – 1948 – On the influence of mineral fertilizers on vegetation of subalpine meadows. *Bot. Zh.* 33: 475–486.

Rabotnov, T. A. – 1950 – Some problems of study of meadow swards' structure. *Bjul. Mosk. O. J. Prir. O. Biol.* 55: 50–71.

Rabotnov, T. A. – 1960 – Some problems in increasing the production of leguminous species in permanent meadows. *Proceed. 8th Intern. Grassl. Congr.*: 260–264.

Rabotnov, T. A. – 1965 – On the dynamics of structure of poly-dominant meadow phytocoenoses. *Bot. Zh.* 50: 1396–1408.

Rabotnov, T. A. – 1966 – Some problems of nitrogen fertilization in flood meadows. Nitrogen and Grassland. *Proceed. 1st general meeting Europ. Grassl. Feder. Wageningen*: 109–119.

Rabotnov, T. A. – 1970 – Some main principles governing influence of fertilizers on meadow plants and meadow phytocoenoses. *Trudy Mosc. Obtsh. Isp. Prir.* 38: 137–153.

Rabotnov, T. A. – 1973 – Influence of mineral fertilizers on meadow plants and meadow phytocoenoses. Moscow, Nauka.

Rabotnov, T. A. – 1974 – Ecology of meadows. Moscow.

Rabotnov, T. A. & A. P. Demin, – 1971 – Number of adventitious roots as parameter of structure of meadow phytocoenoses. *Doklady Akad. Nauk USSR* 201: 246–249.

Rabotnov, T. A. & A. P. Demin, – 1974 – Effect of longterm fertilization on the underground parts of meadow plants and phytocoenoses. *II. Intern. Symp. Ecol. Physiol. Root Systems*: 243–246.

Rabotnov, T. A. & N. P. Krylova, – 1963 – Influence of fertilizers on legumes in sward of short inundated meadow. *Bjul. Mosk. O. J. Prir. O. Biol.* 68 (6): 68–76.

Remy, Th. & J. Vasters, – 1931 – Untersuchungen über die Wirkung steigender Stickstoffgaben auf Rein- und Mischbestände von Wiesen- und Weidenpflanzen. *Landw. Jb.* 73: 521–602.

Richardson, H. L. – 1938 – The nitrogen cycle in grassland soils: with especial reference to the Rhothamsted Park Grass experiment. *J. agric. Sci.* 28: 73–121.

Romashev, P. J. – 1936 – Utilization of legumes' nitrogen by grasses in mixtures. *Khimiz. soc. zemled.* 11: 28- 40.

Romashev, P. J. – 1939a – Utilization of fertilizers by plants in grass-legumes mixtures. *Khimiz. soc. zeml.* 7: 35–39.

Romashev, P. J. – 1939b – Utilization legumes nitrogen in the grassland management. *Pochvovedenie* 4: 99–113.

Romashev, P. J. – 1941 – Effectiveness of nitrogen, phosphate and potassium fertilizers on meadow. *Vestnik Kormodobyvania*, 1: 14–30.

Romashev, P. J. – 1949 – Fertilization of meadows and pastures. Moscow.

Romashev, P. J. – 1963 – Fertilization of meadows and pastures. In: "Prirodn. senokosy i pastbitsha", ed. J. V. Larin, Moscow.

Rumberg, C. B. & C. S. Cooper, – 1961 – Fertilizer-induced changes in botanical composition, yield and quality of native meadow hay. *Agron. Journ.* 53: 255–258.

Sannikova, T. J. – 1963 – Influence of fertilizers on botanical composition of sward of dry meadow. *Bot. Zh.* 48: 631–687.

Sannikova, T. J. – 1965 – Change of vegetation of wet meadows in flood valley of the Seim river under application of fertilizers. *Bot. Zh.* 50: 361–367.

Schulze, E. & H. Mues, – 1961 – Ertragsleistung, Pflanzenbestand und Bewurzelung einer Grasnarbe bei verschiedener Düngungsweise. *Zeitschr. Acker- u. Pflbau*, 112: 141–160.

Schwarz, R. – 1932 – Untersuchungen über den Wasserverbrauch verschiedener Gräser. *Arch. Pflbau* 8: 276–334.

Siebold, M. – 1958 – Der Einfluss langjähriger statischer Düngung auf Pflanzenbestand, Ertrag und Futterwert von Dauerwiesen. *Bayer. landw. Jahrb.* 35 (3): 3–66.

Skripko, G. S. – 1971 – Influence of fertilizers on growth and distribution of leaf area. *Vestnik Mosk. Univers., biolog. i pochvoved.* 1: 65–72.

Smelov, S. P. – 1947 – Biological foundations of meadow management. Moscow.

Smelov, S. P. & K. T. Terekhova, – 1957 – Utilization of natural swards for production of grasses' seeds. *Bjul. nauchn-tekhn. inform. Vsesoyuzn. Inst. Kormav* 2–3: 30–39.

Smelov, S. P. & K. T. Terekhova, – 1959 – Growing of seeds of herbaceous plants on natural meadows. *Vestnik selskokhoz. nauki* 3: 47–51.

Smith, D. – 1964 – Winter injury and the survival of forage plants. *Herb. Abstr.* 34: 203–209.

Speidel, B. – 1962 – Die Artenkombination als Massstab für die Ertragsleistung hessischer Mittelgebirgswiesen. *Ber. Europ. Konf. Naturfutterbau in Berglagen*, Chur: 71–78.

Speidel, B. – 1966 – Änderungen des Pflanzenbestandes von Dauerwiesen bei langjähriger Düngung. – *Bayer. landw. Jahrbuch* 43: 214–222.

Speidel, B. & A. Weiss, – 1971 – Primary production of a meadow (Trisetum flavescentis hercynicum) with different fertilizer treatments – Preliminary Report, in: Integrated experimental ecology, ed. H. Ellenberg, 61–67. Springer, Berlin. Heidelberg. New York.

Speidel, B. & A. Weiss, – 1973 – Zur ober- und unterirdischen Stoffproduktion einer Goldhaferwiese bei verschiedener Düngung. *Angew. Botanik* 46: 75–93.

Stebler, F. G. & C. Schröter, – 1887 – Beiträge zur Kenntnis der Matten und Weiden der Schweiz. II Untersuchungen über den Einfluss der Düngung auf die Zusammensetzung der Grasnarbe. *Landw. Jb. Schweiz* 1: 93–148.

Stebler, F. G. & C. Schröter, – 1892 – Versuch einer Übersicht über die Wiesentypen der Schweiz. *Landw. Jb. Schweiz* 6: 95–212.

Sturm, H. – 1958 – Zehnjähriger Wiesendüngungsversuch auf Oberbayerischen Niedermoor. *Bayer. Landw. Jahrbuch* 35: 530–543.

Tamm, C. O. – 1954 – Some observations on the nutrient turnover in a bog community dominated by *Eriophorum vaginatum* L. *Oikos* 5: 189–194.

Thurston, J. M. – 1869a – The effect of liming and fertilizers on the botanical composition of permanent grassland and on the yield of hay. In: "Ecological aspects of the mineral nutrition of plants", ed. J. N. Rorison 3–9. Blackwell, Oxford a. Edinburgh.

Thurston, J. M. – 1969b – The effect of liming and fertilizers on the botanical composition of permanent grassland. In: "Experimentelle Pflanzensoziologie", ed. R. Tüxen 58–61.

Tyler, G. – 1967 – On the effect of phosphorus and nitrogen, supplied to Baltic shore-meadow vegetation. *Botan. Notiser.* 120: 433–447.

Vartha, E. W. – 1963 – The effects of different levels of Nitrogen und Sulphur on the yield and composition of tussock grassland. *New Zeal. Journ. Agric. Res.* 6: 47–55.

Vorhauer, H. F. – 1958 – Über das Verhalten der Pflanzenbestände und die Entwicklung der Heuerträge von Wiesen und Mähweiden in der Eifel und im Bergischen Land bei steigender Düngungsintensität". *Z. Acker- u. Pflbau* 105: 193–210.

Vries de, D. M. & A. A. Kruijne, – 1960 – The influence of nitrogen fertilization on the botanical composition of permanent grassland. *Stickstoff* 4: 26–36.

Warren, R. G. & A. E. Johnston, – 1964 – The Park Grass experiment. Rothamst. exp. station report for 1963: 240–262.

Widdowson, J. P. & G. W. Jeates & W. B. Healy, – 1973 – The effect of root nematodes on the utilization of phosphorus by white clover on a yellow-brown loam. *New Zeal. J. Agric. Res.* 16: 77–80.

Williams, J. T. – 1968 – The nitrogen relations and other ecological investigations on wet fertilized meadows. *Veröffentlich. geobotan. Inst. Rübel* 41: 71–193.

Wind, G. P. – 1954 – The influence of nitrogen fertilizing on water consumption of grassland. Europ. Grassland Conf. EOOC: 211–214.

Zürn, F. – 1961 – Der Einfluss mineralischer und organischer Düngung auf Ertrag, Pflanzenbestand und chemische Zusammensetzung des Futters von Portionsweiden. *Bayer. landw. Jahrb.* 38: 915–944.

LEAF AREA AND PRODUCTIVITY IN GRASSLANDS

ERIKA GEYGER

Contents

11.1   General relationship among some growth parameters

In searching for causal analyses of the productivity of plants, scientists of various branches of biology have endeavoured to find adequate methods. The basic process of organic matter production, photosynthesis, has become an increasingly interesting target of quantitative investigation; Šesták, Čatský & Jarvis (1971) report on at least nine different methodical approaches. Many of them include measurement of the assimilatory surface, as it is the essential part of the plant body in CO_2-exchange and sun energy fixation. These invisible processes can be detected by the presence of intact Chlorophyll. Active green (chlorophyllous) surfaces are found mainly spread out in flat leaves, but are also present on stems, leaf sheathes, buds, fruits and even under the bark of branches. The amount of green surfaces can be expressed in terms of the ratio of foliage area to unit land area, the "leaf area index (LAI)", introduced by Watson (1947).

Leaf area plays the principle role in intercepting light for photosynthesis. Optimal density of leaf area is achieved by leaf arrangement in different layers and orientation of leaves at different inclination angles, the "architecture of the photosynthetic system", "foliage architecture" or "stand structure". Thus, light is distributed with decreasing intensity from the top to the bottom of the stand until it is almost entirely lacking near the ground.

The reactions of the plants on decreasing light intensity are complicated. On one hand the morphological properties may change from "sun-leaf-structure" to "shade-leaf-structure". On the other hand the activity of the leaf surface in energy fixation, and, hence, organic matter production, may change as well. In order to express this activity quantitatively, the ratio between organic matter production and leaf area is used, the "unit leaf rate" or "leaf efficiency" (E); (Šesták, Čatský & Jarvis, 1971). In terms of CO_2-uptake, other expressions have been frequently used: "net assimilation rate" (NAR) or "net production rate"

* Including results of Solling Project of DFG (IBP), Comm. 166.

(NPR, P_{net}). Detailed measurements are necessary to determine the differences in leaf efficiency among the various species and among the leaves in different layers of the stand. On the whole, however, a certain amount of biomass is produced in a certain period of time by the total sum of green surfaces per unit of land: the "crop growth rate" (CGR). The quotient of CGR and LAI, then, yields the average leaf efficiency of the community within the given period of time.

Because of the changing light conditions in the stand with increasing LAI and during the growth period this ratio is not constant. Research results have led to two divergent concepts (see Fig. 1).

Growth rate of the community (CGR) increases parallel to LAI up to a certain level. According to concept 1 (dotted line), it reaches an optimum, while curve 2 attains a plateau. Brown & Blaser (1968), as well as Tooming & Kallis (1972), and Loomis & Gerakis (1975), present comprehensive discussions of this problem. All of them conclude that recent research attributes more

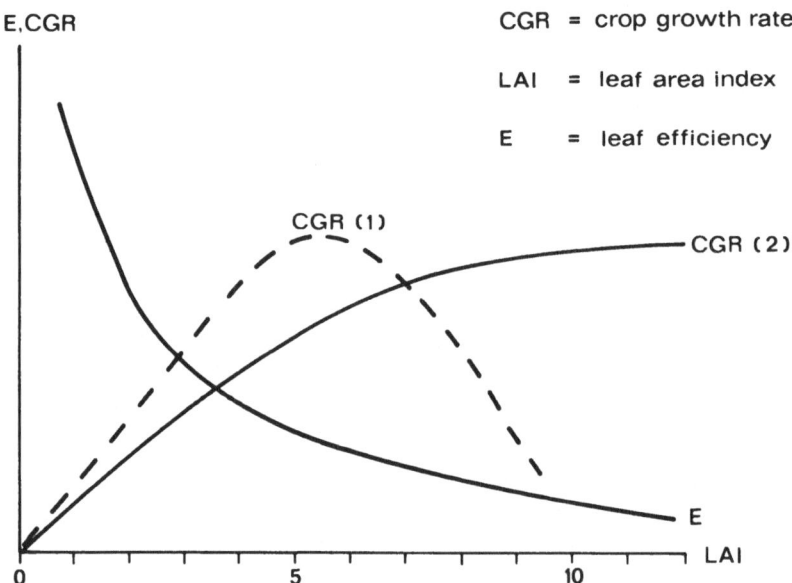

Fig. 1. The relationship of average leaf efficiency (E), LAI and crop growth rate CGR (1) and (2), according to divergent concepts. (from Brown & Blaser 1968, Baeumer 1973).

502

validity to the second concept. E decreases, in any case, with increasing LAI, as a consequence of light reduction due to shade.

Proceeding from CGR to yield production, time must be taken into account. This is done for leaf area with the term "leaf area duration" (LAD), the integral of LAI over the growth period, introduced by Watson (1947).

It seems obvious that high productivity would occur when leaf efficiency (E), total amount of active green area (LAI), and length of time of leaf activity (LAD) are optimally combined, in congruence with the specific combination of the external conditions.

11.2   Leaf area determination

Methods for measuring green surface area have been summarized by Květ & Marshall (1971) with a comprehensive reference list. LAI can be determined by measuring green surface area of part of the green biomass harvested on a certain unit of land and calculated as ratio of total dry matter. LAD is calculated as integral after repeated determinations of LAI at certain time intervals. Non-destructive measurement may be done by the "inclined point quadrat method" (Warren-Wilson, 1963). Measurement of the "cumulative LAI" from top to bottom of the stand is done by "stratified clipping" (Monsi & Saeki, 1953). Graduated leaf area indices can be artificially produced by placing plants – grown up separately in pots – together in different degrees of density (Munakata et al., 1969).

When calculating LAI, it must be decided which plant surfaces should be included. Due to the close relationship of leaf area with light interception (prevailing sunrays hit the leaf on one side only), most authors now use one side of the flat leaves in their calculations (as proposed by Newbould, 1967, for IBP programs). Others (e.g. Geyger, 1971, Rychnovská, Gloser & Petrík, 1973) add the lateral projection of green three-dimensional organs. Warren-Wilson (1963) proposed using the term "foliage area index" instead of LAI when the surfaces of such organs are included in the calculations. For this case I previously used the term "green area index (GAI)" (Geyger, 1971), but meanwhile I prefer the well introduced term LAI, even when surfaces other than leaves are measured. Decisions on the real assimilatory activity of the various green plant surfaces are quite complicated and cannot be discussed in this paper (see Spedding, 1971).

Full light only reaches the leaves of the top layers of a plant stand; light intensity, then, decreases through absorption or reflection when penetrating the various leaf layers downwards.

The respective gradient is expressed as "light extinction coefficient" (K) (Monsi & Saeki, 1953). It is greatly affected by the inclination angles of the leaves: broad, horizontally positioned leaves of dicotyledonous communities absorb much light with only a few leaf layers; the extinction coefficient is high (see Fig. 2). Inversely, K is low in communities with vertically inclined leaves, because light is distributed more uniformly among various leaf layers. As a consequence, leaf area must be larger until all the light is intercepted. Grasslands or grass crops reach LAI values of 10–15 in the field, whereas leaf area indices of up to 36 can be produced experimentally in monocotyledonous communities, as was shown with *Gladiolus* and *Scilla* by Blackman (1968). Monocultures of clover only reach leaf area indices of 3–6 (Brougham, 1958), of about 9 (Alekseenko, 1969) or of maximally 11 (McCree & Troughton, 1966).

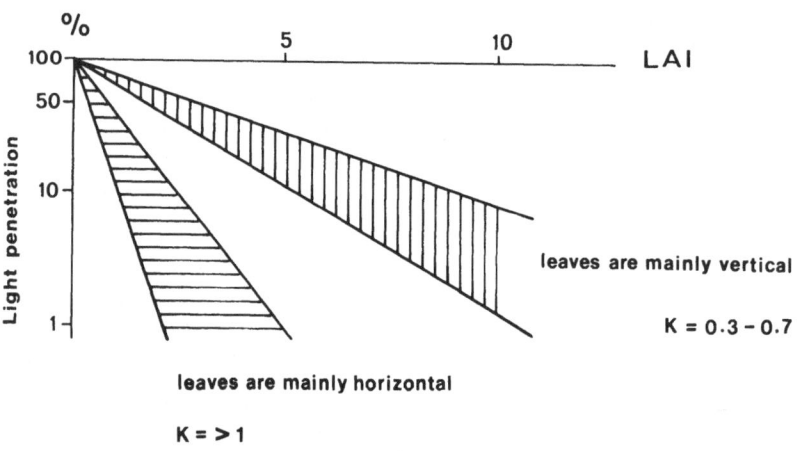

K   = light extinction coefficient

LAI = cumulative leaf area index

Fig. 2. Effect of leaf angles on light extinction (from Saeki 1963, Brown & Blaser 1968, Larcher 1973).

11.4 Adaptation of foliage architecture and LAI for achievement of better production

The relevance of the spatial arrangement of the leaves and especially of leaf angles has been investigated by many authors. Alekseenko (1969) found in experiments with grass and clover species the following LAI maxima:

Monocultures	LAI	Mixtures	LAI
Trifolium pratense	9.1	*Trifolium + Dactylis*	18.5
Dactylis glomerata	9.8	*Trifolium + Phleum*	9.8
Phleum pratense	7.5		

The high LAI of the first mixture is possible because the two species concentrate their leaves in different layers: *Dactylis glomerata* in the lower strata, *Trifolium pratense* at a height of 40–50 cm. *Phleum* competes with *Trifolium* in the same layer. Brown & Blaser (1968) summarize from results of various authors that higher production occurs in communities with vertical orientated leaves. Kuroiwa (1968) has tested with simulation experiments that photosynthesis is highest when the leaves in the upper strata are inclined nearly vertically until a cumulative LAI of 6 has been reached. The adjacent lower leaves must have horizontally orientated leaves and add another 3 LAI to the total.

Various authors (e.g. Brown & Blaser, 1968; Thorne, 1971) proposed breeding new varieties with erect leaves in order to get better yields. Sheehy & Cooper (1973) tested various varieties of the genera *Dactylis, Festuca* and *Phleum*, and proposed the choice of those with prevalently vertical leaf inclination, reaching maximal leaf area indices of up to 15. Tsuno & Kitakado (1969) tested two rice varieties and found that "the straight-leaf variety had a higher optimum leaf area index than the bent-leaf variety", especially "under low light conditions". Brown & Blaser (1968) reported on experiments by various authors and summarized that plants often tend to put their leaves in more vertical position even when only density was increased. This happened with *Medicago sativa, Trifolium subterraneum, Hordeum vulgare* and *Dactylis glomerata*.

De Wit (1968), however, concluded from his simulation experiments and their congruence with field results in the Netherlands that variation in the stand structure has often a disappointingly small influence on net production. The same has been found by Ross (1970), who included different sun elevations in his simulation experiments and stated that stand structure has only little influence on productivity at simulated elevations of 35°-45°. Another statement was that little effect of leaf angles was observed with LAI values between 2 and 3, which is true for most crops of the temperate zone. At higher sun elevations and with higher LAI values, however, well-adapted crop architecture could improve total photosynthesis by as much as 30%. Similar findings are reported by Loomis & Gerakis (1975), who stated that the "erect-leaf-hypothesis" is only valid for high leaf area indices.

On a world scale, Loomis, Williams & Duncan (1967) drew some principle conclusions: "1. In low insolation environments, or with low solar elevations, high efficiency will be obtained with low leaf angles and with low leaf area indices. 2. With high insolation and high solar elevation, greatest efficiency will be achieved with a higher leaf area index and higher leaf angle, particularly in the upper strata".

11.5   Effects of altitude on LAI and yield in some grasslands in Central Europe

Radiation differs not only with geographical latitude, but also with altitude above sea level. A great effect of this site variable on yield production was observed in mountain pastures by Caputa (1966), reported by Dietl in this volume. For comparison I'll try to analyse the effect of altitude on LAI, yield and efficiency. This is not easily done, since numerous other site conditions vary as well, when going from lowlands to mountain regions. Looking only at grasslands under "natural" conditions, LAI does not at all reflect altitude, as is demonstrated through some examples of grasslands going from NW- to SW-Germany (Geyger 1964, 1967, 1971):

Community	Site	Altitude	LAI
Inundated meadows	Elbe marsh	0-10 m	3-13
Permanent meadow	Solling Region	460 m	2-4
Mountain pasture	Menzenschwand Valley	1000 m	4-5
Mountain pasture	Feldberg	1500 m	1-2

Reasons for this variability are numerous. At the given sites, at least one of the natural site variables, nutrition, could be compensated for through fertilization. Comparison shall be done with sites which were fertilized either naturally or artificially: One site was naturally fertilized through short-term inundation (Elbe marsh), four others were fertilized with 120/200 kg N/ha per year and optimal doses of P and K (Speidel & Weiß, 1972, Caputa, 1966).

My investigations concern the following grassland communities:

1. a permanent meadow at the river Gose-Elbe near Hamburg (5 m altitude), a C i r s i u m o l e r a c e u m - A r r h e n a - t h e r u m e l a t i o r -community with *Filipendula ulmaria* and *Glyceria maxima*.

2. a permanent meadow in the Fulda Valley near Hersfeld (200 m altitude), belonging to the order A r r h e n a t h e r e - t a l i a , but lacking association character-species, as it is situated too low for mountain species, but has a climate unfavourable for A r r h e n a t h e r e t u m species (Speidel, unpublished information); an additional impoverishment, here floristically speaking, is due to high, long-term doses of nitrogen. The main grasses are *Festuca rubra, Alopecurus pratensis* and *Poa pratensis*.

3. a permanent meadow on a slight north-western slope in the Solling (460 m altitude). Association: T r i s e t e t u m f l a - v e s c e n t i s h e r c y n i c u m (Speidel 1970). Main grass species are *Festuca rubra, Agrostis tenuis, Holcus lanatus, Anthoxanthum odoratum*.

4. a mountain pasture in a protected valley near Menzenschwand, Black Forest (1000 m altitude). Phytosociologically speaking, it changed during the 4—5 years of the fertilization experiment from a F e s t u c o - G e n i s t e l l e t u m to a highly productive F e s t u c o - C y n o s u r e t u m t r i f o - l i o s u m fazies with *Festuca rubra* and *Agrostis tenuis* (Krause, 1967).

5. a mountain pasture on a wind-exposed ridge near the Feldberg peak, Black Forest (1500 m altitude). It changed through fertilization from a L e o n t o d o n h e l v e t i c u s - N a r - d e t u m to a F e s t u c o - C y n o s u r e t u m (Krause, 1967).

Each of the investigated meadows or pastures was mowed several times a year. LAI was measured every 1—2 weeks or at least on the date of each mowing. Fig. 3 shows average production, LAI and E for several years.

Looking at Fig. 3 from the bottom to the top, three facts are clearly visible:

507

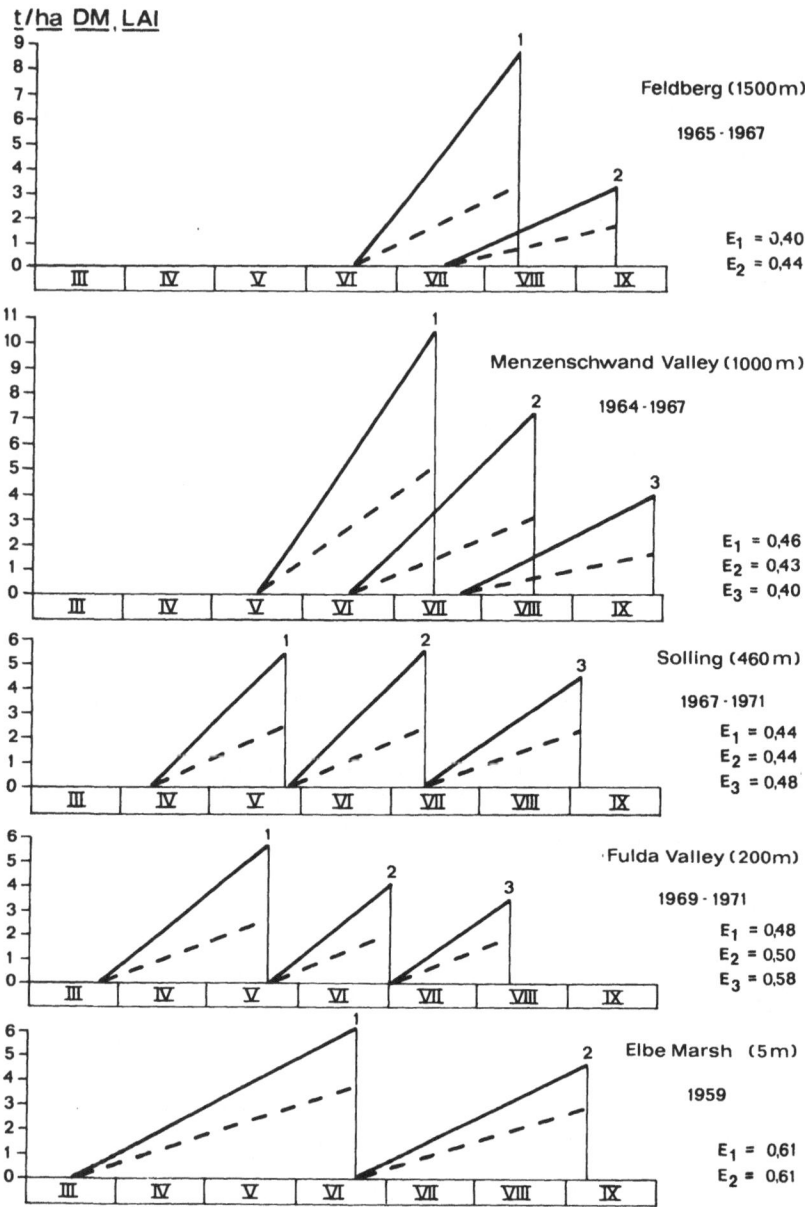

Fig. 3. Average efficiency (E), LAI, and above-ground production in t/ha in consecutive growth periods 1, 2, 3 at different altitudes from Northwest- to Southwest-Germany.

—————— LAI

— — — — t/ha Dry Matter

1. The beginning of the vegetation period is delayed with increasing altitude from March to June.
2. With a later growth start, the development of LAI and the production of biomass is more rapid until the first mowing. The respective LAI gradient increases from 25° near sea level to 53° in the Menzenschwand Valley. At the Feldberg site the LAI gradient is slightly lower.
3. Highest LAI and yield are found at the Menzenschwand site.

These tendencies have created difficulties with the management of the mountain pastures, when attempts have been made to improve them through fertilization. Caputa (1966) reports on results in the Swiss Alps in connection with the same FAO experiments, of which our investigations were a part. Caputa found an "explosive production" of fodder within a short time after intensive NPK fertilization carried out as soon as snow had disappeared. Successive mowings had a far lower level of productivity, especially at high altitudes. As has been shown by Dietl in this volume, p. 430, lower sites required more time for producing the same yield. For purposes of comparison, I calculated my results in the same way (see Table 1):

Table 1

Influence of altitude on length of time required for production of first 1,5 t/ha dry matter and of first LAI 2 in days

1	2	3	4
Site	Altitude	dry matter days	LAI days
Feldberg	1500 m	25	12
Menzenschwand	1000 m	18	11
Solling	460 m	27	16
Fulda Valley	200 m	32	20
Elbe marsh	5 m	38	30

The respective number of days for producing 1,5 t/ha dry matter (see column 3) decreases with elevation. Excluding the Feldberg site because of its unfavourable conditions, the reduction of growth time is about 2 days per 100 m from sea level to 1000 m. For the Alps an average reduction of 1,7 days per 100 m between 400 m and 1200 m altitude and one of 3 days per 100 m between 1200 and 1800 m is calculated from Fig. 8 of Dietl's article, p. 430 in this volume.

Development of LAI in the initial growth period displays a similar trend (see Table 1, column 4): For each 100 m of elevation, the time required for producing LAI 2 is two days less.

A thorough causal analysis of this phenomenon has not yet been attempted. Caputa (1966) presumes that the greater intensity of infrared and ultraviolet radiation might be one reason. In my opinion, the different daylength must also be taken into account, since, in the case of the examples cited here, daylength increases considerably from March to June. In regions of high geographical latitude with extremely long summer days, a similar "explosive" yield occurs in a period of only 4-6 weeks on fertilized meadows or grass crops. In Iceland, 4 t/ha were produced in 1970 during such a period on fertilized grass crops of *Phleum pratense* and *Poa pratensis*, reaching leaf area indices of 4,5 (Geyger, unpublished). This is noteworthy with respect to the relatively low temperatures of this region.

In regard to efficiency per unit leaf area, the higher altitudes are not favoured, on the contrary: The E-values given in Fig. 3 show that average efficiency is lowest at the sites of highest altitude and vice versa.

A comparison of the regrowth tendencies after mowing is complicated because the date when regrowth begins, varies in the given examples from late May to late July. The greatest decrease of yield and LAI between the first and second mowings occurs at the most elevated site, closely followed by the Menzenschwand site.

Efficiency, however, does not at all show such a tendency. The distributions of growth rates of the different mowings for both LAI and yield are much more proportionate at the sites of lower altitude. Here, longtime experience has possibly led to appropriate management of these meadows, while the frequency and amount of fertilization at the more elevated sites had not — during these first experiments — been well adapted to the mountain climate.

11.6   Effects of fertilization on LAI and yield

The investigations presented here are used to demonstrate, too, the differences in LAI at various fertilization levels. Some examples concerning the experiments already discussed are given in Fig. 4 and Table 2.

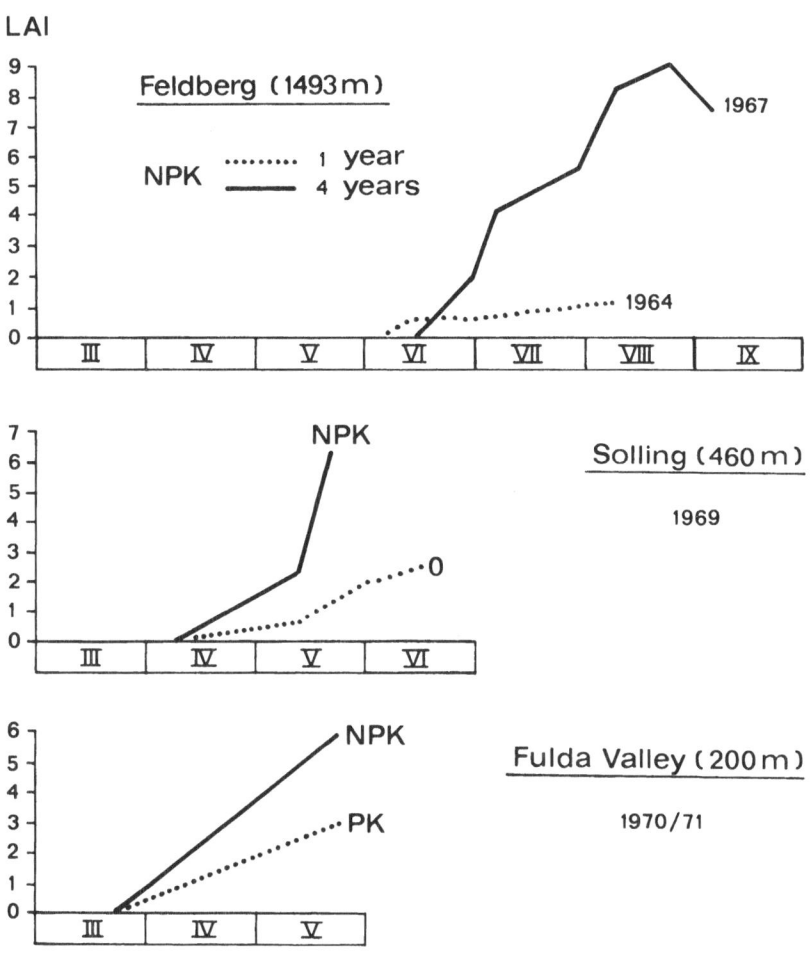

Fig. 4. LAI development at different fertilization levels (first growth period).

Table 2

Production parameters of grasslands at different fertilization levels (first harvest)

1 Site	2 Year	3 Fertilization level	4 LAI	5 Dry matter t/ha	6 E
Feldberg	1964	–	1,2	1,1	0,92
(1500 m)	1967	NPK	9,0	4,7	0,52
Solling		–	1,5	1,1	0,73
(460 m)	1969	NPK	9,0	1,9	0,30
Fulda	1970	PK	3,0	1,8	0,60
Valley	+	NPK	6,0	2,8	0,47
(200 m)	1971				

At the Feldberg site, no control plots were investigated. Therefore comparison has been done between the first and the fourth year of NPK fertilization. The difference is striking: LAI increases by 750%, and yield by 450%. In the Solling experiment, 1969 was the third year of fertilization. The difference between NPK and control was 4:1 for LAI and less than 2:1 for yield. In the Fulda Valley experiment the difference between NPK and PK was 2:1 for LAI and 1.5:1 for yield.

It is evident that the increase in LAI through fertilization is greater than the increase in above-ground yield. As a consequence, average production efficiency per leaf unit decreases. Table 2, column 6 shows the respective values. In all cases the stands with higher LAI have lower E values. These results seem to substanciate once more the validity of Fig. 1, i.e. E decreases with increasing LAI. Furthermore they confirm the statement of Watson (1963) that leaf area is greatly affected by mineral nutrient supply.

Further examples of fertilization experiments in grasslands including LAI measurements are rare. One is the IBP Project in Colorado (Pawnee grassland), where a natural shortgrass prairie in a semi-arid climate with *Bouteloua gracilis* as dominant species was irrigated and fertilized (Knight, 1973). Merely through irrigation, LAI rose from 0.3 to 0.5; with additional fertilization LAI rose to 3. This is ten times greater than under natural site conditions.

The results can be summarized as follows:
1. LAI increases through fertilization with greatest effect when the initial site conditions were previously rather unfavourable.
2. Growth improvement is achieved by increasing the leaf area and not by increasing photosynthetic efficiency.

11.7   Leaf area indices in different vegetation types

The effects of some vital growth factors on LAI and productivity have already been discussed in the previous chapters. Improvement of growth conditions resulted in higher production mainly through increased leaf area, sometimes combined with a change in stand architecture and with a change in the phytosociological composition. All of these variables are linked in a complex way; it is difficult to decide which relationship is closest: LAI to stand structure, to species composition, to productivity or to external conditions.

512

In chapter 3 some references are given for the relationship of LAI, light and foliage architecture. The connection of LAI with productivity is even more complex, as is discussed in various papers (Niciporovic, 1968; Šesták, Čatský & Jarvis, 1971; Spedding, 1971; Knight, 1973; Cooper, 1975; Lieth, 1975). The possibility of describing communities through LAI is mentioned by

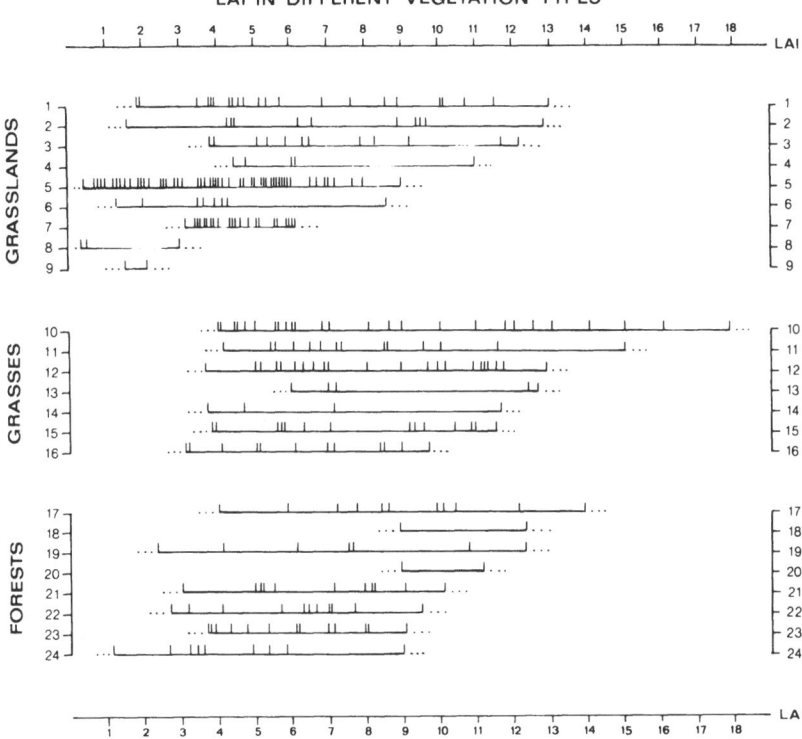

Fig. 5. LAI in different vegetation types.

GRASSLANDS
1. Arrhenatheretalia meadows, Central Europe; 2. Mountain pastures, Black Forest, SW-Germany; 3. Cirsium oleraceum meadows, Elbe marsh, NW-Germany; 4. Salt marsh communities, Baltic Sea; 5. Trisetetum flavescentis hercynicum, Solling; 6. Inundated meadows, Southern Moravia, ČSSR; 7. Miscanthus sinensis community, Japan; 8. Shortgrass-prairie, Colorado; 9. Mixed prairie, Canada.

GRASSES	FORESTS
10. *Zea mays*	17. Spruce
11. *Lolium multiflorum*	18. Douglas fir
12. *Dactylis glomerata*	19. Tropical Rain Forest
13. *Lolium perenne*	20. Alder
14. *Triticum aestivum*	21. Beech
15. *Phragmites communis*	22. Oak
16. *Oryza sativa*	23. Mixed Oak
	24. Pine

513

Loomis, Williams & Duncan, 1967). Many authors stress the significance of environment for leaf area development; only a few references are given here: Ellenberg, 1939; Vareschi, 1951; Donald & Black, 1958; Evans, 1963; Watson, 1963; Geyger, 1964; Alekseenko, 1969, Koblet, 1972; Rychnovská, Gloser & Petrík, 1972; Larcher, 1973; Milthorpe & Moorby, 1974. In this paper a contribution shall be given by comparing the LAI variation in different vegetation types all over the world (Fig. 5).

Each vertical mark along the horizontal lines represents one LAI value found in natural communities, plantations or field experiments (pot experiments were excluded). The graph shows clearly one striking fact. All of the three groups: grasslands, grasses, and forests, show a similar pattern of low and high values; available data lie for grasses between 3 and 18, for grasslands and forests between 1 and 14. (For coniferous forests half of the needle surface was used, as was proposed in the IBP Handbook No. 2, Newbould, 1967).

A similar convergence among the world's vegetation formations is demonstrated by Jordan (1971) for annual production of leaf and litter: The ranges are from 5.5—23 t/ha in tropical rain forests as well as in annual herb communities. Most communities, including all types of forests, produce 3—7 t/ha, so that "leaf and litter production is relatively uniform in perennial herb and grass, and tree communities".

Chlorophyll content is, according to Lieth (1975) convergent among communities, too. He reports chlorophyll contents of 3—9 g/m² ground surface for tropical rain forests, whereas tropical and temperate grasslands, as well as annual crops, reach maxima of 5—6 g/m².

The other conspicuous fact is the great variability within each vegetation type, as is shown by the length of the horizontal lines of Fig. 5. In meadows of the order Arrhenatheretalia (line 1) LAI ranges from 2—13; the same range occurs in tropical rain forests (line 19). The variability of LAI is even greater in some monospecific grass communities, e.g. *Zea mays* (line 10) and *Lolium multiflorum* (line 11). It must be suspected that the shortness of some lines, such as 18 or 20, is only due to the scarcity of values available.

The great variation within the same vegetation type, even the same "association" (see lines 3, 5, 7, 10—16) seems to disprove a close relation of LAI and community composition. What else might be reflected? Let us try to relate the given LAI values with some of the specific site conditions:

514

Most of the lower LAI values of line 1 concern the long-term fertilization experiment in the Fulda Valley (see also Fig. 3 + 4); the lowest leaf area indices were produced by plots without nitrogen or at harvest dates later in summer. Furthermore, the climatic conditions were rather unfavourable. Higher LAI values between 5 and 8 were found on moist sites in Switzerland in fertilized permanent meadows of the association A r r h e n a - t h e r e t u m   e l a t i o r i s (Nösberger, 1970; Koblet, 1972). Values between 8,5—10,5 were found by Medina & Lieth (1964) for   an A r r h e n a t h e r e t u m   e l a t i o r i s   s u b a s s. a l o p e c u r o s u m near Stuttgart. Most of the fertilized meadows of Württemberg, SW-Germany, lie in the same range between 7 and 9 (Lieth, 1972). The highest values of line 1 refer to communities of the Elbe marsh, A r r h e n a t h e r e t u m e l a t i o r i s at moist and naturally well-fertilized sites (Geyger, 1964).

The tendency of LAI within this group of communities seems obvious: Sites rich in nutrients produce high leaf area indices when water is not limiting.

On the other hand, too much water can have an adverse effect on LAI. The lower values of line 3 concern C i r s i u m o l e r a c e u m -communities on permanently wet soils in the valley of the Bille near Hamburg (Geyger, 1964). The higher values concern the same community, but on sandy soils with frequent, but only short-term inundation, and hence, continuous renewal of nutrients.

The remarkable improvement through fertilization already discussed in chapter 6 is reflected by the great range of values of line 2, which represent all of the measured leaf area indices during the experiment in mountain pastures in the Black Forest (see also Fig. 3 + 4). The transformation to other associations was a parallel effect.

A similar improvement has been reached in the IBP fertilization experiment in the Solling Area. Line 5 includes all LAI values of NPK, PK and control plots during the five years of experimentation (see also Fig. 3 + 4). The highest leaf area indices were produced in the first year on the NPK plots, the lowest on the unfertilized control plots in the last year, impoverished after five years of bi-annual harvesting.

In more continental areas inundation improves both the nutrient and the water supply of meadow sites. Line 6 represents from left to right leaf area indices of the following communities, investigated by Rychnovská, Gloser & Petrík (1972) in an alluvial

515

region in Southern Moravia, ČSSR: S e r a t u l o - F e s t u c e -
t u m c o m m u t a t a e = "dry"; G r a t i o l a o f f i c i n a l i s -
C a r e x p r a e c o x - s u z a e -community="moist"; P h a l a r i -
d e t u m a r u n d i n a c e a e = "damp"; G l y c e r i e t u m
m a x i m a e = "wet". The range of annual dry matter production was 1970 between 3.2 − 14.1 t/ha.

Similar conditions with respect to water and nutrient supply
occur in the salt marshes on the Graswarder near Heiligenhafen,
Baltic Sea (Schmeisky, 1974). I measured the highest LAI values
of about 11 (see line 4) at the lowest sites with P l a n t a g i -
n e t u m m a r i t i m a e , whereas at sites of only little more
elevation LAI values of only about 4 were produced in an A r -
t e m i s i e t u m m a r i t i m a e .

The relatively short line 7 includes plots of a semi-natural
grassland at the Japanese Kawatabi IBP Area (Shimada *et al.*,
1975): a M i s c a n t h u s s i n e n s i s -community under
temperate conditions, mown once a year at the end of August.
LAI ranges from 3 to 6; yield is about 7 t/ha.

The extremely low values of the lines 8 and 9 represent leaf
area indices in natural grasslands under unfavourable site conditions. The lowest value of line 8 stands for the natural shortgrass
prairie in Colorado, already mentioned in chapter 6. The higher
values reflect the positive influence of irrigation and fertilization.
Line 9 represents leaf area indices of grasses + sedges of a natural
mixed prairie of the "A g r o p y r o n - K o e l e r i a - type"
(Coupland, Ripley & Robins, 1973), part of the Canadian Matador IBP project in a region where climate is "arid with a cold
season". LAI was about 2 (herbs excluded), annual production
1−1.6 t/ha.

The monospecific communities of grasses (lines 10−16)
show some extremely high values. For *Zea mays* (line 10) the
higher values result from a density experiment with a highly
productive variety (Williams, Loomis & Lepley, 1965). Here surfaces of leaf sheathes were included in the calculations. The great
variation of LAI within this line is due to the world-wide distribution of the respective stands with all their differences in management as well as in edaphic and climatic conditions.

The variability of LAI is great, too, in the third group of
vegetation types, even though most of the lines 17−24 represent
forests with only one tree species. The values cover a wide range
of sites on all continents; most of the authors' references are
given in Geyger (1964) and Art & Marks (1971).

Annual production displays a similar variability in vegetation units all over the world. Lieth (1975) collected available data, of which a few examples are given here:

Vegetation unit	Annual production range, t/ha
tropical rain forests	10—15
tropical grasslands	2—20
temperate grasslands	1—15
swamp and marsh	8—40

It seems evident that the wide ranges of both production and LAI are mainly due to the influences of the world-wide variations of external conditions.

11.8   SUMMARY

Leaf area is known to be essential for production of biomass. The general relationship among leaf area index, leaf efficiency and growth rate as well as the influence of foliage architecture are discussed, including results of simulation experiments by various authors. Greater leaf density is achieved with erect-leaf plants like grasses, allowing a better light distribution and, to a certain extent, especially at high insolation levels, greater net production.

Under similar nutrient conditions, LAI and yield production of some central european grasslands depended significantly on altitude. The main findings were: For every 500 m of elevation growth was delayed by about 4 weeks, and then required 10 days less for producing the same rates of LAI and yield. The higher the altitude, the greater was the improvement of natural grasslands through fertilization. Effects were greater on LAI than on yield.

A comparison of LAI values in different vegetation types all over the world showed convergent ranges for grasslands, grasses, and forests between LAI 1 and 15. On the other hand, a great variability exists within each vegetation type or association. Reasons for this variation in LAI mainly lie in the different environmental conditions.

References

Alekseenko, I. N., – 1969 – Productivity of some meadow phytocenoses as a function of environmental conditions. In "Basic problems of biological productivity", (eds.) D. V. Zalensky & L. E. Rodin, pp. 88–93 (russ.) Soviet IBP, Acad. Sci. USSR.

Art, H. W. & P. L. Marks, – 1971 – A summary table of biomass and net annual production in forest ecosystems of the world. In "Forest biomass studies", (ed.) H. E. Young, pp. 1–34. Univ. of Maine, Life Sci. & Agric. Exp. St., Orono, Maine.

Baeumer, K., – 1973 – Allgemeiner Pflanzenbau. Ulmer, Stuttgart. 264 pp.

Blackman, G. E., – 1968 – The application of the concept of growth analysis to the assessment of productivity. In "Functioning of terrestrial ecosystems at the primary production level, Proceed. Copenhagen Symp. 1965", (ed.) F. E. Eckardt, pp. 243–259. UNESCO, Paris.

Brougham, R. W., – 1958 – Interception of light by the foliage of pure and mixed stands of pasture plants. *Austr. J. Agric. Res.* 9: 39–52.

Brown, R. H. & R. E. Blaser, – 1968 – Leaf area index in pasture growth. *Herb. Abstr.* 38: 1–9.

Caputa, J., – 1966 – Contribution à l'étude de la croissance du gazon des pâturages naturel à différentes altitudes. *Recherche agron. en Suisse* 5: 393–426.

Cooper, J. P., – Control of photosynthetic production in terrestrial systems. In "Photosynthesis and productivity in different environments", (ed.) J. P. Cooper, pp. 593–621. Univ. Press, Cambridge.

Coupland, R. T., E. A. Ripley & P. C. Robins, – 1973 – The Canadian IBP grassland zone programme. Description of site. I. Floristic composition and canopy architecture of the vegetative cover. Matador Project, Techn. Rep. No. 11. Saskatoon. 54 pp.

Dietl, W., – 1976 – Vegetationskunde als Grundlage der Verbesserung des Graslandes in den Alpen. This volume, chapter 9.

Donald, C. M. & J. N. Black, – 1958 – The significance of leaf area in pasture growth. *Herb. Abstr.* 28: 1–6.

Ellenberg, H., – 1939 – Über Zusammensetzung, Standort und Stoffproduktion bodenfeuchter Eichen- und Buchen-Mischwaldgesellschaften Nordwestdeutschlands. *Mitt. florist.-soziol. Arb. gem. Niedersachsen* 5: 3–135.

Evans, L. T. (ed.), – 1963 – Environmental control of plant growth. Proceed. Canberra Symp. 1962. Academic Press, New York. 449 pp.

Geyger, E., – 1964 – Methodische Untersuchungen zur Erfassung der assimilierenden Gesamtoberflächen von Wiesen. *Ber. geobot. Inst. ETH, Stiftg Rübel*, Zürich 35: 41–112.

Geyger, E., – 1967 – Die Bestimmung der assimilierenden Oberfläche des Weidebestandes als Maß für dessen Produktivität. Vortrag 5. Tagung FAO-Studiengruppe Berggrünland 28.–30.6.67 Todtnau/Schw.

Geyger, E., – 1971 – Green area indices of grassland communities and agricultural crops under different fertilizing conditions. Ecol. Studies, Analysis and Synthesis, 2, (ed.) H. Ellenberg, pp. 68–71. Springer, Heidelberg.

Jordan, C. F., – 1971 – Productivity of a tropical forest and its relation to a world pattern of energy storage. *J. Ecol.* 59: 127–142.

Knight, D. H., – 1973 – Leaf area dynamics of a shortgrass prairie in Colorado. *Ecol.* 54: 891–896.

Koblet, R., – 1972 – Über die Entwicklung und die Stoffproduktion von Wiesenpflanzen in Abhängigkeit von der Artenkombination und von Umweltfaktoren. *Angew. Bot.* 46: 59–74.

Krause, W., – 1967 – Die naturgegebenen Grundlagen der Weideverbesserung im Südschwarzwald. Vortrag 5. Tagung FAO-Studiengruppe Bergrünland 28.–30.6.67 Todtnau/Schw.

Kuroiwa, S., – 1968 – Total photosynthesis of a foliage in relation to inclination of leaves. In: "Ecophysiological studies on primary production process in plant communities", Rep. Jap. IBP/PP Photosynthesis Level III Exp. (ed.) M. Monsi, pp. 73–75. Tokyo.

Květ, J. & J. K. Marshall, – 1971 – Assessment of leaf area and other assimilating plant surfaces. In "Plant photosynthetic production. Manual of methods. (eds.) Z. Šesták, J. Čatský & P. G. Jarvis, pp. 517–555. Junk, The Hague.

Larcher, W., – 1973 – Ökologie der Pflanzen. Ulmer, Stuttgart. 320 pp.

Lieth, H., 1 – 1972 – Über die Primärproduktion der Pflanzendecke der Erde. *Angew. Bot.* 46: 1–37.

Lieth, H., – 1975 – Primary production of the mayor vegetation units of the world. In "Primary productivity of the biosphere", (eds.) H. Lieth & R. H. Whittaker, pp. 203–215. Springer, New York.

Loomis, R. S. & P. A. Gerakis, – 1975 – Productivity of agricultural ecosystems. In "Photosynthesis and productivity in different environments", (ed.) J. B. Cooper, pp. 145–172. Univ. Press, Cambridge.

Loomis, R. S., W. A. Williams & W. G. Duncan, – 1967 – Community architecture and the productivity of terrestrial plant communities. In "Harvesting the sun. Photosynthesis in plant life", (eds.) A. San Pietro, F. A. Greer & T. S. Army, pp. 291–308. Academic Press, New York, London.

McCree, K. J. & J. H. Troughton, – 1966 – Non-existence of an optimum leaf area index for the production rate of white clover grown under constant conditions. *Plant Physiol.* 41: 1615–1622.

Medina, E. & H. Lieth, – 1964 – Die Beziehungen zwischen Chlorophyllgehalt, assimilierender Fläche und Trockensubstanzproduktion in einigen Pflanzengemeinschaften. *Beitr. Biol. Pflanzen* 40: 451–494.

Milthorpe, F. L. & J. Moorby, – 1974 – An introduction to crop physiology. Univ. Press, Cambridge, 202 pp.

Monsi, M. & T. Saeki, – 1953 – Über den Lichtfaktor in den Pflanzengesellschaften und seine Bedeutung für die Stoffproduktion. *Jap. J. Bot.* 14: 22–52.

Munakata, K., S. Akita, I. Tanaka & Y. Murata, – 1969 – A mathematical expression for photosynthetic production in the artificial population of rice plant with various leaf area. In "Ecophysiological studies on primary production process in plant communities, Rep. Jap. IBP/PP Photosynthesis Level III exp.", (ed.) M. Monsi, pp. 76–79. Tokyo.

Newbould, P. J., – 1967 – Methods for estimating the primary production of forests. IBP Handbook No. 2, Blackwell, Oxford, 62 pp.

Niciporovic, A. A., – 1968 – Evaluation of productivity by study of photosynthesis as a function of illumination. In "Functioning of terrestrial ecosystems at the primary production level. Proceed. Copenhagen Symp. 1965", (ed.) F. E. Eckardt, pp. 243–259. UNESCO, Paris.

Nösberger, J., – 1970 – Die Analyse der Ertragsbildung von Pflanzen. *Schweiz. Landw. Mon.* 48: 325–345.

Ross, J. – 1970 – Mathematical models of photosynthesis in a plant stand. In "Prediction and measurement of photosynthetic productivity, Proceed. IBP/PP. Techn. Meeting Třeboň 1969", (ed.) I. Šetlik, pp. 29–45. PUDOC, Wageningen.

Rychnovská, M. J. Gloser & B. Petrík, – 1972 – Vertical structure of four inundated meadow stands. In "Ecosystem study on grassland biome in Czechoslovakia, IBP/ PT-PP Rep. No. 2", (ed.) M. Rychnovská, pp. 25–28. Brno.

Schmeisky, H. – 1974 – Vegetationskundliche und ökologische Untersuchungen in Strandrasen des Graswarders vor Heiligenhafen/Ostsee. Diss. Univ. Göttingen, 103 pp.

Šesták, Z., J. Čatský & P. G. Jarvis (ed.), – 1971 – Plant photosynthetic production. Manual of methods. Junk, The Hague, 818 pp.

Sheehy, J. E. & J. P. Cooper, – 1973 – Light interception, photosynthetic activity, and crop growth rate in canopies of six temperate forage grasses. *J. appl. Ecol.* 10: 239–250.

Shimada, Y., H. Iwaki, B. Midorikawa & N. Ohga, – 1975 – Primary productivity of the Miscanthus sinensis community at the Kawatabi IBP area – Standing crop of aboveground parts. In "Ecological studies in japanese grasslands – Productivity of terrestrial communities", (ed.) M. Numata, IBP Synth. 13: 110–114. Univ. Press, Tokyo.

Speidel, B. – 1970 – Grünlandgesellschaften im Hochsolling. *Schriftenr. Veg.-Kunde* 5: 99–114. Bonn-Bad Godesberg.

Speidel, B. & A. Weiss, – 1972 – Zur ober- und unterirdischen Stoffproduktion einer Goldhaferwiese bei verschiedener Düngung. *Angew. Bot.* 46: 75–93.

Spedding, C. R. W., – 1971 – Grassland ecology. Clarendon Press, Oxford. 221 pp.

Thorne, G., – 1971 – Physiological factors limiting the yield of arable crops. In "Potential crop production", (ed.) P. F. Wareing & J. P. Cooper, pp. 14–158. Heinemann, London.

Tooming, H. & A. Kallis, – 1972 – Productivity and growth calculations of plant stands. In "Solar radiation and productivity of plant stand", (ed.) H. Tooming, A. Kallis & J. Ross, pp. 1–121. (russ., engl. summary). Eston. Acad. Sci. Tartu. Contribution to IBP.

Tsuno, Y. & K. Kitakado, – 1969 – Optimum leaf area index for photosynthesis of rice plant population. In "Ecophysiological studies on primary production process in plant communities", Rep. Jap. IBP/PP Photosynthesis Level III Exp. (ed.) M. Monsi, pp. 37–40. Tokyo.

Vareschi, V., – 1951 – Zur Frage der Oberflächenentwicklung von Pflanzengesellschaften der Alpen und der Subtropen. *Planta* 40: 1–35.

Warren-Wilson, J., – 1963 – Estimation of foliage denseness and foliage angle by inclined point quadrats. *Austr. J. Bot.* 11: 95–105.

Watson, D. J., – 1947 – Comparative physiological studies on the growth of field crops. II. The effect of varying nutrient supply on net assimilation rate and leaf area. *Ann. Bot.* 11: 375–407.

Watson, D. J., – 1963 – Climate, weather and plant yield. In "Environmental control of plant growth", (ed.) L. T. Evans, pp. 337–350. Academic Press, New York.

Williams, W. A., R. S. Loomis & C. R. Lepley, – 1965 – Vegetative growth of corn as affected by population density. II. Components of growth, net assimilation rate, and leaf area index. *Crop Sci.* 5: 215–219.

Wit, C. T. de, – 1968 – Plant production. *Misc. Papers Landb. Hogesch.* No. 3: 25–50. Wageningen.

Scientific plant names according to:

Ehrendorfer, F. (ed.), – 1973 – Liste der Gefäßpflanzen Mitteleuropas. 2nd ed. Fischer, Stuttgart.

Purseglove, J. W., – 1972 – Tropical crops. Monocotyledons. Longman, London.

520

INDEX

521

527

530

533